Composite Materials

VOLUME 2

Mechanics of Composite Materials

COMPOSITE MATERIALS

Edited by

LAWRENCE J. BROUTMAN
Illinois Institute of Technology
Chicago, Illinois

AND

RICHARD H. KROCK
P. R. Mallory & Co., Inc.
Laboratory for Physical Science
Burlington, Massachusetts

1. Interfaces in Metal Matrix Composites
   Edited by ARTHUR G. METCALFE
2. Mechanics of Composite Materials
   Edited by G. P. SENDECKYJ
3. Engineering Applications of Composites
   Edited by BRYAN R. NOTON
4. Metallic Matrix Composites
   Edited by KENNETH G. KREIDER
5. Fracture and Fatigue
   Edited by LAWRENCE J. BROUTMAN
6. Interfaces in Polymer Matrix Composites
   Edited by EDWIN P. PLUEDDEMANN
7. Structural Design and Analysis, Part I
   Edited by C. C. CHAMIS
8. Structural Design and Analysis, Part II
   Edited by C. C. CHAMIS

VOLUME 2

Mechanics of Composite Materials

Edited by

G. P. SENDECKYJ

Advanced Composites Branch, Structures Division
Air Force Flight Dynamics Laboratory, Air Force Systems Command
Wright-Patterson Air Force Base, Ohio

ACADEMIC PRESS New York and London 1974

A Subsidiary of Harcourt Brace Jovanovich, Publishers

ACADEMIC PRESS, INC.
111 Fifth Avenue, New York, New York 10003

United Kingdom Edition published by
ACADEMIC PRESS, INC. (LONDON) LTD.
24/28 Oval Road, London NW1

Library of Congress Cataloging in Publication Data
Main entry under title:

Mechanics of composite materials.

(Composite materials, v. 2)
Includes bibliographies.
1. Composite materials. I. Sendeckyj, G. P., ed.
II. Series.
TA418.9.C6M4 620.1'1 74-2392
ISBN 0-12-136502-6

PRINTED IN THE UNITED STATES OF AMERICA

Contents

3 Elastic Behavior of Composites

G. P. Sendeckyj

4 Viscoelastic Behavior and Analysis of Composite Materials

R. A. Schapery

5 Elastoplastic Behavior of Composites

Donald F. Adams

6 Application of Statistical Theories for the Determination of Thermal, Electrical, and Magnetic Properties of Heterogeneous Materials

Mark J. Beran

List of Contributors

Numbers in parentheses indicate the pages on which the authors' contributions begin.

J. D. ACHENBACH (309), The Technological Institute, Northwestern University, Evanston, Illinois

DONALD F. ADAMS (169), Department of Mechanical Engineering, University of Wyoming, Laramie, Wyoming

MARK J. BERAN (209), Towne School of Civil and Mechanical Engineering, University of Pennsylvania, Philadelphia, Pennsylvania

I. M. DANIEL (433), IIT Research Institute, Chicago, Illinois

N. J. PAGANO (1, 23), Non-Metallic Materials Division, Air Force Materials Laboratory, Wright–Patterson Air Force Base, Ohio

ALLEN C. PIPKIN (251), Division of Applied Mathematics, Brown University, Providence, Rhode Island

R. A. SCHAPERY (85), Civil and Aerospace Engineering Departments, Texas A & M University, College Station, Texas

G. P. SENDECKYJ (45), Advanced Composites Branch, Structures Division, Air Force Flight Dynamics Laboratory, Air Force Systems Command, Wright–Patterson Air Force Base, Ohio

EDWARD M. WU (353), Department of Mechanical and Aerospace Engineering, Washington University, St. Louis, Missouri

Foreword

The development of composite materials has been a subject of intensive interest for at least 15 years, but the concept of using two or more elemental materials combined to form the constitutent phases of a composite solid has been employed ever since materials were first used. From the earliest uses, the goals for composite development have been to achieve a combination of properties not achievable by any of the elemental materials acting alone; thus a solid could be prepared from constituents which, by themselves, could not satisfy a particular design requirement. Because physical, chemical, electrical, and magnetic properties might be involved, input from investigators of various disciplines was required. In the various volumes of this treatise references to specific materials have generally only included the man-made or synthetic composites, but certainly the broad definition of composite materials must include naturally occurring materials such as wood. Chapters dealing with analytical studies of course can apply equally to synthetic or naturally occurring composites.

While composites have been used in engineering applications for many years, the severe operating conditions at which materials have to function (in the space age) led to the science of composite materials as we know it today. The efforts of scientists and engineers working on government research and development programs created entirely new materials, fabrication techniques, and analytical design tools within a short span of time to serve a limited market, but one with constantly demanding requirements. At end of the 1960s, a sharp reduction in the level of government expenditures in these areas and redistribution or re-emphasis of much of the personnel and institutions that had been involved in the development of composite materials raised the possibility that no complete reference work would be available to record many of these important developments and techniques. It was also apparent that this great bulk of technology, if properly digested and evaluated, could be employed in industrial and consumer applications for advantages of economy, performance, and design simplicity. For these reasons, the Editors and Academic Press have prepared this treatise detailing the major aspects of the science and technology of composite materials. We believe that the wide representation of contributors and the diversity of subject matters contained in the treatise assure the complete coverage of this field.

We intend that the volumes be used for reference purposes, or for text supplement purposes, but particularly to serve as a bridge in transferring the bulk of composite materials technology to industrial and consumer applications.

The Editors are indebted for the cooperation and enthusiasm they have received from the Editor of each volume and the individual contributors who worked diligently and as a unit to complete this task. We are also grateful to the staff at Academic Press who provided constant support and advice for the project.

Finally, the Editors wish to thank the management of P. R. Mallory and Co., Inc. and the administration of Illinois Institute of Technology for providing a key element in the successful completion of this work through their support and encouragement.

LAWRENCE J. BROUTMAN
RICHARD H. KROCK

Preface to Volume 2

The treatise on *Composite Materials* serves as a state-of-the-art summary of the development of composite materials as of 1973. The present volume, in conjunction with Volumes 5, 7, and 8, deals with the theoretical work on the subject. As the title implies, the volume deals with the prediction of the deformation behavior and strength of composite materials. As originally envisioned, the volume would have contained thirteen chapters, but length limitations forced the shifting of three chapters to other related volumes of the treatise.

The present volume can be visualized as being divided into three parts. The first and by far the largest part, consisting of Chapters 1–8, deals with the deformation behavior of composites. The second part, dealing with strength of composites, consists of Chapter 9 and Chapters 3 and 10 in Volume 5. Additional results dealing with the strength of composites can be found in Volume 5. The last part, consisting of Chapter 10 and Chapter 9 in Volume 8, deals with experimental methods for characterizing composite materials.

Chapter 1 serves as an introduction to the volume. It deals with the basic concepts in micromechanics, definition of effective moduli, and the influence of the number of fibers through-the-thickness within a single composite layer on the effective properties. In Chapter 2, N. J. Pagano gives exact expressions for the effective moduli of laminates. Moreover, he discusses the transition from the exact results to laminated plate theory and the boundary layer phenomenon at free edges in laminates. Chapter 3 summarizes the various approaches for predicting the effective elastic moduli of composite materials. The viscoelastic response of composites is discussed in Chapter 4. Moreover, this chapter serves as an introduction to the theory of viscoelasticity.

After a brief introduction to plasticity, Chapter 5 summarizes the work dealing with the elastoplastic response of composites. Prediction of the effective physical properties of composites is discussed in Chapter 6. This chapter gives an excellent treatment of the statistical approach to the prediction of effective properties. Besides being a state-of-the-art summary, Chapter 7 is an excellent treatment of the finite deformation of ideal fiber-reinforced composites. Ideal fiber-reinforced composites are materials reinforced by inextensible flexible fibers. A typical example of such materials is fiber-reinforced tire cord material. Chapter 8 deals with wave propagation and vibration of composite materials. It is complemented by Chapter 6 in Volume 7.

Chapter 9 by E. M. Wu discusses various empirical strength theories for anisotropic media and, in particular, composite materials. Limitations of the commonly used strength criteria are indicated. Moreover, an experimental data reduction methodology for selecting empirical strength criteria is given. Finally, Chapter 10 summarizes the available photoelastic investigations of composite materials.

G. P. SENDECKYJ

Contents of Previous Volume

Volume 1

1

The Role of Effective Moduli in the Elastic Analysis of Composite Laminates

N. J. PAGANO

Non-Metallic Materials Division, Air Force Materials Laboratory
Wright–Patterson Air Force Base, Ohio

I. Introduction

The purpose of this introductory chapter is to review some of the salient features of micromechanics of composite media. In contrast to the extensive literature reviews given by Hashin (1964) and Chamis and Sendeckyj (1968) which treat the various approaches used in the determination of effective properties of heterogeneous bodies, our emphasis will be placed on the interpretation and application of effective elastic moduli. Physical and mathematical definitions of effective moduli are compared and their use in the analysis of engineering composite laminates will be discussed. Finally, we shall present an approach to treat nonuniform (linear variation) macroscopic membrane stresses within composite laminates.

II. Basic Concepts in Micromechanics

The practical analysis of the mechanical response of composite bodies involves analytical studies on two levels of abstraction. These areas of investigation, in popular terminology, are known as *micromechanics* and *macromechanics*. In micromechanics, one attempts to recognize the fine details of the material structure, i.e., a truly heterogeneous body, consisting of reinforcing elements, such as fibers, particles, or crystals, embedded in a matrix material, is treated. Although the term "composite" includes a wide variety of materials, such as concrete, semicrystalline polymers, paper, leather, bone, etc., our discussion here will be focused on fiber/reinforced bodies. We should clarify that the term "micromechanics" does not usually imply studies on an atomic scale or the use of higher-order stress tensors, such as those occurring in couple stress or strain gradient theories, although some work of this type has been presented, e.g., Sadowsky *et al.* (1967) and Koh (1970).

The essential quantities sought from micromechanics formulation are the effective moduli, which are defined as the constitutive coefficients relating volume averages of the stress and strain components under certain boundary conditions. These boundary conditions can be either of two types (Hashin and Rosen, 1964), i.e., for boundary displacements given by†

$$u_i(S) = \epsilon_{ij}^0 x_j \tag{1}$$

or prescribed tractions of the form

$$T_i(S) = \sigma_{ij}^0 n_j \tag{2}$$

where n_j is the unit outward normal vector on the boundary surface S, x_j are the Cartesian coordinates of that surface, and ϵ_{ij}^0 and σ_{ij}^0 are constants. In a homogeneous body under either set of boundary conditions (1) or (2), the stress and strain components are constant. If (1) is prescribed, the strains are given by ϵ_{ij}^0, while if (2) is prescribed, the stresses take the values σ_{ij}^0. In a nonhomogeneous body, the *volume averaged* strains are given by ϵ_{ij}^0 under (1) and the volume averaged stresses by σ_{ij}^0 under (2). The

† Throughout this work, we shall omit rigid body displacement components.

proof of these statements follows from the identities

$$\int_V \epsilon_{ij}\, dV = \tfrac{1}{2} \int_S (u_i n_j + u_j n_i)\, dS \tag{3}$$

$$\int_V \sigma_{ij}\, dV = \tfrac{1}{2} \int_S (T_i x_j + T_j x_i)\, dS \tag{4}$$

In (3) and (4), V stands for the volume enclosed by surface area S. These relations hold for heterogeneous bodies provided, for (3), that u_i are continuous, and, in (4), the tractions are continuous at all interfaces, while body forces vanish. Therefore, if either (1) or (2) is prescribed, the effective moduli C_{ijkl} are defined by

$$\bar{\sigma}_{ij} = C_{ijkl}\bar{\epsilon}_{kl} \tag{5}$$

where the overbars denote volume averages. It is hypothesized that the effective moduli (5) under boundary conditions (1) and (2) are consistent for statistically homogeneous media (see Chapter 6). The effective moduli are functions of the phase geometry (shape and spatial arrangement of the reinforcing elements and volumetric composition) and the elastic properties of the phase materials.

The approach to calculate the effective moduli of composite materials is based upon the premise of a representative volume element, or an element in which the volume averages of all stress and strain components are identical to the respective quantities in the entire composite. Owing to the mathematical complexity of the problem, the micromechanical formulation is usually accomplished only for relatively simple composites, such as an infinite matrix reinforced by a parallel system of identical fibers arranged in a doubly periodic array, both phases being elastic. An exception to this general approach is the work of Sendeckyj (1970), in which the longitudinal shearing problem of a matrix reinforced by arbitrarily located fibers of arbitrary diameter is solved. Since the mathematical definition of effective moduli given above differs from a physical definition, i.e., one based upon experimentally observable *surface* averages of stress and strain components, it is important to recognize that the two definitions are correlated through micromechanics analysis, as discussed in Section V.

The main product of the micromechanical formulation is the evaluation of the effective moduli of a unidirectional (parallel fibers) composite body, which generally takes the form of a layer in a laminate. This calculation permits one to perform subsequent analytical studies on a new level of abstraction, in which the field variables are the volume averages within representative volume elements, $\bar{\sigma}_{ij}$ and $\bar{\epsilon}_{ij}$. In this way, the actual heterogeneous material is conceptually represented as a homogeneous, anisotropic solid characterized by the computed effective moduli. In turn, there exists a direct correspondence between the "macroscopic" variables $\bar{\sigma}_{ij}$ and $\bar{\epsilon}_{ij}$ and the respective phase responses. The approach is rigorous (Tabaddor, 1972) provided the composite is subjected to boundary conditions (1) or (2), in which cases the macroscopic stress and strain components are uniform. If the latter components are variable, the approach is approximate. We shall make some observations on the degree of this approximation, as well as on the use of effective moduli in laminate calculations, in Sections VI–VIII.

III. Laminate Analysis

Composite structural elements are normally fabricated by a lamination process in which the layers are oriented in a predetermined manner. The field of macromechanics treats the response of such laminates in terms of the (effective) characteristics of the layers. Since engineering laminates are frequently thin members, a common approach in their analysis involves the use of laminated plate or shell theories following the assumption of linear variations of the in-plane displacement components through the thickness (Ashton and Whitney, 1970). More rigorous solutions (e.g., Pagano and Whitney, 1970; Pagano and Hatfield, 1972), which retain the concept of effective layer properties, but satisfy the elasticity field equations within each layer and appropriate interface continuity conditions, have helped to establish the range of validity of this approach. For relatively thick laminates, the assumption of linear displacement is very poor. For example, the thickness distribution of a normalized in-plane displacement on the edge of a particular nine-layer graphite–epoxy laminate under static bending is shown in Fig. 1 (Pagano and Hatfield, 1972). The various curves illustrate the influence of the aspect ratio S (span to thickness) on this displacement. At higher values of S, the deformation is accurately predicted by classical laminated plate theory (CPT), in which the Kirchhoff hypothesis is invoked.

The use of lamination (plate or shell) theory can be considered as serving a dual purpose in composite analysis. Besides giving the approximate field equations for use in the solution of specific boundary value problems, one obtains a constitutive description of the laminate itself. In other words, we can consider an infinitesimal laminate element $dx\,dy\,h$, where h is the laminate thickness, and derive relations between the forces and moments per unit length and the geometric variables, central-plane strains and curvatures. In this way, we can define *laminate* effective moduli, in contrast to *layer* effective moduli, to characterize the constitutive relations at each "point" (x, y) of the laminate. These laminate constitutive equations take the place of generalized Hooke's law in homogeneous, elastic bodies. Assuming homogeneous layers, it can be shown (Chapter 2) that this procedure is exact when the forces and moments per unit length, as well as the tractions on the surfaces $z = \pm h/2$, are constant. This situation is analogous to the requirement governing precise layer effective moduli, i.e., uniform macroscopic stresses.

By use of lamination theories, one can formulate significant boundary value problems, the solutions of which characterize the stiffness and stability of composite laminates. Once a particular boundary value problem is solved in terms of the dependent variables in these theories, i.e., force and moment resultants, the distribution of macroscopic layer stresses can be

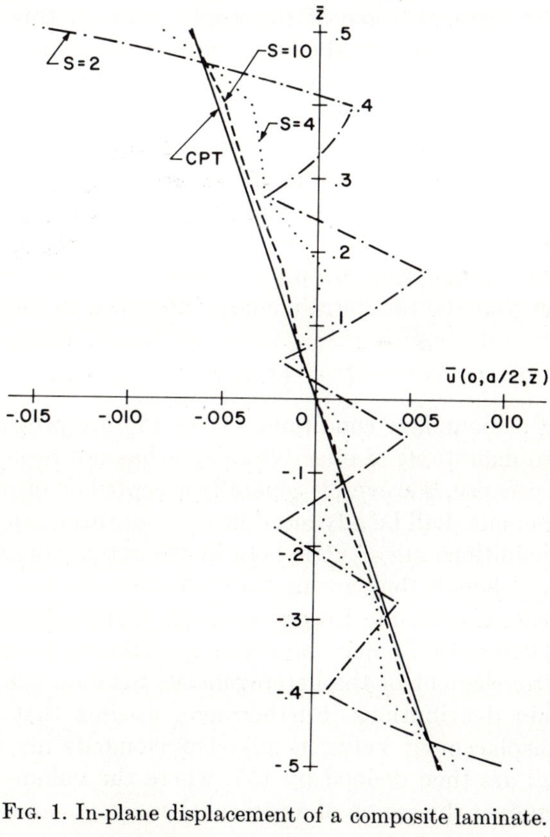

FIG. 1. In-plane displacement of a composite laminate.

determined according to the particular theory being employed. In lieu of the approximate lamination theories, one can attempt an "exact" analysis, as discussed earlier. In this case, the macroscopic layer stresses are the field variables so that the latter step is not needed. In turn, given an approach to treat nonuniform macroscopic stresses (discussed in Section VIII), the stress field within the phase materials can be determined through micromechanics. Hence, micromechanics represents the connecting link between the response of engineering laminates and that of the phase materials, and vice versa.

IV. Comparison of Basic Definitions of Effective Moduli

Since the analytical chain of logic by which one treats the response of composite laminates depends heavily on the concept of effective moduli, it might be well to examine some of the implications of this concept with respect to these calculations. We shall limit our discussion here to composites of uniform phase geometry, i.e., we assume that the fiber array is periodic, at least in one direction. Statistical aspects will be given in Chapter 3.

The fundamental problem associated with the description of effective moduli is represented by the doubly periodic array of fibers, or the infinite array. For example, we consider an infinite matrix reinforced by a square array of circular cylindrical fibers. As discussed earlier, the effective moduli C_{ijkl} are defined through Eq. (5) when either (1) or (2) are prescribed. Despite the fact that the total strain energy absorbed in the body is given by

$$U = \tfrac{1}{2}\bar{\sigma}_{ij}\bar{\epsilon}_{ij}V \tag{6}$$

when either set of boundary conditions (1) or (2) are prescribed, equivalence of the two definitions of effective moduli has not been shown in the literature. The premise, however, is generally accepted on intuitive grounds. A heuristic argument shall be advanced later on in this section.

The above definitions are of little help in the actual computation of effective moduli, although they are useful in the construction of upper and lower bounds, e.g., Hashin and Rosen (1964). A slightly different definition (Adams and Doner, 1967) may be given as follows: Assume that each typical geometric element of the heterogeneous medium exhibits identical stress and strain distributions. Furthermore, assume that continuity of traction and displacement between adjacent elements are satisfied. The effective moduli are then defined by (5), where the volume averages can clearly be taken over the region of a typical element. As an example of this procedure, let us consider boundary conditions on the typical element in the case of a square array that are appropriate for the calculation of the effective extensional moduli, i.e., those which relate the various volume averaged normal stress and strain components. The class of boundary value problems known as *generalized plane strain*, in which the stress and strain components are functions of x_1 and x_2 only and ϵ_{33} is a constant, will suffice for this purpose. The following boundary conditions (see Fig. 2) are prescribed:

$$u_3(x_1, x_2, x_3) = x_3\epsilon_{33}^0 \tag{7a}$$

$$\sigma_{12}(\pm a, x_2) = \sigma_{12}(x_1, \pm a) = 0 \tag{7b}$$

$$u_1(\pm a, x_2) = \pm a\epsilon_{11}^0 \tag{7c}$$

$$u_2(x_1, \pm a) = \pm a\epsilon_{22}^0 \tag{8}$$

where u_i are the x_i displacement components and ϵ_{ij}^0 are constants. These boundary conditions, in conjunction with the planes of symmetry, are sufficient to guarantee the compatibility of adjacent elements. By using (3) and (4), we can show that the volume averaged strains and stresses are given by

$$\bar{\epsilon}_{11} = \epsilon_{11}^0, \qquad \bar{\epsilon}_{22} = \epsilon_{22}^0, \qquad \bar{\epsilon}_{33} = \epsilon_{33}^0 \tag{9}$$

$$\bar{\epsilon}_{12} = \bar{\epsilon}_{13} = \bar{\epsilon}_{23} = 0$$

and

$$\bar{\sigma}_{11} = (1/2a)\int_{-a}^{a} \sigma_{11}(a, x_2)\, dx_2, \qquad \bar{\sigma}_{22} = (1/2a)\int_{-a}^{a} \sigma_{22}(x_1, a)\, dx_1$$

$$\bar{\sigma}_{33} = (1/4a^2)\int_{-a}^{a}\int_{-a}^{a} \sigma_{33}(x_1, x_2)\, dx_1\, dx_2 \tag{10}$$

$$\bar{\sigma}_{12} = \bar{\sigma}_{13} = \bar{\sigma}_{23} = 0$$

Solution of the single inclusion elasticity problem defined by boundary conditions (7) and (8) for various values of ϵ_{ij}^0 and substitution of (9) and (10) into (5) leads to the evaluation of C_{ijkl}.

We shall now present an argument which suggests that the definition of effective moduli given by (1), (2), and (7), (8) are, in fact, equivalent. As an example, we consider the extensional moduli (analogous treatment of shear moduli presents no difficulty) for the square array just discussed. Imagine a large prismatic body which edges parallel to the x_i axes reinforced by a perfect square array of fibers lying in the x_3 direction. According to the solution defined by (7) and (8), the traction σ_{11} on an end-plane x_1 = constant is periodic on an interval of length $2a$ (Fig. 2). On the same plane, if (2) is prescribed, the traction, say $\sigma_{1'}$, is constant. We now set the average value $\bar{\sigma}_{11}$, given by the first of Eqs. (10) equal to $\sigma_{1'}$, and follow the same procedure on the other end-planes of the prism. The boundary tractions in the two problems are, therefore, statically equivalent within each interval of length $2a$. Saint-Venant's principal implies that the respective fields only differ in narrow regions of dimension $2a$ near the boundaries. For large bodies, this difference becomes insignificant in the volume averaging process.

To relate the effective moduli given by the boundary conditions (1) and (7), (8), we conceive of the following hypothetical "solution": Let every typical element be subjected to boundary conditions (7) and (8) with one exception, i.e., those elements adjacent to the boundary surfaces, x_1 = constant and x_2 = constant. On these latter elements (7) and (8) are also applied except on the edges which lie on the boundaries themselves. On these edges, we assume (1) is prescribed. In this "solution," each element deforms into a rectangle. With one exception, all boundary and continuity conditions are satisfied if (1) is prescribed; the exception being that the tangential displacements of the interior edges of the boundary elements do not precisely coincide with those of the adjacent edges of the interior elements.[†] These respective edges are, however, coplanar, and all element corners are coincident. Since only highly localized continuity conditions are violated, we expect that this model only differs from the true solution under (1) in a narrow boundary layer region. Thus, for large bodies, the implication is that the effective moduli defined by (1) and (7), (8) are equivalent to each other, and by comparison, to that defined by (2). Furthermore, the stress and strain fields defined by (7) and (8) are the ones that are postulated to exist away from the boundaries when either (1) or (2) are imposed.

V. Mathematical and Physical Definitions of Effective Moduli

In practice, two distinct definitions of effective moduli are encountered. We might refer to these as "mathematical" and "physical" definitions. The former relates to the description given previously, i.e., the use of (5)

[†] The element adjacent to a corner element in this discussion also lies on the boundary, rather than the interior; however, this does not change the subsequent argument.

in conjunction with (1), (2), or (7), (8), and involves the relationship between volume averaged stress and strain components. The latter relates averages of stress and strain components over *surface* regions, i.e., quantities detectable through routine experimental characterization procedures. In order to compare these definitions, we first observe that certain components of stress and strain on the boundary surface S are defined by the imposed boundary conditions (1) and (2). For example, consider a bounding surface x_1 = constant. If (1) is prescribed, we have

$$\epsilon_{\alpha\beta}(S) = \epsilon_{\alpha\beta}^0 \qquad (\alpha, \beta = 2, 3) \tag{11}$$

while if (2) is prescribed,

$$\sigma_{1j}(S) = \sigma_{1j}^0 \qquad (j = 1, 2, 3) \tag{12}$$

Analogous results are found for surfaces x_2 = constant and x_3 = constant. No such results, however, are available under boundary conditions (7) and (8) aside from the fact that ϵ_{33} is constant everywhere because of the state of generalized plane strain assumed there.

We observe that boundary conditions (1) are physically attractive as the strains (11) correspond to those which can be determined by physical measurements, e.g., strain gages, while the volume averaged stresses can be expressed in terms of applied surface tractions through (4). For example, consider a (conceptual) experiment on a prism whose edges are oriented along x_i with the fibers being parallel to x_3. Let boundary conditions (1) be prescribed, where the only nonvanishing volume averaged strain is $\epsilon_1{}^0$. From (11), $\epsilon_1{}^0$ is the actual value of ϵ_{11} on the boundaries x_2 = constant and x_3 = constant. The volume averaged stress component $\bar{\sigma}_{11}$, from (4), is given by

$$\bar{\sigma}_{11} = (1/S_1) \int_{S_1} \sigma_{11}\, dS_1 = P/S_1 \tag{13}$$

where S_1 is the cross-sectional area in planes x_1 = constant and P is simply the axial force on the cross section. Hence, this conceptual experiment leads to the determination of C_{1111} in terms of physically measurable quantities. (The experiment is viewed as being conceptual since practical restrictions prohibit its execution in the laboratory.)

A similar conceptual experiment to the one just described can be devised in connection with boundary conditions (7) and (8). In this case, consider the same prism under (7) and (8) such that $\epsilon_{22}^0 = \epsilon_{33}^0 = 0$. Equation (13) applies for $\bar{\sigma}_{11}$ here, but ϵ_{11} is not constant on any of the bounding surfaces. However, referring to Fig. 2, the average value of ϵ_{11} over a dimension $2a$

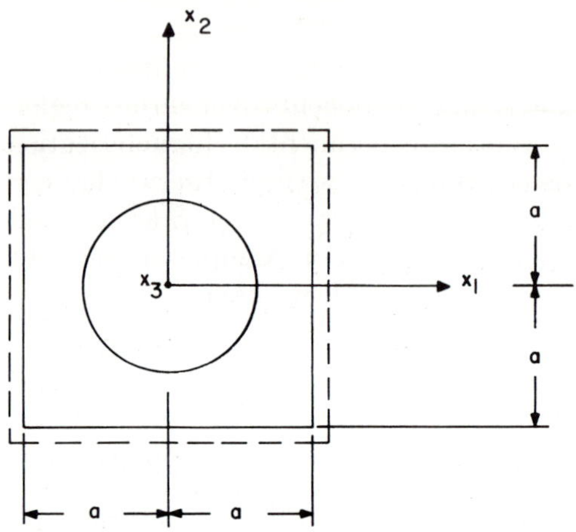

FIG. 2. Typical element of an infinite square array.

on the surface x_2 = constant is given by

$$(1/2a)\int_{-a}^{a} \epsilon_{11}(x_1, x_2)\, dx_1 = [u_1(a, x_2) - u_1(-a, x_2)]/2a = \epsilon_{11}^0 \quad (14)$$

where the last step follows from (7c). The appropriate value of $\bar{\epsilon}_{11}$ in the computation of C_{1111} is, therefore, given by the surface average of ϵ_{11} over a length $2a$, or the fiber spacing. By symmetry, this is the same as the average value over distance a. Alternately, a strain gage of a length equivalent to many fiber spacings will give the desired result.

Finally, we consider the application of (2) to the boundaries of our specimen. Assume that the constants σ_{ij}^0 are adjusted such that the only nonvanishing volume average strain is $\bar{\epsilon}_{11}$. From (12), σ_{11}^0 is the actual value of σ_{11} on the end-planes x_1 = constant; however, using (3), the quantity $\bar{\epsilon}_{11}$ is given by

$$\bar{\epsilon}_{11} = \frac{1}{V}\iint_{S_1} \Delta u_1(x_2, x_3)\, dx_2\, dx_3 \quad (15)$$

where $\Delta u_1(x_2, x_3)$ stands for the difference in displacement of the end-planes and V is the specimen volume. In this case, the volume average $\bar{\epsilon}_{11}$ is a function of the average displacements $\overline{u_1}$ on the end-planes. It will be shown in the next section, however, that $\bar{\epsilon}_{11}$ can be defined from surface measurements over sufficiently long dimensions, as in the previous case.

We have thus seen that the quantities involved in the mathematical and physical definitions of effective moduli are interrelated so that the two types of definitions are, in a sense, equivalent.

VI. Influence of Number of Fiber Rows

The previous section involved a discussion of the effective moduli of an infinite medium, i.e., the phase geometry consists of an indefinitely repeating doubly periodic array of fibers. While this type of model appears to have wide application for certain composite laminates, in others (e.g., boron–epoxy or boron–aluminum) each layer is reinforced by a *single* row† of filaments. Whether or not the effective moduli corresponding to the infinite medium are applicable in such laminates is certainly a valid question. Conversely, one might argue that the appropriate effective moduli to be employed are those corresponding to a strip of matrix material reinforced by a single row of fibers. In this section we shall examine these questions.

Consider the single fiber row problem illustrated in Fig. 3. Let us again assume that each geometrically typical element, which we take to be that represented in Fig. 2, behaves in identical fashion, with the upper and lower surfaces being traction-free. The appropriate boundary conditions are given by Eqs. (7), while (8) is replaced by

$$\sigma_{22}(x_1, \pm a) = 0 \tag{16}$$

By solving the associated single inclusion problem in elasticity theory, one can determine the volume averaged stresses and strains and relate them through equations of form (5). Let the coefficients in these equations be denoted by $C^{(1)}_{ijkl}$. While the $C^{(1)}_{ijkl}$ define the exact relations between $\bar{\sigma}_{ij}$ and $\bar{\epsilon}_{ij}$ in the strip under extensional loading, it is doubtful if they should be considered as true effective material properties, for we might consider an alternate definition of $C^{(1)}_{ijkl}$, in which we impose Eqs. (7) and (8) as boundary conditions, with the stipulation that $\bar{\sigma}_{22}$ vanishes. The effective

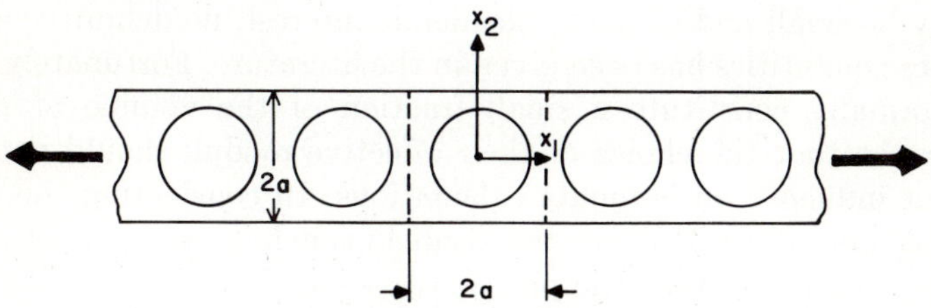

FIG. 3. Single fiber row.

† Despite this fact, the fiber length is considerably greater than its diameter and there are a minimum of approximately 200 fibers per lineal inch of composite. Hence, two dimensions of such a fibrous layer can be considered infinite, but not the third.

moduli so obtained would be precisely those for the infinite array, say $C_{ijkl}^{(\infty)}$, and are presumably different from $C_{ijkl}^{(1)}$. For example, Hulbert and Rybicki (1971) report the value of effective Young's modulus in the x_1 direction for the infinite array $E^{(\infty)} = 1.83 \times 10^6$ psi for boron–epoxy at 50% fiber volume. For the single row of fibers they give a value $E^{(1)} = 1.65 \times 10^6$ psi. Hence, the apparent effective moduli for the single row are dependent on the prescribed boundary conditions [recall that a previous argument has suggested that $C_{ijkl}^{(\infty)}$ are independent of the prescribed boundary conditions (1), (2), or (7), (8)]. Furthermore, the horizontal edges of the single row strip under (16) do not remain parallel after deformation. In other words, certain apparent effective moduli based on a physical definition are functions of position along the strip. However, suppose that the strip is built into a laminate under extensional loading such that it is constrained by other layers at its upper and lower surfaces. Clearly neither (8) nor (16) is generally appropriate in this case, for, if (8) is prescribed, traction continuity will, in general, be violated at the interfaces, while if (16) is prescribed, displacement continuity will not be satisfied. For rigid constraining layers, (8) would be appropriate, while for very soft layers, (16) would seem to be in order. Apparently, the only solution in the literature which bears on this issue is the one by Hulbert and Rybicki (1971) treating the problem of multiple rows of parallel fibers in a square array. This solution indicates, at least for the material properties treated there, that the stress field within interior elements is essentially identical to that corresponding to the infinite array. Sendeckyj (1971) has made a similar observation for the case in which the shear moduli of the fibers and matrix are identical. Since $C_{ijkl}^{(\infty)}$ can be viewed as material properties (they are also well documented in the literature), this further (admittedly meager) evidence supports the use of $C_{ijkl}^{(\infty)}$ to characterize the macroscopic response of interior layers of a laminate. While the difference between $C_{ijkl}^{(1)}$ and $C_{ijkl}^{(\infty)}$ may be small and of purely academic interest, no definitive comparison of these quantities has been given in the literature. Fortunately, surface layers normally constitute a small fraction of the volume of practical laminates, so that the choice of their effective moduli should not have a significant influence on laminate calculations. In considering the problem of multiple fiber rows, Hulbert and Rybicki conclude, at least for the composites treated there, that the phase stresses within the elements along a free surface are independent of the number of fiber rows in the composite. If we assume that this conclusion is general for arbitrary laminates, i.e., that the phase response of a surface element only depends on its averaged strains $\bar{\epsilon}_{ij}$ and the surface boundary conditions, one can determine local (phase) response near a boundary surface by solving a single inclusion problem in which the volume averaged strains $\bar{\epsilon}_{ij}$ and boundary conditions, analogous to (7) and (16), are imposed. The formulation to characterize $\bar{\epsilon}_{ij}$ within the layers of arbitrary laminates is treated in Chapter 2.

VII. Surface Strains

An aspect of micromechanics which has not been emphasized in the literature is its use in predicting surface strains. As noted earlier, the surface strain distribution is not, in general, defined by the boundary conditions. The exceptional case occurs when the tangential surface displacements are prescribed, such as in (1). For the single row problem discussed earlier, Rybicki (1972) gives the distribution of $\epsilon_{11}(x_1, a) = \epsilon_{11}(x_1, -a)$ shown in Fig. 4. Since Eq. (14) applies in this case, the average value of this function over distance a is given by ϵ_{11}^0; however, its distribution is far from being uniform. In typical boron–epoxy composites, the fiber spacing is approximately 0.005 in. so that standard strain gages, as their lengths are many times greater than this distance, would record a value very close to ϵ_1^0. However, Moire grid techniques employing a mesh of 1000 lines/in. are currently being employed in the field of experimental composite mechanics, e.g., Pipes and Daniel (1971). Such a technique can be expected

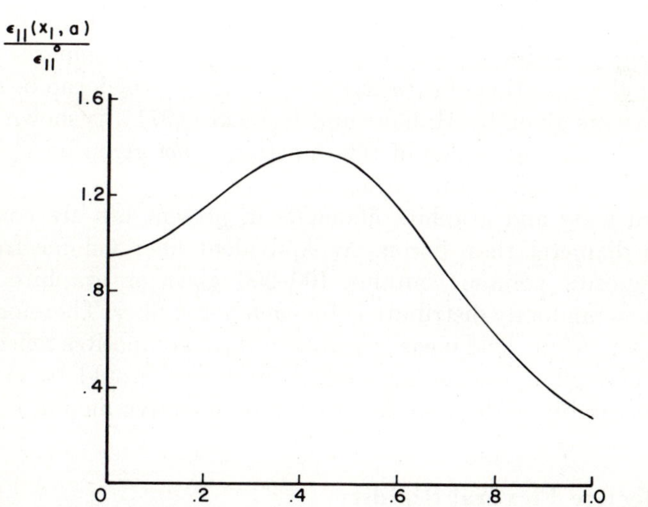

FIG. 4. Tangential strain distribution on free surface (boron–epoxy).

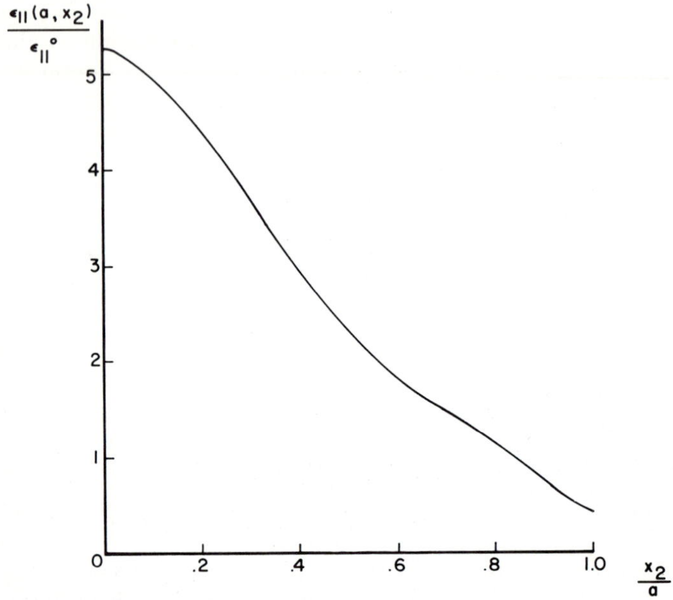

FIG. 5. Normal strain distribution on loaded edge (boron–epoxy).

to lead to a reasonable approximation of the surface strain distribution shown in Fig. 4, assuming uniform phase geometry. For comparison, a curve illustrating the variation of $\epsilon_{11}(a, x_2) = \epsilon_{11}(-a, x_2)$, which can be computed from the results given by Hulbert and Rybicki (1971), is shown in Fig. 5. Note that the average value of this function is *not* given by ϵ_{11}^0 since this component is normal to the surface $x_1 = a$.

Standard glass and graphite filaments in present use are considerably smaller in diameter than boron. At equivalent fiber volume fractions, a given composite volume contains 100–200 glass or graphite filaments (more or less randomly distributed) for each boron fiber. Therefore, even at 1000 lines/in., Moire grid measurements in these composites reflect average strains over regions containing several fibers and would be expected to compare favorably with analysis based upon effective moduli $C_{ijkl}^{(\infty)}$.

VIII. Effective Flexural Rigidity

In the previous sections, we have dealt with problems in which the macroscopic stress field is uniform. That is, in the actual heterogeneous

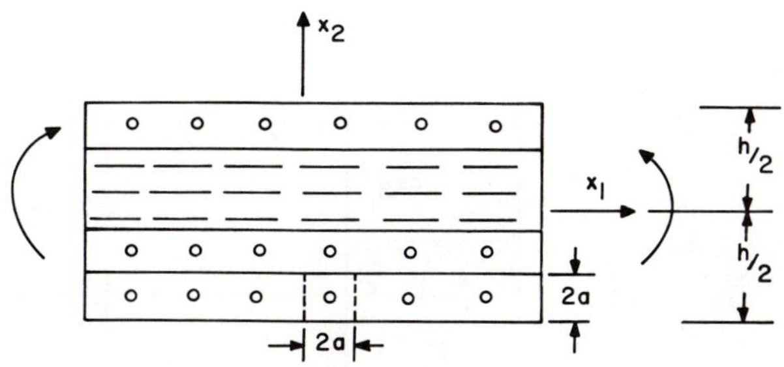

FIG. 6. Bending of N-fiber row strip.

material, the volume averaged stresses over the representative volume element are constant. In the "equivalent" homogeneous material, characterized by the effective moduli of the heterogeneous composite, the state of stress is uniform. However, in many problems of practical interest (e.g. Pagano, 1970; Pipes and Pagano, 1970; Pagano and Whitney, 1970), rather severe macroscopic stress gradients occur. Since the definition of effective moduli is based upon macroscopically uniform stresses, the significance of those results for heterogeneous materials is not clear. To examine this question, we shall develop an approximate treatment for a fiber-reinforced material in the presence of a linear macroscopic stress gradient, and compare the results with an exact solution.

Consider the N-row strip, with the fibers packed in a square array under bending moments M per unit length, as shown in Fig. 6. If the strip were a homogeneous orthotropic material, the only nonvanishing stress component σ_{11} would be given by

$$\sigma_{11} = 12Mx_2/h^3 \tag{17}$$

where h is the thickness of the strip (Fig. 6). The x_1 displacement component is, therefore,

$$u_1 = \kappa x_1 x_2 \tag{18}$$

where κ is the curvature, related to M through

$$M = Eh^3\kappa/12 \tag{19}$$

where E is Young's modulus in the x_1 direction. To define analogous effective material properties for the fiber-reinforced strip, we assume that the following boundary conditions are imposed (Fig. 6): On the planes $x_2 =$ constant,

$$\sigma_{12} = \sigma_{22} = 0 \tag{20}$$

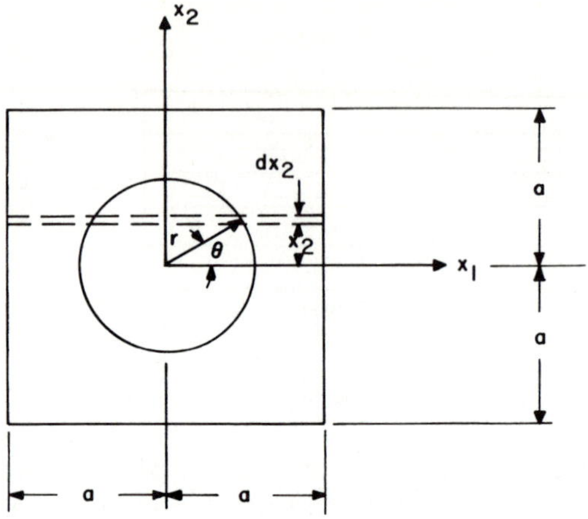

FIG. 7. Typical element of strip.

and on the planes x_1 = constant,

$$\sigma_{12} = 0, \qquad u_1 = \kappa x_1 x_2 \tag{21}$$

while the strip is in a state of plane stress. The effective flexural rigidity F will be defined by the relation

$$M = F\kappa \tag{22}$$

and the macroscopic stress component σ_{11}^* is defined by

$$\sigma_{11}^* = 12Mx_2/h^3 \tag{23}$$

We first consider the single fiber row, $N = 1$, and examine a typical geometric element, in each of which we assume identical stress and strain fields occur (Fig. 7). In order to develop an approximate expression for the constant F, we assume that the stress component σ_{11} within the heterogeneous element is independent of x_1 and that σ_{22} vanishes. The constitutive relation for the dotted strip in Fig. 7 can thus be written as

$$\sigma_{11} = E\kappa x_2 \tag{24}$$

where

$$\frac{1}{E} = \begin{cases} \dfrac{1}{a}\left[\left(\dfrac{1}{E_1} - \dfrac{1}{E_2}\right)(r^2 - x_2{}^2)^{1/2} + \dfrac{a}{E_2}\right] & (|\, x_2 \,| \leq r) \qquad (25a) \\[2ex] \dfrac{1}{E_2} & (|\, x_2 \,| \geq r) \qquad (25b) \end{cases}$$

where E_i are the phase Young's moduli, subscript 1 stands for the fiber and 2 for the matrix. The moment M can now be expressed as

$$M = \int_{-a}^{a} \sigma_{11} x_2 \, dx_2 = \kappa \int_{-a}^{a} E x_2^2 \, dx_2 \tag{26}$$

so that

$$F_1 = \int_{-a}^{a} E x_2^2 \, dx_2 \tag{27}$$

where the subscript on F stands for the number of fiber rows. Transforming into polar coordinates, F_1 can be expressed in the following form, convenient for numerical integration:

$$F_1 = \frac{2}{3} E_2(a^3 - r^3) + 2ar^3 \int_0^{\pi/2} \left[\left(\frac{1}{E_1} - \frac{1}{E_2}\right) r \cos\theta + \frac{a}{E_2}\right]^{-1} \sin^2\theta \cos\theta \, d\theta \tag{28}$$

We now consider the evaluation of F for the strip containing N rows of fibers. Let the nth element $(n = 1, 2, \ldots, N)$, the geometry of which is shown in Fig. 7, be shifted a distance d_n above the x_1 axis. The macrostress distribution on this element under pure bending of the entire strip varies from $\sigma_{11}^*(d_n) + \Delta\sigma_{11}^*$ to $\sigma_{11}^*(d_n) - \Delta\sigma_{11}^*$ at the top and bottom of the element, respectively, where $\sigma_{11}^*(d_n)$ is the macrostress σ_{11}^* at $x_2 = d_n$, the center of the element. Recognizing that the formulation leading to the development of (28) is only approximate, we refrain from a strict interpretation of this procedure. Rather, we superpose two stress fields on the element: (i) the constant stress $\sigma_{11}^*(d_n)$, and (ii) the linear variation defined by the limits $\pm\Delta\sigma_{11}^*$. Hence, the conventional effective moduli are retained in conjunction with (i), while approximation (28) is employed in the description of (ii). Therefore, the moment M_n on this element is given by

$$M_n = \kappa(F_1 + 2a\, d_n^2 E_{\text{ext}}) \tag{29}$$

where E_{ext} is the effective Young's modulus in the x_1 direction. Letting the elements be indexed consecutively from $n = 1$ corresponding to the bottom element to $n = N$, the top element, we can write

$$d_n = (2n - N - 1)a \tag{30}$$

Summing the moments on all elements, and using the identities

$$\sum_{n=1}^{N} n^2 = \frac{N(N+1)(2N+1)}{6}, \qquad \sum_{n=1}^{N} n = \frac{N(N+1)}{2} \tag{31}$$

we obtain

$$M = F_N \kappa \tag{32}$$

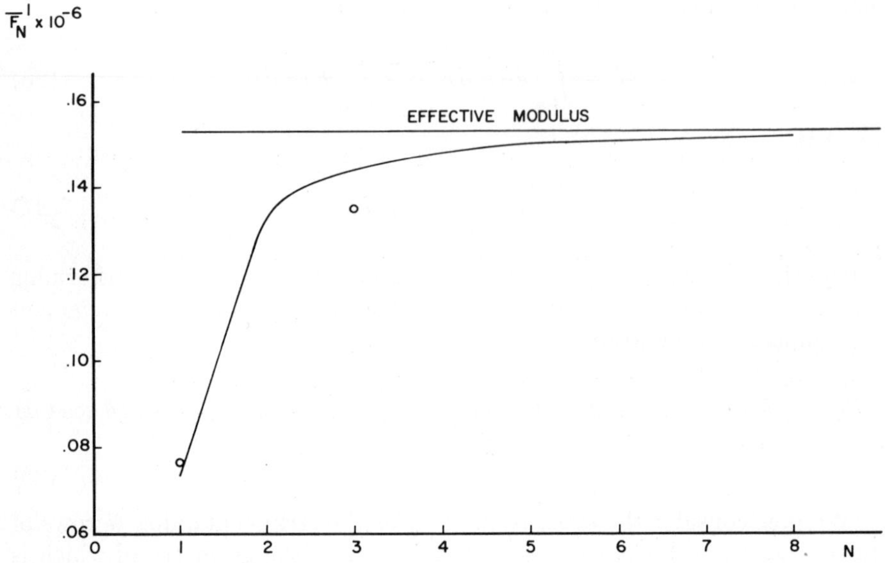

FIG. 8. Effective flexural rigidity (boron–epoxy) for exact results (○) and approximation theory (—).

where

$$F_N = N[F_1 + \tfrac{2}{3}(N^2 - 1)a^3 E_{\text{ext}}] \tag{33}$$

which defines the effective flexural rigidity of the N-row strip under the state of plane stress. For a state of plane strain, it is only necessary to replace E_i by E_i' and E_{ext} by E'_{ext}, where

$$\begin{aligned} E_i' &= E_i/(1 - \nu_i^2) \qquad (i = 1, 2) \\ E'_{\text{ext}} &= E_{\text{ext}}/(1 - \nu_{13}\nu_{31}) \end{aligned} \tag{34}$$

The respective value of effective flexural rigidity is denoted by F_N'. In Eq. (34), ν_i stands for the phase Poisson ratios and ν_{ij} is the effective composite Poisson ratio measuring the normal strain in the x_j direction under uniaxial normal stress in the x_i direction.

A curve showing the dependence of normalized effective flexural rigidity on N is given in Fig. 8 for the state of plane strain. The normalized flexural rigidity $\overline{F}_N'$ is defined by

$$\overline{F}_N' = F_N'/(2Na)^3 \tag{35}$$

where $h = 2Na$ is the thickness of the strip. The composite is boron–epoxy with a fiber volume fraction of 0.5. The phase properties are

$$E_1 = 60 \times 10^6 \quad \text{psi}, \qquad \upsilon_1 = 0.2$$

$$E_2 = 0.5 \times 10^6 \quad \text{psi}, \qquad \upsilon_2 = 0.34$$

The value of E_{ext} used here is that corresponding to the infinite square array. Also shown on the curve are the values determined from the curves given by Hulbert and Rybicki (1971), where an elasticity solution is presented for the one- and three-row composites. The approximate results are seen to be in good agreement with those from the exact solution. The curve marked "effective modulus" gives the result of the calculation based upon the conventional approach, i.e., the assumption of a homogeneous, anisotropic body characterized by the effective moduli of the composite. The approximate results rapidly converge to this value for moderate values of N and approach it asymptotically at large N since

$$\bar{F}_N' \rightarrow \tfrac{1}{12} E'_{\text{ext}} \tag{36}$$

as $N \rightarrow \infty$. For composites such as graphite or glass–epoxy, in which each layer contains many rows of fibers, the "effective modulus" approach is quite accurate.

The distribution of σ_{11} given by (24) is shown in Fig. 9 for the single fiber row (boron–epoxy) and compared to the function given by Hulbert and

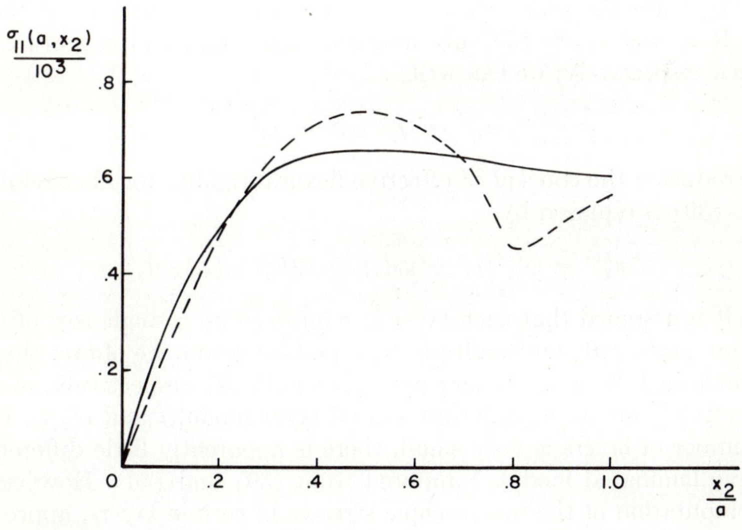

FIG. 9. Matrix stress distribution (boron–epoxy) for exact results (—) and approximation theory (– – –).

Rybicki (1971) along the element edges $x_1 = \pm a$. Reasonable agreement between the exact and approximate functions is again observed. Analogous agreement has been achieved for the $N = 3$ boron–aluminum problem reported in the latter reference.

We have restricted our attention here to the study of bending of a composite layer about an axis parallel to the fiber direction. The analogous problem of bending about an axis normal to the fibers has been considered by Margolin (1967). Furthermore, the approach presented here can be applied in similar fashion to treat the influence of twisting moments. For laminates reinforced by fibers lying in parallel planes, these three problems can be used to define the effective constitutive behavior of the layers under linear macroscopic membrane stresses, including the appropriate transformation equations for the effective flexural rigidities.

Conventional laminated plate theory, as developed by Stavsky (1961) and Dong *et al.* (1962) assumes that each layer is a homogeneous medium. Using standard contracted notation (see Chapter 2), the membrane strain components are assumed in the form

$$\epsilon_i = \epsilon_i{}^0 + z\kappa_i \qquad (i = 1, 2, 6) \tag{37}$$

Thus, the membrane stresses are given by

$$\sigma_i^{(n)} = Q_{ij}^{(n)}(\epsilon_j{}^0 + z\kappa_j) \qquad (i, j = 1, 2, 6) \tag{38}$$

where $Q_{ij}^{(n)}$ are the plane stress reduced stiffness coefficients of the nth layer. If $\epsilon_1{}^0$ and κ_1 are the only nonvanishing strain and curvature components, respectively, we can write

$$\sigma_1^{(n)} = Q_{11}^{(n)}(\epsilon_1{}^0 + z\kappa_1) \tag{39}$$

By introducing the concept of effective flexural rigidity for fiber-reinforced layers, (39) is replaced by

$$\sigma_1^{(n)} = Q_{11}^{(n)}(\epsilon_1{}^0 + \kappa_1 d_n) + 12\bar{F}_1^{(n)}\kappa_1(z - d_n) \tag{40}$$

where it is assumed that each layer is reinforced by a single row of fibers. If we integrate (40) and multiply by z and integrate, we obtain the force per unit length N_1 and moment per unit length M_1, respectively, in terms of effective laminate moduli (integrated layer moduli) and $\epsilon_1{}^0$, κ_1. Unless the number of layers is very small, there is apparently little difference in effective laminated moduli computed from (39) and (40). However, for the computation of the macroscopic stresses in certain layers, appreciable differences can occur, in particular, when $\epsilon_1{}^0 \approx 0$ and $d_n \approx 0$.

IX. Concluding Remarks

We have reviewed some of the fundamental principles in micromechanics with emphasis on the concept of effective elastic moduli and their applicability in the study of the response of fiber-reinforced laminates. A heuristic argument, suggesting the equivalence of various mathematical definitions of effective moduli, has been given. Physical and mathematical definitions are also in agreement provided the physical measurements are taken over sufficiently large surface regions.

Since effective moduli represent the properties of infinite media, we have examined their appropriateness with respect to laminate analysis, where the various layers are of finite thickness. Assuming uniform macroscopic layer stresses and layers consisting of multiple fiber rows, errors in predicting effective laminate properties caused by the use of $C_{ijkl}^{(\infty)}$ appear to be negligible. The use of $C_{ijkl}^{(\infty)}$ to characterize a layer reinforced by a single fiber row is not precise, but the approach seems to be consistent with engineering accuracy and with observed scatter in composite properties. However, only meager evidence on this point is available in the literature. A plausible approach to analyze the response of surface layers has been given, and it has been observed that in boron fibrous composites, contemporary experimental techniques are capable of detecting variations in surface displacements within dimensions smaller than the fiber spacing. Hence, the tools are available to examine some rather fine details in composite response, such as nearly local matrix strain distributions.

Implicit in the definition of effective elastic moduli is the existence of a uniform macroscopic state of stress. To gain some insight into the applicability of such a characterization, which implies that layers can be treated as homogeneous, anisotropic continua, we have presented an approximate formulation to study linear macroscopic membrane stress gradients. The approximate theory is in good agreement with the specific exact solutions developed by Hulbert and Rybicki (1971). As a practical consideration, because of the thinness of typical composite layers, the assumption of linear macroscopic membrane stresses (at least within each layer) is a respectable approximation. Hence, aside from the few exceptions noted at the end of Section VIII, the effective moduli retain their practical utility in the description of the behavior of composite laminates. It must be noted, however, that phase stresses and the linear macroscopic stresses are connected through the solution of a different boundary value problem than that conceived for the determination of the effective moduli, i.e., a linear bending micromechanics solution is also required.

The extension of the above approach to treat laminates in which the interlaminar stresses exhibit severe gradients, such as that described by Pipes and Pagano (1970), is viewed as one of the important subjects for future research in composite mechanics.

We have confined our study here to static problems in composites. Some aspects of the dynamic response of composite bodies are discussed in Chapter 8.

References

Adams, D. F., and Doner, D. R. (1967). *J. Compos. Mater.* **1,** 152.
Ashton, J. E., and Whitney, J. M. (1970). Theory of Laminated Plates. Technomic, Westport, Connecticut.
Chamis, C. C., and Sendeckyj, G. P. (1968). *J. Compos. Mater.* **2,** 332.
Dong, S. B., Pister, K. S., and Taylor, R. L. (1962). *J. Aerosp. Sci.* **29,** 969.
Hashin, Z. (1964). *Appl. Mech. Rev.* **17,** 1.
Hashin, Z., and Rosen, B. W. (1964). *J. Appl. Mech.* **31,** 233.
Hulbert, L. E., and Rybicki, E. F. (1971). *J. Compos. Mater.* **5,** 164.
Koh, S. L. (1970). *In* "Mechanics of Composite Materials" (F. W. Wendt, H. Liebowitz, and N. Peronne, eds.). Pergamon, Oxford.
Margolin, G. G. (1967). *Polym. Mech.* **3,** 492.
Pagano, N. J. (1970). *J. Compos. Mater.* **4,** 20.
Pagano, N. J., and Hatfield, S. J. (1972). *J. AIAA* **10,** 931.
Pagano, N. J., and Whitney, J. M. (1970). *J. Compos. Mater.* **4,** 360.
Pipes, R. B., and Daniel, I. M. (1971). *J. Compos. Mater.* **5,** 255.
Pipes, R. B., and Pagano, N. J. (1970). *J. Compos. Mater.* **4,** 538.
Rybicki, E. F. (1972). Private communication.
Sadowski, M. A., Hsu, Y. C., and Hussain, M. A. (1967). *J. Compos. Mater.* **1,** 174.
Sendeckyj, G. P. (1970). *J. Compos. Mater.* **4,** 500.
Sendeckyj, G. P. (1971). *In* AFML-TR-71-260 (H. Nara, ed.).
Stavsky, Y. (1961). *Proc. Amer. Soc. Civil Eng. J. Eng. Mech. Div.* **87,** 31.
Tabaddor, F. (1972). *J. Appl. Mech.* **39,** 598.

2

Exact Moduli of Anisotropic Laminates

N. J. PAGANO

Non-Metallic Materials Division, Air Force Materials Laboratory
Wright–Patterson Air Force Base, Ohio

I. Introduction

One of the most important aspects of the analysis of the response of composite materials is related to the studies of the overall behavior of laminates composed of anisotropic layers. In this phase of the analysis, we are concerned with the relationship between forces and moments per unit length and the deformations which they produce. These relations may be termed the "effective" laminate constitutive equations since they define the geometric response caused by loads exerted on the laminate, as distinguished from the conventional implication of an elastic constitutive equation, namely the stress–strain relationship of an infinitesimal material element. Once the effective constitutive law is formulated, theories can be developed based upon consideration of the laminate as a unit, rather than resorting to a layer by layer elasticity analysis. Once a particular boundary value problem is solved, however, one can determine the detailed stress distribution throughout the laminate. Another motivation for the study of effective laminate moduli is the popular use of the laminate as a model for the behavior of complex heterogeneous materials, such as composites reinforced with three-dimensional networks of fibers (Halpin and Pagano, 1969; Berkowitz and Cohen, 1970; Halpin *et al.*, 1971). Finally, proper interpretation of data from even the most routine experiments involving composite laminates requires appreciation of an effective constitutive relation.

Hence, in this work, we shall present a formulation to define the effective moduli of a laminated body in which each layer is anisotropic with no special form of elastic symmetry, i.e., each layer is characterized by 21 elastic coefficients. The treatment is limited to the case where the laminate force and moment resultants, as well as the surface forces, are constant. This implies that the interlaminar stresses are also constant. The most general case, where the latter condition is not satisfied, is currently being studied. The description of effective laminate thermal expansion coefficients is also included here. The present formulation is based upon a solution which satisfies the field equations of linear elasticity and which is internally consistent, i.e., all external boundary conditions and interface continuity conditions are satisfied. We shall also show the relationship of other existing laminate constitutive laws, which represent more specific classes of materials, to the present work. Of special importance is our demonstration that the constitutive equations of classical lamination theory, as developed by Stavsky (1961) and Dong *et al.* (1962), as well as those given by Chou *et al.* (1972) and Khoroshun (1966), after correcting some minor errors in the latter, follow directly from the general results presented here for the special loading and symmetry conditions applied in the aforementioned works. Finally, we shall present evidence regarding the validity of our predicted stress field outside of narrow boundary layer regions by reviewing some recent work by Pipes and Pagano (1970), which treats a particular form of boundary layer disturbance, i.e., a traction-free edge.

II. General Formulation of Effective Constitutive Law

Consider the prismatic body shown in Fig. 1a. The edges are parallel to x, y, z coordinate axes and the origin is placed in the horizontal central

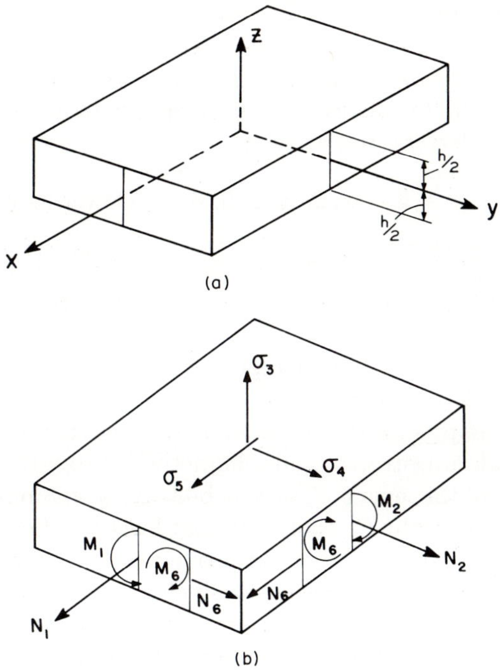

FIG. 1. (a) Coordinate system; (b) applied tractions, forces, and moments per unit length.

plane of the body. It is assumed that the local elastic constants are functions of only one space coordinate, which we take as the z coordinate without loss of generality. It should be noted that the following treatment is appropriate to characterize the response of heterogeneous anisotropic materials, in which the local stiffness coefficients are functions of one space coordinate. The variations in stiffness properties may be either continuous or discontinuous, as in a layered system. We further assume that the dimensions of the body in the x and y directions are much larger than the thickness h.† Under these conditions, the response of the body can be conveniently described in terms of forces and moments per unit length along the edges x = constant, y = constant, and surface tractions on $z = \pm h/2$ (see Fig. 1b). We shall employ standard contracted notation (Hearmon, 1961), where the stress and strain components are represented by single

† The subsequent formulation is independent of this assumption. As discussed later, the "thinness" requirement implies that, under the given loading, the detailed stress field differs from that predicted here only in narrow boundary layer regions.

subscripts, throughout this work. The stress components $\sigma_1, \sigma_2, \sigma_3, \sigma_4, \sigma_5, \sigma_6$, stand for $\sigma_x, \sigma_y, \sigma_z, \tau_{yz}, \tau_{xz}, \tau_{xy}$, respectively, while ϵ_i $(i = 1, 2, \ldots, 6)$ represent the corresponding engineering strain components. It will be helpful to introduce the following ranges for subscripts indicated by lower case, capital, and Greek letters, respectively,

$$i, j, k, l = 1, 2, 3, 4, 5, 6 \qquad \text{(lower case)}$$

$$A, B, C, D = 3, 4, 5 \qquad \text{(capital)}$$

$$\alpha, \beta = 1, 2, 6 \qquad \text{(Greek)}$$

It will also be understood that a repeated subscript requires summation over the range of that subscript.

The material comprising the prism of Fig. 1a is viewed as being locally homogeneous and anisotropic and characterized by the local stiffness coefficients C_{ij}. In the event that the body is a composite, e..g, fiber-reinforced matrix, then the C_{ij} themselves, at least to a first approximation, represent effective moduli, or the relations between volume averages of the stress and strain components within the inclusions and matrix.† The local stiffnesses C_{ij} in this case can be computed through micromechanics analysis as discussed in Chapters 3 and 6. Hence, the local material behavior is given by the elastic constitutive law

$$\sigma_i = C_{ij}\epsilon_j \qquad \text{where} \quad C_{ij} = C_{ij}(z) \tag{1}$$

and C_{ij} may be a discontinuous function of z as in the case of a laminated body. No special material symmetry will be imposed aside from the symmetry requirement implied by the existence of a strain energy potential, i.e.,

$$C_{ij} = C_{ji} \tag{2}$$

Therefore, in general, the C_{ij} matrix can contain 21 independent elastic constants (at a given value of z).

Let us now discuss the solution of a boundary value problem in heterogeneous elasticity theory which leads to the prediction of the effective moduli of our material. The body considered is the rectangular prism shown in Fig. 1a. The governing field equations in terms of the stress and strain components are (1), where the stiffness distribution is of form (2), as well as the conventional stress equilibrium and strain compatibility equations

† Tabaddor (1972) discusses the approximate nature of this approach in the presence of arbitrary surface tractions and/or displacements. (See also Chapter 1.)

of homogeneous isotropic elasticity. Since the latter relations are given in standard textbooks on isotropic elasticity, they will not be reproduced here. It will suffice to indicate that (stress) field variables of the form

$$\sigma_\alpha = \sigma_\alpha(\xi), \qquad \sigma_A = \sigma_A{}^0 \tag{3}$$

where $\sigma_A{}^0$ are constants and

$$\xi = z/h \tag{4}$$

satisfy the equilibrium equations identically. Furthermore, the constitutive law (1) and compatibility relations require that

$$\epsilon_A = \epsilon_A(\xi), \qquad \epsilon_\alpha = \epsilon_\alpha{}^0 + \xi\kappa_\alpha \tag{5}$$

where $\epsilon_\alpha{}^0$ and κ_α, which may be referred to as *central plane strains* and *curvatures*, respectively, are constants. We now suppose that the body is subjected to applied tractions on each of its boundary surfaces, where the tractions are such that they are consistent with (3), and denote the average value of any function $f(\xi)$ by an asterisk, i.e.,

$$f^* = \int_{-1/2}^{1/2} f(\xi)\, d\xi \tag{6}$$

Our goal is to establish the relations connecting applied forces and moments to curvatures and volume averaged strains, i.e.,

$$\sigma_i{}^* = \bar{C}_{ij}\epsilon_j{}^* + \bar{B}_{i\alpha}\kappa_\alpha \tag{7a}$$

$$(\xi\sigma_\beta)^* = \bar{E}_{\beta j}\epsilon_j{}^* + \bar{D}_{\beta\alpha}\kappa_\alpha \tag{7b}$$

where, for notational convenience, we have put

$$\sigma_\alpha{}^* = N_\alpha/h \tag{8}$$

and

$$(\xi\sigma_\beta)^* = M_\beta/h^2 \tag{9}$$

We also observe, from (3) and (5), that

$$\sigma_A{}^* = \sigma_A{}^0, \qquad \epsilon_\alpha{}^* = \epsilon_\alpha{}^0 \tag{10}$$

According to (7), we have defined effective stiffness coefficients relating certain "overall" response functions, which incidentally, can be observed in an experiment. These quantities are the effective extensional matrix $\bar{C}_{ij}$, flexural rigidities $\bar{D}_{\beta\alpha}$, and bending extensional coupling matrices $\bar{B}_{i\alpha}$ and $\bar{E}_{\beta j}$. We now proceed to examine the explicit forms of these matrices.

We begin by substituting Eqs. (5) into (1), which gives, after a slight rearrangement,

$$\sigma_i = C_{i\alpha}(\epsilon_\alpha{}^0 + \xi\kappa_\alpha) + C_{iA}\epsilon_A \tag{11}$$

which leads to the expression

$$C_{BA}\epsilon_A = \sigma_B - C_{B\alpha}(\epsilon_\alpha{}^0 + \xi\kappa_\alpha) \tag{12}$$

Equation (12) may now be inverted to yield

$$\epsilon_C = T_{CB}[\sigma_B - C_{B\alpha}(\epsilon_\alpha{}^0 + \xi\kappa_\alpha)] \tag{13}$$

where

$$T_{CB}C_{BA} = \delta_{CA} \tag{14}$$

and δ_{CA} is the usual Kronecker delta. In other words, T_{CB} is the inverse of the matrix

$$\begin{pmatrix} C_{33} & C_{34} & C_{35} \\ C_{34} & C_{44} & C_{45} \\ C_{35} & C_{45} & C_{55} \end{pmatrix}$$

Hence, T_{CB} is a symmetric matrix. We now integrate (13) through the normalized thickness to establish the result

$$\epsilon_C{}^* = T^*_{CB}\,\sigma_B - (T_{CB}C_{B\alpha})^*\epsilon_\alpha{}^0 - (\xi T_{CB}C_{B\alpha})^*\kappa_\alpha \tag{15}$$

where definition (6) has been applied. Introducing another symmetric matrix F_{AC}, defined by

$$F_{AC}T^*_{CB} = \delta_{AB} \tag{16}$$

into (15), we obtain

$$\sigma_B = F_{BA}[\epsilon_A{}^* + (T_{AC}C_{C\alpha})^*\epsilon_\alpha{}^0 + (\xi T_{AC}C_{C\alpha})^*\kappa_\alpha] \tag{17}$$

Now substituting (17) into (13) and in turn, the resulting expression for ϵ_C into (11), we find that

$$\sigma_i = C_{i\alpha}(\epsilon_\alpha{}^0 + \xi\kappa_\alpha) + C_{iC}T_{CB}\{F_{BA}[\epsilon_A{}^* + (T_{AD}C_{D\alpha})^*\epsilon_\alpha{}^0 + (\xi T_{AD}C_{D\alpha})^*\kappa_\alpha] - C_{B\alpha}(\epsilon_\alpha{}^0 + \xi\kappa_\alpha)\} \tag{18}$$

Finally, integrating (18) and multiplying (18) by ξ and integrating yields the results in the desired form,

$$\sigma_i{}^* = \bar{C}_{i\alpha}\epsilon_\alpha{}^0 + \bar{C}_{iA}\epsilon_A{}^* + \bar{B}_{i\alpha}\kappa_\alpha \tag{19a}$$

$$(\xi\sigma_\alpha)^* = \bar{E}_{\alpha\beta}\epsilon_\beta{}^0 + \bar{E}_{\alpha A}\epsilon_A{}^* + \bar{D}_{\alpha\beta}\kappa_\beta \tag{19b}$$

where

$$\bar{C}_{i\alpha} = C^*_{i\alpha} + F_{BA}(C_{iC}T_{CB})^*(T_{AD}C_{D\alpha})^* - (C_{iC}T_{CB}C_{B\alpha})^* \tag{20}$$

$$\bar{C}_{iA} = F_{BA}(C_{iC}T_{CB})^* \tag{21}$$

$$\bar{B}_{i\alpha} = (\xi C_{i\alpha})^* + F_{BA}(C_{iC}T_{CB})^*(\xi T_{AD}C_{D\alpha})^* - (\xi C_{iC}T_{CB}C_{B\alpha})^* \tag{22}$$

$$\bar{E}_{\alpha\beta} = (\xi C_{\alpha\beta})^* + F_{BA}(T_{AD}C_{D\beta})^*(\xi C_{\alpha C}T_{CB})^* - (\xi C_{\alpha C}T_{CB}C_{B\beta})^* \tag{23}$$

$$\bar{E}_{\alpha A} = F_{BA}(\xi C_{\alpha C}T_{CB})^* \tag{24}$$

$$\bar{D}_{\alpha\beta} = (\xi^2 C_{\alpha\beta})^* + F_{BA}(\xi C_{\alpha C}T_{CB})^*(\xi T_{AD}C_{D\beta})^* - (\xi^2 C_{\alpha C}T_{CB}C_{B\beta})^* \tag{25}$$

It can be noted from Eqs. (20)–(25) that all of the effective stiffness matrices are completely defined by the distribution $C_{ij}(\xi)$.

Observing, from (5), that $\epsilon_\alpha{}^* = \epsilon_\alpha{}^0$, and using (14), we see that (20) and (21) can be written as

$$\bar{C}_{ij} = C^*_{ij} + F_{BA}(C_{iC}T_{CB})^*(T_{AD}C_{Dj})^* - (C_{iC}T_{CB}C_{Bj})^* \tag{26}$$

whereas (19a) takes the form (7a). Similar considerations indicate that

$$\bar{E}_{\alpha j} = (\xi C_{\alpha j})^* + F_{BA}(T_{AD}C_{Dj})^*(\xi C_{\alpha C}T_{CB})^* - (\xi C_{\alpha C}T_{CB}C_{Bj})^* \tag{27}$$

so that (19b) takes the form (7b). Furthermore, the symmetry of $C_{ij}(\xi)$, $T_{AB}(\xi)$, and F_{AB} implies that $\bar{C}_{ij}$ and $\bar{D}_{\alpha\beta}$ are also symmetric, i.e.,

$$\bar{C}_{ij} = \bar{C}_{ji}, \qquad \bar{D}_{\alpha\beta} = \bar{D}_{\beta\alpha} \tag{28}$$

while it also follows that

$$\bar{E}_{\alpha j} = \bar{B}_{j\alpha} \tag{29}$$

Therefore, the effective constitutive law, (7) or (19), can be simplified as

$$\sigma_i{}^* = \bar{C}_{ij}\epsilon_j{}^* + \bar{B}_{i\alpha}\kappa_\alpha \tag{30a}$$

$$(\xi\sigma_\alpha)^* = \bar{B}_{i\alpha}\epsilon_i{}^* + \bar{D}_{\alpha\beta}\kappa_\beta \tag{30b}$$

where the effective stiffness coefficients are given by (22), (25), and either (26) or (20) and (21). Thus, in general, i.e., when no internal material symmetry is present, there exist 21 extensional stiffness coefficients $\bar{C}_{ij}$, 18 coupling coefficients $\bar{B}_{i\alpha}$, and 6 flexural stiffnesses $\bar{D}_{\alpha\beta}$, required to characterize the medium under the loading shown in Fig. 1b, all effective stiffnesses being completely defined by the distribution $C_{ij}(\xi)$. We observe that $\bar{B}_{i\alpha}$ vanishes in the event that $C_{ij}(\xi)$ is an even function. Furthermore, the use of (30), (3), and (5) defines the stress and strain distributions, $\sigma_i(\xi)$ and $\epsilon_i(\xi)$. This process may be envisioned as one of treating the medium as being homogeneous (however, possessing a rather sophisticated constitutive behavior) by use of the effective constitutive law (30), and then returning to define the detailed response by recognizing the actual material structure given by $C_{ij}(\xi)$.

Equation (26) has also been derived by Khoroshun (1966), except that apparent typographical errors are present in that work, namely an exponent (-1), signifying matrix inversion, has been consistently omitted from Khoroshun's equations (2.6)–(2.9), as well as (3.6)–(3.9). We might also clarify that the function E_{j1k1}^{-1} given by Khoroshun should be defined as the inverse of E_{j1k1}, rather than the inverse of E_{jikl} with $i = l = 1$.

A. *Effective Thermal Expansion Coefficients*

We now consider the case where the heterogeneous medium is subjected to a constant temperature rise T, in addition to the "loads," $\sigma_i{}^*$ and $(\xi\sigma_\alpha)^*$, in order to define its effective thermal expansion coefficients. Let the local thermal coefficients of expansion be denoted by $\alpha_i = \alpha_i(\xi)$, where we note that, for material anisotropy of the most general form, all six strain components result from a change in ambient temperature. Thus, under a uniform change in temperature T, a homogeneous anisotropic body is in the state of strain, $\epsilon_i = \alpha_i T$, when no surface tractions are imposed. Let these "free expansion" strains† be denoted by e_i, so that

$$e_i = \alpha_i T \tag{31}$$

Superposing the latter strains $e_i(\xi)$ and those caused by stress, our local constitutive law becomes

$$\epsilon_i = S_{ij}\sigma_j + e_i \tag{32}$$

where $S_{ij}(\xi)$ represents the local compliance matrix, i.e.,

$$S_{ij}C_{jk} = \delta_{ik} \tag{33}$$

Using Eq. (33), we can write (32) in the form

$$\sigma_i = C_{ij}(\epsilon_j - e_j) \tag{34}$$

† The strains induced in a homogeneous body under zero surface tractions and body forces have been called "expansional strains" by Halpin and Pagano (1970). They may also be produced, for example, by absorption of a fluid (swelling). The formulation given here is valid for any such environmental effect, as long as it produces no change in the elastic moduli.

Employing relations (3) and (5), and following the same procedure which led to the development of (18), we obtain

$$\begin{aligned}\sigma_i = {} & (C_{i\alpha} - C_{iA}T_{AB}C_{B\alpha})(\epsilon_\alpha{}^0 + \xi\kappa_\alpha) \\ & + C_{iC}T_{CA}F_{AD}[\epsilon_D{}^* + (T_{DB}C_{B\alpha})^*\epsilon_\alpha{}^0 + (\xi T_{DB}C_{B\alpha})^*\kappa_\alpha] \\ & + (C_{iA}T_{AB}C_{Bj} - C_{ij})e_j - C_{iC}T_{CA}F_{AD}(T_{DB}C_{Bj}e_j)^* \end{aligned} \tag{35}$$

Finally, integrating σ_i and $\xi\sigma_\alpha$, and using (20)–(22), and (25) yields

$$\sigma_i{}^* = \bar{C}_{ij}\epsilon_j{}^* + \bar{B}_{i\alpha}\kappa_\alpha + [(C_{iA}T_{AB}C_{Bj} - C_{ij})e_j]^* - \bar{C}_{iA}(T_{AB}C_{Bj}e_j)^* \tag{36}$$

and

$$\begin{aligned}(\xi\sigma_\beta)^* = {} & \bar{B}_{i\beta}\epsilon_i{}^* + \bar{D}_{\alpha\beta}\kappa_\alpha \\ & + [\xi(C_{A\beta}T_{AB}C_{Bj} - C_{\beta j})e_j]^* - \bar{B}_{A\beta}(T_{AB}C_{Bj}e_j)^* \end{aligned} \tag{37}$$

Hence, the effective expansional strains E_i and curvatures K_α are defined by putting $\sigma_i{}^* = (\xi\sigma_\beta)^* = 0$ in (36) and (37), which leads to the following simultaneous equations:

$$\begin{aligned}\bar{C}_{ij}E_j + \bar{B}_{i\alpha}K_\alpha &= [(C_{ij} - C_{iA}T_{AB}C_{Bj})e_j]^* + \bar{C}_{iA}(T_{AB}C_{Bj}e_j)^* \\ \bar{B}_{i\beta}E_i + \bar{D}_{\alpha\beta}K_\alpha &= [\xi(C_{\beta j} - C_{A\beta}T_{AB}C_{Bj})e_j]^* + \bar{B}_{A\beta}(T_{AB}C_{Bj}e_j)^* \end{aligned} \tag{38}$$

The effective constitutive equations can now be expressed as

$$\begin{aligned}\sigma_i{}^* &= \bar{C}_{ij}(\epsilon_j{}^* - E_j) + \bar{B}_{i\alpha}(\kappa_\alpha - K_\alpha) \\ (\xi\sigma_\beta)^* &= \bar{B}_{i\beta}(\epsilon_i{}^* - E_i) + \bar{D}_{\alpha\beta}(\kappa_\alpha - K_\alpha) \end{aligned} \tag{39}$$

where E_j and K_α follow from the solution of Eqs. (38).

B. Monoclinic Symmetry

We now proceed to examine the forms of the effective stiffness matrices for a special class of material symmetry, namely, we assume that each material particle possesses a single plane of elastic symmetry normal to the z axis. This is known as *monoclinic symmetry*. As before, the local stiffness coefficients may vary either continuously or abruptly through the thickness. The latter situation is typical of most engineering composite laminates, which consist of layers of fiber-reinforced material such that the fibers in the various layers lie in parallel planes. For monoclinic symmetry, it can be shown (Lehknitskii, 1963) that

$$C_{14} = C_{15} = C_{24} = C_{25} = C_{34} = C_{35} = C_{46} = C_{56} = 0 \qquad (40)$$

for the case treated here, i.e., the plane of elastic symmetry is normal to the z axis. Therefore, the local stiffness matrix has the appearance

$$C_{ij} = \begin{pmatrix} C_{11} & C_{12} & C_{13} & 0 & 0 & C_{16} \\ C_{12} & C_{22} & C_{23} & 0 & 0 & C_{26} \\ C_{13} & C_{23} & C_{33} & 0 & 0 & C_{36} \\ 0 & 0 & 0 & C_{44} & C_{45} & 0 \\ 0 & 0 & 0 & C_{45} & C_{55} & 0 \\ C_{16} & C_{26} & C_{36} & 0 & 0 & C_{66} \end{pmatrix} \qquad (41)$$

containing 13 nonzero coefficients. Hence, in this case, we have

$$C_{3A} = \delta_{3A} C_{33} \qquad (42a)$$

$$C_{A\alpha} = \delta_{3A} C_{3\alpha} \qquad (42b)$$

By substituting (42a) into (14), we find that

$$T_{3A} = \delta_{3A}/C_{33} \qquad (43)$$

and by use of this result, in conjunction with (16), we obtain

$$F_{3A} = \delta_{3A}/(1/C_{33})^* \qquad (44)$$

By putting (43) into (14), we can compute the remaining nonvanishing elements of T_{AB} as

$$T_{44} = C_{55}/C, \qquad T_{45} = -C_{45}/C, \qquad T_{55} = C_{44}/C \qquad (45)$$

while, from (44), (16), and (45), we have

$$F_{44} = (1/T)(C_{44}/C)^* \qquad F_{45} = (1/T)(C_{45}/C)^*$$
$$F_{55} = (1/T)(C_{55}/C)^* \qquad (46)$$

where

$$C = C_{44}C_{55} - C_{45}^2 \qquad (47)$$

and

$$T = (C_{44}/C)^*(C_{55}/C)^* - (C_{45}/C)^{*2} \qquad (48)$$

We now substitute (42)–(44) into (20), which establishes the result

$$\bar{C}_{i\alpha} = C^*_{i\alpha} + \frac{1}{(1/C_{33})^*}\left(\frac{C_{3\alpha}}{C_{33}}\right)^*\left(\frac{C_{3i}}{C_{33}}\right)^* - \left(\frac{C_{3i}C_{3\alpha}}{C_{33}}\right)^* \tag{49}$$

Similarly, putting (42)–(44) into (21) with $i = \alpha$ gives

$$\bar{C}_{\alpha A} = \frac{\delta_{3A}}{(1/C_{33})^*}\left(\frac{C_{3\alpha}}{C_{33}}\right)^* \tag{50}$$

and with $i = D$, Eq. (21) yields

$$\bar{C}_{DA} = F_{BA}(\delta_{3D}\delta_{3B} + C_{4D}T_{4B} + C_{5D}T_{5B})^* \tag{51}$$

Finally, substituting (44)–(46) into (51) leads to the remaining nonzero coefficients of $\bar{C}_{ij}$, viz.,

$$\bar{C}_{33} = 1/(1/C_{33})^* \tag{52}$$

and

$$\bar{C}_{AB} = (1/T)(C_{AB}/C)^* \qquad (A, B = 4, 5 \text{ only}) \tag{53}$$

The preceding analysis also indicates that

$$\bar{C}_{14} = \bar{C}_{15} = \bar{C}_{24} = \bar{C}_{25} = \bar{C}_{34} = \bar{C}_{35} = \bar{C}_{46} = \bar{C}_{56} = 0 \tag{54}$$

which, in comparison with (40) shows that the effective extensional stiffness matrix is also monoclinic. We can also observe that Eqs. (49)–(54) are identical to those given by Chou *et al.* (1972), aside from the fact that finite summations replace the present integrals since a layered system was specifically treated there. In the latter reference, however, only extensional response was considered with the implicit assumption of symmetric lamination geometry, i.e., $C_{ij}(\xi) = C_{ij}(-\xi)$, hence the matrices $\bar{B}_{i\alpha}$ and $\bar{D}_{\alpha\beta}$ were not discussed. It is also interesting to note that other approaches to define effective laminate stiffnesses for media composed of isotropic layers, such as White and Angona (1955), Postma (1955), Rytov (1956), Salamon (1968), and Behrens (1967), as pointed out by Chou *et al.* (1972), as well as Khoroshun (1966), lead to the same results as those given here when the assumption of local isotropy is invoked.

For future reference, we shall now record the forms of $\bar{B}_{i\alpha}$ and $\bar{D}_{\alpha\beta}$ for the case of local monoclinic symmetry. Following the same procedure as in the previous paragraph, i.e., utilizing Eqs. (42)–(44), the statements

$$T_{34} = T_{35} = F_{34} = F_{35} = 0 \tag{55}$$

along with results (22) and (25), we find that

$$\bar{B}_{\beta\alpha} = (\xi C_{\alpha\beta})^* + \frac{1}{(1/C_{33})^*}\left(\frac{C_{3\beta}}{C_{33}}\right)^*\left(\frac{\xi C_{3\alpha}}{C_{33}}\right)^* - \left(\frac{\xi C_{3\alpha}C_{3\beta}}{C_{33}}\right)^* \tag{56a}$$

$$\bar{B}_{3\alpha} = \frac{1}{(1/C_{33})^*}\left(\frac{\xi C_{3\alpha}}{C_{33}}\right)^* \tag{56b}$$

$$\bar{B}_{4\alpha} = \bar{B}_{5\alpha} = 0 \tag{56c}$$

and

$$\bar{D}_{\alpha\beta} = (\xi^2 C_{\alpha\beta})^* + \frac{1}{(1/C_{33})^*}\left(\frac{\xi C_{3\alpha}}{C_{33}}\right)^*\left(\frac{\xi C_{3\beta}}{C_{33}}\right)^* - \left(\frac{\xi^2 C_{3\alpha}C_{3\beta}}{C_{33}}\right)^* \tag{57}$$

for the case of monoclinic symmetry. Following the same procedure, and using the fact that

$$e_4 = e_5 = 0 \tag{58}$$

for the monoclinic case, we find that (38) reduces to

$$\bar{C}_{\beta\alpha}E_\alpha + \bar{C}_{3\beta}E_3 + \bar{B}_{\beta\alpha}K_\alpha = (C_{\beta\alpha}e_\alpha)^* - (C_{3\beta}C_{3\alpha}e_\alpha/C_{33})^* + \bar{C}_{3\beta}(C_{3j}e_j/C_{33})^* \tag{59a}$$

$$\bar{C}_{3\alpha}E_\alpha + \bar{C}_{33}E_3 + \bar{B}_{3\alpha}K_\alpha = \bar{C}_{33}(C_{3j}e_j/C_{33})^* \tag{59b}$$

$$\bar{B}_{\alpha\beta}E_\alpha + \bar{B}_{3\beta}E_3 + \bar{D}_{\alpha\beta}K_\alpha = (\xi C_{\alpha\beta}e_\alpha)^* - (\xi C_{3\beta}C_{3\alpha}e_\alpha/C_{33})^* + \bar{B}_{3\beta}(C_{3j}e_j/C_{33})^* \tag{59c}$$

$$E_4 = E_5 = 0 \tag{59d}$$

In the situation where the material structure is symmetric with respect to the central plane, i.e., $C_{ij}(\xi) = C_{ij}(-\xi)$, the coupling matrix $\bar{B}_{i\alpha}$ vanishes so that, from (59), K_α also vanish. Hence, in this case, the effective expansional strains E_i satisfy monoclinic symmetry.

III. Reduction to Classical Lamination Theory

We now turn our attention to the most familiar theory being employed in the analysis of composite laminates, classical laminated plate theory (CPT), as developed by Stavsky (1961), Dong *et al.* (1962), and utilized extensively by Ashton and Whitney (1970). CPT is an approximation theory employed to predict the response of laminated composites normally consisting of monoclinic layers. Invoking the Kirchhoff hypothesis (undeformed normals) as in standard isotropic thin plate theory, one can develop effective constitutive equations which, in terms of the present notation for normalized stress and moment resultants, can be expressed as

$$\begin{aligned} \sigma_\alpha{}^* &= A_{\alpha\beta}\epsilon_\beta{}^0 + B_{\alpha\beta}\kappa_\beta - (Q_{\alpha\beta}e_\beta)^* \\ (\xi\sigma_\alpha)^* &= B_{\alpha\beta}\epsilon_\beta{}^0 + D_{\alpha\beta}\kappa_\beta - (\xi Q_{\alpha\beta}e_\beta)^* \end{aligned} \tag{60}$$

where

$$A_{\alpha\beta} = Q^*_{\alpha\beta}, \qquad B_{\alpha\beta} = (\xi Q_{\alpha\beta})^*, \qquad D_{\alpha\beta} = (\xi^2 Q_{\alpha\beta})^* \tag{61}$$

Also,

$$Q_{\alpha\beta} = Q_{\alpha\beta}(\xi) = C_{\alpha\beta} - (C_{3\alpha}C_{3\beta}/C_{33}) \tag{62}$$

are the (layer) plane stress reduced stiffness coefficients. We will now prove that, under the conditions where $\sigma_\alpha{}^*$ and $(\xi\sigma_\alpha)^*$ are independent of x and y and $\sigma_3 = 0$, these equations are exact for monoclinic layers.

For monoclinic symmetry, because of (54), (56c), and (59d), we can easily show that the first of Eqs. (39) reduces to

$$\sigma_\alpha{}^* = \bar{C}_{\alpha\beta}(\epsilon_\beta{}^* - E_\beta) + \bar{C}_{3\alpha}(\epsilon_3{}^* - E_3) + \bar{B}_{\alpha\beta}(\kappa_\beta - K_\beta) \tag{63a}$$

$$\sigma_3 = \bar{C}_{3\beta}(\epsilon_\beta{}^* - E_\beta) + \bar{C}_{33}(\epsilon_3{}^* - E_3) + \bar{B}_{3\beta}(\kappa_\beta - K_\beta) \tag{63b}$$

$$\sigma_A = \bar{C}_{AB}\epsilon_B{}^* \qquad (A, B = 4, 5 \text{ only}) \tag{63c}$$

Putting $\sigma_3 = 0$ and eliminating $(\epsilon_3{}^* - E_3)$ from (63a, b), we obtain

$$\sigma_\alpha{}^* = \left(\bar{C}_{\alpha\beta} - \frac{\bar{C}_{3\alpha}\bar{C}_{3\beta}}{\bar{C}_{33}}\right)(\epsilon_\beta{}^* - E_\beta) + \left(\bar{B}_{\alpha\beta} - \frac{\bar{B}_{3\beta}\bar{C}_{3\alpha}}{\bar{C}_{33}}\right)(\kappa_\beta - K_\beta) \tag{64}$$

However, from (49), (50), (52), and (62), we can establish that

$$\bar{C}_{\alpha\beta} - (\bar{C}_{3\alpha}\bar{C}_{3\beta}/\bar{C}_{33}) = Q^*{}_{\alpha\beta} \tag{65}$$

while (50), (52), (56a, b), and (62) imply that

$$\bar{B}_{\alpha\beta} - (\bar{B}_{3\beta}\bar{C}_{3\alpha}/\bar{C}_{33}) = (\xi Q_{\alpha\beta})^* \tag{66}$$

Hence, (64) becomes simply

$$\sigma_\alpha{}^* = Q^*_{\alpha\beta}(\epsilon_\beta{}^* - E_\beta) + (\xi Q_{\alpha\beta})^*(\kappa_\beta - K_\beta) \tag{67}$$

In the same manner, we can eliminate $(\epsilon_3{}^* - E_3)$ in Eqs. (59a, b), which yields the result

$$Q^*_{\alpha\beta}E_\beta + (\xi Q_{\alpha\beta})^*K_\beta = (Q_{\alpha\beta}e_\beta)^* \tag{68}$$

Finally, substituting (68) into (67) gives

$$\sigma_\alpha{}^* = Q^*_{\alpha\beta}\epsilon_\beta{}^* + (\xi Q_{\alpha\beta})^*\kappa_\beta - (Q_{\alpha\beta}e_\beta)^* \tag{69}$$

Similarly, the reduced form of the second of Eqs. (39) for the monoclinic case with $\sigma_3 = 0$ becomes

$$(\xi\sigma_\alpha)^* = (\xi Q_{\alpha\beta})^*\epsilon_\beta^* + (\xi^2 Q_{\alpha\beta})^*\kappa_\beta - (\xi Q_{\alpha\beta}e_\beta)^* \tag{70}$$

We observe then that the results of (69) and (70) are identical to those of (60) and (61). We can, therefore, conclude that under the assumed conditions, the approximate theory (CPT) is exact. The assumed conditions are: (i) elastic monoclinic layers (or monoclinic heterogeneity such that $C_{ij} = C_{ij}(\xi)$ and $e_i = e_i(\xi)$, (ii) the stresses are independent of x and y, with $\sigma_3 = 0$ and σ_4, σ_5 constant. Examples are uniform bending and/or twisting, uniform membrane loading, pure interlaminar shear, as well as combinations of these elementary loadings. This observation, in conjunction with a few key exact (elasticity) solutions for boundary value problems in which these conditions are not satisfied, leads to a rather thorough understanding of the elastic response of composite flat laminates. The aforementioned exact solutions and comparison with CPT have been discussed by Pagano (1969, 1970a, b), Pagano and Wang (1971), and Pagano and Hatfield (1972). It is of interest to note that the conditions under which classical lamination theory is valid are precisely those which lead to rigorous solutions in the theory of homogeneous isotropic plates. We can also point out that the elements of $A_{\alpha\beta}$, $B_{\alpha\beta}$, and $D_{\alpha\beta}$ defined by (61) are each components of symmetric† fourth-order tensors in two-space since $Q_{\alpha\beta}$ is of this character (Tsai and Pagano, 1968), hence they all transform according to the relation

$$A'_{ijkl} = \alpha_{im}\alpha_{jn}\alpha_{kp}\alpha_{lq}A_{mnpq} \tag{71}$$

where all subscripts have the range 1, 2, and α_{im} is the direction cosine of the angle between axes x_i' and x_m. Of course, standard Cartesian tensor notation has been used in (71). The correspondence between the latter (4 subscript) and contracted (2 subscript) notations is shown by replacing mn and pq by single subscripts, say α and β, respectively, where if

$$m = n, \quad \alpha = m \qquad m \neq n, \quad \alpha = 6$$

and (72)

$$p = q, \quad \beta = p \qquad p \neq q, \quad \beta = 6$$

It is also understood that x_i' and x_i lie in the xy plane.

For convenience, we shall list here the expressions for the various components of $Q_{\alpha\beta}$ in terms of engineering constants for an orthotropic material,

† Note, however, that although $B_{\alpha\beta}$ is symmetric, $\bar{B}_{\alpha\beta}$ is not, in general, symmetric, e.g., see Eqs. (22) and (56a).

such as a unidirectional fiber-reinforced layer of a laminate. With respect to the x, y, z axes lying in the planes of elastic symmetry (one of the axes is parallel to the fibers in a unidirectional composite), we have the following (Tsai and Pagano, 1968):

$$\begin{aligned} &Q_{11} = E_{11}/(1 - \nu_{12}\nu_{21}), \qquad Q_{22} = E_{22}/(1 - \nu_{12}\nu_{21}) \\ &Q_{12} = \nu_{12}Q_{22} = \nu_{21}Q_{11}, \qquad Q_{66} = G_{12} \\ &Q_{16} = Q_{26} = 0 \end{aligned} \tag{73}$$

That is, we have the elastic constitutive law

$$\sigma_1 = Q_{11}\epsilon_1 + Q_{12}\epsilon_2, \qquad \sigma_2 = Q_{12}\epsilon_1 + Q_{22}\epsilon_2, \qquad \sigma_6 = Q_{66}\epsilon_6 \tag{74}$$

under plane stress. In (73) and (74), E stands for Young's modulus, G for shear modulus, and ν_{12} is the Poisson ratio measuring the strain in the y direction under uniaxial normal stress in the x direction. Similarly, the complete stiffness matrix C_{ij} with respect to the same x, y, z axes is given by (Tsai, 1966)

$$\begin{aligned} &C_{11} = (1 - \nu_{23}\nu_{32})HE_{11}, \quad C_{22} = (1 - \nu_{31}\nu_{13})HE_{22}, \quad C_{33} = (1 - \nu_{12}\nu_{21})HE_{33} \\ &C_{12} = (\nu_{21} + \nu_{23}\nu_{31})HE_{11}, \quad C_{13} = (\nu_{31} + \nu_{21}\nu_{32})HE_{11}, \quad C_{23} = (\nu_{32} + \nu_{12}\nu_{31})HE_{22} \\ &C_{44} = G_{23}, \qquad C_{55} = G_{13}, \qquad C_{66} = G_{12} \\ &H = (1 - \nu_{12}\nu_{21} - \nu_{23}\nu_{32} - \nu_{31}\nu_{13} - 2\nu_{12}\nu_{23}\nu_{31})^{-1} \end{aligned} \tag{75}$$

where the constitutive law is given by (1) and (2). The elements of C_{ij} that are not listed in (75) vanish for the axes being considered here, i.e., the orthotropic case.

IV. Boundary Layer Response

The preceding formulation gives the equations which define the effective moduli of a laminated system comprised of anisotropic layers. Once these effective moduli have been determined, the detailed stress field can be described by returning to the local constitutive law (11). Clearly, the latter, in conjunction with (3) and (4), require that the stress components σ_α be linear functions of the thickness coordinate within each layer, while σ_A are constant. Thus, in a prism having finite boundaries, and boundary tractions which precisely follow these distributions, the solution is exact. However, this situation will rarely occur in actual laminates, e.g., at a free

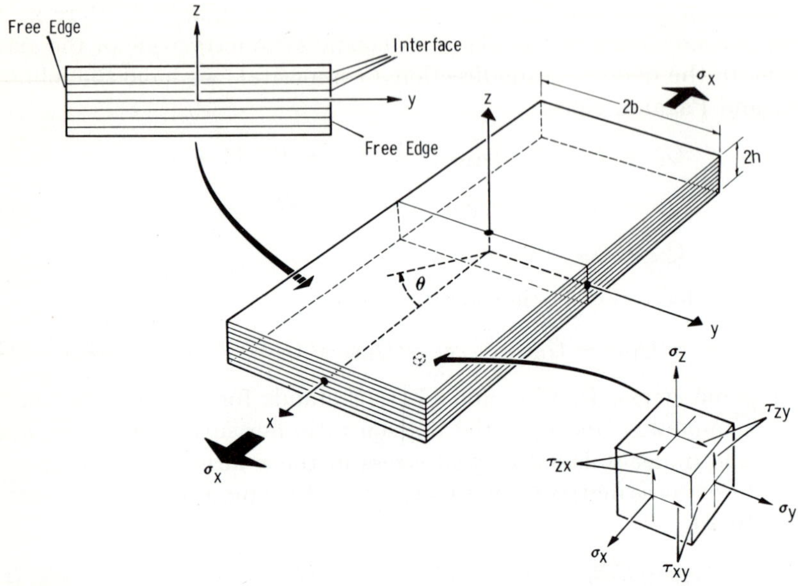

FIG. 2. Laminate geometry.

boundary, the surface tractions vanish at each point, while the present solution can only guarantee zero force and moment *resultants*. Therefore, one may question the meaning of the stress field given by the present analysis. Recent work by Pipes and Pagano (1970) and Rybicki (1971), however, has illustrated that in all cases treated in these publications, the stress field only suffers disturbances in boundary layer regions of a dimension comparable to the laminate thickness. These disturbances are attributed to high localized interlaminar stresses σ_A, which vanish in interior regions. Furthermore, the interior stress field agrees with that given in the present formulation. Hence, for a laminate in which the thickness is much smaller than its other dimensions and which the loading functions N_α, M_α, and σ_A are constant, we have strong evidence to support the premise that the present work yields accurate results except in narrow boundary layer regions.†

As an example of the nature of the boundary layer disturbance discussed above, we present some of the features of the elasticity solution for the laminate shown in Fig. 2. The body consists of 4 anisotropic unidirectional

† Timoshenko (1925) has made a similar observation in his work on the bimetallic strip problem, although no proof was given.

composite layers in which the fiber orientations are given by $[-\theta, \theta, \theta, -\theta]$ in the various layers, from top to bottom (or vice versa), respectively. In the example chosen, $\theta = 45°$. The layer engineering moduli are those representative of a high modulus graphite–epoxy fiber-reinforced material,

$$E_{\mathrm{l}} = 20 \times 10^6 \quad \text{psi}, \qquad E_{\mathrm{t}} = E_z = 2.1 \times 10^6 \quad \text{psi}$$

$$\nu_{\mathrm{lt}} = \nu_{\mathrm{l}z} = \nu_{\mathrm{t}z} = 0.21$$

$$G_{\mathrm{lt}} = G_{\mathrm{l}z} = G_{\mathrm{t}z} = 0.85 \times 10^6 \quad \text{psi}$$

where l is the fiber direction and t is the transverse direction (in the plane of the layer). The laminate is subjected to a constant uniaxial strain ϵ_x while the boundaries $z = \pm 2h_0$ and $y = \pm \mathrm{b}$ are traction-free. The problem was solved numerically by use of the finite-difference procedure. Results giving the distribution of the interlaminar stresses $\sigma_A(h_0, y)$ and membrane stresses $\sigma_\alpha(h_0, y)$ are shown in Fig. 3. The boundary layer phenomena and

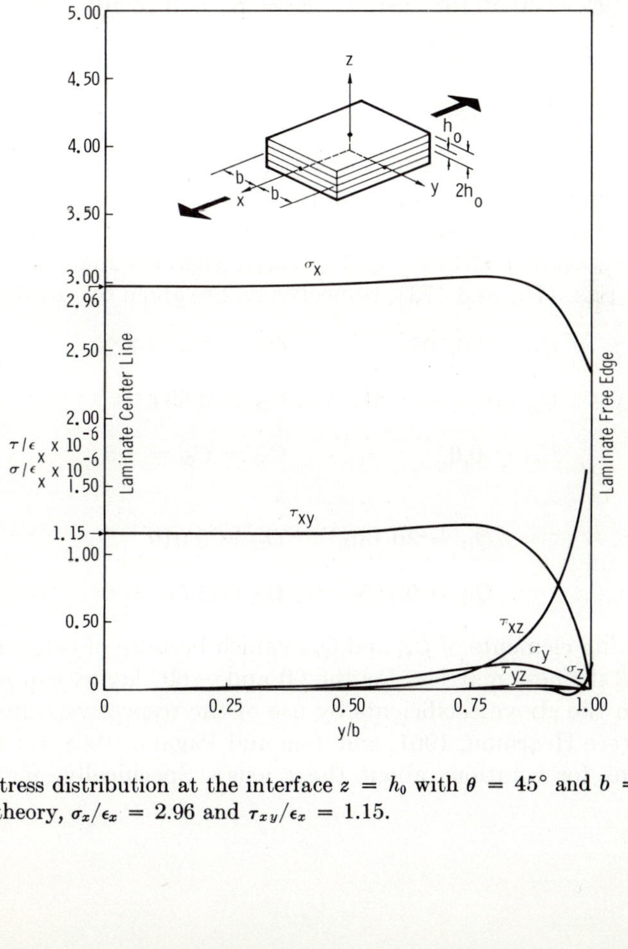

FIG. 3. Stress distribution at the interface $z = h_0$ with $\theta = 45°$ and $b = 8h$. Using lamination theory, $\sigma_x/\epsilon_x = 2.96$ and $\tau_{xy}/\epsilon_x = 1.15$.

agreement with the present formulation in regions away from the free edges are quite evident.

V. Illustrative Problem

As an illustration of the present formulation, we consider the effective moduli of a laminate consisting of 3 layers of unidirectional boron–epoxy. The layers are of equal thickness $h/3$ and the fiber orientations are -60, 0, and 60° with respect to the x axis for the bottom, middle, and top layers, respectively. Here, the positive angle from the x axis to the fiber direction is measured in the clockwise direction, as viewed from the positive extension of the z axis, and the fibers in each layer lie in planes parallel to xy. The layer elastic moduli are taken as

$$E_{\rm l} = 30 \times 10^6 \quad \text{psi}, \qquad E_{\rm t} = E_z = 3 \times 10^6 \quad \text{psi}$$

$$\nu_{\rm lt} = \nu_{{\rm l}z} = \nu_{{\rm t}z} = 0.25$$

$$G_{\rm lt} = G_{{\rm l}z} = 1.5 \times 10^6 \quad \text{psi}, \qquad G_{{\rm t}z} = 0.6 \times 10^6 \quad \text{psi}$$

The fact that the Poisson ratios are selected to be equal is incidental. The general formulation makes no assumption specifying relationships among any of the local elastic coefficients.

The corresponding stiffness and reduced stiffness matrices for the 0° layer, from Eqs. (75) and (73), respectively, are given by (in 10^6 psi units)

$$C_{11} = 30.508, \qquad C_{22} = C_{33} = 3.234$$

$$C_{12} = C_{13} = 1.017, \qquad C_{23} = 0.834$$

$$C_{44} = 0.6, \qquad C_{55} = C_{66} = 1.5$$

and

$$Q_{11} = 30.189, \qquad Q_{22} = 3.019$$

$$Q_{12} = 0.755, \qquad Q_{66} = 1.5$$

The remaining elements of C_{ij} and $Q_{\alpha\beta}$ vanish because of orthotropic symmetry. The stiffness matrices for the 60 and $-60°$ layers can now be obtained from the above coefficients by use of the respective transformation equations (see Hearmon, 1961, and Tsai and Pagano, 1968, for the simplified versions for rotations about the z axis). Specifically, for the $\pm 60°$

layers, we have (10^6 psi)

$$
\begin{aligned}
&C_{11} = 5.232, && C_{22} = 18.869, && C_{33} = 3.234\\
&C_{12} = 5.837, && C_{13} = 0.880, && C_{23} = 0.971\\
&C_{16} = \mp 3.122, && C_{26} = \mp 8.688, && C_{36} = \mp 0.079\\
&C_{44} = 1.275, && C_{55} = 0.825, && C_{45} = \mp 0.390\\
&C_{66} = 6.320
\end{aligned}
$$

and

$$
\begin{aligned}
&Q_{11} = 4.993, && Q_{22} = 18.578, && Q_{12} = 5.573\\
&Q_{16} = \mp 3.101, && Q_{26} = \mp 8.664, && Q_{66} = 6.318
\end{aligned}
$$

The remaining elements of C_{ij} all vanish, and the upper and lower signs refer to the +60 and −60° layers, respectively. Because of these sign reversals, we observe that

$$C^*_{16} = C^*_{26} = C^*_{36} = C^*_{45} = Q^*_{16} = Q^*_{26} = 0$$

while it is easily shown that C_{33} and $C_{44}C_{55} - C^2_{45}$ are invariants with respect to rotation about the z axis for monoclinic layers of the same material. Hence, in this case, Eqs. (49), (52), and (53), along with definition (62), reduce to

$$
\begin{aligned}
&\bar{C}_{\beta\alpha} = Q^*_{\beta\alpha} + (C^*_{3\beta}C^*_{3\alpha}/C_{33}) \qquad (\alpha = 1, 2)\\
&\bar{C}_{3\alpha} = C^*_{3\alpha} \qquad (\alpha = 1, 2)\\
&\bar{C}_{66} = Q^*_{66}, \qquad \bar{C}_{33} = C_{33}, \qquad \bar{C}_{44} = C/C^*_{55}, \qquad \bar{C}_{55} = C/C^*_{44}
\end{aligned}
\tag{76}
$$

where

$$C = C_{44}C_{55} - C^2_{45} \tag{77}$$

while the remaining elements of $\bar{C}_{ij}$ vanish. In similar fashion, Eqs. (56) lead to the following nonvanishing components of $\bar{B}_{i\alpha}$:

$$
\begin{aligned}
&\bar{B}_{\alpha 6} = (\xi Q_{\alpha 6})^* + (C^*_{3\alpha}/C_{33})(\xi C_{36})^* \qquad (\alpha = 1, 2)\\
&\bar{B}_{6\alpha} = (\xi Q_{\alpha 6})^* \qquad (\alpha = 1, 2)\\
&\bar{B}_{36} = (\xi C_{36})^*
\end{aligned}
\tag{78}
$$

Finally, Eqs. (57) lead to the nonvanishing elements of $\bar{D}_{\alpha\beta}$,

$$
\begin{aligned}
&\bar{D}_{\alpha\beta} = (\xi^2 Q_{\alpha\beta})^* \qquad (\alpha, \beta = 1, 2)\\
&\bar{D}_{66} = (\xi^2 Q_{66})^* + (\bar{B}^2_{36}/C_{33})
\end{aligned}
\tag{79}
$$

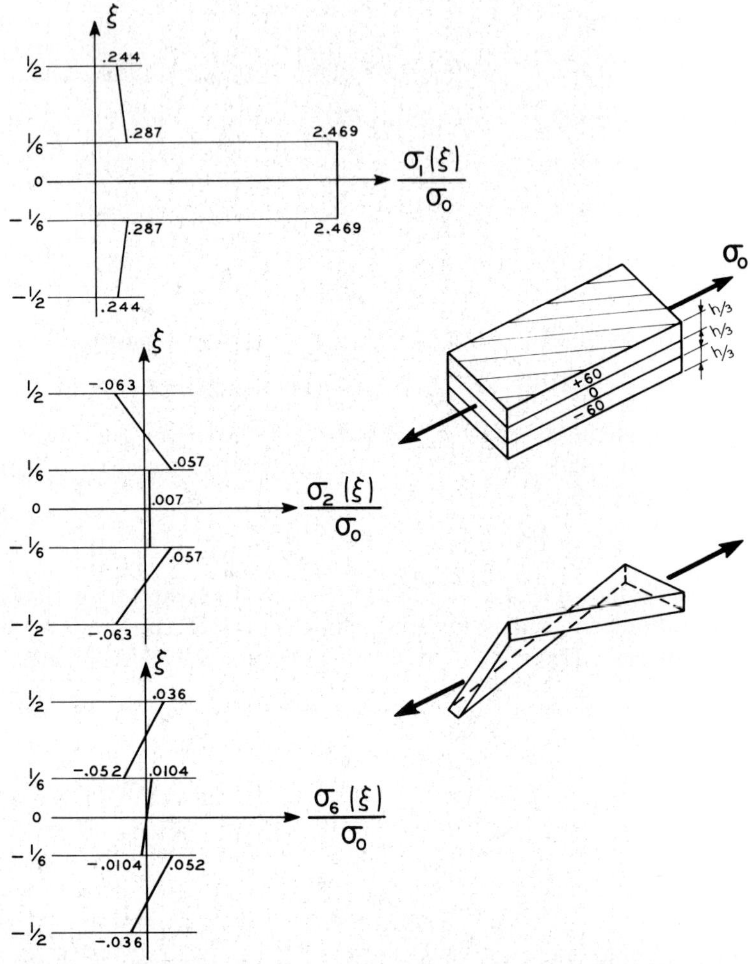

FIG. 4. Stress distribution in laminate under uniaxial load.

In Eqs. (76)–(79), any starred function, say f^*, in the present example, is given by

$$f^* = \int_{-1/2}^{-1/6} f^{(1)}(\xi)\, d\xi + \int_{-1/6}^{1/6} f^{(2)}(\xi)\, d\xi + \int_{1/6}^{1/2} f^{(3)}(\xi)\, d\xi \tag{80}$$

where the numbers (1), (2), (3) stand for the -60, 0, and 60° layers, respectively. We now substitute the respective values of C_{ij} and $Q_{\alpha\beta}$ into

(76)–(80) to arrive at the effective stiffness matrices (10^6 psi),

$$\bar{C}_{11} = \bar{C}_{22} = 13.656, \qquad \bar{C}_{33} = 3.234$$
$$\bar{C}_{12} = 4.232, \qquad \bar{C}_{13} = \bar{C}_{23} = 0.925$$
$$\bar{C}_{44} = \bar{C}_{55} = 0.857, \qquad \bar{C}_{66} = 4.712$$

and

$$\bar{B}_{16} = -0.694, \qquad \bar{B}_{61} = -0.689$$
$$\bar{B}_{26} = -1.930, \qquad \bar{B}_{62} = -1.925$$
$$\bar{B}_{36} = -0.018$$

and

$$\bar{D}_{11} = 0.494, \qquad \bar{D}_{22} = 1.500$$
$$\bar{D}_{12} = 0.450, \qquad \bar{D}_{66} = 0.512$$

Thus, according to (30), the above effective stiffnesses characterize the elastic constitutive behavior of the present laminate. Inversion of (30) leads to the determination of the geometric variables ϵ_i^* and κ_α. Thus, under a given set of prescribed "loads," σ_i^* and $(\xi\sigma_\alpha)^*$, one can return to (18) and compute the detailed distribution of stresses $\sigma_i(\xi)$. Suppose, for example, that the uniaxial loading

$$\sigma_1^* = \sigma_0, \qquad \sigma_2^* = \sigma_6^* = \sigma_A^* = (\xi\sigma_\alpha)^* = 0$$

is imposed on the laminate. Inversion of (30), or more simply in this case, (69) and (70), leads to

$$\epsilon_1 = 0.0823 \times 10^{-6}\sigma_0, \qquad \epsilon_2^* = -0.0184 \times 10^{-6}\sigma_0,$$
$$\kappa_6 = 0.0416 \times 10^{-6}\sigma_0, \qquad \epsilon_6^* = \kappa_1 = \kappa_2 = 0$$

Substituting into the reduced form of (18), namely

$$\sigma_\alpha = Q_{\alpha\beta}(\epsilon_\beta^* + \xi\kappa_\beta), \qquad \sigma_A = 0 \tag{81}$$

the stress field corresponding to the given loading can be determined. The resulting distributions of $\sigma_\alpha(\xi)$ are shown in Fig. 4.

VI. Summary

In summary, we have presented an approach to define the effective coefficients, $\bar{C}_{ij}$, $\bar{B}_{i\alpha}$, $\bar{D}_{\alpha\beta}$, or the extensional stiffnesses, coupling matrix, and flexural rigidities, respectively, as well as the effective expansional coefficients, for anisotropic laminates or materials in which the elastic constants vary in one coordinate direction. The formulation is rigorous within the framework of three-dimensional heterogeneous elasticity theory. No local material symmetry has been assumed, i.e., at each point in the medium, the elastic constitutive law may involve 21 independent moduli. This work unifies several contributions to the literature which have treated similar, but more specialized, classes of materials. It also clarifies the approximations involved in classical laminated plate theory. Finally, aside from highly localized boundary layer regions under loadings which satisfy σ_A = constant, $(\xi\sigma_\alpha)^*$ = constant, one can compute the detailed stress field in prismatic bodies of the type treated here.

References

Ashton, J. E., and Whitney, J. M. (1970). Theory of Laminated Plates. Technomic, Westport, Connecticut.

Behrens, E. (1967). *J. Acoust. Soc. Amer.* **42**, 378.

Berkowitz, H. M., and Cohen, L. J. (1970). Elastic Properties of Multi-directionally Reinforced Composites. McDonnell Douglas Tech. Paper 10064.

Chou, P. C., Carleone, J., and Hsu, C. M. (1972). *J. Compos. Mater.* **6**, 80.

Dong, S. B., Pister, K. S., and Taylor, R. L. (1962). *J. Aerosp. Sci.* **29**, 969.

Halpin, J. C., and Pagano, N. J. (1969). *J. Compos. Mater.* **3**, 720.

Halpin, J. C., and Pagano, N. J. (1970). *In* "Recent Advances in Engineering Science" (A. C. Eringen, ed.), pp. 33–46. Gordon and Breach, New York.

Halpin, J. C., Jerina, K., and Whitney, J. M. (1971). *J. Compos. Mater.* **5**, 36.

Hearmon, R. F. S. (1961). "Applied Anisotropic Elasticity." Oxford, London and New York.

Khoroshun, L. P. (1966). *Sov. Appl. Mech.* **2**, 9.

Lehknitskii, S. G. (1963). "Theory of Elasticity of an Anisotropic Elastic Body." Holden-Day, San Francisco, California.

Pagano, N. J. (1969). *J. Compos. Mater.* **3**, 398.

Pagano, N. J. (1970a). *J. Compos. Mater.* **4**, 20.

Pagano, N. J. (1970b). *J. Compos. Mater.* **4**, 330.

Pagano, N. J., and Hatfield, S. J. (1972). *J. AIAA* **10**, 931.

Pagano, N. J., and Wang, A. S. D. (1971). *J. Compos. Mater.* **5**, 521.

Pipes, R. B., and Pagano, N. J. (1970). *J. Compos. Mater.* **4**, 538.

Postma, G. W. (1955). *Geophysics* **20**, 780.

Rybicki, E. F. (1971). *J. Compos. Mater.* **5**, 354.

Rytov, S. M. (1956). *Sov. Phys. Acoust.* **2**, 68.

Salamon, M. D. G. (1968). *Int. J. Rock Mech. Min. Sci.* **5**, 519.

Stavsky, Y. (1900). *Proc. Amer. Soc. Civil Eng. J. Eng. Mech. Div.* **87**, 31.

Tabaddor, F. (1972). *J. Appl. Mech.* **39**, 598.

Timoshenko, S. (1925). *J. Opt. Soc. Amer.* **11**, 233.

Tsai, S. W. (1966). AFML-TR-66-149, Part II.

Tsai, S. W., and Pagano, N. J. (1968). *In* "Composite Materials Workshop" (S. W. Tsai, J. C. Halpin, and N. J. Pagano, eds.), pp. 233–253. Technomic, Westport, Connecticut.

White, J. E., and Angona, F. A. (1955). *J. Acoust. Soc. Amer.* **27**, 311.

3

Elastic Behavior of Composites

G. P. SENDECKYJ

Advanced Composites Branch, Structures Division
Air Force Flight Dynamics Laboratory, Air Force Systems Command
Wright–Patterson Air Force Base, Ohio

I. Introduction

The literature dealing with composite materials has undergone a phenomenal growth since the early 1960s at which time glass–epoxy composites came into prominent use. Further impetus was provided by the introduction of boron and graphite fibers and their composites. The specific properties (such as uniaxial modulus and strength) were sufficiently greater than those of monolithic metals to make application of these materials very attractive for weight and stiffness sensitive structures. This led to extensive material development funded by the Air Force Materials Laboratory.

When attempts were made to predict the behavior of the newly developed materials, it was soon found that the available theoretical methods were far from satisfactory. As a result, theoretical work dealing with the prediction of composite properties from constituent properties was funded profusely. This had the natural effect of leading to considerable duplication of effort and a plethora of contractor reports which seldom found their way into the open literature. The *Journal of Composite Materials* was founded in 1967 as an organ for researchers working in the area of composite materials. This journal was rapidly followed by two more journals, *Composites* and *Fibre Science and Technology*.

Considering the growth of the subject, it is not surprising that survey articles (Hashin, 1964; Chamis and Sendeckyj, 1968) became obsolete on the date of publication. The literature dealing with the elastic behavior of composite materials has recently matured to the point where monographs have appeared stressing particular aspects of the subject (Ashton *et al.*, 1969, Hashin, 1972; Frantsevich and Karpinos, 1972). Ashton *et al.* (1969) present an excellent introduction to the subject and it should be mandatory reading for those not intimately familiar with the theoretical work dealing with composite materials. On the other hand, Hashin (1972) and Frantsevich and Karpinos (1972) give detailed treatments of special topics.

Obviously this chapter cannot deal as deeply with all aspects of the elastic behavior of composites as might be desired. At best it is an attempt to cull the best results from the literature, indicating the approaches used and their limitations. Special attention is paid to the formulation of the various theories.

A. Types of Composite Materials

For the purpose of this work, a *composite material* is defined to be any material consisting of two or more distinct constituents (or phases) that are insoluble in one another. Besides materials that are commonly considered to be composites, this definition includes dispersion-hardened alloys, concrete, coated metals, etc. Polycrystalline solids, which consist of a single constituent, can be included within the scope of the definition if the grains are considered to be distinct phases for each grain orientation.

For the sake of convenience, one of the phases will be referred to as the *matrix,* while the others as the *reinforcement.* Even though reinforcement implies strengthening of the material, the term will be used here to denote any phase that is imbedded in the matrix. Thus, voids are included within this term.

Composite materials are commonly characterized by the shape of the reinforcement. The main types of reinforcement are particles, chopped (or discontinuous) fibers, continuous fibers (or filaments), and flakes.

Composites, reinforced by particles having all dimensions of the same order of magnitude, are referred to as *particulate composites.* Materials that can be classified as particulate composites can vary widely from dispersion-hardened alloys and syntactic foams to neutron-irradiated metals (containing dispersed vacancies). Polycrystalline solids can also be considered to be particulate composites containing a zero matrix volume content. Even though recent work dealing with composites has emphasized fibrous composites, particulate composites enjoy the unique position of being the material for which most of the analysis methods have been first used.

Composites, reinforced by particles having two dimensions much larger than the third, are referred to as *flake-reinforced materials.* Even though these materials have the attractive property of being isotropic in a plane containing the flakes, little theoretical work has been done for this type of reinforcement. Hence, this composite will be omitted from consideration in this chapter.

By far the most extensively analyzed composite material is the fiber-reinforced composite of which there are two main types, namely, the continuous fiber and short (chopped) fiber composites. Continuous fiber-reinforced composites can be made either by aligning all the fibers or by weaving a cloth and then impregnating the resulting structure with a matrix material. Since the weaving process leads to fiber damage resulting in inferior composites, emphasis will be placed on the uniaxial continuous fiber (filamentary) composites.

While filamentary composites have exceptional strength and stiffness properties in the direction of the fibers, their transverse properties are rather poor. The overall properties are normally improved by laminating single plies (in which all fibers have the same orientation). This leads to what is called *laminated composites* or *laminates*. These are discussed in Chapter 2.

Since filamentary composites are most commonly encountered, it is useful to examine the constituents of these materials. The most commonly used fibers are glass, boron (with a tungsten substrate), and carbon (graphite). The glass fibers are circular in cross section. Boron fibers are also circular with surface irregularities resembling corn on the cob. Graphite fibers can be either circular or extremely irregular in shape depending on the manufacturer.

Both polymers and metals have been used as the matrix material. Among the polymeric matrices, we have the epoxies and polyimides; while, aluminum and nickel have been used as the metallic matrices. The choice of the matrix material depends on the potential application of the composite. For example, fiber-reinforced epoxy would be used in low temperature environments, while metal matrix composites would be used at high temperatures.

Materials containing short fibers whose length is about a hundred times the other two dimensions are called *chopped fiber-reinforced composites*. The short fibers can either be aligned or arbitrarily distributed. Furthermore, fiber lengths can vary anywhere from about an inch down to a microinch (for whiskers). Since the strength of the fibers increases with decreasing fiber length and diameter, short fiber-reinforced composites have a tremendous potential. Attempts to exploit this potential have led to whisker- (elongated single crystals) reinforced composites and metal–metal composites made by eutectic decomposition.

B. Modeling of Composite Materials

Considering the variety of possible composite materials, it is apparent that no single all inclusive theoretical treatment is possible. The present chapter is restricted to the linearly elastic behavior of composites under static loading conditions. (Elastoplastic, viscoelastic, dynamic, and finite deformations are considered in Chapters 5, 4, 8, and 7, respectively.) This implies that macroscopic behavior of the composite is linearly elastic to failure. Furthermore, the constituent properties are also linear; that is, the composite is assumed to be a heterogeneous linearly elastic solid.

Since all materials are heterogeneous if a small enough scale is chosen, the heterogeneity of a composite needs some clarification. To illustrate what is meant by the heterogeneity of a composite, consider a two-phase solid. On the atomic level, the phases are heterogeneous. As the size of the region being examined increases, a point is reached at which the atomic structure loses importance and the phases can be considered to be homogeneous elastic solids. This puts a lower limit on the size of the reinforcing particles. For particles below this limiting size, interatomic forces must be taken into account and little can be done by way of analysis. Fortunately, reinforcing particles (fibers, flakes, etc.) are large enough to be considered homogeneous.

Over distances of the order of the particle size, the composite is heterogeneous. If the distances over which quantities such as stresses are defined are much greater than the particle size (as measured by some characteristic dimension), the composite can be viewed as being homogeneous. This is exactly the situation encountered in metals which are treated as being homogeneous for the purpose of stress analysis while actually being heterogeneous.

Even though the above discussion indicates how a composite material can be treated as being both heterogeneous and homogeneous at the same time, it fails to provide us with a criterion that would tell us when a composite material can be assumed to be homogeneous. Such a criterion would be useful in establishing the ranges of validity of various theoretical treatments of composite materials. At present, little work has been done to establish such a criterion. The common practice has been, and still is, to assume that the heterogeneous composite can be replaced by an equivalent homogeneous solid with properties established experimentally. Theoretical estimates of the properties of the equivalent solid are then made on the basis of models that are analyzable but not necessarily reflecting the structure of the composite. These estimates are compared with experimental data. If reasonable agreement is obtained, the model is accepted and further work stops. The various models that have been used are the subject of this chapter.

Before turning to the discussion of the models and resulting estimates for the elastic properties of the composite, it is convenient to introduce the concept of a representative volume. By a *representative volume*, we mean any region of the composite whose minimum dimension is much larger than the characteristic dimensions of the phases. The term *macroscopic* will be used to refer to regions that are larger than the representative volume; while, the term *microscopic* will refer to regions that are of the order of the characteristic dimensions of the phases. For example, macroscopic stresses refer to stresses defined over regions greater in size than the representative volume.

For the purpose of this chapter, we assume that the macroscopic stresses vary sufficiently slowly with position so that they can be considered to be uniform over a representative volume. If the variation is sufficiently fast so that the macroscopic stresses are not uniform over the representative volume, the analyses presented below break down. We assume further that the phases are perfectly bonded; that is, the microscopic tractions and displacements are continuous as we pass from one phase to another.

C. Scope of the Chapter

As indicated above, the present work is restricted to the linearly elastic behavior of composites under static loading conditions. With the exception of Section II where some general results are derived, only particulate and fiber-reinforced composites are considered. The reader is referred to Chapter 2 for a discussion of layered composites.

Section II presents the derivation of exact expressions for the effective (macroscopic) elastic moduli of composites in terms of the phase volume fractions and elastic moduli of the constituents. These expressions depend on the average stress and strain concentration factors for the phases. Furthermore, it is shown that rudimentary information about the phase geometry allows one to derive relations between the effective elastic moduli that depend only on the volume fractions and moduli of the phases but not on average stress and strain concentration factors.

If we assume that the shear moduli of the phases are identical, exact expressions for the bulk modulus of a particulate composite and for the transverse planar bulk modulus of a fiber-reinforced composite can be derived for arbitrary phase geometries. The results of these derivations are presented in Section II,C. If the phase geometry is known *a priori*, the microscopic stress distribution can also be established and an explicit solution is presented for the transverse microscopic stresses for a fiber-reinforced composite modeled as an arbitrary array of circular inclusions in an unbounded matrix.

Section III summarizes various approximations for the effective elastic moduli. In particular, the predicted effective elastic moduli are given for:

(i) the Reuss and Voigt models,
(ii) sphere (cylinder) incased in a spherical (cylindrical) shell,
(iii) sphere (cylinder) incased in a spherical (cylindrical) shell which in turn is incased in an unbounded matrix possessing the effective moduli of the composite, and
(iv) Hill's self-consistent model.

The variational approach for deriving bounds on the effective moduli is described in Section IV. Derivations based on classical variational principles are used to illustrate this approach. Best bounds, based on new variational principles, are also presented.

The extensive literature dealing with fiber-reinforced composites, modeled as doubly periodic arrays of identical fibers in an unbounded matrix, is discussed in Section V. The statistical approach for predicting effective moduli is presented in Section VI. The derivation of the basic equations is given in considerable detail and approximations necessary for their solution are indicated. Explicit results for particulate composites are presented and discussed. No results are given for fiber-reinforced composites due to their complexity and crudeness.

Section VII contains a brief discussion of various other approaches for predicting the effective elastic moduli. Among these we have the method of long waves and the mechanics of materials method. Work dealing with the effect of fiber curvature and twist, random distribution of fibers, and chopped fiber composites is also discussed.

Finally, the Appendix presents a brief treatment of the effective coefficients of thermal expansion.

II. Some Exact Results

A major problem in micromechanics consists of deriving relations between the effective elastic moduli of the composite and the properties of the constituents (or phases). In addition, the stress and strain distributions within the phases have to be determined for a complete understanding of material behavior. Considering the complicated structure of real composites, it is apparent that only limited theoretical results can be derived. The present section is restricted to some fundamental results, due to Hill (1963, 1964), relating the effective elastic moduli of the composite to the phase volume fractions and elastic moduli of the constituents.

With this in mind, consider a composite manufactured from homogeneous phases which can have any shape and distribution. If the phases are anisotropic, it will be assumed that the material axes of each constituent are aligned. For a given macroscopic stress distribution imposed on the composite, the stresses and strains are complicated functions of the volume fraction v, distribution, shape, and elastic moduli of the constituents. Some universally valid results, relating the effective moduli of the composite and constituent properties, can be derived by a careful averaging of the stress and strain fields within the phases. These results are presented in this section. Even though the results are valid for any number of phases, the derivations are given for the case of a two phase composite.

Averaging the stresses and strains in the composite yields

$$\langle \sigma_{ij} \rangle = v_1 \langle \sigma_{ij}^1 \rangle + v_2 \langle \sigma_{ij}^2 \rangle \tag{1}$$

$$\langle e_{ij} \rangle = v_1 \langle e_{ij}^1 \rangle + v_2 \langle e_{ij}^2 \rangle \tag{2}$$

where the volume average is defined as

$$\langle \phi \rangle = (1/v) \int_v \phi \, dv \tag{3}$$

Subscripts on the volume fractions and superscripts on elastic quantities are used to refer to the phases. In each phase, the stresses and strains are related by the generalized Hooke's law

$$\sigma_{ij}^k = C_{ijmn}^k e_{mn}^k \tag{4}$$

where C_{ijmn} are elastic moduli.† For a discussion of the symmetry properties of C_{ijmn} the reader is referred to any standard text on anisotropic elasticity. Upon averaging, Eq. (4) can be written as

$$\langle \sigma_{ij}^k \rangle = C_{ijmn}^k \langle e_{mn}^k \rangle \tag{5}$$

For a representative volume (defined in Section I) with a uniform imposed stress state, there exists a unique dependence of the macroscopic stresses on the macroscopic strains which can be written as

$$\langle \sigma_{ij} \rangle = C_{ijmn}^* \langle e_{mn} \rangle \tag{6}$$

Herein, C_{ijmn}^* are the effective elastic moduli of the composite. The main problem of this section consists of finding "exact" relationships between C_{ijmn}^*, C_{ijmn}^k, and v_k. This problem can be dealt with as follows.

A. *Effective Elastic Moduli in Terms of Average Strain and Stress Concentration Factors*

Following Hill (1963), we assume the existence of a unique dependence of the average strains in the phases on the overall strains in the composite which can be written as

$$\langle e_{ij}^k \rangle = a_{ijmn}^k \langle e_{mn} \rangle \tag{7}$$

where a_{ijmn}^k are constants defining the average strain concentration factors

† The Einstein summation convention, in which repeated subscripts are summed over from one to three, is used throughout this chapter.

for the kth phase and $\langle e_{mn} \rangle$ are the uniform macroscopic strains corresponding to the uniform macroscopic stresses $\langle \sigma_{ij} \rangle$. It should be noted that the constants a^k_{ijmn} are not independent. In view of Eq. (2) they are related by

$$v_1 a^1_{ijmn} + v_2 a^2_{ijmn} = I_{ijmn} = \delta_{ij}\delta_{mn} + \delta_{im}\delta_{jn} \tag{8}$$

where δ_{ij} is the Kronecker delta

$$\delta_{ij} = \begin{cases} 1 & \text{if} \quad i = j \\ 0 & \text{if} \quad i \neq j \end{cases} \tag{9}$$

The tensors a^k_{ijmn} depend on the volume fractions, moduli, distribution, and shape of the phases.

Substitution of Eq. (7) into Eq. (5) and the resulting expressions into Eq. (1) yields Eq. (6) with

$$C^*_{ijmn} = v_1 C^1_{ijkl} a^1_{klmn} + v_2 C^2_{ijkl} a^2_{klmn} \tag{10}$$

Equation (10) is an *exact* expression for the effective elastic moduli in terms of the average strain concentration factors, moduli, and volume fractions of the constituents.

Similarly, by writing the generalized Hooke's law for the composite and phases in terms of the elastic compliances S_{ijmn} and repeating the derivations, we obtain

$$S^*_{ijmn} = v_1 S^1_{ijkl} b^1_{klmn} + v_2 S^2_{ijkl} b^2_{klmn} \tag{11}$$

where b^k_{ijmn} are the average phase stress concentration factors defined by

$$\langle \sigma^k_{ij} \rangle = b^k_{ijmn} \langle \sigma_{mn} \rangle \tag{12}$$

Equations (10) and (11) define the effective elastic properties of the composite. Since a^k_{ijmn} and b^k_{ijmn} are not known *a priori*, these results are of limited practical value. If reasonable consistent estimates for a^k_{ijmn} and b^k_{ijmn} can be obtained, then Eqs. (10) and (11) will give the effective elastic moduli. Various approximations for the average strain and stress concentration factors are possible and the simplest of these are summarized in Section III for particulate and filamentary composites when the phases are isotropic.

Alternately, if the effective elastic moduli of the composite are known (e.g., from experimental data), then Eqs. (10) and (11) can be used to determine the average strains and stresses in the phases.

B. *Relations between Effective Elastic Moduli*

The results presented in Section II,A are weak, in that knowledge of average strain and stress concentration factors is required. If further

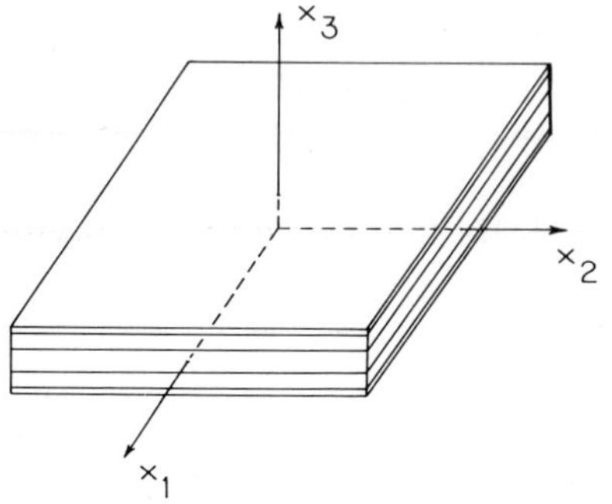

FIG. 1. Coordinate system used in analysis of layered composite.

information about the structure of the composite is available, stronger results can be derived. To illustrate the derivations, assume that the composite is made by alternately stacking two orthotropic layers (or plies). For convenience take the coordinate axes as shown in Fig. 1.

The assumption of orthotropic layers implies that some of the constants C^k_{ijmn} are identically zero, i.e.,

$$C^k_{1123} = C^k_{2223} = C^k_{1131} = C^k_{2231} = C^k_{1223} = C^k_{1231} = C^k_{3312} = 0 \quad (13)$$

By appropriately choosing the in-plane coordinates, additional elastic constants can be made to vanish for one of the plies; but, this does not lead to any simplifications.

The effective moduli of the layered composite (laminate) are formally defined by Eq. (6). Upon using Eq. (1), Eq. (6) can be written in terms of strain quantities as

$$C^*_{ijmn}\langle e_{mn}\rangle = v_1 C^1_{ijmn}\langle e^1_{mn}\rangle + v_2 C^2_{ijmn}\langle e^2_{mn}\rangle \quad (14)$$

Now let us perform an armchair experiment in which the laminate is loaded by smooth rigid planes so that

$$\langle e_{11}\rangle = \langle e^k_{11}\rangle, \qquad \langle e_{22}\rangle = \langle e^k_{22}\rangle, \qquad \langle e^k_{ij}\rangle = 0 \qquad (i \neq j) \quad (15)$$

For such loading conditions, Eq. (14) can be rewritten as

$$v_1(C^*_{ijmn} - C^1_{ijmn})\langle e^1_{ij}\rangle + v_2(C^*_{ijmn} - C^2_{ijmn})\langle e^2_{ij}\rangle = 0 \quad (16)$$

which gives three equations in four unknown strain quantities ($\langle e_{11}\rangle$, $\langle e_{22}\rangle$, $\langle e^1_{33}\rangle$, $\langle e^2_{33}\rangle$). Without loss of generality the loading planes can be adjusted so that two of the average strains vanish. For example, let $\langle e_{11}\rangle = \langle e_{22}\rangle = 0$, which corresponds to compression or extension of the composite in the x_3 direction between constraining walls parallel to the x_1x_2 axes. In this case, Eqs. (16) reduce to three equations in two strains. Solving each of these equations for $\langle e^1_{33}\rangle / \langle e^2_{33}\rangle$ and equating the results yields

$$\frac{C^*_{1133} - C^2_{1133}}{C^*_{1133} - C^1_{1133}} = \frac{C^*_{2233} - C^2_{2233}}{C^*_{2233} - C^1_{2233}} = \frac{C^*_{3333} - C^2_{3333}}{C^*_{3333} - C^1_{3333}} \tag{17}$$

Equation (17) gives two relations between the effective elastic moduli of the composite. Further relations follow by performing additional armchair experiments. For these the reader is referred to the work of Sendeckyj (1972).

Hill (1964) carried out similar derivations for a fiber-reinforced composite. In his work, which is of a fundamental nature, he assumed that the phases and resulting composite are transversely isotropic. The reader is referred to that paper for further results.

C. Exact Results for Equal Phase Shear Moduli

As was indicated above, knowledge of the most rudimentary information about the structure of a composite permits one to derive universal relations between the effective elastic moduli. It is also possible to derive exact expressions for the effective elastic moduli if the elastic properties of the phases are restricted. In fact, the assumption that the shear moduli of the phases of a composite made from isotropic constituents are equal leads to exact expressions for the bulk modulus.

For a particulate composite, Hill (1963) derived the expression

$$k^*/k_R = (3k_1k_2 + 4\mu k_V)/(3k_1k_2 + 4\mu k_R) \tag{18}$$

where

$$1/k_R = (v_1/k_1) + (v_2/k_2), \qquad k_V = v_1k_1 + v_2k_2 \tag{19}$$

Herein, k and μ are the bulk and shear modulus, respectively. Expressions for Poisson's ratio ν and Young's modulus E follow from Eq. (18) upon using well-known relations. Thus,

$$\begin{aligned} \nu^* &= (v_1\nu_1 + v_2\nu_2 - \nu_1\nu_2)/(1 - v_1\nu_2 - v_2\nu_1) \\ E^* &= (v_1E_1 + v_2E_2 - E_1E_2/4\mu)/[1 - (v_1E_2 + v_2E_1)/4\mu] \end{aligned} \tag{20}$$

For details of the derivation, see the work of Hill (1963).

For a fiber-reinforced composite, Hill (1964) gives

$$1/E_l^* = (v_1/E_{l1}) + (v_2/E_{l2}) \tag{21}$$

$$1/(K^* + \mu) = [v_1/(K_1 + \mu)] + [v_2/(K_2 + \mu)] \tag{22}$$

$$1/(E_t^*/4 - \mu) = v_1/(E_{t1}/4 - \mu) + v_2/(E_{t2}/4 - \mu) \tag{23}$$

$$1/(1 - \nu^*) = [v_1/(1 - \nu_1)] + [v_2/(1 - \nu_2)] \tag{24}$$

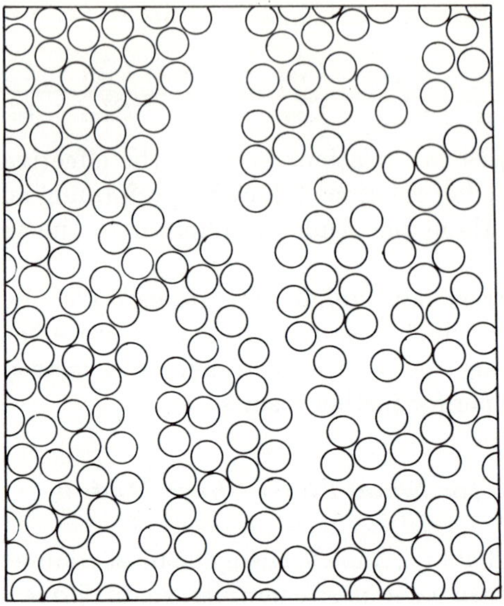

FIG. 2. Arbitrary distribution of fibers in a typical glass–epoxy composite.

where E_l, E_t, ν, μ, and K are the longitudinal Young's modulus, transverse Young's modulus, major Poisson's ratio, shear modulus, and plane bulk modulus, respectively.

In addition to the effective elastic moduli, simple analytical solutions giving the stress distribution within the composite can be derived. Sendeckyj and Yu (1971) give the solution for a fiber-reinforced composite loaded by a uniform stress ($\sigma_{xx} = 1$, $\sigma_{xy} = \sigma_{yy} = 0$) normal to the fiber direction. In the analysis, they assumed the fibers to be circular and distributed arbitrarily (see Fig. 2). Their results in terms of complex potentials (Muskhelishvili, 1963) are

$$\begin{aligned} \phi_m &= Az/4, & \psi_m &= A[-z + \gamma B(z)]/2 \\ \phi_f &= A(1+\gamma)z/4, & \psi_f &= A[-z + \gamma B^*(z)]/2 \end{aligned} \tag{25}$$

where

$$B(z) = \sum_{n=1}^{\infty} R_n^2 (z - a_n)^{-1} \tag{26}$$

$$A = (1 + v_f\gamma)^{-1}, \qquad \gamma = (\kappa_m - \kappa_f)/(1 + \kappa_f) \tag{27}$$

$$\kappa = \begin{cases} 3 - 4\nu & \text{for plane strain} \\ (3-\nu)/(1+\nu) & \text{for (generalized) plane stress} \end{cases} \tag{28}$$

where R_n and a_n are the radius and location of the center of the nth inclusion, respectively, $z = x_1 + ix_2$ is a complex variable, and subscripts f and m refer to the fibers and matrix. Furthermore, the asterisk on $B(z)$ denotes that the term singular inside the ith inclusion is to be omitted from the potentials for that inclusion.

For a square array of fibers of unit radius (Fig. 3), $R_n = 1$ and the centers of the inclusions are at the points

$$z = 2a(m + in) \qquad (m, n = 0, \pm 1, \pm 2, \ldots) \tag{29}$$

where a is a parameter specifying the spacing between the fibers. The fiber volume fraction v_f is given by $v_f = \pi/4a^2$. The stresses are readily computed by using well-known relations and at the interface they are given by

$$\begin{aligned}
\sigma_{rr} &= \tfrac{1}{2}A\left[1 + \gamma + (\cos 2\theta/A) + \gamma \sum_{n=2}^{\infty} c_n a^{-2n} \cos 2n\theta\right] \\
\sigma_{r\theta} &= -\tfrac{1}{2}\left[\sin 2\theta + A\gamma \sum_{n=2}^{\infty} c_n a^{-2n} \sin 2n\right] \\
\sigma_{\theta\theta} &= A - \sigma_{rr}
\end{aligned} \tag{30}$$

for the matrix. Within the inclusion σ_{rr} and $\sigma_{r\theta}$ have the form given by Eqs. (30), while

$$\sigma_{\theta\theta} = (1 + \gamma)A - \sigma_{rr} \tag{31}$$

In Eqs. (30), $c_2 = 0.590852$, $c_3 = 0$, and

$$c_n = [3/(2n + 1)(n - 3)] \sum_{m=2}^{n-2} c_m c_{n-m} \qquad (n \geq 4) \tag{32}$$

It should be noted that Eqs. (30) and (31) are *exact*.

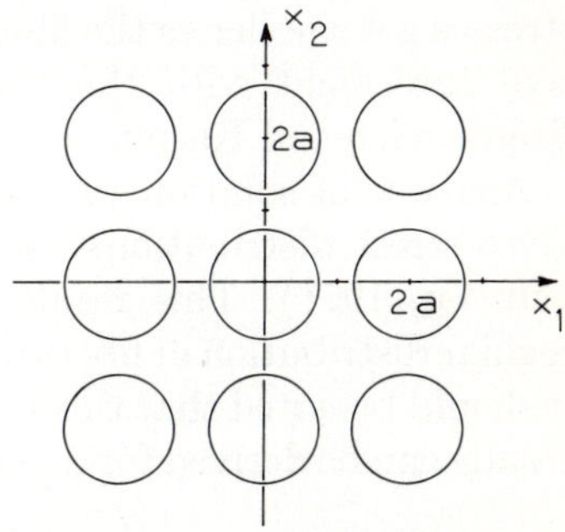

FIG. 3. Square array of circular fibers of unit radius.

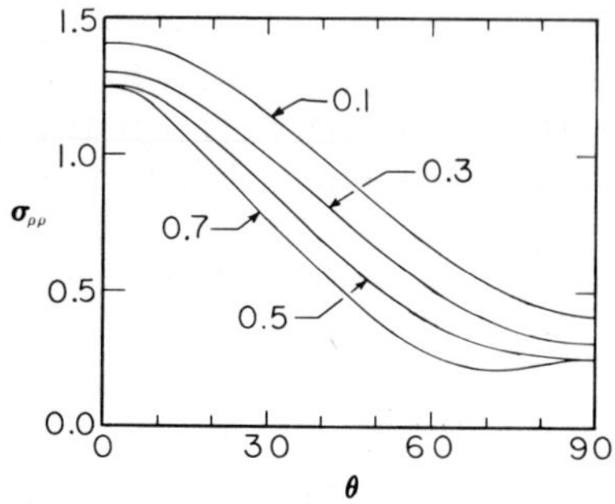

FIG. 4. Stress $\sigma_{\rho\rho m}$ at the interface versus θ for $\gamma = 1$ and discrete values of v for the case of transverse uniaxial tension.

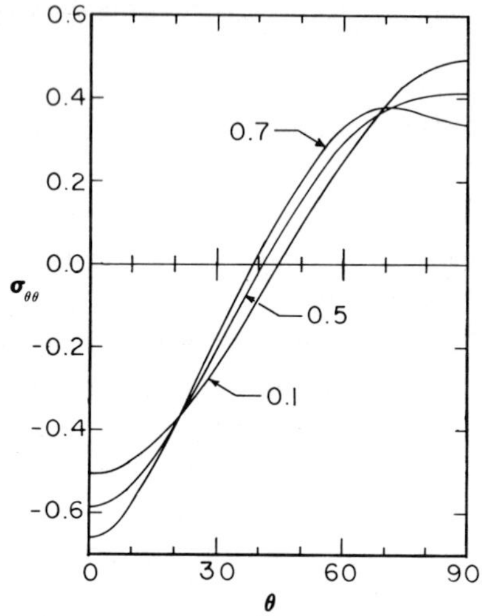

FIG. 5. Stress $\sigma_{\theta\theta m}$ at the interface versus θ for $\gamma = 1$ and discrete values of v for the case of transverse uniaxial tension.

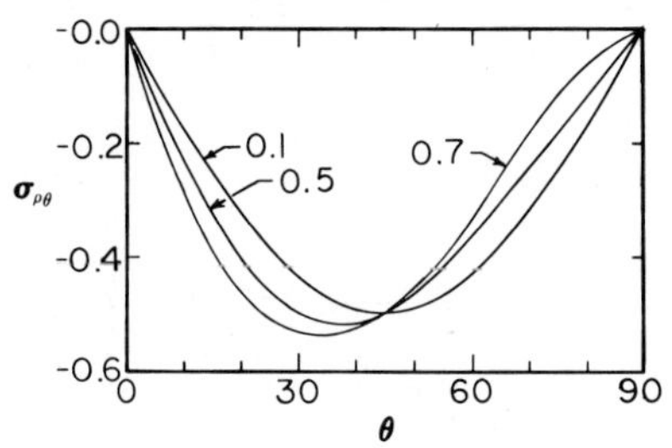

FIG. 6. Stress $\sigma_{\rho\theta m}$ at the interface versus θ for $\gamma = 1$ and discrete values of v for the case of transverse uniaxial tension.

The stresses at the interface for points in the matrix are shown in Figs. 4–6 for discrete values of v_f and $\gamma = 1$, which is the largest admissible positive value of γ. This value of γ corresponds to $\nu_m = 0.0$ and $\nu_f = 0.5$ which implies that the Young's modulus of the fibers is 1.5 times the Young's modulus of the matrix. The general trend of the results is that the stresses get smaller as the fiber volume fraction increases. This is in qualitative agreement with the results of Van Fo Fy (first nine chapters of Frantsevich and Karpinos, 1972) and Foye (1968).

Analysis of solution (25) shows that an arbitrary distribution of fibers gives stress distributions similar to those shown in Figs. 4–6 (Sendeckyj and Yu, 1971). This result tends to justify using the assumption of a regular distribution of fibers made by some authors (see Section V). Finally, it should be noted that this is the only case in which simple exact analytical results can be derived for a realistic fiber-reinforced composite. Even though analytical solutions that do not make the assumption of equal phase shear moduli can be derived in principle (Sendeckyj, 1970, 1971; Yu, 1971), the results are sufficiently complicated that no information can be extracted from them.

III. Approximations for the Effective Elastic Moduli

With the exception of laminates (Chapter 2), Section II exhausts the exact results for real composites. Considering the material presented above, it is obvious that other approaches must be used to gain useful information about composites. One such approach is to estimate the effective stress and strain concentration factors required to use Eqs. (10) and (11). Various estimates have been proposed in the literature for particulate and fiber-reinforced composites. These are summarized below.

A. Particulate Composites

When the composite consists of spheroidal isotropic inclusions embedded in an isotropic matrix in such a manner that it is macroscopically isotropic, Eqs. (10) and (11) simplify considerably. In this case, Eqs. (10) and (11) are replaced by

$$k^* = v_1 k_1 a_1 + v_2 k_2 a_2, \qquad 1/k^* = (v_1 b_1/k_1) + (v_2 b_2/k_2) \tag{33}$$

$$\mu^* = v_1 \mu_1 c_1 + v_2 \mu_2 c_2, \qquad 1/\mu^* = (v_1 d_1/\mu_1) + (v_2 d_2/\mu_2) \tag{34}$$

where a_i (c_i) and b_i (d_i) are the average volumetric (deviatoric) strain and stress concentration factors satisfying a constraint of the form

$$a_1 v_1 + a_2 v_2 = 1 \tag{35}$$

A crude estimate, suggested by Voigt (Hill, 1963a), is to take the strains throughout the composite to be uniform. Thus, upon setting a_i and c_i equal to unity, the first of Eqs. (33) and (34) reduce to

$$k_V^* = v_1 k_1 + v_2 k_2, \qquad \mu_V^* = v_1 \mu_1 + v_2 \mu_2 \tag{36}$$

Similarly by taking b_k and d_k equal to unity, the second of Eqs. (33) and (34) reduce to an estimate suggested by Reuss (Hill, 1963a):

$$1/k_R^* = (v_1/k_1) + (v_2/k_2), \qquad 1/\mu_R^* = (v_1/\mu_1) + (v_2/\mu_2) \tag{37}$$

Another estimate of the stress and strain concentration factors is obtained by assuming that the phases form two perfectly bonded concentric spheres (Hill, 1963a). Letting subscripts 1 and 2 refer to the spherical shell and inner sphere, respectively, and assuming the composite sphere to be loaded by a uniform external pressure yields

$$k^* = k_R[4\mu_2 k_V + 3k_1 k_2]/[4\mu_2 k_R + 3k_1 k_2] \tag{38}$$

for the effective bulk modulus. If the subscripts are interchanged in Eq. (38), one obtains an estimate based on phase 1 forming the inner sphere. It should be noted that Eq. (38) reduces to Eq. (18) when the two phases have the same shear modulus. Thus estimate (38) is reasonable since it does reduce to the exact expression in the limit $\mu_1 = \mu_2 = \mu$.

The next estimate in order of complexity is based on modeling the composite as a sphere of reinforcing material encased in a spherical shell of matrix material which in turn is encased in an unbounded medium possessing the as yet unknown properties of the composite. The inner r_i and outer r_0 radii of the matrix spherical shell are normally fixed by requiring that the volume fraction of the reinforcement be r_i^3/r_0^3 (Frohlich and Sack, 1946; Mackenzie, 1950; Kerner, 1956). Upon imposing simple boundary conditions at infinity and solving simple problems in the three-dimensional theory of elasticity, one obtains

$$k^* = \frac{[k_1 v_1/(3k_1 + 4\mu_1)] + [k_2 v_2/(3k_2 + 4\mu_1)]}{[v_1/(3k_1 + 4\mu_1)] + [v_2/(3k_2 + 4\mu_1)]} \tag{39}$$

$$\mu^* = \mu_2[\alpha\mu_2 v_2 + v_1]/[\alpha\mu_1 v_2 + v_2] \tag{40}$$

where

$$\alpha = 15(1 - \nu_1)/[(7 - 5\nu_1)\mu_1 + (8 - 10\nu_1)\mu_1] \tag{41}$$

Herein, subscripts 1 and 2 refer to the spherical shell and inner inclusion, respectively.

Finally, consider the estimate based on the self-consistent model (Hill, 1965b) in which the composite is modeled as a spherical reinforcing particle in a homogeneous medium possessing the effective properties of the composite. After considerable computation, one has

$$3/(3k^* + 4\mu^*) = [3v_1/(3k_1 + 4\mu^*)] + [3v_2/(3k_2 + 4\mu^*)] \tag{42}$$

$$[3v_1 k_1/(3k_1 + 4\mu^*)] + [3v_2 k_2/(3k_2 + 4\mu^*)] + 5\{[v_1\mu_2/(\mu^* - \mu_2)] + [v_2\mu_1/(\mu^* - \mu_1)]\} + 2 = 0 \tag{43}$$

Equation (43) is a quartic which must be solved for μ^*.

In conclusion, it should be noted that Eqs. (38)–(43) reduce to the result given by Eshelby (1957) for a dilute dispersion of spherical particles in an isotropic matrix. Other pertinent works, not explicitly discussed here, are those by Hashin (1955, 1959), Budiansky (1965), Wu (1966), Furuse (1967), and Dewey (1947).

B. Fiber-Reinforced Composites

The approximate models proposed for determining the average stress and strain concentration factors and, hence, effective moduli of fiber-reinforced composites are basically the same as those used when dealing with particulate composites. Since the analysis is more complicated because of the larger number of effective elastic moduli characterizing fiber-reinforced materials (assumed to be transversely isotropic), only results are presented here. For the sake of convenience, subscripts L and T will be used to distinguish the effective elastic moduli. A single subscript L (T) refers to the Young's modulus due to extension along (perpendicular to) the fibers. The subscripts on shear modulus μ denote the plane in which the shear deformation occurs. Thus μ_{TT}^* is the effective shear modulus for a shearing deformation in a plane perpendicular to the fibers. By definition ν_{LT}^* (ν_{TT}^*) is the negative ratio of the strain in a transverse direction to the strain in the longitudinal (transverse) direction for a stress imposed in the longitudinal (transverse) direction. An alternate definition of the symbol ν_{LT}^* is used by some authors and the reader is cautioned to make sure that his interpretation of ν_{LT}^* is the same as that of the authors'. The remaining Poisson's ratio ν_{TL}^* is related to ν_{LT}^* and the Young's moduli by the universal relation

$$\nu_{LT}^*/E_L^* = \nu_{TL}^*/E_T^* \tag{44}$$

The Reuss and Voigt type of estimates for the effective moduli have been made by many authors (see Chamis and Sendeckyj, 1968). The pertinent results are

$$\begin{aligned} E_L^* &= v_f E_f + v_m E_m, & \nu_{TT}^* &= v_f \nu_f + v_m \nu_m \\ 1/E_T^* &= v_f/E_f + v_m/E_m, & 1/\mu_{LT}^* &= v_f/\mu_f + v_m/\mu_m \end{aligned} \tag{45}$$

where subscripts f and m refer to the fibers and matrix, respectively, which are both assumed to be isotropic. Estimates (45) are in reasonable agreement with experimental results.

The concentric cylinder model, consisting of a fiber embedded in a concentric matrix cylinder, was used by Hill (1964), Whitney and Riley

(1965, 1966), and Hermans (1967). The predicted effective moduli are

$$E_L{}^* = v_f E_f + v_m E_m + \frac{2(\nu_f - \nu_m)^2 E_m E_f v_f (1 - \nu_f)}{E_m(1 - v_m)L_f + [L_m v_m + (1 + \nu_m)]E_f} \tag{46}$$

$$\nu_{LT}^* = \nu_m - \frac{2(\nu_m - \nu_f)(1 - \nu_m{}^2)E_f v_f}{E_m(1 - v_f)L_f + [L_m v_f + (1 + \nu_m)]E_f} \tag{47}$$

$$K^* = [(K_f + \mu_m)K_m - (K_f - K_m)v_f\mu_m]/[(K_f + \mu_m) - v_f(K_f - K_m)] \tag{48}$$

$$\mu_{LT}^* = [(1 + v_f)\mu_f + (1 - v_f)\mu_m]\mu_m/[\mu_f + \mu_m - v_f(\mu_f - \mu_m)] \tag{49}$$

$$E_T{}^* = [2K^*(1 - \nu_{TT}^*)E_L{}^*]/[E_L{}^* + 4K^*\nu_{LT}^{*2}] \tag{50}$$

where K^* is the transverse plane bulk modulus and

$$L_m = 1 - \nu_m - 2\nu_m{}^2, \qquad L_f = 1 - \nu_f - 2\nu_f{}^2 \tag{51}$$

Equation (46) predicts an effective longitudinal Young's modulus that is only slightly greater than that given by Eq. (45).

The next model in complexity, consisting of a fiber embedded in a matrix cylinder which in turn is embedded in an unbounded solid possessing the effective moduli of the composite, was used by Kilchinskii (1965, 1966) and Hashin and Rosen (1964). Hashin and Rosen formulated, but did not explicitly solve, the elastic boundary value problems for determining the effective moduli. Hashin (1972) indicates that explicit solutions have been obtained; but, these results have not been published.

Even though mathematically the models used by Kilchinski and Hashin and Rosen are identical, the work of Hashin and Rosen is superior in that they show that the results are *exact* for an ideal composite which they refer to as the *cylinder assemblage*. For details of the model see the work of Hashin (1972).

Finally, Hill (1965a) used a self-consistent model to estimate the transverse plane bulk and shear moduli and the longitudinal shear modulus. His estimates for isotropic phases are

$$K^* = \{[v_f/(K_f + \mu_{TT}^*)] + [v_m/(K_m + \mu_{TT}^*)]\}^{-1} - \mu_{TT}^* \tag{52}$$

$$[v_f K_f/(K_f + \mu_{TT}^*)] + [v_m K_m/(K_m + \mu_{TT}^*)] = 2\{[v_f\mu_m/(\mu_m - \mu_{TT}^*)] + [v_m\mu_f/(\mu_f - \mu_{TT}^*)]\} \tag{53}$$

$$[v_f/(\mu_{LT}^* - \mu_m)] + [v_m/(\mu_{LT}^* - \mu_f)] = 1/(2\mu_{LT}^*) \tag{54}$$

The remaining effective moduli follow upon using the exact relations between the effective moduli derived by Hill (1964).

IV. Variational Approach

As shown above, various estimates for the effective elastic moduli can be derived. Comparison with experimental data has to be used in discriminating between the estimates. Since a large number of estimates is possible and scatter in experimental data is considerable, selection of the best estimates is difficult.

Some of the estimates can be shown to be unreasonable by deriving bounds on the effective moduli by using variational principles of the theory of elasticity. The derivation of bounds is first illustrated in Section IV,A for particulate composites. Then results for filamentary composites are presented in Section IV,B.

A. Bounds on Effective Moduli for Particulate Composites

Variational principles of the classical theory of elasticity were apparently first applied to particulate composites by Paul (1960). Hashin and Shtrikman (1961, 1962, 1963), Hashin (1962, 1965b, 1967, 1970), and Hill (1963) published extensive results based on the use of the variational formulation. Use of the classical principles will be illustrated in this section.

The strain energy density W is defined as

$$W = \tfrac{1}{2}\sigma_{ij}e_{ij} = \tfrac{1}{2}C_{ijmn}e_{ij}e_{mn} = \tfrac{1}{2}S_{ijmn}\sigma_{ij}\sigma_{mn} \tag{55}$$

For a representative volume under uniform loading, the strain energy density is given in terms of the average stresses and strains by

$$\langle W \rangle = \tfrac{1}{2}\langle\sigma_{ij}\rangle\langle e_{ij}\rangle = \tfrac{1}{2}C^*_{ijmn}\langle e_{ij}\rangle\langle e_{mn}\rangle = \tfrac{1}{2}S^*_{ijmn}\langle\sigma_{ij}\rangle\langle\sigma_{mn}\rangle \tag{56}$$

where C^*_{ijmn} and S^*_{ijmn} are the effective elastic moduli.

The principle of minimum potential energy for an elastic medium states that the actual strain energy in the composite does not exceed the energy corresponding to any fictitious (unequilibrated) state of deformation provided displacement boundary conditions are satisfied. Thus upon assuming the strains to be uniform (guaranteeing satisfaction of the displacement boundary conditions), the variational principle states

$$C^*_{ijmn}\langle e_{ij}\rangle\langle e_{mn}\rangle \leq \langle e_{ij}\rangle(v_1C^1_{ijmn} + v_2C^2_{ijmn})\langle e_{mn}\rangle \tag{57}$$

for every $\langle e_{ij}\rangle$. Thus

$$C^*_{ijmn} \leq v_1C^1_{ijmn} + v_2C^2_{ijmn} \tag{58}$$

For isotropic phases, Eq. (58) implies

$$k^* \leq k_V, \qquad \mu^* \leq \mu_V$$

where k_V and μ_V are the Voigt estimates, Eq. (36).

Similarly, the principle of minimum complementary energy implies that

$$S^*_{ijmn}\langle\sigma_{ij}\rangle\langle\sigma_{mn}\rangle \leq \langle\sigma_{ij}\rangle(v_1S^1_{ijmn} + v_2S^2_{ijmn})\langle\sigma_{mn}\rangle \tag{59}$$

for every uniform state of stress satisfying the stress boundary conditions. This yields

$$S^*_{ijmn} \leq v_1S^1_{ijmn} + v_2S^2_{ijmn} \tag{60}$$

which for isotropic phases gives

$$k^* \geq k_R, \qquad \mu^* \geq \mu_R \tag{61}$$

where k_R and μ_R are the Reuss estimates, Eqs. (37).

Closer bounds can be derived by using Hill's (1963b) strengthening theorem: For any heterogeneous elastic solid, the strain energy increases under fixed surface constraints when the material is "strengthened" in any nonuniform way. By "strengthening," Hill means that in each changed element of material the energy density after strengthening would exceed that before strengthening if the local strains were the same. By applying this theorem, Hill showed that improved upper and lower bounds on the bulk modulus are given by Eq. (18) upon setting μ equal to the largest and smallest shear modulus of the two phases. That these are the best possible bounds was shown by comparing the results with the concentric spheres model.

An alternate derivation of these bounds was presented by Hashin and Shtrikman (1961), who used the so-called *polarization theorems* (Hashin, 1967). Their bounds for the shear modulus are

$$\mu_1 + \frac{v_2}{\dfrac{1}{\mu_2 - \mu_1} + \dfrac{6}{5}\dfrac{(k_1 + 2\mu_1)v_1}{(3k_1 + 4\mu_1)\mu_1}} \leq \mu^* \leq \mu_2 + \frac{v_1}{\dfrac{1}{\mu_1 - \mu_2} + \dfrac{6}{5}\dfrac{(k_2 + 2\mu_2)v_2}{(3k_2 + 4\mu_2)\mu_2}} \tag{62}$$

The bounds for the bulk modulus coincide with Hill's bounds. Other pertinent results based on variational principles are given by Walpole (1966, 1969), Wu (1965), Yeh (1970, 1971), and Rubenfeld and Keller (1969).

B. Bounds on Effective Moduli for Filamentary Composites

Bounds on the effective moduli of fiber-reinforced composites were derived almost simultaneously by Hill (1964), Hashin and Rosen (1964), and Hashin (1965, 1972). Since the derivations are rather lengthy, only the final results for the best bounds are presented. These are (Hashin, 1972)

$$K_m + \frac{v_f}{[1/(K_f - K_m)] + [v_m/(K_m + \mu_m)]}$$

$$\leq K_{TT}^* \leq K_f + \frac{v_m}{[1/(K_m - K_f)] + [v_f/(K_f + \mu_f)]} \tag{63}$$

$$\mu_m + \frac{v_f}{[1/(\mu_f - \mu_m)] + [v_m(K_m + 2\mu_m)/2\mu_m(K_m + \mu_m)]}$$

$$\leq \mu_{TT}^* \leq \mu_f + \frac{v_m}{[1/(\mu_m - \mu_f)] + [v_f(K_f + 2\mu_f)/2\mu_f(K_f + \mu_f)]} \tag{64}$$

$$\frac{\mu_m[\mu_m v_m + \mu_f(1 + v_f)]}{\mu_m(1 + v_f) + \mu_f v_m} \leq \mu_{LT}^* \leq \frac{\mu_f[\mu_f v_f + \mu_m(1 + v_m)]}{\mu_f(1 + v_m) + \mu_m v_f} \tag{65}$$

where K is the transverse planar bulk modulus. Bounds on the remaining moduli follow from Eqs. (63)–(65) upon using the exact relations indicated in Section II. Furthermore, Hill (1964) showed that

$$E_L^* \geq E_m v_m + E_f v_f \tag{66}$$

with equality for $\nu_f = \nu_m$. The above bounds are best possible for arbitrary phase geometry.

Upon comparing the estimates of Section III with the bounds, it can be shown that the concentric cylinder model gives estimates that coincide with the best bounds and that Hill's self-consistent model gives results within the bounds. Finally, it should be noted that the bounds are far apart for high-performance composites (boron–epoxy and graphite–epoxy).

V. Regular Array Models

Although the results based on variational principles are exact, they suffer from a major weakness, namely, the bounds are far apart. Attempts to narrow the bounds by using statistical information have met with limited success (see Section VI). If one were to examine typical micrographs of boron–epoxy and boron–aluminum fiber-reinforced composites, it would be apparent that the structure can be modeled as a doubly periodic array of identical inclusions in an unbounded matrix.

Thus let us consider an unbounded matrix containing a doubly periodic array of identical aligned circular inclusions as shown in Fig. 3. For the sake of convenience both the fibers and matrix are normally assumed to be isotropic. The composite is assumed to be loaded by macroscopically uniform stresses and the resulting elastic boundary value problems are solved to give the exact stress and strain distributions within the composite. These are then used to obtain expressions for the effective elastic moduli. It should be noted that this is the only approach that gives the actual stresses within the composite on the microscale level.

There are basically two different approaches for carrying out the necessary analysis. These consist of (a) using doubly periodic functions and (b) reducing the problem by symmetry arguments to an elastic boundary value problem for a finite region. The two approaches are discussed below.

A. Solution by Doubly Periodic Functions

Doubly periodic functions have been used by the Soviet writers to determine the stress distributions within the composite. Since this approach requires a background in the theory of the Weierstraus elliptic and special meromorphic functions, a detailed discussion of this approach is beyond the scope of the present work. For details of the derivations, the reader is referred to the extensive literature cited below.

The systematic treatment of fiber-reinforced composites by the use of doubly periodic functions can be traced to Kurshin and Fil'shtinskii (1961), who considered the problem of an isotropic plate weakened by a doubly periodic array of circular cavities. Subsequently, this work was extended to composite materials by Fil'shtinskii (1963), Grigolyuk *et al.* (1965) and Grigolyuk and Fil'shtinskii (1966, 1970). In 1965, Van Fo Fy and his co-workers began systematically applying this approach. The initial papers (Van Fo Fy and Savin, 1965; Van Fo Fy, 1965a) presented the general formulation of the problem. Both exact and approximate expressions for the shear moduli are given by Van Fo Fy (1965b). Expressions for the other effective moduli are also given by Van Fo Fy (1966, 1967). Composites containing hollow fibers are considered in other works of Van Fo Fy (1966b, 1966d). Stress distributions are presented by Van Fo Fy and Klyavlin (1967), Klyavlin (1970), and Frantsevich and Karpinos

(1972, Chaps. 1–9). Finally, it should be noted that Meijers (1967) also presented some results for composites containing rigid fibers.

B. Solution by Reduction to Simpler Problems

An alternate approach was adopted in this country. The doubly periodic problem was reduced by symmetry arguments to a problem for a rectangular or hexagonal prism (containing a single fiber), called the *fundamental repeating region,* and then solved by various numerical methods. Apparently, Herrmann and Pister (1962, 1963) were the first to adopt this approach. They considered a rectangular array of circular fibers and used a series formulation to solve the resulting problem for the rectangular fundamental repeating region. Subsequently, rectangular arrays of fibers were considered by Wilson and Hill (1965) who used a complex variable method and Adams and Doner (1967) who used a finite-difference scheme.

Hexagonal arrays of isotropic fibers were considered by Pickett (1965, 1968), Chen and Cheng (1967), Clausen and Leissa (1967), and Leissa and Clausen (1968). More recently, Chen and Cheng (1970) and Chen (1970, 1971) considered composites reinforced by a doubly periodic array of anisotropic fibers. Other pertinent results are due to Piehler (1967), Haener (1966), Haener and Ashbaugh (1967), Haener *et al.* (1968, 1969), Heaton (1968, 1970), Chen and Lin (1969), Stevenson (1970), Bloom and Adams (1969), Bloom and Wilson (1967), Foye (1968, 1970, 1972), and Hulbert and Rybicki (1971).

The results derived by the above mentioned authors agree with the analysis of Van Fo Fy with the major trends being:

(1) Estimates for the effective elastic moduli based on the doubly periodic array models fall slightly above the best lower bounds with the hexagonal array being nearest to the lower bound. This, in conjunction with the cylinder assemblage model, led Hashin (1972) to claim that the lower bounds give the true values of most of the elastic moduli.

(2) The maximum stresses decrease as the fiber volume content increases from zero, attaining minimum values for $v_f \approx 0.5$. For higher fiber volume fractions, they increase rapidly.

VI. Statistical Formulation

Even though the analyses based on the doubly periodic array model give reasonable estimates for the effective elastic moduli for some fiber-reinforced materials, they fail for others. This failure can be traced to the fact that the model does not accurately represent the microstructure of the material. The statistical formulation is an attempt to circumvent this difficulty. It apparently has been proposed independently by groups of researchers. Extensive Soviet work in this area is due to Khoroshun (1966, 1967, 1968), Savin and Khoroshun (1967), Levin (1967, 1970), Darinski and Shermergor (1965), Fokin and Shermergor (1967, 1968, 1969), Lomakin (1965, 1966, 1970), Lomakin and Sheinin (1970), Volkov (1962, 1969, 1970), Volkov and Klinskikh (1965), Stavrov *et al.* (1967), Stavrov and Volkov (1968), Volkov and Dolgikh (1967, 1968), Volkov and Kossimarova (1968), Stavrov and Fomina (1968, 1969), Van Fo Fy *et al.* (1969), Bolotin and Moskalenko (1968, 1969), and Moskalenko (1969). English language work is due to Beran (1965, 1968, 1970, 1971), Beran and McCoy (1970), Beran and Molyneux (1966), Kroner (1967), Molyneux and Beran (1965), and Davies (1971).

The formulation of the problem employed by the above mentioned authors is the same; but, they deviate in subsequent treatment. By far the most complete expositions of this approach are due to Beran (1968, 1970, 1971), Kroner (1967), and Lomakin (1970b). The reader is encouraged to refer to these works for a discussion of the underlying assumptions and difficulties encountered in the solution of the problem. The present work will be restricted to the derivation of the basic equations (without paying attention to fine details) and indication of the various solution methods adopted. With this in mind consider an elastic heterogeneous medium with Hooke's law given by

$$\sigma_{ij} = C_{ijmn}e_{mn} \tag{67}$$

where the C_{ijmn} are random functions of position. This implies that the stresses and strains will also be random functions of position. Formally decomposing the elastic quantities into "constant" and fluctuating parts, we have

$$C_{ijmn} = \langle C_{ijmn} \rangle + C'_{ijmn}, \qquad e_{mn} = \langle e_{mn} \rangle + e'_{mn} \tag{68}$$

where C'_{ijmn} and e'_{mn} are random functions of position such that the statistical average is zero. If the region under consideration is large enough and the variation of the macroscopic elastic fields is sufficiently smooth, then the statistical average can be replaced by the simple volume average, i.e.,

$$\langle e_{mn} \rangle = \frac{1}{v} \int e_{mn}\, dv \tag{69}$$

Substituting Eq. (68) into Eq. (67) and the resulting expression into the equilibrium equations yields

$$\frac{\partial}{\partial x_j}[(\langle C_{ijmn}\rangle + C'_{ijmn})(\langle e_{mn}\rangle + e'_{mn})] = 0 \tag{70}$$

which can be written as

$$\langle C_{ijmn}\rangle \frac{\partial}{\partial x_j} e'_{mn} = -\langle C_{ijmn}\rangle \frac{\partial}{\partial x_j}\langle e_{mn}\rangle - \frac{\partial}{\partial x_j}[C'_{ijmn}\langle e_{mn}\rangle] - \frac{\partial}{\partial x_j}[C'_{ijmn}e'_{mn}] \tag{71}$$

By assumption the random variation of the elastic quantities occurs over much smaller regions than the macroscopic variation of the stresses and strains. Thus the operation of volume (or statistical) averaging can be interchanged with differentiation with respect to x_j.

Taking a volume average of Eq. (71) gives

$$\langle C_{ijmn}\rangle \frac{\partial}{\partial x_j}\langle e_{mn}\rangle + \frac{\partial}{\partial x_j}\langle C'_{ijmn}e'_{mn}\rangle = 0 \tag{72}$$

as the governing equation for $\langle e_{mn}\rangle$. Substituting this equation into Eq. (71) gives

$$\langle C_{ijmn}\rangle \frac{\partial}{\partial x_j} e'_{mn} = -\frac{\partial}{\partial x_j}[C'_{ijmn}\langle e_{mn}\rangle] - \frac{\partial}{\partial x_j}[C'_{ijmn}e'_{mn} - \langle C'_{ijmn}e'_{mn}\rangle] \tag{73}$$

as the equation for e'_{mn}. Equation (73) is given by Beran (1971). In his derivations, Kroner (1967) assumed that $\langle e_{mn}\rangle$ was constant and, hence, obtained Eq. (73) without the last term. No exact solution is available for Eqs. (72) and (73).

Before considering the various methods used to obtain approximate solutions of the equations, let us see how they enter into the definition of the effective elastic constants. Upon averaging Eq. (67), we obtain

$$\langle \sigma_{ij}\rangle = \langle C_{ijmn}\rangle\langle e_{mn}\rangle + \langle C'_{ijmn}e'_{mn}\rangle \tag{74}$$

which may be rewritten as

$$\langle \sigma_{ij}\rangle = C^*_{ijmn}\langle e_{mn}\rangle \tag{75}$$

where

$$C^*_{ijmn} = \langle C_{ijmn}\rangle + \langle C'_{ijkl}a_{klmn}\rangle \tag{76}$$

and a_{ijmn} is defined by

$$e'_{ij} = a_{ijmn}\langle e_{mn}\rangle \tag{77}$$

Thus it is seen that the quantity $C'_{ijmn}e'_{mn}$ appears in the definition of the effective elastic moduli as a correction to the rule of mixtures value. Equations (72) and (73) must be solved to determine e'_{mn} and, hence, $\langle C'_{ijmn}e'_{mn}\rangle$.

One approach for determining $\langle C'_{ijmn}e'_{mn}\rangle$ consists of multiplying Eq. (73) by C'_{pqrs} and then averaging to give

$$\langle C_{ijmn}\rangle \frac{\partial}{\partial x_j} \langle C'_{pqrs}e'_{mn}\rangle + \langle C'_{ijmn}C'_{pqrs}\rangle \frac{\partial}{\partial x_j} \langle e_{mn}\rangle + \frac{\partial}{\partial x_j} \langle C'_{pqrs}C'_{ijmn}e'_{mn}\rangle = 0 \tag{78}$$

Upon multiplying this equation by the inverse of $\langle C_{ijmn}\rangle$, one has a system of partial differential equations for $\langle C'_{pqrs}e'_{mn}\rangle$ in terms of second- and third-order correlation functions

$$\langle C'_{ijmn}C'_{pqrs}\rangle \quad \text{and} \quad \langle C'_{pqrs}C'_{ijmn}e'_{mn}\rangle \tag{79}$$

It only remains to set $r = m$ and $s = n$ to complete the derivation. The correlation function $\langle C'_{ijmn}C'_{pqrs}\rangle$ can be determined from the statistics of the problem. Unfortunately, $\langle C'_{pqrs}C'_{ijmn}e'_{mn}\rangle$ can only be obtained by multiplying Eq. (73) by $C'_{abcd}C'_{pqrs}$ and repeating the averaging process. Since this gets out of hand very quickly, only crude approximations have been derived using this approach.

An alternate approach is based on the fact that Eq. (73) can be transformed into an integral equation by using the free space Green's function. After the transformation, the integral equation is solved by the iteration method. Again the result depends on all the correlation functions of C'_{ijmn}.

Normally, the correlation functions can only be determined experimentally. Since measurement of the correlation functions is tedious, only limited results are available. In fact, only second- and third-order correlation functions have been measured. For the limited experimental data, the reader is referred to the work of Stavrov and Fomina (1969), Van Fo Fy *et al.* (1969), and Frantsevich and Karpinos (1972).

Most of the exact work in this area is based on the assumption of statistical independence. Under this assumption, the higher-order correlation functions can be written in terms of simple averages of the elastic moduli of the constituents for a two-phase solid. Thus, for example, Stavrov *et al.* (1967) obtained series expressions, which were summed by Sendeckyj (1969a), for the effective bulk and shear moduli of a two-phase particulate composite:

$$\begin{aligned} \mu^* &= \langle \mu\rangle \left\{1 - \frac{v_f(1 - v_f)fg^2}{1 + fg(1 - 2v_f) - f^2g^2(1 - v_f)v_f}\right\} \\ k^* &= \langle k\rangle \left\{1 - \frac{v_f(1 - v_f)FG^2}{1 + FG(1 - 2v_f) - F^2G^2(1 - v_f)v_f}\right\} \end{aligned} \tag{80}$$

where

$$f = 2(4 - 5\langle\nu\rangle)/15(1 - \langle\nu\rangle), \qquad F = (1 + \langle\nu\rangle)/3(1 - \langle\nu\rangle)$$
$$g = (\mu_f - \mu_m)/\langle\mu\rangle, \qquad G = (k_f - k_m)/\langle k\rangle \tag{81}$$

and subscripts f and m refer to the reinforcement and matrix, respectively. For small fluctuations of the moduli ($|g| \ll 1$ and $|G| \ll 1$), expressions (80) reduce to the results of Khoroshun (1966b, 1967), Krivoglaz and Cherevko (1959), Fokin and Shermergor (1967a), and Beran (1968). Comparison of results with experimental data and variational bounds for glass–epoxy shows that the moduli predicted by Eqs. (80) could violate the variational bounds (Sendeckyj, 1969b). Also Moskalenko (1969) showed theoretically that the assumption of statistical independence leads to inconsistent results unless the fluctuations of the phase moduli are small.

The results for fiber-reinforced composites are extremely approximate and, hence, they will not be presented here. It suffices to say that they are based on small fluctuations in phase properties.

Bounds using statistical information have been derived by Beran (1965, 1970), McCoy (1970) and Miller (1969). These are improvements on the purely variational results. Since the computations are rather involved, results will not be presented here.

VII. Additional Results

A. Other Approaches

Two other approaches for predicting the effective elastic moduli of composites are worth mentioning in addition to those presented above. These are the method of long waves due to Behrens (1967, 1969, 1971) and the mechanics of materials approach. Behrens' approach consists of studying the propagation of waves in a composite material and comparing the results with those for wave propagation in homogeneous solids. The analysis is performed on simplified models of the composite. Effective moduli are derived by passing to the limit of infinitely long wavelengths which is the static case.

The mechanics of materials approach belongs to the early history of composite materials. It is based on rather crude models of the composite with analyses making all the assumptions made in typical strength of materials derivations. Estimates based on this approach have been superceded by results based on the more sophisticated models discussed above.

For references to this approach see the work of Chamis and Sendeckyj (1968) and Sendeckyj (1969b).

B. *Effect of Fiber Curvature and Twist*

The results for fiber-reinforced composites are based on the assumption of perfectly aligned straight fibers. A number of attempts have been made to relax this assumption. Bazant (1968), Nosarev (1967), and Tarnopol'ski *et al.* (1967) assumed that the fibers were not straight, but bent slightly. Their analysis showed that the longitudinal Young's modulus is a strong function of fiber curvature. Slight amounts of bent fibers can reduce Young's modulus by as much as 10%. Whitney (1966) considered the effect of fiber twist with similar results for the case of compressive loading.

C. *Random Distribution of Fibers*

Attempts to model a fiber-reinforced composite material as an arbitrary array of fibers have been made by Adams and Tsai (1968, 1969), Stevenson (1970), and Sendeckyj (1970). Adams and Tsai assumed that the composite can be modeled by a repeating structure of rectangular cells containing a small number of arbitrarily placed fibers. Due to mathematical difficulties, they assumed the fibers within the cells to have a square cross section. This had the effect of reducing the problem to manageable proportions permitting them to obtain some numerical results. Based on the analysis they concluded that the effective transverse Young's modulus lies somewhat above the best lower bound.

Stevenson (1970) attempted to perform a more realistic treatment of the problem. His starting point was the construction of unique finite elements consisting of a circular fiber incased within a square matrix region with the stresses, strains, and displacements within the element being obtained by a finite difference scheme. In particular, he generated elements with different diameter fibers placed off center and at the center of the square. Forming various combinations of these elements with square elements containing no fibers, he was able to solve some rather significant problems. Among these we have an array of 25 elements containing both central and noncentral fibers. These results correspond to an improvement over the model used by Adams and Tsai. Considering the long computational times on the UNIVAC 1107 and 1108 computers, he was not able to run all the examples that one would like to see analyzed.

Sendeckyj (1970) modeled the composite as an arbitrary distribution of circular fibers. Assuming that the composite was loaded by shearing stresses parallel to the fibers, he solved the elastic boundary value problem exactly by a series method. He also derived an exact expression for the longitudinal shear modulus. Unfortunately, the convergence of the series is extremely slow, relegating the solution to the realm of results that are of academic interest only.

D. Chopped Fiber Composites

Derivations of estimates for the effective elastic moduli of chopped fiber-reinforced composite materials are extremely complex due to the nature of the material. For aligned chopped fiber composites, estimates have been developed by Halpin (1969) and Halpin and Pagano (1970) by using the Halpin–Tsai equations. Letting subscripts 1 and 2 refer to the fiber and in-plane transverse directions, respectively, the estimates are

$$E_1^*/E_m = (1 + \xi\eta v_f)/(1 - \eta v_f) \tag{82}$$

$$\mu_{12}^*/\mu_m = (1 + \lambda v_f)/(1 - \lambda v_f) \tag{83}$$

$$E_2^*/E_m = (1 + \eta v_f)/(1 - \eta v_f) \tag{84}$$

$$\nu_{12}^* = \nu_f v_f + \nu_m v_m \tag{85}$$

where

$$\eta = [(E_f/E_m) - 1]/[(E_f/E_m) + \xi] \tag{86}$$

$$\lambda = [(\mu_f/\mu_m) - 1]/[(\mu_f/\mu_m) + 1] \tag{87}$$

$$\xi = 2(L/D) \tag{88}$$

where L/D is the ratio of fiber length to diameter.

Estimates for the effective moduli of composites reinforced by randomly oriented chopped fibers were derived by Nielsen and Chen (1968) and Halpin and Pagano (1969). Nielsen and Chen (1968) averaged the Young's modulus in an arbitrary direction for a unidirectionally reinforced composite over all possible directions to obtain the desired expression for the in-plane Young's modulus of a composite containing randomly oriented fibers. Since the computation is extremely tedious, they did not derive an explicit expression for the effective Young's modulus; but, they did present extensive graphical results.

Halpin and Pagano (1969) assumed that the composite could be represented by a quasi-isotropic laminate. Using laminated plate theory they obtained simple analytical expressions for the moduli of composites reinforced by randomly oriented fibers. In their computations, Halpin and

Pagano assumed that the unidirectional moduli are given by the Halpin–Tsai equations. Their predictions are higher than those of Nielsen and Chen.

VIII. Discussion and Recommendations for Future Work

As can be seen from the results presented, many methods have been used for predicting the elastic behavior of composite materials. By far the theoretically soundest results are the bounds based on the variational theorems of elasticity and the exact expressions based on Hill's fundamental work. The variational approach gives upper and lower bounds for the effective elastic moduli between which all other estimates must fall. For the purpose of experiment–theory comparison, the basic assumptions made in deriving the bounds must be kept in mind. The bounds were derived for a perfect composite containing homogeneous phases possessing identical and distinct elastic properties. Thus all reinforcing particles must have the same physical properties. Furthermore, a perfect bond must exist between the reinforcement and the matrix material. These conditions are seldom met in real composite materials. A perfect bond seldom exists at all reinforcement–matrix interfaces as a result of the manufacturing process. Furthermore, the reinforcement, be it particles or fibers, has elastic properties varying about a mean value. It is possible that the properties of the reinforcement in a particular composite component may differ considerably from the mean values for that reinforcement material. Thus, it is possible that the experimental data will fall outside the best variational bounds. For example, the longitudinal Young's modulus of a unidirectional fiber-reinforced composite could be below the lower bound due to fiber damage during the manufacturing process.

The main use of the variational bounds is in assessing the quality of the various other theoretical estimates which must fall between the bounds. If the estimates fall outside the bounds, the assumptions made in deriving the estimates are probably wrong and the estimate should be discarded. One area for improvement, indicated by the above discussion, is the derivation of bounds which take into account the actual distribution of the constituent properties. This would aid in interpreting experimental results.

Another area where tremendous improvements can be made is the statistical approach for predicting the effective elastic moduli of composites. In principle, the approach should give the best theoretical results. Unfortunately, at present the methods for solving partial differential equations with random coefficients are not sufficiently well developed to give reliable results. It is necessary to develop an approach for solving these equations for the case of large fluctuations in the random coefficients (that is, large variations in the elastic moduli of the phases). Furthermore, it is necessary to determine statistical information describing the actual distribution of the reinforcement within the composite. With the exception of some limited results, no systematic attempts have been made to generate this information.

Finally, the behavior of composite materials under rapidly varying macroscopic stress fields has been virtually unexplored. Since most fiber-reinforced composites contain a relatively small number of fibers in the thickness direction, it is expected that the macroscopic elastic fields will not be uniform over the representative volume, that is, the assumption underlying all the analyses discussed in this chapter might not hold. Analyses of the effect of rapidly varying macroscopic stresses would be of considerable help in assessing whether predictions based on the concept of a representative volume are valid. Furthermore, knowledge of the actual stress distributions within the composite under rapidly varying macroscopic stress fields would shed some light on the seemingly anamolous strength behavior of composite materials.

Appendix: Coefficients of Thermal Expansion

Since composites are often used in thermal environments, their coefficients of thermal expansion play an important role in design considerations. Hence, this chapter will be rounded off by a brief discussion of the theoretical estimates for this property.

Considering their importance, it is at first surprising that only a small number of papers have been written dealing with prediction of the effective coefficients of thermal expansion of composite materials. Apparently the first attempt at prediction was made by Turner (1946) who gave the following estimate:

$$\alpha^* = \langle \alpha k \rangle / \langle k \rangle \tag{89}$$

for the effective volumetric coefficient of thermal expansion, where α_i and k_i are the coefficients of thermal expansion and bulk moduli of the phases. Equation (89) is valid for any number of phases. Subsequently, various estimates were derived by Kerner (1956), Van Fo Fy (1966a, d), Van Fo Fy and Savin (1965), Khoroshun (1966c), and Kilchinskii (1965, 1966).

Levin (1967b) and Schapery (1968), working independently, derived exact expressions for the coefficients of thermal expansion in terms of the effective moduli of composites. Since Schapery's work is more general, his results will be presented here. For an n-phase particulate composite, Schapery gives the following expressions for the upper and lower bounds for α^*:

$$\alpha_+{}^* = 3\langle\alpha\rangle + \frac{3[(\langle k\alpha\rangle/\langle k\rangle) - \langle\alpha\rangle](1/k_R - 1/k^*)}{(1/k_R - 1/k_V)} + \Delta\alpha^* \tag{90}$$

$$\alpha_-{}^* = 3\langle\alpha\rangle + \frac{3[(\langle k\alpha\rangle/\langle k\rangle) - \langle\alpha\rangle](1/k_R - 1/k^*)}{(1/k_R - 1/k_V)} - \Delta\alpha^* \tag{91}$$

where

$$\Delta\alpha^* = 3\left(\frac{1}{k^*} - \frac{1}{k_V}\right)^{1/2}\left(\frac{1}{k_R} - \frac{1}{k^*}\right)^{1/2}$$

$$\times \frac{\{[\langle k\alpha^2\rangle - (\langle k\alpha\rangle^2/\langle k\rangle)][(1/k_R) - (1/k_V)] - [(\langle k\alpha\rangle/\langle k\rangle) - \langle\alpha\rangle^2]\}^{1/2}}{(1/k_R) - (1/k_V)}$$

$$\geq 0 \tag{92}$$

For a two-phase solid $\Delta\alpha^* = 0$ and expressions (90) and (91) coincide. They can be rewritten as

$$(\alpha^* - 3\alpha_2)/3(\alpha_1 - \alpha_2) = k_1(k_2 - k^*)/[k^*(k_2 - k_1)] \tag{93}$$

which is identical with the expression given by Levin (1967b). It should be noted that Eqs. (90)–(93) depend on the effective bulk modulus of the composite. If only upper and lower bounds on the effective bulk modulus are available, Eq. (93) gives only bounds on α^*.

Similarly, bounds on the linear coefficients of thermal expansion have the form

$$\alpha_{1-}^* = \langle\alpha\rangle + \frac{[(\langle k\alpha\rangle/\langle k\rangle) - \langle\alpha\rangle][(1/E_R) - (1/E_1{}^*)]}{(1/E_R) - (1/E_V{}^*)} - \Delta\alpha_1{}^* \tag{94}$$

$$\alpha_{1+}^* = \langle\alpha\rangle + \frac{[(\langle k\alpha\rangle/\langle k\rangle) - \langle\alpha\rangle][(1/E_R) - (1/E_1{}^*)]}{(1/E_R) - (1/E_V{}^*)} + \Delta\alpha_1{}^* \tag{95}$$

where

$$\Delta\alpha_1^* = 3\left(\frac{1}{E_1^*} - \frac{1}{E_V}\right)^{1/2}\left(\frac{1}{E_R} - \frac{1}{E_1^*}\right)^{1/2}$$

$$\times \frac{\{[\langle k\alpha^2\rangle - (\langle k\alpha\rangle^2/\langle k\rangle)][(1/E_R) - (1/E_V)] - \frac{1}{9}[\langle\alpha\rangle - (\langle k\alpha\rangle/\langle k\rangle)^2]\}^{1/2}}{(1/E_R) - (1/E_V)} \tag{96}$$

Herein, the subscript 1 denotes the direction in which the uniaxial properties are measured.

For fiber-reinforced composites, the longitudinal coefficient of thermal expansion, denoted by α_L^* (as given by Levin, 1967), is

$$(\alpha_L^* E_L^*/\langle E\rangle) - (\langle E\alpha\rangle/\langle E\rangle) = [\alpha_1(1+\nu_1) - \alpha_2(1+\nu_2)][E_L^* - \langle E\rangle]/[\langle E\rangle(\nu_1 - \nu_2)] \tag{97}$$

The transverse coefficient of thermal expansion, as given by Van Fo Fy (1966a), is

$$\alpha_T^* = \alpha_1 + (\alpha_1 - \langle\alpha_L^*\rangle)\nu_{21}^* - [(\alpha_1 - \alpha_2)(1+\nu_2)(\nu_1 - \nu_{21}^*)]/(\nu_1 - \nu_2) \tag{98}$$

Expressions (97) and (98) are exact depending only on the effective elastic moduli of the composite. Thus the problem of predicting the effective coefficients of thermal expansion has been reduced to the problem of estimating the effective moduli of the composite, which is the main subject of the present chapter. This explains the lack of extensive literature dealing with the prediction of the effective coefficients of thermal expansion.

References

Adams, D. F., and Doner, D. R. (1967a). *J. Compos. Mater.* **1**, 4.

Adams, D. F., and Doner, D. R. (1967b). *J. Compos. Mater.* **1**, 152.

Adams, D. F., and Tsai, S. W. (1968). The Influence of Random Filament Packing on the Elastic Properties of Composite Materials. The RAND Corp. Memorandum RM-5608-PR.

Adams, D. F., and Tsai, S. W. (1969). *J. Compos. Mater.* **3**, 368.

Ashton, J. E., Halpin, J. C., and Petit, P. H. (1969). Primer on Composite Materials: Analysis. Technomic, Westport, Connecticut.

Bazant, Z. P. (1968). *Mekh. Polim.* No. 2, 314.

Behrens, E. (1967a). *J. Acoust. Soc. Amer.* **42,** 367.
Behrens, E. (1967b). *J. Acoust. Soc. Amer.* **42,** 378.
Behrens, E. (1969a). *J. Acoust. Soc. Amer.* **45,** 102.
Behrens, E. (1969b). *J. Acoust. Soc. Amer.* **45,** 1567.
Behrens, E. (1971). *J. Appl. Mech.* **38,** 1062.
Beran, M. (1965). *Trans. Soc. Rheol.* **1,** 339.
Beran, M. (1968). "Statistical Continuum Theories." Wiley (Interscience), New York.
Beran, M. (1970). *In* "Mechanics of Composite Materials" (F. W. Wendt, H. Leibowitz, and N. Perrone, eds.). pp. 243–254. Pergamon, New York.
Beran, M. (1971). *Phys. Status Solidi (a)* **6,** 365.
Beran, M. J., and McCoy, J. J. (1970a). *Int. J. Solids Struct.* **6,** 1035.
Beran, M. J., and McCoy, J. J. (1970b). *Int. J. Solids Struct.* **6,** 1267.
Beran, M. J., and McCoy, J. J. (1970c). *Quart. Appl. Math.* **28,** 245.
Beran, M. J., and Molyneux, J. (1966). *Quart. Appl. Math.* **24,** 107.
Bloom, J. M., and Adams, D. F. (1969). *J. Compos. Mater.* **3,** 186.
Bloom, J. M., and Wilson, H. B., Jr. (1967). *J. Compos. Mater.* **1,** 268.
Bolotin, V. V., and Moskalenko, V. N. (1968a). *Prikl. Mekh. Teor. Fiz.* No. 1, 67.
Bolotin, V. V., and Moskalenko, V. N. (1968b). *Sov. Phys.—Dokl.* **13,** 73.
Bolotin, V. V., and Moskalenko, V. N. (1969). *Inzh. Zh. Mekh. Tverd. Tela* No. 3, 106.
Budiansky, B. (1965). *J. Mech. Phys. Solids* **13,** 223.
Chamis, C. C., and Sendeckyj, G. P. (1968). *J. Compos. Mater.* **2,** 332.
Chen, C. H. (1970). *J. Appl. Mech.* **37,** 198.
Chen, C. H. (1971). *J. Appl. Mech.* **38,** 710.
Chen, C. H., and Cheng, S. (1967). *J. Compos. Mater.* **1,** 30.
Chen, C. H., and Cheng, S. (1970). *J. Appl. Mech.* **37,** 186.
Chen, P. E., and Lin, J. M. (1969). *Mater. Res. Std.* No. 8, 29.
Clausen, W. E., and Leissa, A. W. (1967). Stress and Deflection Analysis of Fibrous Composite Materials under External Load. AFML-TR-67-151.
Darinski, B. M., and Shermergor, T. D. (1965). *J. Appl. Mech. Tech. Phys.* No. 4, 79.
Davies, W. E. A. (1971a). *J. Phys. D Appl. Phys.* **4,** 1176.
Davies, W. E. A. (1971b). *J. Phys. D Appl. Phys.* **4,** 1325.
Dewey, J. M. (1947). *J. Appl. Phys.* **18,** 578.
Eshelby, J. D. (1957). *Proc. Roy. Soc. London* **A241,** 376.
Fil'shtinskii, L. A. (1963). *J. Appl. Math. Mech.* **28,** 530.
Fokin, A. G., and Shermergor, T. D. (1967a). *Inzh. Zh. Mekh. Tverd. Tela* No. 1, 129.
Fokin, A. G., and Shermergor, T. D. (1967b). *Inzh. Zh. Mekh. Tverd. Tela* No. 2, 93.
Fokin, A. G., and Shermergor, T. D. (1968a). *Prikl. Mat. Mekh.* **32,** 660.
Fokin, A. G., and Shermergor, T. D. (1968b). *Prikl. Mekh. Teor. Fiz.* No. 3, 37.
Fokin, A. G., and Shermergor, T. D. (1968c). *Prikl. Mekh. Teor. Fiz.* No. 4, 39.
Fokin, A. G., and Shermergor, T. D. (1968d). *Inzh. Zh. Mekh. Tverd. Tela* No. 3, 73.
Fokin, A. G., and Shermergor, T. D. (1968e). *Inzh. Zh. Mekh. Tverd. Tela* No. 4, 93.
Fokin, A. G., and Shermergor, T. D. (1968f). *Mekh. Polim.* No. 4, 624.
Fokin, A. G., and Shermergor, T. D. (1969). *Prikl. Mekh. Teor. Fiz.* No. 1, 51.
Foye, R. L. (1968). Advanced Design Concepts for Advanced Composite Airframes. AFML-TR-68-91, Vol. 1.
Foye, R. L. (1970). *J. Compos. Mater.* **4,** 562.
Foye, R. L. (1972). *J. Compos. Mater.* **6,** 293.
Frantsevich, I. N., and Karpinos, D. M. (eds.) (1972). Fibrous Composites. NASA TT F-675.

Frohlich, H., and Sack, R. (1946). *Proc. Roy. Soc. London* **A185,** 415.
Furuse, H. (1967). *J. Phys. Soc. Japan* **23,** 1163.
Grigolyuk, E. I., Kurshin, L. M., and Fil'shtinskii, L. A. (1965). *Prikl. Mekh.* **1,** No. 1, 22.
Grigolyuk, E. I., and Fil'shtinskii, L. A. (1966). *Prikl. Mekh.* **2,** No. 9, 1.
Grigolyuk, E. I., and Fil'shtinskii, L. A. (1970). *Inzh. Zh. Mekh. Tverd. Tela* No. 1, 75.
Haener, J. (1966). Micromechanical Behavior of Fiber Reinforced Plastics. Whittaker Corp., USAAVLABS Tech. Rep. 66–22.
Haener, J., and Ashbaugh, N. (1967). *J. Compos. Mater.* **1,** 54.
Haener, J., Puppo, A., and Feng, M.-Y. (1968). Oblique Loading of Unidirectional Fiber Composites: Shear Loading. Whittaker Corp., USAAVLABS Tech. Rep. 68–81.
Haener, J., Puppo, A., and Feng, M.-Y. (1969). Transverse Loading of Unidirectional Fiber Composites. Whittaker Corp., USAAVLABS Tech. Rep. 69-45.
Halpin, J. C. (1969). *J. Compos. Mater.* **3,** 732.
Halpin, J. C., and Pagano, N. J. (1969). *J. Compos. Mater.* **3,** 720.
Halpin, J. C., and Pagano, N. J. (1970). Stiffness and Expansion Estimates for Oriented and Random Fibrous Composites. AFML-TR-69-341.
Hashin, Z. (1955). *Bull. Res. Counc. Isr. Sect. C* **5,** 46.
Hashin, Z. (1959). *In* "Non-Homogeneity in Elasticity and Plasticity" (W. Olszak, ed.), pp. 463–478. Pergamon, New York.
Hashin, Z. (1962). *J. Appl. Mech.* **29**E, 143. Discussion by B. Paul, *ibid.* **29**E, 765.
Hashin, Z. (1964). *Appl. Mech. Rev.* **17,** No. 1, 1.
Hashin, Z. (1965a). *Trans. Soc. Rheol.* **9,** 381.
Hashin, Z. (1965b). *J. Mech. Phys. Solids* **13,** 119.
Hashin, Z. (1967). *Int. J. Eng. Sci.* **5,** 213.
Hashin, Z. (1970). *In* "Mechanics of Composite Materials" (F. W. Wendt, H. Leibowitz, and N. Perrone, eds.), pp. 201–242. Pergamon, New York.
Hashin, Z. (1972). Theory of Fiber Reinforced Materials. NASA CR-1974.
Hashin, Z., and Rosen, B. W. (1964). *J. Appl. Mech.* **31,** 223.
Hashin, Z., and Shtrikman, S. (1961). *J. Franklin Inst.* **271,** 336.
Hashin, Z., and Shtrikman, S. (1962a). *Proc. U. S. Nat. Congr. Appl. Mech., 4th,* **1962,** 619.
Hashin, Z., and Shtrikman, S. (1962b). *J. Mech. Phys. Solids* **10,** 335.
Hashin, Z., and Shtrikman, S. (1962c). *J. Mech. Phys. Solids* **10,** 343.
Hashin, Z., and Shtrikman, S. (1963). *J. Mech. Phys. Solids* **11,** 127.
Heaton, M. D. (1968). *J. Phys. D Appl. Phys.* **1,** 1039.
Heaton, M. D. (1970). *J. Phys. D Appl. Phys.* **3,** 672.
Hermans, J. J. (1967). *Proc. Kon. Ned. Akad. Wetensch. Amsterdam* **B70,** No. 1, 1.
Herrmann, L. R., and Pister, K. S. (1962). Composite Properties of Filament-Resin Systems. Aerojet-General Corp. Tech. Paper 119SRP.
Herrmann, L. R., and Pister, K. S. (1963). ASME Paper No. 63-WA-239.
Hill, R. (1963a). *J. Mech. Phys. Solids* **11,** 357.
Hill, R. (1963b). *In* "Progress in Applied Mechanics" (The Prager Anniv. Vol.), pp. 99–106. Macmillan, New York.
Hill, R. (1964). *J. Mech. Phys. Solids* **12,** 199.
Hill, R. (1965a). *J. Mech. Phys. Solids* **13,** 189.
Hill, R. (1965b). *J. Mech. Phys. Solids* **13,** 213.
Hulbert, L. E., and Rybicki, E. F. (1971). *J. Compos. Mater.* **5,** 164.
Kerner, E. H. (1956). *Proc. Phys. Soc. London* **B69,** 808.
Khoroshun, L. P. (1966a). *Prikl. Mekh.* **2,** No. 2, 14.

Khoroshun, L. P. (1966b). *Mekh. Polim.* **2,** 365.
Khoroshun, L. P. (1966c). *Prikl. Mekh.* **2,** No. 9, 99.
Khoroshun, L. P. (1967). *Prikl. Mekh.* **3,** No. 9, 12.
Khoroshun, L. P. (1968a). *Mekh. Polim.* No. 1, 78.
Khoroshun, L. P. (1968b). *Prikl. Mekh.* **4,** No. 7, 8.
Khoroshun, L. P. (1968c). *In* "Kontsentratsiya Napryazhenii" (G. N. Savin, ed.), Vol. 2, p. 232. Naukova Dumka, Kiev.
Kilchinskii, A. A. (1965). *Prikl. Mekh.* **1,** No. 12, 65.
Kilchinskii, A. A. (1966). *In* "Thermal Stresses in Elements of Construction" (A. D. Kovalenko, ed.), Vol. 6, pp. 123–136. Naukova Dumka, Kiev.
Klyavlin, V. V. (1970a). *Prikl. Mekh.* **6,** No. 7, 123.
Klyavlin, V. V. (1970b). *Probl. Proch.* **2,** No. 6, 69.
Krivoglaz, M. A., and Cherevko, A. S. (1959). *Phys. Met. Metall.* **8,** No. 2, 1.
Kroner, E. (1967). *J. Mech. Phys. Solids* **15,** 319.
Kurshin, L. M., and Fil'shtinskii, L. A. (1961). *Izv. Akad. Nauk USSR Mekh. Mash.* No. 6, 100.
Leissa, A. W., and Clausen, W. E. (1968). *In* "Fundamental Aspects of Fiber Reinforced Plastic Composites" (R. T. Schwartz and H. S. Schwartz, eds.), pp. 29–44. Wiley (Interscience), New York.
Levin, V. M. (1967a). *In* "Issledovaniya po Uprugosti i Plastichnosti" (L. M. Kachanov, ed.), Vol. 6, pp. 58–71. Leningrad Univ. Press, Leningrad.
Levin, V. M. (1967b). *Inzh. Zh. Mekh. Tverd. Tela* No. 1, 88.
Levin, V. M. (1970). *Inzh. Zh. Mekh. Tverd. Tela* No. 5, 83.
Lomakin, V. A. (1965). *Prikl. Mat. Mekh.* **29,** 1048.
Lomakin, V. A. (1966a). *Inzh. Zh. Mekh. Tverd. Tela* No. 3, 72.
Lomakin, V. A. (1966b). *Inzh. Zh. Mekh. Tverd. Tela* No. 6, 68.
Lomakin, V. A. (1966c). *Prikl. Mat. Mekh.* **30,** 1035.
Lomakin, V. A. (1970a). *Vestn. Mosk. Univ. Mat. Mekh.* No. 2, 107.
Lomakin, V. A. (1970b). "Statistical Problems of Deformable Solid Bodies." Nauka, Moscow.
Lomakin, V. A., and Sheinin, V. I. (1970). *Inzh. Zh. Mekh. Tverd. Tela* No. 4, 124.
Mackenzie, J. K. (1950). *Proc. Phys. Soc. London* **B63,** 2.
McCoy, J. J. (1970). *In* "Recent Advances in Engineering Science" (A. C. Eringen, ed.), Vol. 5, Part 2, pp. 235–254. Gordon and Breach, New York.
Meijers, P. (1967). Doubly-Periodic Distributions in Perforated Plates. PhD Dissertation, Hogeschool Delft.
Miller, M. (1969). *J. Math. Phys.* **10,** 2005.
Molyneux, J., and Beran, M. (1965). *J. Math. Mech.* **14,** 337.
Moskalenko, V. N. (1969). *Izv. Akad. Nauk Arm. SSR Mekh.* **22,** No. 6, 45.
Muskhelishvili, N. I. (1963). "Some Basic Problems of the Mathematical Theory of Elasticity." Noordhoff, Groningen.
Nielsen, L. E., and Chen, P. E. (1968). *J. Mater.* **2,** 352.
Nosarev, A. V. (1967). *Mekh. Polim.* No. 5, 858.
Paul, B. (1960). *Trans. Metall. Soc. AIME* **218,** 36.
Pickett, G. (1965). Analytical Procedures for Predicting the Mechanical Properties of Fiber Reinforced Composites. AFML-TR-65-220.
Pickett, G. (1968). *In* "Fundamental Aspects of Fiber Reinforced Plastic Composites" (R. T. Schwartz and H. S. Schwartz, eds.), pp. 13–27. Wiley (Inerscience), New York.
Piehler, H. R. (1967). *In* Fiber-Strengthened Metallic Composites, ASTM STP 427, p. 3. Amer. Soc. Testing Mater., Philadelphia, Pennsylvania.

Rubenfeld, L. A., and Keller, J. B. (1969). *SIAM J. Appl. Math.* **17,** 495.

Savin, G. N., and Khoroshun, L. P. (1967). *Dokl. Akad. Nauk Ukr. SSR,* Series A, No. 9, 804.

Schapery, R. A. (1968). *J. Compos. Mater.* **2,** 380.

Sendeckyj, G. P. (1969a). *J. Compos. Mater.* **3,** 177.

Sendeckyj, G. P. (1969b). *Proc. Annu. Tech. Conf. Reinforced Plast./Compos. Div., 24th, SPI.* Sect. 14-E.

Sendeckyj, G. P. (1970). *J. Compos. Mater.* **4,** 500.

Sendeckyj, G. P. (1971). *J. Compos. Mater.* **5,** 82.

Sendeckyj, G. P. (1972). *J. Compos. Mater.* **6,** 160.

Sendeckyj, G. P., and Yu, I.-W. (1971). *J. Compos. Mater.* **5,** 533.

Stavrov, V. P., and Fomina, T. G. (1968). *Mekh. Polim.* No. 4, 631.

Stavrov, V. P., and Fomina, T. G. (1969). *Mekh. Polim.* No. 2, 362.

Stavrov, V. P., and Volkov, S. D. (1968). *Mekh. Polim.* No. 1, 86.

Stavrov, V. P., Dolgikh, V. Ya., and Volkov, S. D. (1967). *Mekh. Polim.* No. 2, 259.

Stevenson, J. F. (1970). Discrete-Element Microstress Analysis of Unidirectional Composites. Ph.D. Dissertation, Case Western Reserve Univ.

Tarnopol'ski, Yu. M., Portnov, G. G., and Zhigun, I. G. (1967). *Mekh. Polim.* No. 2, 243.

Turner, P. S. (1946). *J. Res. Nat. Bur. Std.* **37,** 239.

Van Fo Fy, G. A. (1965a). *Prikl. Mekh.* **1,** No. 2, 110.

Van Fo Fy, G. A. (1965b). *Pirkl. Mekh.* **1,** No. 5, 111.

Van Fo Fy, G. A. (1966a). *Sov. Phys.—Dokl.* **11,** No. 2, 176.

Van Fo Fy, G. A. (1966b). *Prikl. Mekh.* **2,** No. 7, 1.

Van Fo Fy, G. A. (1966c). *Mekh. Polim.* No. 4, 593.

Van Fo Fy, G. A. (1966d). *Mekh. Polim.* No. 5, 763.

Van Fo Fy, G. A. (1967a). *In* "Fiziko-Khimiya i Mekhanika Orientirovannikh Steklo-plastikov" (G. D. Andreevskaya, ed.), pp. 110–115. Nauka, Moscow.

Van Fo Fy, G. A. (1967b). *Prikl. Mekh.* **3,** No. 2, 106.

Van Fo Fy, G. A., and Klyavlin, V. V. (1967). *Mekh. Polim.* No. 4, 667.

Van Fo Fy, G. A., and Savin, G. N. (1965). *Polym. Mech.* **1,** No. 1, 106.

Van Fo Fy, G. A., Klyavlin, V. V., and Gordienko, V. P. (1969). *Mekh. Polim.* No. 2, 282.

Volkov, S. D. (1962). *Fiz. Metal. Metalloved.* **13,** No. 1, 10.

Volkov, S. D. (1969). *Mekh. Polim.* No. 1, 63.

Volkov, S. D. (1970). *Mekh. Polim.* No. 4, 676.

Volkov, S. D., and Dolgikh, V. Ya. (1967). *Mekh. Polim.* No. 3, 413.

Volkov, S. D., and Dolgikh, V. Ya. (1968). *Mekh. Polim.* No. 3, 438.

Volkov, S. D., and Klinskikh, N. A. (1965). *Fiz. Metal. Metalloved.* **19,** No. 1, 25.

Volkov, S. D., and Komissarova, M. L. (1968). *Mekh. Polim.* No. 4, 618.

Walpole, L. J. (1966a). *J. Mech. Phys. Solids* **14,** 151.

Walpole, L. J. (1966b). *J. Mech. Phys. Solids* **14,** 289.

Walpole, L. J. (1969). *J. Mech. Phys. Solids* **17,** 235.

Whitney, J. M. (1966). *Textile Res. J.* **36,** 765.

Whitney, J. M., and Riley, M. B. (1965). Elastic Stress-Strain Properties of Fiber Reinforced Composite Materials. AMFL-TR-65-238.

Whitney, J. M., and Riley, M. B. (1966). *AIAA J.* **4,** 1537.

Wilson, H. B., Jr. and Hill, J. L. (1965). Mathematical Studies of Composite Materials. Rohm and Haas Special Rep. No. S-50, AD 468569.

Wu, T. T. (1965). *J. Appl. Mech.* **32,** 211.

Wu, T. T. (1966). *Int. J. Solids Struct.* **2,** 1.

Yeh, R. H. T. (1970). *J. Appl. Phys.* **41,** 3353.

Yeh, R. H. T. (1971). *J. Appl. Phys.* **42,** 1101.
Yu, I.-W. (1971). Multiple Circular Inclusion Problems in Plane Elastostatics. Ph.D. Dissertation, Case Western Reserve Univ.

Bibliography

The following references were not used in this chapter, but may be of interest to the reader.

Abolin'sh, D. S. (1965). *Polym. Mech.* **1,** No. 4, 28.
Abolin'sh, D. S. (1966). *Mekh. Polim.* **2,** No. 3, 372.
Beer, F. (1959). *VDI-Z.* **101,** 463.
Biderman, V. L. (1965). *In* "Raschety na Prochnost'" (L. N. Danilov and Yu. N. Kanin, eds.), Vol. 11, pp. 3–30. Mashinostroenie, Moscow.
Bolotin, V. V. (1966a). *Mekh. Polim.* No. 1, 11.
Bolotin, V. V. (1966b). *In* "Raschety na Prochnost,'" Vol. 12, p. 3. Mashinostroenie, Moscow.
Bolotina, K. S. (1966). *Izv. Vyssh. Ucheb. Zav. Mashinostr.,* No. 12, 23.
Borzova, T. V. (1965). *Inzh. Zh.* **5,** 972.
Brandmaier, H. E. (1969). *J. Compos. Mater.* **3,** 728.
Brown, C. B. (1965). *J. Franklin Inst.* **279,** 189.
Bryzgalin, G. I. (1964). *Sov. Plast.* No. 7, 65.
Chamis, C. C. (1968). *Proc. SAMPE Nat. Meeting, 14th.*
Chen, P. E., and Nielsen, L. E. (1969). *Kolloid-Z. Z. Polym.* **235,** 1174.
Chow, T. S., and Hermans, J. J. (1969). *J. Compos. Mater.* **3,** 382.
Cohen, L. J., and Ishai, O. (1967). *J. Compos. Mater.* **1,** 390.
Corten, H. T. (1967). *In* "Modern Composite Materials" (L. J. Broutman and R. H. Krock, eds.), pp. 27–105. Addison-Wesley, Reading, Massachusetts.
Cox, H. L. (1952). *Brit. J. Appl. Phys.* **3,** 72.
Cuevas, J. E. (1968). *J. Compos. Mater.* **2,** 113.
Date, E. H. F. (1970). *J. Phys. D Appl. Phys.* **3,** 778.
Eimer, Cz. (1967). *Arch. Mech. Stosowanej* **19,** 521.
Eimer, Cz. (1968). *Arch. Mech. Stosowanej* **20,** 87.
Frisch, H. L. (1965). *Trans. Soc. Rheol.* **9,** 293.
Gillis, P. P. (1970). *Fibre Sci. Technol.* **2,** 193.
Greszczuk, L. B. (1964). Elastic Constants and Analysis Methods for Filament Wound Shell Structures. Missile and Space Syst. Div., Douglas Aircraft Co., Douglas Rep. SM-45849.
Greszczuk, L. B. (1966). *Proc. ASCE J. Eng. Mech. Div.* **92,** No. EM6, 63.
Greszczuk, L. B. (1970). *Proc. AIAA/ASME Struct., Struct. Dynamics Mater. Conf., 11th Denver, Colorado.*
Guz, A. N. (1967). *Dopov. Akad. Nauk Ukr. SSR* No. 12, 1098.
Halpin, J. C., and Thomas, R. L. (1968). *J. Compos. Mater.* **2,** 488.
Hashin, Z., Brull, M. A., Chu, T. Y., and Zudans, Z. (1967). U. S. Army Aviation Res. Lab., USAAVLABS Tech. Rep. 67-41.
Hershey, A. V. (1954). *J. Appl. Mech.* **21,** 236.

Hewitt, R. L., and deMalherne, M. C. (1970). *J. Compos. Mater.* **4,** 280.
Hill, R. (1952). *Proc. Phys. Soc. London* **A65,** 349.
Holister, G. S., and Thomas, C. (1966). "Fibre Reinforced Materials." Elsevier, Amsterdam.
Kausch-Blecken von Schmeling, H. H. (1967). *J. Appl. Phys.* **38,** 4213.
Kneer, G. (1965). *Phys. Status Solidi* **9,** 825.
Koh, S. L. (1970). *In* "Mechanics of Composite Materials" (F. W. Wendt, H. Liebowitz, and N. Perrone, eds.), pp. 387–402. Pergamon, Oxford.
Krock, R. H., and Broutman, L. J. (1967). *In* "Modern Composite Materials" (L. J. Broutman and R. H. Krock, eds.), pp. 3–26. Addison-Wesley, Reading, Massachusetts.
Kroner, E., Datta, B. K., and Kessel, D. (1966). *J. Mech. Phys. Solids* **14,** 21.
Lee, K. J., and Westmann, R. A. (1970). *J. Compos. Mater.* **4,** 242.
Lees, J. K. (1968). *Polym. Eng. Sci.* **8,** 186.
Lomakin, V. A., and Koltunov, M. A. (1965). *Polym. Mech.* **1,** No. 2, 79.
Margolin, G. G. (1967). *Mekh. Polim.* No. 4, 737.
Morris, P. R. (1970). *Int. J. Eng. Sci.* **8,** 49.
Nielsen, L. E. (1967). *J. Compos. Mater.* **1,** 100.
Nielsen, L. E. (1970). *J. Appl. Phys.* **41,** 4626.
Nosarev, A. V. (1967). *Mekh. Polim.* **3,** No. 2, 266.
Novozhilov, V. V. (1968). *Prikl. Mat. Mekh.* **24,** 59.
Rabinovich, A. L. (1962). *Issled. Mekh. Prikl. Mat., Tr. Mosk. Fiz.-Tekh. Inst.* **9,** 54–68.
Rabinovich, A. L. (1964). *Sov. Phys.—Dokl.* **8,** 1233.
Rabinovich, A. L. (1967). *In* "Fiziko-Khimiya i Mekhanika Orientirovannikh Steklo-plastikov" (G. D. Andreevskaya, ed.), pp. 101–110. Nauka, Moscow.
Rabinovich, A. L., and Verkhovskii, I. A. (1964). *Inzh. Zh.* **4,** No. 1, 90.
Rossi, R. C. (1968). *J. Amer. Ceram. Soc.* **51,** 433.
Sendeckyj, G. P. (1970). *Fibre Sci. Technol.* **2,** 211.
Shaffer, B. W. (1964). *AIAA J.* **2,** 348.
Somoreoff, A. (1970). *Quart. Appl. Math.* **28,** 219.
Symm, G. T. (1970). *J. Compos. Mater.* **4,** 426.
Tsai, S. W. (1963). Structural Behavior of Composite Materials. NASA CR-52410.
Tsai, S. W. (1966). Mechanics of Composite Materials, Part I. AFML-TR-66-149.
Walpole, L. J. (1970). *J. Mech. Phys. Solids* **18,** 343.
Whitney, J. M. (1967a). *J. Compos. Mater.* **1,** 188.
Whitney, J. M. (1967b). *Textile Res. J.* **37,** 1108.

4

Viscoelastic Behavior and Analysis of Composite Materials

R. A. SCHAPERY

Civil and Aerospace Engineering Departments
Texas A & M University
College Station, Texas

I. Introduction

Composite materials which have one or more polymeric constituents exhibit a significant amount of time-dependent mechanical behavior in many service environments. This behavior, termed *viscoelasticity*, manifests itself in different ways; creep under constant load, stress relaxation under constant deformation, damping of dynamic response, and dependence of stress on strain rate are just some of the examples of viscoelastic phenomena. The efficient design and utilization of such composites in load bearing members and in other structurally related applications demands a good understanding of viscoelasticity.

This chapter is concerned primarily with (i) the analytical characterization of linearly viscoelastic stress–strain behavior of polymeric composites and their constituents, and (ii) the prediction of effective mechanical properties and structural response in terms of constituent properties. However, recognizing that composites may be nonlinearly viscoelastic under design stresses, this area is reviewed in Section VI. Although we have restricted the discussion on behavior to polymeric composites, it should be realized that the linear theory itself is not limited to these materials, but can be used whenever the linearity condition is at least approximately satisfied.

Much of the chapter is a summary and review of the literature on viscoelastic behavior and analysis of composites, especially the most recent developments. Some new results are also given which deal with the prediction of upper and lower bounds for effective complex moduli and compliances, and with the dynamic analysis of structures; included is a simple method for extending dynamic elasticity solutions with microstructural effects to viscoelasticity.

A large number of papers on viscoelasticity of composites has appeared in the last ten years. As a result, most of our references to the literature were necessarily limited to this time period. The general rule we followed was to reference only the most recent papers, especially if they themselves provided a review of earlier work. Hopefully, those included plus recent books on viscoelasticity (Christensen, 1971; Ferry, 1970; Flügge, 1967;

Pipkin, 1972; Ward, 1971) and surveys on composites (Beckwith, *et al.*, 1970, 1971; Krokosky, 1967) will enable the interested reader to make an in-depth study.

II. Linear Behavior

The analytical representation of linear constitutive behavior is discussed in this section. Some examples of actual material response are given for illustrative purposes and for later use in the sections on analysis.

Linear theory is, of course, only an approximation to real behavior. If it is to be properly used, one should know its range of applicability. Therefore, we first give the definition of linearity in a form which can be readily used in conjunction with experimental data to establish the range for which the theory is sufficiently accurate. We believe it is important to elaborate on this definition because there are many published studies in which linearity is claimed, but only *one* of the two criteria which comprise the definition of linear behavior is verified. A proper verification is of considerable practical importance because, as will be shown in Section VI, some important composite materials satisfy the one criterion which is often used as a supposed check on linearity, and yet are highly nonlinear.

A. Definition of Linearity; Hereditary Law

The assumption that a viscoelastic body is linear is sufficient to establish explicit single-integral expressions connecting responses to inputs, without having to specify a priori its physical makeup and the physical significance of input and response quantities. Thus, for later use in both material characterization and analysis, we shall view the body in this subsection simply as a "black box," which may be, for example, a bar under uniaxial loading or a complex composite structure under several loads.

Consider first a response R (e.g., internal stress or displacement) to one input (e.g., boundary force or displacement). We write

$$R = R\{I\} \tag{1}$$

where the curly brackets $\{\ \}$ are used to denote the fact that the current value of R is a function of the *history* of I, and not just its current value; the response R is called a *functional* of I.

This functional is defined to be linear if and only if it satisfies the following two criteria:

(i) homogeneity (or proportionality):

$$R\{cI\} = cR\{I\}; \qquad c = \text{constant} \tag{2}$$

(ii) superposition:

$$R\{I_a + I_b\} = R\{I_a\} + R\{I_b\} \tag{3}$$

where I_a and I_b may be either the same or different input histories. Any material or structure which satisfies Eqs. (2) and (3) is termed *linearly viscoelastic.*

The first property states that a proportional change in input causes the same proportional change in response (e.g., if the input is doubled, the response doubles). On the other hand, the second one is not restricted to proportional changes, and states that the response due to the input $I_a + I_b$ is equal to the sum of the responses due to each input acting separately. It is the first criterion which, when satisfied by experimental data, has been used by some investigators as the only basis for concluding that a material is linearly viscoelastic; that there are some composite materials which satisfy Eq. (2), but not Eq. (3), will be covered in Section VI. Although superposition is not implied by the homogeneity condition, the opposite relationship is true for all practical purposes; viz., one can easily show that the homogeneity condition for all rational values of c (including $c = 0$) is automatically satisfied if it is known that the body obeys the superposition criterion.

When the response functional is linear, it can be written as an integral containing the actual input and the response to *one* preselected input history. In characterizing viscoelastic materials it is customary (and convenient) to select the one input as a unit step function H, where

$$H(t - t') \equiv \begin{cases} 0, & t < t' \\ 1, & t > t' \end{cases} \tag{4}$$

The response to H is denoted by R_H, and, in general, is a function of the current time t *and* the time at which the unit input is applied t'; viz.,

$$R_H = R_H(t, t') \tag{5}$$

Obviously, from physical considerations, R_H vanishes when $t < t'$. In many applications, the response of the viscoelastic body depends only upon the time elapsed since application of the input; viz.,

$$R_H = R_H(t - t') \tag{6}$$

Viscoelastic materials or structures which satisfy Eq. (6) are called *nonaging* or *time-translation invariant.*

The function R_H is called, for example, the *creep compliance* when the input is stress and the response is strain; however, for general linear systems the name *indicial admittance* is used (von Kármán and Biot, 1940).

The response to a time-varying input is found by considering this input to be the limit of a sum of step inputs and by using Eqs. (2) and (3) (e.g., see Pipkin, 1972). There results,

$$R = \int_{-\infty}^{t} R_H(t, t') \frac{dI}{dt'} dt' \tag{7}$$

and for a nonaging body,

$$R = \int_{-\infty}^{t} R_H(t - t') \frac{dI}{dt'} dt' \tag{8}$$

The type of relation connecting input and response in Eqs. (7) and (8) is called a *hereditary law.* For a nonaging body, several names are commonly used for the integral representation, Eq. (8): (i) superposition integral, (ii) Boltzmann superposition integral, (iii) Duhamel's integral, and (iv) convolution integral.

The lower integration limit in the hereditary law is taken as $-\infty$ in order to allow for the possibility of input application prior to the time origin; this choice facilitates, for example, the prediction of steady-state vibration response (e.g., Flügge, 1967).

Generalization of the above forms to several responses (e.g., six components of the stress tensor at a point, displacements at many points, etc.) and several input quantities (e.g., six components of the strain tensor at a point, concentrated applied forces, etc.) is easily accomplished when the response of the body is linear with respect to all inputs. For example, for each response R_α of a nonaging system,

$$R_\alpha = \int_{-\infty}^{t} R_{H\alpha\beta}(t - t') \frac{dI_\beta}{dt'} dt' \tag{9}$$

in which $R_{H\alpha\beta}(t)$ is the response R_α due to a unit input $I_\beta = H(t)$, with all other inputs zero. Also, the summation convention is used throughout *Sections II–IV,* in which a repeated index (β here) is to be summed out over its range unless indicated otherwise.

B. Isothermal Constitutive Equations

The so-called "effective modulus" (or "effective compliance") theory can often be used to characterize and analyze the linear viscoelastic response of composite materials. Just as for elastic composites, this theory is valid for static or quasi-static problems when the dimensions which define the microstructural inhomogeneity are small compared to (i) dimensions of the specimen or structure of interest, and (ii) the distances over which forces and/or displacements are applied; if, for example, concentrated loads are applied and/or the body contains cracks or any sharp notches, the theory cannot be expected to yield accurate predictions of stresses and strains in the regions of high stress concentration. In the case of dynamic problems, the wavelengths must be very long compared to the microstructural inhomogeneity dimensions.

In the remainder of Section II, it will be assumed that the effective modulus theory is applicable. Later, in Section V,B, we shall examine some problems in which the microstructure is explicitly taken into account.

General linear relations between stresses σ_{ij} $(i, j = 1, 2, 3)$ and strains ϵ_{ij} [referred to an orthogonal set of Cartesian axes (x_1, x_2, x_3)] for a *non-aging* viscoelastic material with an arbitrary degree of anisotropy follow immediately from Eq. (9). In terms of relaxation moduli $C_{ijkl}(t)$ $(i, j, k, l = 1, 2, 3)$, we write

$$\sigma_{ij} = \int_{-\infty}^{t} C_{ijkl}(t - t')\, \partial\epsilon_{kl}/\partial t'\, dt' \tag{10}$$

The inverse relations, in terms of creep compliances S_{ijkl} are

$$\epsilon_{ij} = \int_{-\infty}^{t} S_{ijkl}(t - t')\, \partial\sigma_{kl}/\partial t'\, dt' \tag{11}$$

Explicit dependence of the stresses, strains, and material properties on coordinates x_i is not shown. However, this omission is made for notational simplicity only; the stresses, strains, moduli, and compliances may vary spacially as long as the variation is not so large as to invalidate the effective modulus theory.

The moduli and compliances in Eqs. (10) and (11) are response functions for unit inputs applied at $t = 0$, as may be seen by noting that, apart from a notational difference, they are indicial admittances $R_{H\alpha\beta}$ in Eq. (9). The significance of these material properties may also be brought out directly from Eqs. (10) and (11). For example, consider a generalized relaxation test in which, by definition,

$$\epsilon_{ij} = \epsilon'_{ij} H(t) \tag{12}$$

where all ϵ'_{ij} are constant. Substituting strains, Eq. (12), into Eq. (10), and recognizing that dH/dt' is the Dirac delta function $\delta(t')$, yields

$$\sigma_{ij} = C_{ijkl}(t)\epsilon'_{kl} \tag{13}$$

Similarly, for a generalized creep test,

$$\sigma_{ij} = \sigma'_{ij} H(t) \tag{14}$$

where all σ'_{ij} are constant, we find from Eq. (11) that

$$\epsilon_{ij} = S_{ijkl}(t) \sigma'_{kl} \tag{15}$$

The relaxation moduli and creep compliances are components of fourth-order tensors and, just as for elastic materials, they are symmetric with respect to an interchange of the index i with j and k with l; this symmetry is a consequence of the symmetry of the stress and strain tensors (Fung, 1965).

Symmetry with respect to an interchange of the first two and last two indices, viz.,

$$C_{ijkl}(t) = C_{klij}(t) \qquad \text{and} \qquad S_{ijkl}(t) = S_{klij}(t) \tag{16}$$

has been derived by Biot (1954) from the thermodynamics of stable, irreversible systems; this theory is based on the first and second laws of thermodynamics and on Onsager's principle (Fung, 1965). In contrast, the symmetry in Eq. (16) for *elastic* materials follows from an application of only the first and second laws to reversible processes. [Modulus and compliance tensors which satisfy Eq. (16) are said to be *completely* symmetric.]

Inasmuch as the theoretical basis for irreversible thermodynamics is not as well established as for the classical equilibrium theory (although no violations have been found experimentally for systems not far from equilibrium (DeGroot and Mazur, 1962), it is desirable to have an independent verification of the symmetry property. Without appealing to irreversible thermodynamics, Schapery (1967) has shown that the effective modulus and compliance tensors for *composites* satisfy Eq. (16) if these tensors for the *phases* satisfy Eq. (16). This result implies that the symmetry property is rigorously valid for most composites of technological interest; the symmetry of phase properties is a direct consequence of: (i) geometric symmetry (rather than irreversible thermodynamics) for elastic and viscoelastic phases which are either isotropic or cubic (Rogers and Pipkin, 1963), and (ii) equilibrium thermodynamics for elastic phases with an arbitrary degree of anisotropy.

The constitutive equations for viscoelastic, composite (or monolithic) media with particular types of mechanical symmetry can be derived from the anisotropic relations, Eqs. (10) or (11), by following the same procedure used for elastic materials; viz., the relaxation modulus or creep compliance tensor is required to be invariant under appropriate time-independent coordinate transformations (Sokolnikoff, 1956).

In order to illustrate the implications of symmetry condition Eq. (16) and its experimental verification, we consider next a type of geometric symmetry common to many technological composites: transverse isotropy. For discussion of more general types the reader is referred to (i) the paper by Hayes and Morland (1969) in which a systematic series of 24 tests is proposed for determining all 36 relaxation moduli $C_{ijkl}(t)$ when the symmetry condition, Eq. (16), is *not* assumed, and (ii) the anisotropic elasticity literature when the modulus and compliance tensors are known a priori to be completely symmetric.

C. Isothermal Characterization of Transversely Isotropic Media

We now assume, for simplicity, that the axes x_i are aligned with the axes of material symmetry, and the x_2x_3 plane is the plane of isotropy; for example, a fiber-reinforced plastic with isotropic phases, in which all fibers are in the x_1 direction and are distributed randomly in the x_2x_3 plane, fits the assumed situation. Using single-index notation (Sokolnikoff, 1956), Eq. (15) for the generalized creep test becomes (Rogers and Pipkin, 1963):

$$\begin{aligned}\epsilon_1 &= S_{11}\sigma_1 + S_{12}\sigma_2 + S_{12}\sigma_3\\ \epsilon_2 &= S_{21}\sigma_1 + S_{22}\sigma_2 + S_{23}\sigma_3\\ \epsilon_3 &= S_{21}\sigma_1 + S_{23}\sigma_2 + S_{22}\sigma_3\end{aligned} \tag{17a}$$

$$\gamma_4 = S_{44}\tau_4, \qquad \gamma_5 = S_{66}\tau_5, \qquad \gamma_6 = S_{66}\tau_6 \tag{17b}$$

where

$$S_{44} \equiv 2(S_{22} - S_{23}) \tag{17c}$$

and the following definitions have been used:

$$\sigma_i \equiv \sigma_{ii}, \qquad \epsilon_i \equiv \epsilon_{ii}, \qquad i = 1, 2, 3 \quad (i \text{ not summed}) \tag{18a}$$

and

$$\tau_4 \equiv \tau_{23}, \qquad \tau_5 \equiv \tau_{13}, \qquad \tau_6 \equiv \tau_{12} \tag{18b}$$

$$\gamma_4 \equiv 2\epsilon_{23}, \qquad \gamma_5 \equiv 2\epsilon_{13}, \qquad \gamma_6 \equiv 2\epsilon_{12} \tag{18c}$$

Also, the stresses in Eqs. (17) are timewise constant (the prime has been dropped for notational simplicity), but all strains and creep compliances may, in general, be functions of time.

It is to be noted that geometric symmetry does *not* imply $S_{21} = S_{12}$ and, therefore, in writing Eqs. (17) we have allowed for the possibility of these compliances being unequal; on the other hand, $S_{23} = S_{32}$ on account of the geometric symmetry involved. Thus, at this point there are *six* independent creep compliances in Eqs. (17).

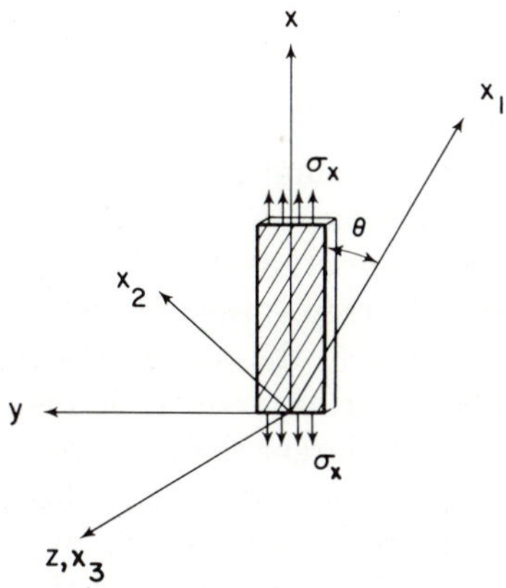

FIG. 1. Anisotropic tensile specimen.

Let us now consider the problem of evaluating these six properties by means of a series of tensile tests on the specimen shown in Fig. 1. First, the tests used by Darlington and Saunders (1971) in their study of oriented, low-density polyethylene will be described. (They measured the axial strain ϵ_x and thickness strain ϵ_3 for several values of the angle θ between the draw direction x_1 and the specimen's axis x.) The axial compliance

$$D_\theta \equiv \epsilon_x/\sigma_x \tag{19}$$

is related to the principal compliances in Eqs. (17) by using standard second-order tensor transformation relations to find the strains in Eqs. (17) in terms of ϵ_x and σ_x; the result is

$$D_\theta = S_{11}\cos^4\theta + (S_{21} + S_{12} + S_{66})\sin^2\theta\cos^2\theta + S_{22}\sin^4\theta \tag{20}$$

Also, define the Poisson's ratio for the thickness direction as

$$\nu_{t\theta} \equiv -\epsilon_3/\epsilon_x \tag{21}$$

Then, again from tensor transformation relations,

$$\nu_{t\theta} = -(S_{21}\cos^2\theta + S_{23}\sin^2\theta)/D_\theta \tag{22}$$

Observe that when $\theta = 0°$, Eqs. (20) and (22) become

$$D_0 = S_{11}, \qquad \nu_{t0} = -S_{21}/S_{11} \tag{23}$$

and when $\theta = 90°$,

$$D_{90} = S_{22}, \qquad \nu_{t90} = -S_{23}/S_{22} \tag{24}$$

Also, when $\theta = 45°$, Eq. (20) can be written as

$$S_{12} + S_{66} = -S_{11} - S_{22} - S_{21} + 4D_{45} \tag{25}$$

Thus, the four creep compliances S_{11}, S_{22}, S_{23}, and S_{21} can be found from Eqs. (23) and (24) and measurements of axial ϵ_x and thickness ϵ_3 strains for $\theta = 0$ and 90°. The sum $S_{12} + S_{66}$ is then found from Eq. (25) and a $\theta = 45°$ test. Darlington and Saunders (1971) assumed $S_{12} = S_{21}$, and therefore were able to find S_{66} from Eq. (25). They mention that the shear compliance S_{66} derived from the above tests was in good agreement with that obtained from torsional creep measurements on a $\theta = 0°$ specimen; the latter results are not shown in the paper. The agreement would seem to imply that $S_{12} = S_{21}$; however, this conclusion is open to question since $4D_{45}$ in Eq. (25) is significantly larger than S_{11}, S_{22}, and S_{21}, which means, of course, that the values of S_{66} predicted from this equation probably are not very sensitive to the difference between S_{12} and S_{21}, if any exists.

A more critical test of the symmetry condition can be achieved by measuring the transverse Poisson's ratio for $\theta = 0$ and 90° under uniaxial stress σ_x. When $\theta = 0°$, this Poisson's ratio is

$$\nu_{12} \equiv -\epsilon_y/\epsilon_x = -\epsilon_2/\epsilon_1 \tag{26}$$

and from Eq. (17a) with $\sigma_1 \equiv \sigma_x$ (and all other stresses zero)

$$\nu_{12} = -S_{21}/S_{11} \tag{27}$$

which is called the *major* creep Poisson's ratio; note that this ratio is identical to ν_{t0}, Eq. (23), as a result of the transverse isotropy assumption. When $\theta = 90°$, the Poisson's ratio is

$$\nu_{21} \equiv -\epsilon_y/\epsilon_x = -\epsilon_1/\epsilon_2 \tag{28}$$

and setting $\sigma_2 = \sigma_x$ in Eq. (17a) we find

$$\nu_{21} = -S_{12}/S_{22} \tag{29}$$

Hence, in terms of experimentally determined quantities,

$$S_{21} - S_{12} = \nu_{21}D_{90} - \nu_{12}D_0 \tag{30}$$

and therefore the compliance tensor is completely symmetric if

$$\nu_{21}D_{90} = \nu_{12}D_0 \tag{31}$$

which is identical to the symmetry condition for elastic materials (where $E_0 \equiv D_0^{-1}$ and $E_{90} \equiv D_{90}^{-1}$), except all properties in the equation may be functions of time. Equation (31) has been verified (within experimental error) by Halpin and Pagano (1968) for a nylon fiber-reinforced rubber with a high ratio of longitudinal (fiber direction) to transverse stiffness, and by Lou and Schapery (1971) for a glass fiber-reinforced epoxy resin.

In both of these studies the fibers were essentially elastic, while the matrix phase was viscoelastic and isotropic. Therefore, that Eq. (31) was verified does *not* provide a check on irreversible thermodynamic theory, and Onsager's principle in particular; for, as indicated in Section II,B, complete symmetry of the composite compliance tensor is a consequence of geometric symmetry of the phase properties. Only if at least one of the phases is significantly viscoelastic *and* anisotropic does the experimental determination of symmetry provide information on the validity of thermodynamics for viscoelastic solids.

Anisotropic (oriented) polymers, used either by themselves or as constituents in composites, should provide an indication of the correctness of the irreversible thermodynamic theory under certain conditions. For example, by cold drawing low-density polyethylene, various amounts of anisotropy and viscoelasticity are induced (Darlington and Saunders, 1970, 1971). The thermodynamic basis for symmetry can be checked as long as a significant portion of the overall viscoelasticity is due to creep in the crystalline regions. Motion in these regions is believed to produce the so-called "α-transition" in polyethylene (Huseby and Matsuoka, 1967), and, in turn, the existence of this transition is dependent on the draw ratio, temperature, and time scale (or frequency). In view of the observations made by Darlington and Saunders (1971), intracrystalline creep is usually most significant at low draw ratios; fortunately, the amount of anisotropy can be quite large in slightly drawn specimens (e.g., at a draw ratio approximately equal to 1.2, Darlington and Saunders, 1971).

The author is not aware of any data on drawn polyethylene in the α-transition, or any other anisotropic material, which are sufficient to either confirm or deny the irreversible thermodynamic basis for symmetry of the modulus and compliance tensors. However, until these data become available, acceptance of the thermodynamic theory is not much of a gamble, considering the large amount of experimental verification already available on other than viscoelastic processes (e.g., DeGroot and Mazur, 1962).

Prediction of the strains due to time-varying stresses is now readily accomplished by evaluating the convolution integrals in Eq. (11) after expressing the compliances S_{ijkl} in terms of the principal compliances S_{ij} appearing in Eqs. (17). For example, if the coordinate axes are assumed to be aligned with the axes of material symmetry and the double index notation for stress and strains is used, the normal strain ϵ_{11} in Eq. (17a) is generalized to read

$$\epsilon_{11} = \int_{-\infty}^{t} S_{11}(t - t')\, \partial\sigma_{11}/\partial t'\, dt' + \int_{-\infty}^{t} S_{12}(t - t')\, \partial(\sigma_{22} + \sigma_{33})/\partial t'\, dt' \tag{32a}$$

and the equation for shear strain γ_4, for example, becomes

$$\epsilon_{23} = \int_{-\infty}^{t} [S_{22}(t - t') - S_{23}(t - t')]\, \partial\tau_{23}/\partial t'\, dt' \tag{32b}$$

In terms of the experimentally defined quantities these equations become

$$\epsilon_{11} = \int_{-\infty}^{t} D_0(t - t')\, \partial\sigma_{11}/\partial t'\, dt' - \int_{-\infty}^{t} D_{90}(t - t')\nu_{21}(t - t')\, \partial(\sigma_{22} + \sigma_{33})/\partial t'\, dt' \tag{33a}$$

$$\epsilon_{23} = \int_{-\infty}^{t} D_{90}(t - t')[1 + \nu_{t90}(t - t')]\, \partial\tau_{23}/\partial t'\, dt' \tag{33b}$$

When the coordinate and principal material axes are *not* parallel, the compliance tensor S_{ijkl} is found in terms of S_{ij} by using the standard fourth-order tensor transformation

$$S_{ijkl} = l_{mi}l_{nj}l_{pk}l_{ql}S'_{mn\,pq} \tag{34}$$

where $S'_{mn\,pq}$ and S_{ijkl} are the compliance tensors in the (x_1', x_2', x_3') and (x_1, x_2, x_3) rectangular Cartesian coordinate systems, respectively; this transformation equation is valid for media with an arbitrary degree of anisotropy, and is *not* restricted to tensors which satisfy the symmetry condition, Eq. (16). Also, l_{ij} is the cosine of the angle between the x_i' axis and x_j axis. If the axes (x_1', x_2', x_3') are taken as the principal material axes, then for a transversely isotropic material,

$$\begin{aligned}
S'_{1111} &\equiv S_{11} \\
S'_{2222} &= S'_{3333} \equiv S_{22} \\
S'_{1122} &= S'_{1133} \equiv S_{12} \\
S'_{2211} &= S'_{3311} \equiv S_{21} \\
S'_{3322} &= S'_{2233} \equiv S_{23} \\
S'_{2323} &\equiv (S_{22} - S_{23})/2 \\
S'_{1313} &= S'_{1212} \equiv S_{66}/4
\end{aligned} \tag{35}$$

All other components of S'_{ijkl} are identically zero. Equations (34) and (35), together with Eq. (11), complete the three-dimensional constitutive formulation for a transversely isotropic material.

The problem of calculating the relaxation moduli C_{ijkl} in Eq. (10), given the creep compliances S_{ijkl} and vice versa, will be taken up later in Section II,G.

It is to be noted that the equations for isotropic media are easily recovered from the above equations. Transformation relation (34) is not needed of course, as the properties are independent of the coordinate orientation. Let $D = D(t)$ and $\nu_c = \nu_c(t)$ be the creep compliance and creep Poisson's ratio, respectively, for uniaxial loading. Then,

$$D_0 = D_{90} \equiv D$$

$$\nu_{12} = \nu_{21} = \nu_{t0} = \nu_{t90} \equiv \nu_c \tag{36a}$$

Also, by using the usual symbol $J = J(t)$ for the shear creep compliance, where $J \equiv S_{44} \equiv S_{66}$, Eq. (17c) yields

$$J = 2(1 + \nu_c)D \tag{36b}$$

which is identical with that for an elastic material except the functions in this equation may be time-dependent.

D. *Nonisothermal Constitutive Equations*

In order to characterize the effect of temperature on viscoelastic behavior, one must first distinguish between reversible and irreversible effects. By definition the latter catagory consists of those cases in which permanent changes are produced, such as primary bond rupture and weight loss; this catagory will be discussed later in Section II,E on aging and thermal degradation.

The present section is concerned with the reversible phenomena consisting of thermal expansion (or contraction) and changes in those material properties which appear in the mechanical constitutive equations and depend on just the instantaneous temperature. Consideration of only these phenomena is sufficient for most engineering applications at moderate and low temperatures (the values of which depend, of course, on the particular material of interest).

We should add that there is very little experimental data in the literature on reversible thermal effects in viscoelastic composites, especially with anisotropy. Consequently, the constitutive equations which are given in this section, while they have some theoretical and experimental support, should be considered as tentative.

1. Uniaxial Behavior

Reversible thermal effects will be introduced by considering, for simplicity, the problem of characterizing the viscoelastic behavior of a bar under a uniform uniaxial stress σ_x and a spacewise constant temperature. In principle, the degree of anisotropy is not restricted.

A completely general approach to this problem (within the context of linear theory) consists of using Eq. (7) with

$$I \equiv \sigma_x, \qquad R \equiv \epsilon_{x\sigma}, \qquad R_H(t, t') \equiv D(t, t') \tag{37}$$

where $\epsilon_{x\sigma}$ is the axial strain due to stress (i.e., total strain less that due to thermal expansion) and $D(t, t')$ is the creep compliance at time t due to a constant stress applied at time t'; thus

$$\epsilon_{x\sigma} = \int_{-\infty}^{t} D(t, t') \frac{d\sigma_x}{dt'} dt' \tag{38}$$

If the creep compliance is independent of temperature, which in practice is approximately true for only *small* temperature changes, and/or the temperature is timewise constant, then Eq. (38) reduces to that for a nonaging material; viz., $D = D(t - t')$.

When these two conditions do not exist, and it is desired to find experimentally $D(t, t')$ without making any a priori assumptions about its time or temperature dependence, a large number of tests may be needed, especially if the temperature varies cyclically or randomly in time, for what one must do is subject the specimen(s) to the actual time-varying temperature and then, starting at different times $t_1', t_2', \ldots, t_N'$, conduct N creep tests and record the resulting strains until the greatest time of interest is reached; the number of tests N must be large enough to be able to determine the continuous dependence of compliance on t'.

In addition, in order to complete the constitutive formulation, the thermal expansion must be found. This problem is not trivial for composites if the thermal expansion is affected appreciably by a release of residual stresses as the material is heated, and is, therefore, thermal-history-dependent; in this case, the correct thermal expansion would be found by subjecting an unloaded specimen to the actual thermal history.

In view of the above comments, a general approach to the characterization problem may be quite impractical. Fortunately, there is experimental evidence which implies viscoelastic characterization for *transient* temperature applications can be accomplished by performing tests at a set of different *constant* temperatures. Although the published data which support this statement are for only amorphous polymers, we shall assume this simplification is applicable in the remainder of this chapter.

(a) The Thermorheologically Simple Material (TSM). The simplest, realistic viscoelastic constitutive equation for which response under constant temperatures can be used to predict response under transient temperatures is that for a so-called "thermorheologically simple material" (TSM) (Morland and Lee, 1960). By definition, the one-dimensional equation in terms of compliance is

$$\epsilon_{x\sigma} = \int_0^{\xi} D(\xi - \xi') \frac{d\sigma_x}{d\xi'} d\xi' \tag{39a}$$

or, equivalently,

$$\epsilon_{x\sigma} = \int_0^{t} D(\xi - \xi') \frac{d\sigma_x}{dt'} dt' \tag{39b}$$

where

$$\xi = \xi(t) \equiv \int_0^t d\tau/a_{\mathrm{T}}, \qquad \xi' \equiv \xi(t') = \int_0^{t'} d\tau/a_{\mathrm{T}} \tag{40}$$

It has been assumed that $\sigma_x = \epsilon_{x\sigma} = 0$ when $t < 0$, which enables later use of Laplace transform theory; no actual generality is lost since the time origin always can be selected so as to satisfy this condition. Also, ξ is called *reduced time* and

$$a_{\mathrm{T}} = a_{\mathrm{T}}[T(\tau)] \tag{41}$$

is the so-called *temperature shift factor*.

The only temperature-dependent material property is this shift factor; it reflects the influence of temperature on internal viscosity and consequently is quite sensitive to temperature changes.

The inverse of Eq. (39), in which stress is expressed as a functional of strain, can be derived from Eq. (39) by first introducing the Laplace transform (LT) with respect to reduced time,

$$\hat{f} \equiv \int_0^{\infty} e^{-s\xi} f(\xi)\, d\xi \tag{42}$$

where f is a function of reduced time and $\hat{f}$ is its LT. [Although not shown explicitly, the lower limit in Eq. (42) is assumed to be 0^- throughout this chapter in order to account for a possible infinity in f at $t = 0$, such as exists in the (ideal) creep and relaxation tests.] Next, transform Eq. (39a), use the familiar rule for the LT of convolution integrals (e.g., Flügge, 1967), and then solve for $\hat{\sigma}_x$; thus

$$\hat{\sigma}_x = s\hat{E}\hat{\epsilon}_{x\sigma} \tag{43}$$

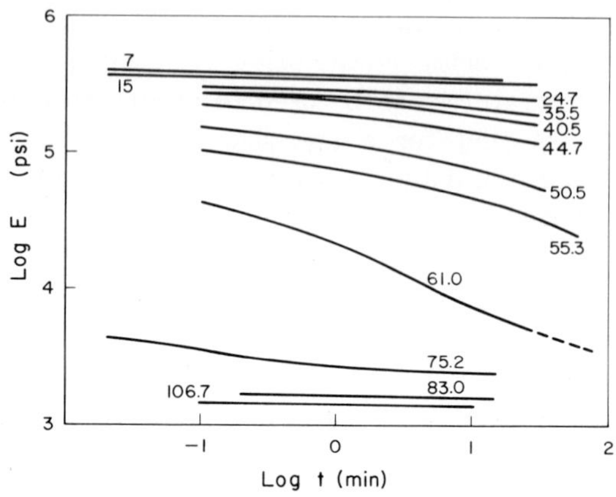

FIG. 2. Compression–relaxation moduli at different temperatures (°C) for epoxy resin (after Moehlenpah *et al.*, 1969).

where, by definition,

$$s\hat{E} \equiv (s\hat{D})^{-1} \tag{44}$$

Laplace inversion of Eq. (43) by means of the convolution rule yields

$$\sigma_x = \int_0^{\xi} E(\xi - \xi') \frac{d\epsilon_{x\sigma}}{d\xi'} \, d\xi' \tag{45a}$$

Equivalently,

$$\sigma_x = \int_0^{t} E(\xi - \xi') \frac{d\epsilon_{x\sigma}}{dt'} \, dt' \tag{45b}$$

Conversely, it can be shown in a similar manner that if Eq. (45) is known to be valid for a given material, then Eq. (39) also will be true.

The experimental bases for Eqs. (39) and (45) under *constant* temperature conditions can be demonstrated by considering isothermal creep and relaxation tests, respectively. Namely, for the creep test [$\sigma_x = \sigma_x' H(t)$, where σ_x' is constant], Eq. (39b) yields

$$\epsilon_{x\sigma} = D(\xi)\sigma_x \tag{46}$$

and for the relaxation test [$\epsilon_{x\sigma} = \epsilon_{x\sigma}' H(t)$, where $\epsilon_{x\sigma}'$ is constant], Eq. (45b) yields

$$\sigma_x = E(\xi)\epsilon_{x\sigma} \tag{47}$$

For both types of isothermal tests,

$$\xi = t/a_{\mathrm{T}} \tag{48a}$$

Alternatively (where $\log \equiv \log_{10}$),

$$\log \xi = \log t - \log a_T \tag{48b}$$

Equation (48b) shows that the effect of temperature on compliance or modulus for a TSM produces only horizontal translations when the property is plotted against $\log t$; this behavior is illustrated in Fig. 2 for compression–relaxation tests on a plasticized epoxy resin. Conversely, if it is found that constant temperature curves, such as those in Fig. 2, can be superposed so as to form a single curve (which is called the *master curve*) by means of only rigid, horizontal translations, then the relaxation modulus depends on time and temperature through the one parameter ξ; Schwarzl and Staverman (1952) originally defined a TSM as one for which the data are superposable in this way.

Master curves of relaxation and constant strain rate moduli in tension, compression, and flexure have been obtained for the epoxy resin whose compression moduli are shown in Fig. 2 by Moehlenpah *et al.* (1969). They later obtained master curves for composites of the same epoxy resin with glass beads, unidirectional glass fibers, and air (foam) tested in tension, compression, and flexure (Moehlenpah *et al.*, 1971). Experimental points which define the flexure relaxation master curves are shown in Fig. 3, and the horizontal translations ($\log a_T$) required to form these curves are given in Fig. 4. The relaxation modulus at 50°C was arbitrarily selected as the reference data, in that data at other temperatures were shifted horizontally until they superposed on this modulus.

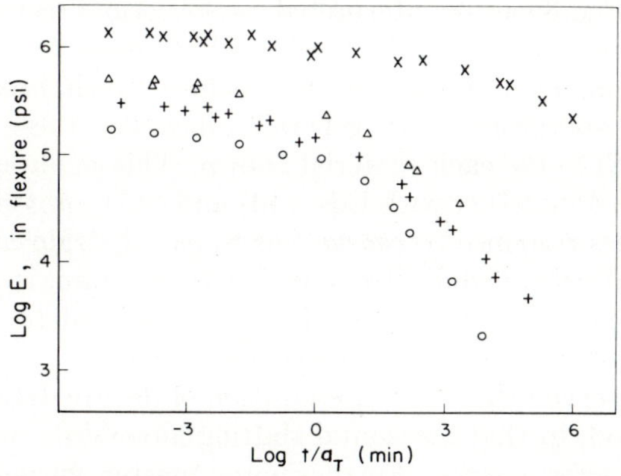

FIG. 3. Flexural stress–relaxation master curves for epoxy composites T_R = 50°C (after Moehlenpah *et al.*, 1971). The points plotted are: ×, continuous transverse; △, particle filled; +, unfilled, ○, foam.

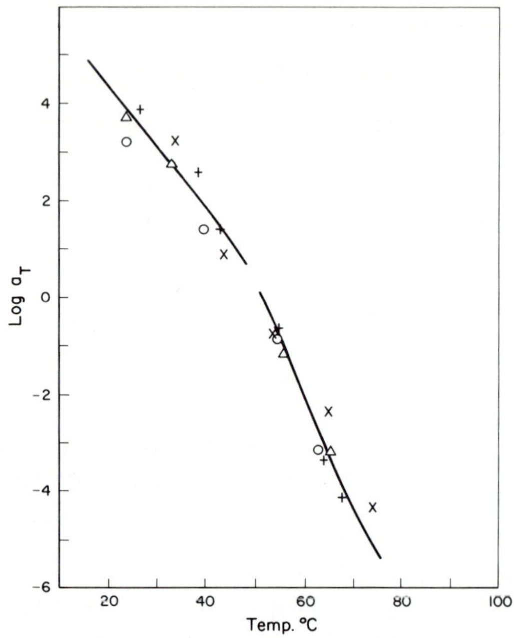

FIG. 4. Flexural stress–relaxation shift factors for epoxy composites $T_R = 50°C$ corresponding to the data in Fig. 3 (after Moehlenpah *et al.*, 1971). The points plotted are: ○, foam; +, unfilled; △, particulate; ×, transversely reinforced with glass fibers.

The data in Fig. 3 can be interpreted as isothermal relaxation moduli at the reference temperature of 50°C (for which $a_T \equiv 1$); they define the relaxation moduli over a time scale which is much broader than that covered in the laboratory. More generally, however, these data define the function $E = E(\xi)$ for each material shown. This modulus and the shift factor a_T in Fig. 4, together with Eqs. (40) and (45), provide the means for calculating stress response to *constant* or *transient* strain and temperature histories. Furthermore, when stress, rather than strain, is given, Eq. (44) or an approximation (see Section II,G) can be used to find $D = D(\xi)$, after which the strain is calculated from Eqs. (39).

It is to be observed that the superposition of flexure data in Fig. 3 is not particularly good, in that horizontal shifting alone does not form smooth, well-defined master curves. Furthermore, master curves for the compression data on composites (Moehlenpah *et al.*, 1971) and that for the resin in Fig. 2, are even less well defined (probably due, in part, to the relatively stubby specimens which were used for compression testing). However, by using vertical *and* horizontal translations, we have found that superposition of data is very good for all cases. Extension of the constitutive theory to this type of behavior is made in the next subsection.

In applications of the constitutive theory, it is usually necessary to express it in terms of total strain ϵ_x rather than the strain due to stress $\epsilon_{x\sigma}$. For an unstressed specimen whose length depends on instantaneous temperature, and not temperature history, the following relation applies:

$$\epsilon_{x\sigma} = \epsilon - \alpha\,\Delta T \tag{49}$$

where α is the linear thermal expansion coefficient (which may depend on temperature) and $\Delta T \equiv T - T_{\mathrm{R}}$ is the difference between the current temperature T and a reference value T_{R}.

On the other hand, for some materials, such as amorphous polymers close to their glass transition temperature T_{g} and for certain composites, the thermal expansion depends on temperature history; in this case, the thermal expansion may be found experimentally by applying the actual thermal history, or found theoretically by using, for example, a nonlinear model for near T_{g} behavior (Ferry, 1970) or the integral expressions in Subsection II,D,2,a.

Finally, it should be recognized that even if isothermal relaxation or creep data can be horizontally translated to form a master curve, this superposability is not sufficient, in itself, to justify use of Eqs. (39) and (45) for transient temperatures; indeed, nonisothermal expressions for reduced time can be selected which are different from Eq. (40), but yet reduce to t/a_{T} under isothermal conditions. The integral form in Eq. (40) was originally postulated by Morland and Lee (1960). Later, Schapery (1964) derived this form and the constitutive equations [e.g., Eqs. (39) and (45)] from irreversible thermodynamic theory. Experimental studies directed at checking these equations for transient temperatures above T_g are very limited; they consist of (i) creep tests on an epoxy resin (Johnson *et al.*, 1970), and (ii) controlled deformation tests on solid propellant (Farris, 1971a; Leeming, 1972). Good agreement was obtained between theory, Eqs. (39) and (45), and experiment for low stress levels. The following generalization is usually needed when $T < T_g$.

(b) THE THERMORHEOLOGICALLY COMPLEX MATERIAL (TCM). Any viscoelastic material whose temperature dependence cannot be characterized by Eqs. (39) and (45) will be called *thermorheologically complex* (TCM).

One type of TCM is a composite material consisting of two or more TSM phases which are viscoelastic over some common temperature range and which have *different* shift factors a_{T}. We shall designate this type as TCM-1;

it will be discussed in Section IV,C on the prediction of effective composite properties.

A second type TCM-2 will be covered in this subsection because it is potentially applicable to monolithic and composite materials. Thermodynamic theory has been used to derive the constitutive equation for linear and nonlinear materials (Schapery, 1969a). Here only some of the pertinent results for uniaxial, linear behavior will be reviewed.

By definition, a TCM-2 is one whose uniaxial behavior is described by the following equation for constant or transient temperatures (assuming $\sigma_x = \epsilon_{x\sigma} = 0$ for $t < 0$):

$$\epsilon_{x\sigma} = D_{\mathrm{I}}\sigma_x + \int_0^t \Delta D(\xi - \xi') \frac{d}{dt'}\left(\frac{\sigma_x}{a_{\mathrm{G}}}\right) dt' \tag{50}$$

where ξ and ξ' are still defined by Eq. (40). Also, $D_{\mathrm{I}} = D_{\mathrm{I}}(T)$ is the initial value of creep compliance. The significance of the functions $a_{\mathrm{G}} = a_{\mathrm{G}}(T)$ and $\Delta D = \Delta D(\xi)$ can be brought out by applying Eq. (50) to an isothermal creep test; the creep compliance $D_{\mathrm{T}} \equiv \epsilon_{x\sigma}/\sigma_x$ is predicted to be

$$D_{\mathrm{T}} = D_{\mathrm{I}}(T) + \Delta D(\xi)/a_{\mathrm{G}}(T) \tag{51}$$

where $\xi = t/a_{\mathrm{T}}$. Since D_{I} is the initial compliance, it follows that $\Delta D(0) = 0$. Furthermore, by assuming $a_{\mathrm{G}} = a_{\mathrm{T}} = 1$ at an arbitrarily selected reference temperature T_{R} (which is only a normalization condition, and not an actual restriction), we see that

$$\Delta D(t) = D_{\mathrm{T}}(t, T_{\mathrm{R}}) - D_{\mathrm{I}}(T_{\mathrm{R}}) \tag{52}$$

is the transient component of compliance at temperature T_{R}.

In order to relate a_{T} and a_{G} to experimental data, write Eq. (51) in the form

$$\log[D_{\mathrm{T}} - D_{\mathrm{I}}] = \log \Delta D - \log a_{\mathrm{G}} \tag{53a}$$

Also, recall that

$$\log \xi = \log t - \log a_{\mathrm{T}} \tag{53b}$$

These equations show that a plot of $\log[D_{\mathrm{T}} - D_{\mathrm{I}}]$ vs $\log t$ at a generic temperature T will be identical to that at T_{R}, apart from rigid horizontal and vertical translations of magnitudes $|\log a_{\mathrm{T}}|$ and $|\log a_{\mathrm{G}}|$, respectively; when $a_{\mathrm{T}} < 1$ ($a_{\mathrm{G}} < 1$), the curve at T is left of (above) the reference curve, and when $a_{\mathrm{T}} > 1$ ($a_{\mathrm{G}} > 1$) the curve at T is right of (below) the reference curve. In analogy with the TSM case, the final reference curve, which consists of the original data at T_{R} and all segments obtained from the graphical shifting process, is called the *master curve*.

This master curve, which is actually the function $\Delta D(\xi)$, and the temperature-dependent functions D_{I}, a_{T}, and a_{G} form a complete set of material properties for predicting nonisothermal, viscoelastic behavior.

One important special case of Eq. (50) is that for which

$$D_{\mathrm{I}} = D_{\mathrm{I}}(T_{\mathrm{R}})/a_{\mathrm{G}}(T) \tag{54}$$

which implies the creep compliance, Eq. (51), reduces to

$$D_{\mathrm{T}} = D(\xi)/a_{\mathrm{G}}(T) \tag{55a}$$

where

$$D(\xi) = D_{\mathrm{I}}(T_{\mathrm{R}}) + \Delta D(\xi) \tag{55b}$$

By using Eq. (52), with the argument t replaced by ξ, and (55b), the constitutive Eq. (50) is rendered as

$$\epsilon_{x\sigma} = \int_0^t D(\xi - \xi') \frac{d}{dt'}\left(\frac{\sigma_x}{a_{\mathrm{G}}}\right) dt' \tag{56}$$

The inverse of this equation may be obtained by recognizing that Eq. (56) is identical with Eq. (39b) except for a_{G}. Thus, by analogy with Eq. (45b), we write

$$\sigma_x = a_{\mathrm{G}} \int_0^t E(\xi - \xi') \frac{d\epsilon_{x\sigma}}{dt'} dt' \tag{57}$$

Hence, the relaxation modulus, $E_{\mathrm{T}} \equiv \sigma_x/\epsilon'_{x\sigma}$, for $\epsilon'_{x\sigma}$ constant, is

$$E_{\mathrm{T}} = a_{\mathrm{G}}(T)E(\xi) \tag{58}$$

where ξ is given by Eq. (40).

Equations (55) and (58) show that, when applicable, master curves of $D = D(\xi)$ and $E = E(\xi)$ can be formed by making horizontal ($\log a_{\mathrm{T}}$) and vertical ($\log a_{\mathrm{G}}$) shifts of the experimentally derived compliance D_{T} and modulus E_{T}.

Equation (50) has been verified experimentally for *transient* temperatures, but the results are very limited; see studies by Watkins (1973) for an epoxy resin and by Schapery *et al.* (1973) for graphite fiber-reinforced epoxy. However, there are significant amounts of *isothermal* data on amorphous and semicrystalline polymers and on metals with which Eqs. (51), (55), and (58) agree. This point can be demonstrated by referring to four different techniques which have been used by various investigators to form master curves of creep compliance. These so-called *superposition procedures* are listed below; we define them in terms of the temperature-dependent functions which comprise the compliance, Eq. (51). The reader is referred to the paper by McCrum and Pogany (1970) for concise reviews on the background of these procedures and on the way in which experimental data are reduced (or "normalized") in order to form master curves.

(i) Tobolsky–Ferry procedure:

$$D_{\mathrm{I}}(T) \equiv D_{\mathrm{I}}(T_{\mathrm{R}})/a_{\mathrm{G}}(T), \qquad a_{\mathrm{G}}(T) \equiv (T/T_{\mathrm{R}})(\rho/\rho_{\mathrm{R}}) \tag{59}$$

where ρ and ρ_{R} are densities at absolute temperature T and T_{R}, respectively.

(ii) Ferry–Fitzgerald procedure:

$$D_{\mathrm{I}}(T) \equiv D_{\mathrm{I}}(T_{\mathrm{R}}), \qquad a_{\mathrm{G}} \equiv (T/T_{\mathrm{R}})(\rho/\rho_{\mathrm{R}}) \tag{60}$$

(iii) Kê procedure:

$$D_{\mathrm{I}}(T) \equiv D_{\mathrm{I}}(T_{\mathrm{R}})/a_{\mathrm{G}}(T), \qquad a_{\mathrm{G}}(T) \equiv D_{\mathrm{T}}(0, T_{\mathrm{R}})/D_{\mathrm{T}}(0, T) \tag{61}$$

(iv) McCrum–Morris procedure:

$$D_{\mathrm{I}}(T) \equiv D_{\mathrm{T}}(0, T), \qquad a_{\mathrm{G}}(T) \equiv \frac{D_{\mathrm{T}}(\infty, T_{\mathrm{R}}) - D_{\mathrm{T}}(0, T_{\mathrm{R}})}{D_{\mathrm{T}}(\infty, T) - D_{\mathrm{T}}(0, T)} \tag{62}$$

The Tobolsky–Ferry and Ferry–Fitzgerald procedures are based on the kinetic theory of rubber elasticity which indicates that the long-time (rubbery) compliance is inversely proportional to absolute temperature and density (Ferry, 1970). The former method assumes this temperature dependence is applicable to all times and temperatures, while the latter one applies it only to the transient component of compliance; inasmuch as $D_{\mathrm{T}}(\infty, T)/D_{\mathrm{T}}(0, T) \gg 1$ for amorphous polymers, both procedures are consistent with the kinetic theory at sufficiently long times and/or high temperatures. However, neither method properly accounts for the temperature dependence of the initial compliance.

The Kê procedure, which was originally applied to aluminum, uses the initial compliance to define the temperature dependence of the coefficient a_{G}. It implies that the long-time compliance has this temperature dependence, which is not true for cross-linked, amorphous polymers above their glass transition temperature.

The McCrum–Morris procedure contains the first three as special cases in that it allows for arbitrary variations of the initial and long-time compliances with temperature.

McCrum and Pogany (1970) have applied the four procedures to the creep compliance of an epoxy resin at temperatures between 24 and 190°C, where $T_{\mathrm{g}} \approx 140$°C. A smooth master curve was obtained for all temperatures by the McCrum–Morris procedure, and for the range $24 \leq T \leq 100$°C by the Kê method. Superposition of data by the other two methods was good only above T_{g}.

It is to be noted that the Kê and McCrum–Morris procedures are, in effect, based on the assumption that the forms of the creep compliance are those shown in Eqs. (55) and (51), respectively, without further restrictions. In principle, therefore, the graphical shift methods described earlier in connection with these equations should yield the same function $a_G(T)$ and master curve $\Delta D(\xi)$ as found by these two procedures. However, according to Eqs. (61) and (62), it is necessary to know, or be able to estimate, the initial compliance for both methods and the long-time compliance for the McCrum–Morris method before the master curve can be found. In view of these requirements and the usual variability of experimental data, along with the fact that experimental factors often do not allow, or else make difficult, the direct determination of the limiting compliances, some differences in the results can be expected. It is believed that the graphical shift methods usually will provide the best "average" material properties and be easier to apply, especially when the experimental data are insufficient to determine directly the limiting values of compliance.

Observe that the long-time compliance is not needed ahead of time to evaluate graphically the properties in Eq. (51), and neither limiting value is needed when Eq. (55) is applicable. Furthermore, even when D_I cannot be determined from the original data, the graphical method may still be used; for, one can make a preliminary estimate of D_I at some given temperature (using, for example, data at the earliest available time), and then, by guessing some nearby values for D_I and forming master curves $\Delta D(\xi)$, select as the "best" function $D_I(0, T)$ the one that leads to the smoothest master curve. This determination of initial compliance is accomplished even more simply when it is known that ΔD obeys a power law (see Section II,F).

The procedure of forming a master curve by means of rigid horizontal and vertical translations of log (compliance), which, as noted above, may be viewed as a form of the Kê procedure, is not restricted to amorphous polymers. For example, it has been applied by Nagamatsu *et al.* (1958) to vibration data on a semicrystalline polymer (polyethylene); the variation of a_G in this case is believed due to the effect of temperature on the rigidity of the crystalline regions.

Finally, we note that the strain $\epsilon_{x\sigma}$ in this subsection can be converted to total strain by using the techniques mentioned previously for a TSM.

2. Multiaxial Behavior

(a) The Thermorheologically Simple Material. Three-dimensional, nonisothermal constitutive equations for the anisotropic TSM are given by

$$\sigma_{ij} = \int_0^t C_{ijkl}(\xi - \xi')\, \partial\epsilon_{kl}/\partial t'\, dt' - \int_0^t \beta_{ij}(\xi - \xi')\, \partial\Delta T/\partial t'\, dt' \quad (63)$$

with the inverse representation being

$$\epsilon_{ij} = \int_0^t S_{ijkl}(\xi - \xi')\, \partial\sigma_{kl}/\partial t'\, dt' + \int_0^t \alpha_{ij}(\xi - \xi')\, \partial\Delta T/\partial t'\, dt' \qquad (64)$$

where the modulus and compliance tensors are identical with the isothermal tensors in Eqs. (10) and (11), respectively, except for the change in argument from physical time to reduced time [where reduced time is defined in Eq. (40)]. The second-order tensors β_{ij} and α_{ij} are associated with the thermal expansion characteristics of the TSM and define, respectively, thermal stresses in a completely constrained body and thermal strains in the absence of applied stress.

Equations (63) and (64) can be viewed as being a direct result of the linear hereditary theory in Section II,A, where (i) the input variables are the strains (or stresses) and temperature change, which are not applied until $t = 0$, and (ii) the response to unit inputs is assumed to be of the nonaging type when expressed in terms of the "material" time scale ξ.

These equations also have been derived from thermodynamic theory by Schapery (1964), which, as noted previously, predicts complete symmetry of the modulus and compliance tensors. Furthermore, when they are specialized to the equations for isotropic media, the constitutive theory proposed originally by Morland and Lee (1960) is recovered.

(b) The Thermorheologically Complex Material. In view of the discussion in Section II,D,1, it may be necessary to use the constitutive equations for a TCM, rather than a TSM, in the nonisothermal analysis of semicrystalline polymers and of amorphous polymers at temperatures near and below their glass transition value. One specific equation based on thermodynamic theory was given earlier [Eq. (50)], and was shown to be consistent with some constant temperature data. However, it must be emphasized that experimental data on the effect of temperature are much more limited than for the TSM, and are almost nonexistent in the case of transient temperatures.

Moreover, even if it is known that a composite material has one or more phases which are described by this particular form of constitutive equation, the form of the effective constitutive equations for the composite will, in general, be different (see Section IV).

In view of the above comments, we shall restrict our present considerations to the three-dimensional version of Eq. (50) for isotropic media. This version, as well as other more general expressions, follow from the thermodynamic theory (Schapery, 1969a).

The equations can be written in terms of any two independent creep compliances for isotropic materials. We shall use the uniaxial creep compliance, as given by Eq. (51), and the shear creep compliance:

$$J_{\mathrm{T}} = J_{\mathrm{I}}(T) + \Delta J(\xi)/a_{\mathrm{G}}(T) \tag{65}$$

where notation analogous to that for D_{T}, Eq. (51), is used. Thus,

$$\epsilon_{11\sigma} = D_{\mathrm{I}}\sigma_{11} - \left(\frac{J_{\mathrm{I}}}{2} - D_{\mathrm{I}}\right)(\sigma_{22} + \sigma_{33}) + \int_0^t \Delta D(\xi - \xi') \frac{\partial}{\partial t'}\left(\frac{\sigma_{11}}{a_{\mathrm{G}}}\right) dt'$$
$$- \int_0^t \left[\frac{\Delta J(\xi - \xi')}{2} - \Delta D(\xi - \xi')\right] \frac{\partial}{\partial t'}\left(\frac{\sigma_{22} + \sigma_{33}}{a_{\mathrm{G}}}\right) dt' \tag{66}$$

plus two companion equations for $\epsilon_{22\sigma}$ and $\epsilon_{33\sigma}$. Also,

$$2\epsilon_{23} = J_{\mathrm{I}}\tau_{23} + \int_0^t \Delta J(\xi - \xi') \frac{\partial}{\partial t'}\left(\frac{\tau_{23}}{a_{\mathrm{G}}}\right) dt' \tag{67}$$

plus two companion equations for ϵ_{13} and ϵ_{12}.

The strain due to stress and total strain are related as follows:

$$\epsilon_{11\sigma} = \epsilon_{11} - \alpha_{\mathrm{I}}\,\Delta T - \int_0^t \Delta\alpha(\xi - \xi') \frac{\partial}{\partial t'}\left(\frac{a_{\mathrm{H}}\,\Delta T}{a_{\mathrm{G}}}\right) dt' \tag{68}$$

plus two companion equations for $\epsilon_{22\sigma}$ and $\epsilon_{33\sigma}$, where α_{I}, $\Delta\alpha$, and a_{H} are defined by the expression for thermal expansion coefficient under a constant temperature change applied at $t = 0$:

$$\alpha_{\mathrm{T}} = \alpha_{\mathrm{I}}(T) + [a_{\mathrm{H}}(T)/a_{\mathrm{G}}(T)]\,\Delta\alpha(\xi) \tag{69}$$

In many cases, it will be acceptable to assume that the thermal expansion is independent of temperature history, which implies

$$\epsilon_{ii\sigma} = \epsilon_{ii} - \alpha\,\Delta T \qquad (i \text{ not summed}) \tag{70}$$

where α is at most a function of the current temperature.

Under isothermal conditions, the Poisson's ratio must satisfy Eq. (36b); hence, at each temperature

$$\nu_{\mathrm{c}} = (J_{\mathrm{T}}/2D_{\mathrm{T}}) - 1 \tag{71}$$

Furthermore, the bulk creep compliance $B_{\mathrm{T}} \equiv 3\epsilon_{ii\sigma}/\sigma_{ii}$, where σ_{ii} is constant, is found by summing Eq. (66) and its two companions; there results,

$$B_{\mathrm{T}} = 3[3D_{\mathrm{T}} - J_{\mathrm{T}}] \tag{72}$$

Since B_{T} is a linear combination of D_{T} and J_{T}, its form is the same as that for the uniaxial and shear compliances [viz., Eq. (51)].

The constitutive equations for a TSM are immediately recovered from Eqs. (66) and (67) by setting $a_{\mathrm{G}} = a_{\mathrm{H}} = 1$ and assuming J_{I}, D_{I}, and α_{I} are constants.

E. Additional Environmental and Aging Effects

The mechanical response of composite materials and their constituents is subject to changes under a variety of environmental and chemical aging influences, with temperature and water (or water vapor) being especially important factors in many applications of polymeric composites (e.g., Fried, 1970; Steel, 1965; Tsai, 1970). These effects are often irreversible (permanent) and lead to time-dependent degradation of the properties. Our only concern here is with the way in which such influences are incorporated into the viscoelastic constitutive equations. Detailed consideration of the physical and chemical mechanisms which produce the changes, and of their mathematical representations, is beyond the scope of this chapter.

As in the previously considered case of reversible thermal effects, the influence of other environmental and aging factors can be explicitly taken into account through the indicial admittances $R_{\mathrm{H}\alpha\beta}(t, t')$ in general, and the creep compliances and relaxation moduli in particular, and by modifying the expression for strain due to stress when shrinkage or expansion is involved. For example, the axial strain response to a uniaxial stress is still given by Eq. (38) if $D(t, t')$ is obtained from specimens in the presence of all significant influences.

From the point of view of viscoelastic analysis, the simplest situation is when the applied loading acts over times that are short compared to the time scale over which changes due to environmental factors and/or aging occur. In this case, one can calculate the response over the same short period by using the nonaging form of the constitutive equations, Eqs. (10) and (11), for the age or state of the material at the time of load application.

In some cases, the viscoelastic properties may assume a form which is analogous to that for the TSM and TCM cases discussed in Section II,D. In particular, Halpin (1968) has indicated that the effect of water vapor (as one example of a swelling agent) in some cases is reversible and is completely analogous to the effect of temperature in a TSM; following his suggestion, we would simply allow for the dependence of a_{T} in Eq. (40) on the instantaneous concentration of swelling agent, and add expansion terms which are analogous to those involving ΔT in Eqs. (63) and (64). A similar generalization of the TCM equations [e.g., Eqs. (66) and (67)] could be made, if necessary; one or more of the temperature-dependent properties might also depend on concentration of the swelling agent.

It should be possible to characterize *irreversible* changes, such as the thermal degradation described by Tsai (1970), in a similar manner. However, for this situation the properties (e.g., a_G and a_T) would have to be expressed in terms of temperature *history* or, more directly, in terms of the time t' for a given thermal history; it is interesting to observe that Tsai's analysis of the effect of elevated temperature on a phenolic–glass composite suggests that the properties (e.g., a_G and a_T) depend on a single parameter, which is the reduced time t'/a_D; a_D is a function of temperature, and is obtained by horizontally shifting curves of weight loss versus log t.

The creep compliance of aging concrete seems to have the same form as that for the TCM. For example, the creep compliance reported by the U.S. Bureau of Reclamation (Anon, 1955) and McHenry (1943) has the form

$$D(t, t') = D_I(t') + [\Delta D(t - t')/a'_G(t')] \tag{73}$$

We have restricted our attention in this section to the use of hereditary integrals in characterizing environmental and aging effects; this has been done because it is believed that this formulation usually will be more convenient to apply to engineering problems than a differential equation representation. If, however, the material can be represented by a low-order differential equation (which is *not* the case for most polymers), then this alternative formulation may prove to be more easily used in some applications (Lubliner and Sackman, 1966).

F. *Analytical Representation of Viscoelastic Properties*

In many applications, including micromechanics analysis and the problem of interrelating viscoelastic properties (see Section II,G), it is desirable to have analytical expressions for the viscoelastic functions. Some commonly used ones are given in this section.

1. *Time-Dependent Properties*

According to thermodynamic theory (Biot, 1954) the creep and relaxation functions in Eqs. (10) and (11) have the following time dependence:

$$C_{ijkl}(t) = C^{\infty}_{ijkl} + \sum_s C^{(s)}_{ijkl} e^{-t/\rho_s} \tag{74a}$$

$$S_{ijkl}(t) = S^{I}_{ijkl} + S^{F}_{ijkl} t + \sum_s S^{(s)}_{ijkl}(1 - e^{-t/\tau_s}) \tag{74b}$$

where the range of the summation index s depends on the particular material, ρ_s is a relaxation time (positive constant), and τ_s is a retardation time (positive constant). Also, each fourth-order tensor on the right-hand side of Eqs. (74) is completely symmetric and constant; these tensors are individually positive semidefinite, while the full creep compliance and relaxation modulus tensors are positive definite (Schapery, 1964, 1967). These positive definite and semidefinite properties are very important in that they assure uniqueness of viscoelasticity solutions and provide the basis for obtaining upper and lower bounds on certain effective composite properties (see Section IV). It should also be mentioned that the tensor S^{F}_{ijkl} produces steady flow under constant stresses, which vanishes if the elastic modulus tensor C^{∞}_{ijkl} is positive definite; for example, this steady flow behavior does not exist with a crosslinked polymer.

The thermal relaxation and creep tensors have similar forms; viz. (Schapery, 1964, 1967),

$$\beta_{ij}(t) = \beta_{ij\infty} + \sum_s \beta_{ij}^{(s)} e^{-t/\rho_s} \tag{75a}$$

$$\alpha_{ij}(t) = \alpha_{ij\mathrm{I}} + \sum_s \alpha_{ij}^{(s)} (1 - e^{-t/\tau_s}) \tag{75b}$$

where each second-order tensor on the right-hand side of Eqs. (75) is symmetric as a result of symmetry of the stress and strain tensors, and is *not* restricted by thermodynamics to be positive semidefinite or definite.

For many materials, including most polymers, there are a large number of relaxation and retardation times which are closely spaced. This implies that the summations (over s) in Eqs. (74) and (75) can be approximated by integrals over continuously varying relaxation and retardation times (e.g., Biot, 1958; Ferry, 1970).

Although, in an exact sense, there may be a large number of time constants, the creep and relaxation functions often can be approximated by exponential series consisting of relatively few terms (typically 10–20), in which the time constants are selected without regard to the values predicted by thermodynamics (e.g., Schapery, 1961). Criteria which assure positiveness of the exponential coefficients found by the collocation method of curve fitting (Schapery, 1961) have been given by Gradowczyk and Moavenzadeh (1969); it should be added that the thermodynamically derived coefficients in Eqs. (74) and (75) are *not* required by the positive semidefinite condition to be nonnegative unless they appear in the diagonal elements, $C_{ijij}(t)$ and $S_{ijij}(t)$ (i, j not summed).

The fact that all of the above time-dependent properties consist of sums of exponentials implies they can be represented by mechanical model analogs consisting of springs and dashpots (e.g., Ferry, 1970). However, except for the diagonal elements mentioned above, the spring moduli and dashpot viscosities may be negative as long as the corresponding time constants are positive.

We have used real time t as the argument in the properties, Eqs. (74) and (75). However, these functions are readily converted to the notation used in Section II,D on nonisothermal characterization of the TSM and TCM; the real time argument is simply replaced by reduced time ξ.

As an alternative representation of time dependence, various forms of power laws have been fit to creep and relaxation data. Specifically, for rigid plastics with or without reinforcement (e.g., McCammond, 1967; Ward, 1971):

$$D(t) = D_{\mathrm{I}} + D_1 t^n, \qquad 0 \le n \le 1 \tag{76}$$

where D_{I}, D_1, and n are independent of time. Whenever D_{I} can be either neglected or determined directly from short-time experimental data, evaluation of D_1 and n is readily accomplished by plotting $\log(D - D_{\mathrm{I}})$ vs $\log t$. Recovery data are useful for finding D_1 and n when there are insufficient creep data available to accurately determine these constants (Lou and Schapery, 1971). The so-called *modified power law*,

$$E(t) = E_\infty + \frac{E_{\mathrm{I}} - E_\infty}{(1 + t/\tau_0)^m} \tag{77a}$$

where E_∞, E_{I}, τ_0, and m are independent of time, often provides an excellent representation of the relaxation modulus of amorphous polymers *above* their glass transition temperature (Williams, 1964). Note that when $t/\tau_0 \gg 1$, Eq. (77a) reduces to a form which is analogous to Eq. (76); viz.,

$$E(t) = E_\infty + E_1 t^{-m} \tag{77b}$$

For a TSM, temperature dependence is incorporated into Eqs. (76) and (77) by replacing t with ξ. In addition, for the TCM discussed in Section II,D, the parameters D_{I}, D_1, E_∞, and E_{I} may depend on temperature.

The power laws, Eqs. (76) and (77b), lead to a particular problem when isothermal experimental data are used to evaluate the temperature-dependent functions of a TCM. Consider, for example, the power law creep compliance expressed in the TCM form Eq. (51); viz.,

$$D_{\mathrm{T}} = D_{\mathrm{I}} + D_1' \xi^n / a_{\mathrm{G}} \tag{78a}$$

where D_1' and n are constants, and $\xi = t/a_{\mathrm{T}}$; thus

$$D_{\mathrm{T}} = D_{\mathrm{I}} + D_1' t^n / (a_{\mathrm{T}}^n a_{\mathrm{G}}) \tag{78b}$$

This last result shows that the product $a_{\mathrm{T}}^n a_{\mathrm{G}}$, but not a_{T} and a_{G} by themselves, can be found from isothermal data; however, constant temperature rate plus isothermal tests are sufficient to obtain the individual functions (Schapery *et al.*, 1973; Watkins, 1973).

2. Temperature-Dependent Properties

Analytical representation of the shift factor a_T for polymers has received much greater attention than any of the other temperature-dependent functions. When $T < T_g$, the familiar Arrhenius relation is usually applicable (Ferry, 1970; Halpin, 1968):

$$\log a_T = \frac{\Delta F}{2.303R}\left(\frac{1}{T} - \frac{1}{T_R}\right) \tag{79}$$

where ΔF is a constant activation energy (per mole) and $R = 1.987$ (cal/mole K) is the universal gas constant. Thus, if $\log a_T$ is plotted versus the reciprocal of absolute temperature $1/T$, a straight line results. The shift factor in Fig. 4 in the range 20–40°C roughly fits this description; for this range, Eq. (79) is (approximately) a straight line when plotted against temperature in degrees Kelvin or Celsius.

When $T > T_g$, the so-called WLF equation normally applies (Ferry, 1970):

$$\log a_T = -c_1(T - T_R)/(c_2 + T - T_R) \tag{80}$$

where c_1 and c_2 are constants. One can show that even after Eq. (80) is fit to experimental data using one choice of reference temperature, its form does not change when another value of T_R is used. The values of c_1 and c_2, however, do depend on T_R. It has been found that c_1 and c_2 do not vary appreciably from material to material when T_R is approximately 50°C above T_g; for this case they have the values

$$c_1 = 8.86, \qquad c_2 = 101.6 \quad (\mathrm{K}) \tag{81}$$

Another form of a_T, which is valid when $T > T_g$, is the power law (Schapery, 1967)

$$a_T = [(T_R - T_a)/(T - T_a)]^\mu \tag{82}$$

where T_a and μ are constants. This equation closely matches the WLF equation over a wide range when $\mu = 12$; typically, $T_a \approx T_g - 10$°C. This power law is useful for some problems because it enables reduced time, Eq. (40), to be evaluated analytically under a constant rate of change of temperature, whereas the WLF equation does not.

Little established information exists on the shift factor a_G in the neighborhood of and below T_g. A form derived from kinetic theory by Schapery and Martin (1972) for this temperature range is

$$a_G = \frac{\gamma \ln(1+z)}{\exp[\gamma z/(1+z)] - 1} \tag{83}$$

where $\gamma = \Delta H/RT_R$, ΔH is the constant activation enthalpy for hole formation, and $z = (T - T_R)/T_R$; typically, $1 < \gamma < 2$ when $T_R = T_g$. Recall that the simple form of a_G in Eqs. (59) and (60) is well established for crosslinked amorphous polymers significantly *above* T_g.

Using high-frequency shear tests on supercooled liquids, initial compliance J_I was found by Barlow *et al.* (1969) to be a linearly increasing function of temperature in the range $T_g < T < T_g + 20°C$. McCrum and Pogany (1970) assumed that the logarithm of initial compliance increases linearly with temperature on the basis of 10 sec creep data in shear.

G. Viscoelastic Property Interrelationships

One can find a considerable amount of information on exact and approximate interrelations among viscoelastic functions in the literature (e.g., Williams, 1964; Ferry, 1970). We cannot hope to reproduce even a representative sampling of these many results in the limited space available to us. Therefore, our attention will be confined to recording only some relations which will be needed in later sections on analysis. Furthermore, it will be sufficient to use the notation of isothermal viscoelasticity; the interrelationships will be valid regardless of whether or not the viscoelastic functions depend on temperature, as long as it is timewise constant. Explicit incorporation of temperature effects is accomplished in the same way as described in Section II,F.

The LT used in this section is defined in terms of real time as follows:

$$\bar{f} = \int_0^\infty e^{-st} f(t)\, dt \tag{84}$$

which may be compared to the LT defined in terms of reduced time, Eq. (42). The Carson transform $\tilde{f}$ also will be used, where

$$\tilde{f} \equiv s\bar{f} \tag{85}$$

1. Uniaxial Relationships

Some relations connecting axial compliance and modulus for uniaxial loading are given here; application to other simple states of stress (e.g., simple shear and hydrostatic pressurization) can be made by simply changing the symbols used for compliance and modulus.

The exact relation between Carson transforms of creep compliance and relaxation modulus is [see Eq. (44)]

$$\tilde{E} = \tilde{D}^{-1} \tag{86}$$

The response to a steady-state vibrational stress of frequency ω is defined by the complex compliance or modulus; viz., with $\sigma_x = \sigma_A \exp(i\omega t)$ and $\epsilon_x = \epsilon_A \exp(i\omega t)$, where σ_A and ϵ_A are real or complex constants,

$$D^* \equiv \epsilon_x/\sigma_x \qquad \text{or} \qquad E^* \equiv \sigma_x/\epsilon_x, \qquad i = (-1)^{1/2} \tag{87a}$$

Thus,

$$E^* = D^{*-1} \tag{87b}$$

It is also well known that steady-state vibration functions and Carson transforms are simply related (e. g., Pipkin, 1972):

$$E^*(\omega) = \tilde{E}\,|_{s\to i\omega}\,, \qquad D^*(\omega) = \tilde{D}\,|_{s\to i\omega} \tag{88}$$

By using the power law creep compliance, Eq. (76), we find

$$\tilde{D} = D_{\text{I}} + D_1\Gamma(1+n)s^{-n} \tag{89}$$

where $\Gamma(1+n)$ is the gamma function with argument $(1+n)$. From Eq. (88),

$$D^* = D' - iD'' \tag{90a}$$

where the real part (also called the *storage compliance*) D' is

$$D' = D_{\text{I}} + D_1\Gamma(1+n)\cos(n\pi/2)\omega^{-n} \tag{90b}$$

and the imaginary part (also called the *loss compliance*) D'' is

$$D'' = D_1\Gamma(1+n)\sin(n\pi/2)\omega^{-n} \tag{90c}$$

The "loss tangent" (also called the *loss factor*) is defined by

$$\tan\phi \equiv D''/D' \tag{90d}$$

where ϕ is the phase angle between stress and strain. When $D_{\text{I}} \ll D'$, the above results imply

$$\phi = n\pi/2 \tag{90e}$$

When D_{I} in Eq. (89) is negligible, the LT inversion of Eq. (86) can be accomplished analytically to find

$$D(t)E(t) = \sin n\pi/n\pi \tag{91}$$

which implies

$$E(t) \approx D(t)^{-1} \tag{92}$$

if $0 < n \lesssim 0.25$, with a maximum error of approximately 10%. Ferry (1970) has shown that Eq. (91) is very accurate for amorphous polymers above T_g even when $D(t)$ does not obey a power law; in this case n is the instantaneous log–log slope of D (or E).

Equation (92) is called the *quasi-elastic approximation.* It too is not restricted to power laws, and is valid when a plot of D (or E) vs log t has small curvature (Schapery, 1962, 1967).

Results similar to the preceding obviously are obtained when the power law modulus Eq. (77b) is used instead of Eq. (76).

2. Multiaxial Relationships

By applying the LT to constitutive Eqs. (10) and (11) (after taking the lower integration limit as 0^-) and multiplying through by s, we obtain the exact result

$$\tilde{C}_{ijkl} = [\tilde{S}_{ijkl}]^{-1} \tag{93}$$

where the right-hand side is the inverse of the matrix whose elements are the Carson transforms $\tilde{S}_{ijkl}$. This same relation holds for the complex moduli and compliances, which are obtained by making the substitution $s \rightarrow i\omega$; viz.,

$$C^*_{ijkl} = [S^*_{ijkl}]^{-1} \tag{94}$$

The quasi-elastic approximation for transient properties

$$C_{ijkl}(t) \approx [S_{ijkl}(t)]^{-1} \tag{95}$$

often is quite accurate (Schapery, 1967).

Equations (93)–(95) provide the means for relating moduli and compliances in the s, ω, and t planes, regardless of the degree of anisotropy. Existence of the inverse tensors in Eqs. (93)–(95) is guaranteed by thermodynamics because of the positive-definite and semidefinite character of the tensors in Eqs. (74) (Schapery, 1964). An important point to remember here is that because the tensors in the above equations are completely symmetric (which is provided by thermodynamics), Eqs. (93)–(95) are identical with those for elastic media, except the moduli and compliances are constant in the latter case.

For elementary types of material symmetry, such as transverse isotropy and isotropy, explicit interrelationships are easily deduced in terms of engineering properties. Consider, for example, a few consequences of Eq. (33) (plus companions) after it has been specialized to an isotropic body, in accordance with Eq. (36a). Some exact, time-dependent relations involving creep Poisson's ratio $\nu_c(t)$, uniaxial compliance $D(t)$, shear compliance $J(t)$, and bulk compliance $B(t)$, have already been given in Eqs. (36b) and (72). Still assuming the body is undisturbed for $t < 0$, transform Eqs. (33) and write the result in a form which is analogous to that used in engineering elasticity; viz.,

$$\bar{\epsilon}_{11} = (1/\tilde{E})[\bar{\sigma}_{11} - \tilde{\nu}_r(\bar{\sigma}_{22} + \bar{\sigma}_{33})] \tag{96a}$$

plus two more analogous equations, and

$$2\bar{\epsilon}_{23} = \bar{\tau}_{23}/\tilde{G} \tag{96b}$$

plus two more analogous equations, where we have used the definitions,

$$\tilde{\nu}_r \equiv (\widetilde{\nu_c D})\tilde{E} \tag{97a}$$

$$\tilde{G} \equiv [2\tilde{D} + 2(\widetilde{D\nu_c})]^{-1} = \tilde{J}^{-1} \tag{97b}$$

Note that these two definitions yield the elasticlike relation

$$\tilde{G} = \tilde{E}/2(1 + \tilde{\nu}_r) \tag{97c}$$

Physical interpretation of $\tilde{\nu}_r$ and $\tilde{G}$ is accomplished by applying Eqs. (96) to relaxation tests (constant strain) in uniaxial tension and simple shear, respectively. In the first case, all stresses (and their LT) are zero except for σ_x; we find from Eq. (96a) and its companion for $\bar{\epsilon}_{22}$,

$$\nu_r(t) = -\epsilon_{22}(t)/\epsilon_{11} \tag{98a}$$

which establishes ν_r as the relaxation Poisson's ratio. For constant ϵ_{23}, Eq. (96b) yields $G(t)$ as the shear relaxation modulus,

$$G(t) = \tau_{23}(t)/2\epsilon_{23} \tag{98b}$$

There is an important conclusion that can be drawn from Eq. (97a), which will be used later in Section IV. Specifically, let us suppose ν_c is independent of time. Then by virtue of the definition of the LT,

$$\widetilde{\nu_c D} = \nu_c\tilde{D} \tag{99}$$

Since $\tilde{D}\tilde{E} = 1$, we find $\tilde{\nu}_r = \nu_c$ from Eq. (97a); upon inverting, there results

$$\nu_r = \nu_c = \text{constant} \tag{100}$$

Figure 5 shows the creep and relaxation Poisson's ratios for an epoxy covering the glassy to rubbery range of behavior. Although Theocaris (1964) does not state a value of T_g for this material, it is estimated from his modulus data that the material is near to or below T_g when $\log(t/a_T) \lesssim 12$.

First, it is of interest to observe that the ratio ν_r/ν_c is very close to unity at all times; this behavior is predicted from Eq. (97a) and the direct method of LT inversion, Eq. (120); thus

$$\nu_r(t) \approx \nu_c(t)D(t)E(t) \approx \nu_c(t) \tag{101}$$

Because of the low value of shear modulus in the rubbery region, the Poisson's ratios should approach a value of $\frac{1}{2}$ at long times; this is not quite the case for the data in Fig. 5, probably due to experimental factors.

The behavior below T_g, or at short times $[\log(t/a_T) \lesssim 12]$, is of much interest because this is the use range for many structural applications of polymers and their composites. It is seen that $0.36 \lesssim (\nu_r, \nu_c) \lesssim 0.39$ in this range. There is a considerable amount of additional evidence that the Poisson's ratio of epoxy and other rigid plastics is very nearly constant (with values in the range of 0.38 ± 0.03) even when the uniaxial and shear moduli are significantly time- and frequency-dependent (McCammond, 1967; Theocaris and Hadjijoseph, 1965; Wada *et al.*, 1959); for the epoxy of Fig. 5, the uniaxial modulus is approximately 40% lower at $\log(t/a_T) = 12$ than at $\log(t/a_T) = 0$ (Theocaris, 1964).

III. Correspondence Principles and Approximate Methods of Analysis

For many viscoelasticity problems, taking the Laplace (or Fourier) transform of the governing field and boundary equations with respect to

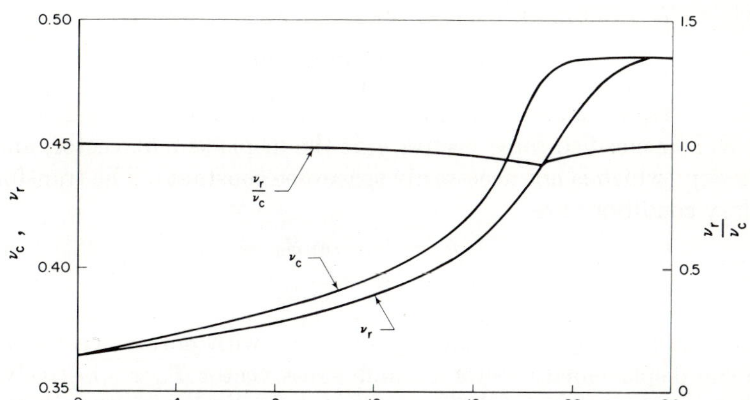

FIG 5. Master curves of Poisson's ratio in creep and relaxation for a cold-setting epoxy, $T_R = 25°C$ (after Theocaris, 1964).

real or reduced time reduces them so that they are mathematically equivalent to those for elasticity problems. This analogy is called a "correspondence principle" (CP), and implies that elastic analysis methods can be used to derive transformed viscoelasticity solutions. The CP for anisotropic media was originally established by Biot (1958).

First, we shall examine the basis for a CP when the temperature is constant in time (but not necessarily spacewise constant) and indicate its implications for steady-state vibration problems. Extension to transient temperature problems, assuming reversible thermal behavior, is then discussed. Finally, approximate methods of analysis are briefly described, primarily for problems in which the correspondence principle is applicable.

Temperature is the only environmental factor that will be taken explicitly into account in this section. However, in view of the comments in Section II,E on swelling agents, for some situations the effects of water vapor may be analyzed using methods which are entirely identical to those for thermal effects.

A. The Correspondence Principle When Temperature Is Timewise Constant

Assuming the material is undisturbed when $t < 0$, the Laplace transformed equilibrium and strain displacement equations are, respectively,

$$\frac{\partial \bar{\sigma}_{ij}}{\partial x_j} + \bar{F}_i = \rho s^2 \bar{u}_i \tag{102}$$

$$\bar{\epsilon}_{ij} = \frac{1}{2}\left(\frac{\partial \bar{u}_j}{\partial x_i} + \frac{\partial \bar{u}_i}{\partial x_j}\right) \tag{103}$$

where F_i is a specified force vector, u_i is the displacement vector, and ρ is the density (which is not necessarily spacewise constant). The transformed boundary conditions are

$$\bar{u}_i = \bar{U}_i \quad \text{on } S_{\mathrm{u}} \tag{104}$$

$$\bar{\sigma}_{ij} n_j = \bar{T}_i \quad \text{on } S_{\mathrm{T}} \tag{105}$$

where S_{u} and S_{T} are the boundary surfaces (with normal vector n_i) on which the displacement vector U_i and stress vector T_i, respectively, are specified; combinations of different stress and displacement vector components can also be considered. In order to be able to LT the boundary conditions, it is necessary to assume that the boundary surfaces S_{u} and S_{T} are independent of time (apart from small displacements). Finally, application of the LT to constitutive Eq. (11), taking 0^- as the lower limit of integration, yields

$$\bar{\epsilon}_{ij} = \tilde{S}_{ijkl}\bar{\sigma}_{kl} \tag{106}$$

and a similar result when Eq. (10) is used. (The LT of thermal expansion $\tilde{\alpha}_{ij}\,\overline{\Delta T}$ can be added to Eq. (106) if needed, without affecting the correspondence principle as long as the temperature is timewise constant.)

Under the assumption that $S_{ijkl}(t)$ is completely symmetric, Eqs. (102)–(106) are identical in form to the equations for an elastic body having the same geometry as the viscoelastic body and having compliances S_{ijkl} instead of $\tilde{S}_{ijkl}$; the name *associated elastic problem* is used for the problem defined by Eqs. (102)–(106). Thus, one can immediately deduce the LT of a viscoelastic solution from the LT of an elastic solution by making the substitution $S_{ijkl} \rightarrow \tilde{S}_{ijkl}$ (or a similar substitution using moduli). It should be added that this last observation may be viewed as an alternative statement of the correspondence principle; it is valid for dynamic and quasi-static problems.

It is often possible to obtain the solution to the associated elastic problem directly from the elasticity literature. This is especially true if the space and time dependence of all nonzero prescribed functions appear as separate factors; viz. (i not summed),

$$U_i = U_i'(x_j)U_i''(t), \qquad T_i = T_i'(x_j)T_i''(t), \qquad F_i = F_i'(x_j)F_i''(t) \tag{107}$$

When these inputs are Laplace transformed in order to obtain the prescribed functions for the associated elastic problem, the spacial distributions (e.g., U_i') are not changed; therefore, apart from constant coefficients (e.g., $\bar{U}_i''$), the LT of the inputs corresponds to the same physical loading condition as in the time plane. Prediction of effective moduli and compliances forms an important class of problems for which Eq. (107) applies.

There are many problems in quasi-static, isotropic theory for which the associated elasticity solution is independent of material properties. The viscoelastic and elastic solutions in the time plane are then identical.

A correspondence principle does not exist when the boundary surface S_u and/or S_T is a function of time, since the boundary conditions cannot be transformed; the integration in the LT is defined only when the spacial location x_i is constant. The spherical indentor problem (e.g., Christensen, 1971) is an example in which a portion of the surface is S_T part of the time and S_u at other times. Specialized methods have been used to solve such problems.

After the associated elastic solution is obtained, one inverts the LT in order to obtain the time-dependent solution. This step is often very difficult, if not impossible, to perform using exact methods; Section III,C,1 describes two approximate methods of LT inversion which are easily applied to numerical or analytical solutions of the associated elastic problem.

In the case of steady-state vibrations, the time-dependent solution for dynamic and quasi-static problems is readily obtained if an *analytical* solution to the associated elastic problem exists. Namely, one makes the substitutions (e.g., Christensen, 1971):

$$\tilde{S}_{ijkl} \rightarrow S^*_{ijkl}, \qquad \bar{u}_i \rightarrow u_i \equiv u_{i\mathrm{A}} e^{i\omega t}$$
$$\bar{\sigma}_{ij} \rightarrow \sigma_{ij} \equiv \sigma_{ij\mathrm{A}} e^{i\omega t}, \qquad \bar{\epsilon}_{ij} \rightarrow \epsilon_{ij} \equiv \epsilon_{ij\mathrm{A}} e^{i\omega t} \tag{108}$$

where the subscript A denotes a time-independent amplitude and ω is the frequency. Also $S^*_{ijkl} = S^*_{ijkl}(\omega)$ is the complex compliance tensor, which appeared in Eq. (94). Also, as noted previously, this complex tensor is related simply to the Carson transform of the creep compliance tensor; viz. (e.g., Pipkin, 1972),

$$S^*_{ijkl} = \tilde{S}_{ijkl}\,|_{s\to i\omega} \tag{109}$$

B. *Correspondence Principles When Temperature Is Spacewise Constant*

Correspondence principles exist if temperature is transient, but it must be spacewise constant *and* the body must be either a TSM or a restricted type of TCM; moreover, the problem must be quasi-static (i.e., $\rho\, \partial^2 u_i/\partial t^2$ in the equilibrium equations has to be negligible). These correspondence principles provide the basis for predicting effective, nonisothermal composite properties, as described in Section IV, C.

Let us first consider the correspondence principle for a TSM. Now, the four independent variables are taken as ξ and x_i, instead of t and x_i. By taking the LT with respect to reduced time ξ, Eq. (42), of the equilibrium and strain-displacement equations and the given boundary conditions, we obtain equations which are identical to Eqs. (102)–(105), except the caret (e.g., $\hat{\sigma}_{ij}$) now appears instead of the overbar (e.g., $\bar{\sigma}_{ij}$) on all quantities in these equations. The LT of the constitutive Eq. (64) yields

$$\hat{\epsilon}_{ij} = \tilde{S}^{\mathrm{R}}_{ijkl} \hat{\sigma}_{kl} + \tilde{\alpha}_{ij}\, \widehat{\Delta T} \tag{110}$$

where

$$\tilde{S}^{\mathrm{R}}_{ijkl} \equiv \tilde{S}_{ijkl}\,|_{\mathrm{T}=\mathrm{T_R}} \tag{111}$$

The governing set of equations is now identical to that for an elastic body with temperature-independent compliances $\tilde{S}^{\mathrm{R}}_{ijkl}$. Once a solution is obtained to this so-called *associated elastic problem*, a LT inversion is made, which yields a solution in terms of reduced time $\xi = \xi(t)$ [Eq. (40)] and coordinates x_i.

When the material is a TCM whose constitutive behavior is defined by the three-dimensional (isotropic or anisotropic) version of Eq. (56), a correspondence principle exists if the temperature is spacewise constant. The principle is analogous to that for a TSM, except all stresses and body forces are divided by the temperature-dependent property a_G. Correspondence principles do not appear to exist for more involved types of TCM.

Moreover, correspondence principles do not exist for either the TSM or TCM discussed above if the temperature is *both* transient and spacewise nonuniform, for, on one hand, if the independent variables t and x_i are used, transformation of the constitutive equations does *not* convert them into elasticlike relations. On the other hand, if the independent variables are ξ and x_i, the spacial derivatives in the equilibrium and strain-displacement equations must be changed according to (Morland and Lee, 1960)

$$\frac{\partial}{\partial x_i} \to \frac{\partial}{\partial x_i} + \frac{\partial \xi}{\partial x_i}\frac{\partial}{\partial \xi} \tag{112}$$

which destroys the elasticlike character of these equations. For this situation, it is usually necessary to resort to direct numerical methods (e.g., Oden, 1972).

C. *Approximate Methods of Analysis*

1. *Laplace Transform Inversion*

The correspondence principles enable one to calculate viscoelastic solutions from elastic solutions. An essential step in this analysis procedure is Laplace transform inversion, and, as noted previously, an exact analytical inversion is not often feasible. In many problems the elastic solution is known only numerically, or is so analytically complex that standard methods are not adequate. The use of realistic relaxation and creep functions adds further complexity to analytical methods of inversion.

In view of such complications, Schapery (1962) proposed two approximate methods of transform inversion which use transformed solutions on only the positive real s axis. For this range of s, $\tilde{C}_{ijkl}$ and $\tilde{S}_{ijkl}$ are real and possess the same positive-definite character exhibited by elastic moduli (Schapery, 1964). Therefore, only numerical values of elastic solutions are required to calculate the time-dependent viscoelastic solutions. These two methods are outlined below.

Several methods of inversion using the positive real s axis have been examined in detail by Cost (1964) and Cost and Becker (1970) and, therefore, we shall not attempt here to evaluate the numerous methods. Suffice it to say that the following techniques have been found to be both accurate and easy to apply in quasi-static viscoelasticity problems; difficulties which may arise in their application to dynamic problems will be noted.

The problem we now consider is to find an approximate solution $f(t)$ to the integral equation

$$\bar{f} = \int_0^\infty e^{-st} f(t)\, dt \tag{113}$$

where $\bar{f}$ is known at least numerically for all real values of the transform parameter $s \geq 0$. This function may depend on coordinates x_i, but explicit indication is omitted for notational simplicity. Although the discussion is in terms of real time, the methods are equally valid when reduced time is used; see Eq. (42).

The two inversion methods are most accurate and most easily applied when f is the quasi-static response to prescribed quantities (U_i, T_i, F_i) which are step functions in time; viz., when f is an indicial admittance, R_{H} or $R_{\mathrm{H}\alpha\beta}$. This represents no practical limitation since superposition integrals, Eq. (9), can be used to generate solutions to inputs which vary in time and space; it is *not* essential that the inputs have the product form in Eq. (107).

(a) Collocation Method. When the viscoelastic solution is finite at $t = \infty$, we write it in the form

$$f(t) = \Delta f(t) + f_\infty \tag{114a}$$

where, by definition,

$$\Delta f(\infty) \equiv 0, \qquad f_\infty \equiv f(\infty) \tag{114b}$$

The collocation method deals with the inversion of $\Delta\bar{f}$.

The constant f_∞ is the long-time response, assumed finite. It can be evaluated directly from the associated elastic solution $\bar{f}$ by using the well-known final value theorem (Van der Pol and Bremmer, 1955),

$$s\bar{f}\,|_{s=0} = f_\infty \tag{115a}$$

If $f(t)$ represents a displacement or strain, it may contain a term proportional to t (steady flow). The coefficient f', say, can be evaluated easily by using a similar relation,

$$s^2\bar{f}\,|_{s=0} = f/t\,|_{t=\infty} \equiv f' \tag{115b}$$

and the remaining problem is one of calculating $\Delta f \equiv f - f't - f_\infty$.

In order to invert $\Delta\bar{f}$, we assume that the solution can be approximated by a finite Dirichlet (or Prony) series

$$\Delta f_{\mathrm{D}} \equiv \sum_{i=1}^{N} S_i e^{-\lambda_i t} \tag{116}$$

where λ_i and S_i are real constants. The coefficients S_i are found by minimizing the square error

$$E^2 \equiv \int_0^\infty [\Delta f - \Delta f_{\mathrm{D}}]^2 \, dt \tag{117}$$

with respect to each coefficient. This condition leads to the following system of equations after multiplying through by s:

$$\sum_{j=1}^{N} [1 + (\lambda_j/\lambda_i)]^{-1} S_j = s\Delta\bar{f}\,|_{s=\lambda_i} \tag{118}$$

It is seen that the total square error in time is minimized by collocating the Carson transformed Dirichlet series and the Carson transformed solution at N real numbers $s = \lambda_i$. With an elastic solution known numerically for $s > 0$, suitable values of λ_i can be prescribed by inspection of a plot of $s\bar{f}$ vs log s, as shown previously (Schapery, 1962, 1967); one-decade spacing of the λ_i is often adequate.

Although this method is usually approximate for $N < \infty$, in principle the error can be made arbitrarily small when N is sufficiently large and the spacing of λ_i is sufficiently small; this observation follows from the completeness property of the Dirichlet series for quadratically integrable functions (e.g., Schapery, 1967). In practice, however, accuracy of the inversion is limited by the smoothness of the LT data; round-off error, which is inherent in any numerical representation of the LT, and other errors (due to interpolation techniques, for example, in the case of associated elastic solutions obtained by the finite difference or finite element method) determine a lower limit on the square error in the time plane (Cost and Becker, 1970; Schapery, 1962, 1967). One finds that, for a given set of numerical values of the LT, the square error E^2 initially diminishes as the spacing between λ_i values is decreased, but eventually grows as smaller and smaller spacing is used; this growth in E^2 usually reflects oscillations of increasing magnitude in the function $\Delta f_{\mathrm{D}}(t)$.

In an effort to overcome this difficulty with the collocation method, Cost and Becker (1970) proposed a related technique, called the *multidata method*; an exponential series in the time plane is still used, but the coefficients S_j are found by minimizing the square error in the transform plane, in which an arbitrary number of LT data points are used. They found that for sufficiently large random error in the LT data, the square error in the time plane was larger for the collocation method than the multidata method. It is the author's experience, however, that the accuracy of the collocation method for quasi-static problems is more than adequate for engineering purposes. Moreover, even if there are some oscillations in the approximation Δf_D, the fact that the method is based on minimizing the square error in the time plane means a smooth curve drawn through the mean values of oscillations normally will be close to the actual solution [see, e.g., the orthotropic cylinder problem (Schapery, 1967)].

The collocation method has been applied to one- and two-dimensional viscoelastic wave propagation problems in isotropic media by Arenz (1964, 1965). Oscillations in the time-dependent solution were observed at points relatively far from the surface of load application, and were attributed to dispersion as a result of the frequency dependence of the complex modulus. Later, Knauss (1968) solved the same one-dimensional problem by the Fourier method and did not predict oscillations. Some of the author's unpublished research has also verified that oscillations reported by Arenz are indeed of a numerical origin, and, in particular, due to round-off error. By examining a sequence of solutions with less and less round-off error, we were able to eliminate the oscillations at successively greater distances from the surface of load application; the need for reduced round-off error with increasing distance from this surface was found to be directly related to the growth in curvature of the function $s\Delta\bar{f}$ vs log s with increasing distance.

The accuracy of the multidata method for wave problems has not yet been studied. More importantly, neither the collocation nor the multidata method has been applied to dynamic problems in composite media. In this case, actual oscillations in the solution may exist with sufficiently short wavelengths due to dispersion produced by the microstructure.

(b) Direct Method. The direct method is easier to apply than the collocation method but, in contrast, offers little means of reducing error and is restricted to functions $s\bar{f}$ vs log s having small curvature. An important feature, however, is that it provides a direct relation between the time-dependent and transformed solutions. As will be shown later, this is important for physical interpretation of upper and lower bounds on effective transformed moduli and compliances.

Schapery (1962) showed that if the solution $f(t)$ has small curvature when plotted against log t, then

$$s\bar{f} \approx f(t)\,|_{t=\beta/s} \tag{119}$$

where $\beta = e^{-c} \approx \frac{1}{2}$ (c is Euler's constant). This equation is exact if f is proportional to $\log t$. A modification of the factor β was also given for functions in which $\log f$ has small curvature with respect to $\log t$; if the log–log slope L is in the range $-0.3 \lesssim L \lesssim 0.1$, then $\beta \approx \frac{1}{2}$.

The direct inversion method is obtained by writing Eq. (119) in the equivalent form

$$f(t) \approx s\bar{f}\,|_{s=1/2t} \tag{120}$$

Notice that Eq. (120) was derived using an assumption about the exact inversion $f(t)$, rather than $s\bar{f}$. Since in practice the exact $f(t)$ will not be known, one must assume Eq. (120) is valid if $s\bar{f}$ satisfies the small curvature condition. As to what is meant by "small" is illustrated in Fig. 6 in terms of a supposed exact indicial admittance. A study of the LT shows that if the solution has constant curvature over a generic interval 2.0 decades wide, the relative error in the direct method at the center of the interval (point A in Fig. 6) is equal to the ratio of solid to shaded area; this observation obviously provides a rule of thumb for estimating the accuracy of the direct method (by examining a plot of $s\bar{f}$ vs $\log s$) in those intervals for which the curvature is at least approximately constant.

It can be shown that the approximate inversion lies above the exact solution at points where the plot of R_H has positive curvature, with just the opposite being true when the solution has negative curvature (Schapery, 1962). This fact, together with an error estimate based on Fig. 6, enables

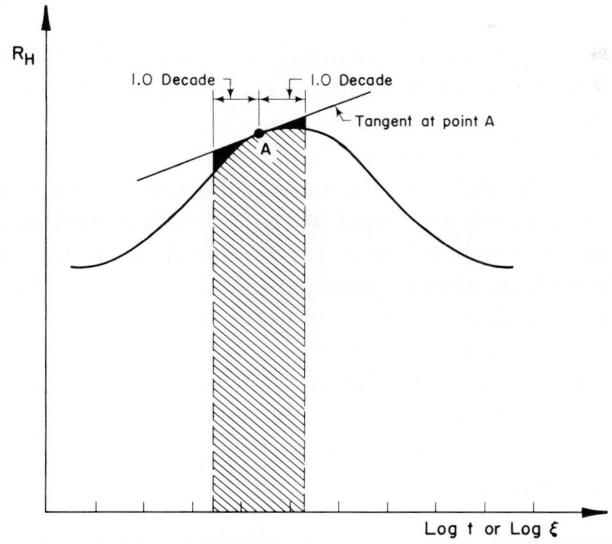

FIG. 6. Estimation of error in the direct method of Laplace transform inversion.

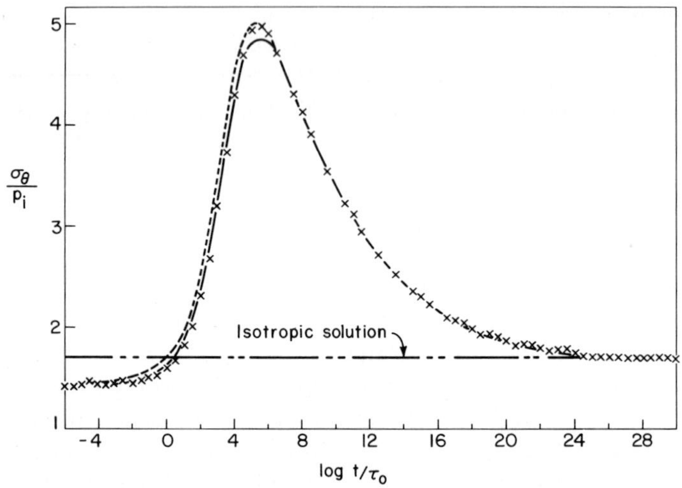

FIG. 7. Stress at inner boundary of cylinder calculated by three methods (after Schapery, 1967: ×, collocation method (5 and 7 digit accuracy input data); —, direct method; – – –, quasi-elastic method.

us to improve the accuracy of the direct solution if the error is small to begin with. Consider, for example, the direct solution in Fig. 7, which is the indicial admittance of hoop stress in an internally pressurized, orthotropic cylinder. Now, if one computes the ratio of solid to shaded areas for points all along the direct solution, and then adds or subtracts (depending on the sign of the curvature) the product of this ratio and the ordinate values, the result is found to coincide with the mean of the collocation solution in the same figure; this correction is obviously negligible except in the neighborhood of the maximum stress. Considering the large number of terms which were used in the collocation solution (Schapery, 1967), it can be concluded that the corrected solution is graphically indistinguishable from an exact inversion.

Finally, we note that the curvature of f vs $\log t$ may not be small in dynamic problems, as shown by Arenz (1964, 1965). In spite of this, he found that the direct method, while not very accurate, predicted the general behavior correctly.

2. *Quasi-Elastic Method*

In the quasi-elastic method, a viscoelastic solution (i.e., indicial admittance) is approximated by an elastic solution wherein all elastic constants are replaced by the corresponding time-dependent relaxation moduli or creep compliances (Schapery, 1965). While this method is based on approximate inversion, Eq. (120), and therefore is limited to quasi-static problems, it actually enables one to completely bypass transform theory in deriving the indicial admittance. In its most general form, the method is equivalent to approximating constitutive Eqs. (10) and (11) by, respectively,

$$\sigma_{ij}(t) \approx C_{ijkl}(t)\epsilon_{kl}(t) \tag{121}$$

and

$$\epsilon_{ij}(t) \approx S_{ijkl}(t)\sigma_{kl}(t) \tag{122}$$

These two equations yield, for example, the previously discussed approximations, Eqs. (95) and (101). However, in certain cases it may be desirable to use this method more restrictively, as there are some problems in which one of Eq. (121) or (122) is significantly more accurate than the other; this point will be elaborated upon in Section V,A.

IV. Prediction of Effective Viscoelastic Properties

A. Exact and Approximate Isothermal Properties

A direct result of the correspondence principle is that the Carson transformed effective moduli $\tilde{C}_{ijkl}$ and compliances $\tilde{S}_{ijkl}$ for a composite material are equal to those for an elastic material with the same phase geometry, but with Carson transformed phase properties in place of elastic constants. Moreover, in view of Eqs. (108) and (109), effective complex properties are obtained by making the substitution $s \to i\omega$ in the Carson transform of the phase properties or, more directly, by replacing phase elastic constants by corresponding complex properties. These elastic–viscoelastic relationships were pointed out by Hashin for the Carson transformed properties (1965, 1966), which he called "transform domain" properties, and for the complex properties (1970a).

The time-dependent effective creep and relaxation properties can be found either by inversion of the Carson transforms or by the quasi-elastic method. In the latter case, all one does to obtain these time-dependent properties is replace phase elastic constants by the appropriate creep compliances or relaxation moduli. In view of the mathematical basis for this method, as well as the experience gained in its application (Schapery, 1965, 1967; Sims 1972), we believe that its accuracy will be well within usual engineering requirements in most cases.

Effective composite properties can be derived by starting with either exact or approximate elastic solutions. On the basis of the accuracy of the quasi-elastic method, it can be concluded that the "quality" of composite creep and relaxation functions derived from *approximate elastic solutions* will be about the same as that of the elastic solutions themselves.

It is not possible to make as general a statement about the quality of effective complex properties; however, a direct relationship between the quality of the elastic and viscoelastic solutions does exist when damping is small or when Eq. (131) is valid. Inasmuch as damping is small in the use range for many structural composites (Hashin, 1970a), this case is of considerable practical importance, and therefore we shall use it to develop a simplified method for calculating complex viscoelastic properties from elastic solutions. For the sake of argument let F represent any one effective elastic modulus or compliance. Then, we write this elastic property as $F = F(p_j)$ for a composite with N phase properties $p_j (j = 1, 2, \ldots, N)$; these latter properties form the set of all elastic constants of all phases which are needed to predict the effective property F. The effective complex viscoelastic property is

$$F^* = F(p_j^*) \tag{123}$$

where F is still the elasticity solution, but its arguments are now complex phase properties. Also, write (j not summed),

$$p_j^* = p_j'[1 \pm i \tan \phi_j] \tag{124}$$

where $+$ is used if p_j^* is a modulus and $-$ is used if it is a compliance (e.g., see Ferry, 1970). Let us now assume that F can be expanded in a multiple, complex Taylor series, using the real part of the complex phase properties p_j' as the set of values about which the expansion is made. Assuming all constituent loss tangents are sufficiently small, we drop second- and higher-order terms in $\tan \phi_j$ and write the result in the form [see also Eq. (163)]

$$F^* = F'[1 \pm i \tan \phi] \equiv F' \pm iF'' \tag{125a}$$

where the real part F' is the elastic solution in terms of the real part of the phase properties; viz.,

$$F' = F(p_j') \tag{125b}$$

and the effective loss tangent is

$$\tan \phi = \sum_{j=1}^{N} \partial \log F / \partial \log p_j \,|_{p_j = p_j'} \tan \phi_j \tag{125c}$$

The logarithmic derivative in Eq. (125c) can be written in terms of the regular derivative of the elastic solution,

$$\partial \log F/\partial \log p_j = (p_j/F)\,\partial F/\partial p_j \tag{125d}$$

It is to be noted that Eq. (125c) reduces immediately to that given by Hashin (1970a) when it is specialized to the shear modulus of a lowly damped particulate composite with one viscoelastic and one elastic phase.

Equations (125) show that with small damping the effective complex properties can be derived directly from analytical or *numerical* elasticity solutions. Obviously, if an elasticity solution is approximate, error in the real part F' of the viscoelastic property is identical with that in the elastic solution; on the other hand, the fact that the loss tangent depends on derivatives of the elastic solution implies the relative error in $\tan\phi$ could be larger. Furthermore, even if one or more of the constituent loss tangents is not small, numerical elasticity solutions can, in principle, still be used. However, as it would be necessary to retain second- and possibly higher-order terms in the Taylor series expansion, the resulting equation for the effective complex property would be much more involved; also, the process of forming derivatives from a numerical solution introduces error, which places a practical upper limit on the number of terms in the expansion that can be evaluated with acceptable accuracy.

As an application of the quasi-elastic method, Halpin (1969) used the approximate Halpin–Tsai elasticity equations (Ashton *et al.*, 1969) to predict the principal shear creep compliance S_{66} and transverse creep compliance S_{22} (loading perpendicular to the fibers) for a (viscoelastic) rubber reinforced with (elastic) unidirectional, nylon fibers. These predictions, along with experimental results, are shown in Fig. 8.

Sims (1972) used the Halpin–Tsai equations to calculate relaxation moduli of unidirectional (elastic) graphite and boron fiber-reinforced (viscoelastic) epoxy composites. Both the quasi-elastic method and collocation method of LT inversion were used, with very good agreement achieved between the two methods. The epoxy's bulk modulus was assumed constant, while its shear compliance was represented by a power law of the form in Eq. (76); in view of our comments in Section II,G,2, the use of a constant Poisson's ratio, rather than constant bulk modulus, would have been more realistic because the epoxy was below T_g. When the exact relation $\tilde{G} = \tilde{J}^{-1}$ was used to compute $G(t)$, before applying the quasi-elastic method, the average difference between quasi-elastic and collocation solutions for the orthotropic moduli was approximately 2%, with the instantaneous error never exceeding 4.1%; use of the additional approximation $G(t) \approx J(t)^{-1}$ resulted in net errors in the effective moduli of approximately twice these values.

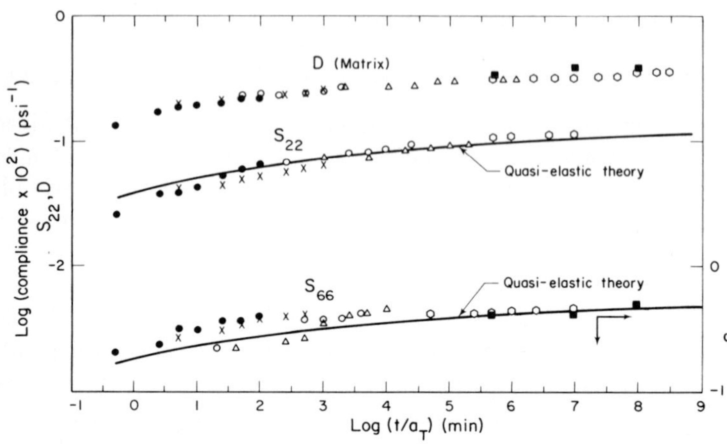

FIG. 8. Master curves of creep compliances for unidirectional nylon fiber-reinforced rubber (after Halpin, 1969). ●, −40°C; ×, −35°C; ○, −30°C; △, −20°C; ⬡, 0°C; ■, 20°C.

Some examples of effective complex modulus prediction have been given by Hashin for particulate (1970a) and fibrous (1970b, c) composites, with and without the assumption of small damping. Figures 9 and 10 show the real and imaginary parts of the complex axial shear modulus $G_A^* = G_A' + iG_A''$ for polyisobutylene (above T_g) reinforced with parallel rigid fibers; the complex shear modulus variation of polyisobutylene (plots for $v_2 = 0$) have been taken from the master curves given by Tobolsky and Catsiff (1956). These predictions were made by starting with the elastic shear modulus G_A for the so-called "composite cylinder assemblage model" (e.g., Hashin, 1966)

$$G_A = \frac{1 + v_2}{1 - v_2} G_1 \tag{126}$$

where G_1 is the matrix shear modulus. Upon replacing G_1 by the complex modulus $G_1^* = G_1' + iG_1''$, and G_A by G_A^*, the real and imaginary parts of G_A^* are found as

$$G_A' = \frac{1 + v_2}{1 - v_2} G_1', \qquad G_A'' = \frac{1 + v_2}{1 - v_2} G_1'' \tag{127}$$

Observe that the loss tangent G_A''/G_A' for the composite is equal to that of the matrix. Moreover, since G_A^* is a product of a frequency-independent term $(1 + v_2)/(1 - v_2)$ and G_1 itself, the only effect of fiber volume fraction

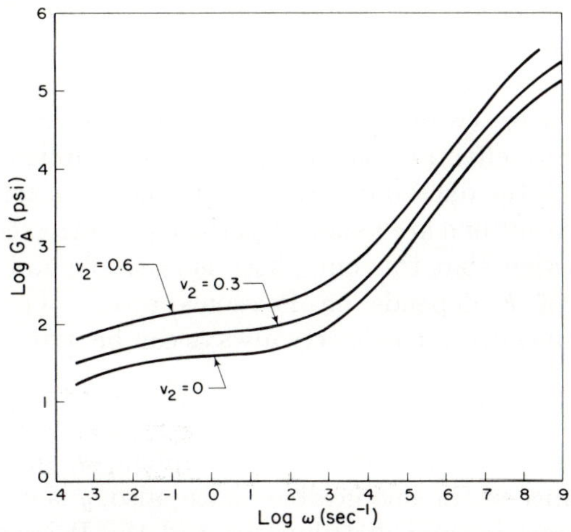

FIG. 9. Real part of complex axial shear modulus of polyisobutylene with rigid fibers (after Hashin, 1970c), where v_2 is the volume fraction of fibers.

is to produce a vertical shift of plots of log G'_A and log G''_A, as may be seen in Figs. 9 and 10.

The fact that the effective modulus in Eq. (126) is dependent on and proportional to a single matrix property is not unusual. Indeed, there are many technically important isotropic and anisotropic composites for which

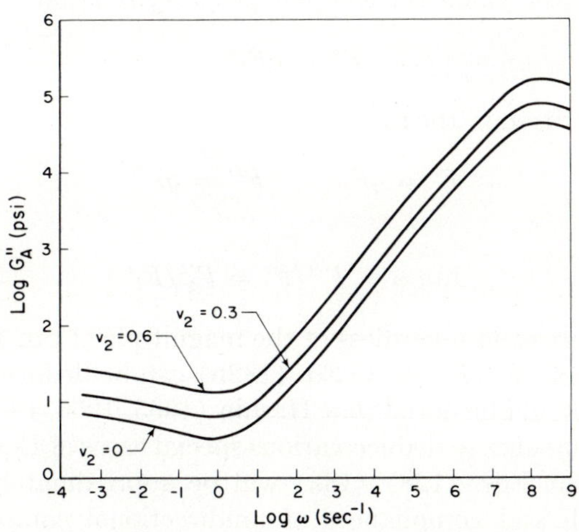

FIG. 10. Imaginary part of complex axial shear modulus of polyisobutylene with rigid fibers (after Hashin, 1970c), where v_2 is the volume fraction of fibers.

this representation is at least approximately true. We shall discuss the basis for this behavior because it has important implications in the derivation of viscoelasticity solutions and calculation of upper and lower bounds on effective properties (which will be shown in the next subsection).

Consider first any effective elastic modulus or compliance F (say) of a composite in which the overall deformation is due essentially to one phase (i.e., all phases except one are assumed perfectly rigid) except possibly for holes. Assume further that this one phase is isotropic and has a constant Poisson's ratio (if F depends on Poisson's ratio). Then, dimensional analysis implies the effective property always can be written in the form

$$F = gF_1 \tag{128}$$

where F and F_1 are both either moduli or compliances, and F_1 is the elastic modulus/compliance of the one flexible phase; also, g is *at most* a function of the phase geometries and distributions and the Poisson's ratio of the flexible phase (ν_1, say). The correspondence principle yields

$$\tilde{F} = g\tilde{F}_1 \tag{129}$$

where g is identical with that in the elastic solution. Divide through by the LT parameter s and invert

$$F(t) = gF_1(t) \tag{130}$$

where $F(t)$ and $F_1(t)$ are both either relaxation moduli or creep compliances. Equation (129) also yields the complex property relation

$$F^* = gF_1^* \tag{131}$$

which may be written in the form

$$F' = gF_1', \qquad F'' = gF_1'' \tag{132a}$$

Also,

$$\tan\phi \equiv F''/F' = F_1''/F_1' \tag{132b}$$

Equation (132) is valid regardless of the magnitude of $\tan\phi$; obviously for the special case of $|F''/F'| \ll 1$, Eq. (132b) can be deduced directly from Eq. (125c). It should be noted that Hashin (1965, 1966, 1970a–c) has used dimensional arguments to deduce various special cases of Eqs. (128)–(132).

We expect that Eqs. (128)–(132) will be approximately valid for the principal moduli and compliances of unidirectional composites, such as epoxy with boron or graphite fibers, except for the effective axial modulus. Nevertheless, the axial modulus E_A of M-phase unidirectional composites with relatively stiff fibers is normally very well represented by

$$E_A = \sum_{i=1}^{M} v_i E_i \tag{133}$$

where the subscripts $i = 1$ and $i > 1$ refer to the matrix and fibers, respectively. The correspondence principle yields

$$\tilde{E}_A = \sum_{i=1}^{M} v_i \tilde{E}_i \tag{134}$$

from which analogous expressions can be derived for the relaxation and complex moduli; if the fibers are elastic, $\tilde{E}_i = E_i$ for $i > 1$. The uniaxial (and shear) moduli and compliances of rubber and asphalt matrices filled with relatively hard particles have been observed to obey Eqs. (130) and (132) (Krokosky, 1967).

When Eq. (128) is valid for all principal creep compliances, the compliance for uniaxial loading will have this form regardless of the orientation of the load vector [see Eq. (120)]. This behavior was reported by Kaye and Saunders (1964) for a glass fiber-reinforced epoxy laminate and by Darlington and Saunders (1970) for oriented polyethylene; crystalline regions in the polyethylene seem to behave as fibers.

B. Bounds on Isothermal Properties

Numerous results exist on the upper and lower bounds on effective elastic properties (e.g., Hashin, 1970c). They provide limits on the possible range of variation of effective properties when information on only phase volume fractions and moduli is known. As pointed out by Schapery (1967), these bounds carry over to the Carson transforms of effective moduli and compliances when s is real and nonnegative; the bases for this statement are (i) the correspondence principle, and (ii) the positive definite character of these Carson transforms for this range of s. However, additional considerations are needed to establish bounds on the relaxation, creep, and complex properties.

First, assume the phases are either perfectly rigid or have zero rigidity (holes) except for one phase which is isotropic with a constant Poisson's ratio. Then, dimensional considerations imply the exact effective properties *and* their upper and lower bounds have the form in Eq. (128). Elastic analysis yields, in effect, the bounds on g after dividing through by F_1,

$$g^{(-)} \leq g \leq g^{(+)} \tag{135}$$

As the same function g appears in Eqs. (130) and (132), bounds on time-dependent and complex properties are readily deduced; viz., multiply Eq. (135) by $F_1(t)$

$$g^{(-)}F_1(t) \le F(t) \le g^{(+)}F_1(t) \tag{136}$$

Similarly, from Eqs. (132a) and (135),

$$g^{(-)}F_1' \le F' \le g^{(+)}F_1' \tag{137a}$$

$$g^{(-)}F_1'' \le F'' \le g^{(+)}F_1'' \tag{137b}$$

It was assumed that F_1, $F_1(t)$, F_1', and F_1'' are all positive in deriving Eqs. (135)–(137). Since these properties can be negative over some or all times and frequencies (unless F_1 is on the main diagonal of the material property tensor), it may be necessary to reverse one or more of the above inequalities.

There are situations for which Eq. (128) is not applicable but for which bounds on transient and complex properties can still be deduced. In particular, Schapery (1967) has used the quasi-elastic method to show that

$$F^{(-)}(t) \lesssim F(t) \lesssim F^{(+)}(t) \tag{138}$$

where $F^{(+)}(t)$ and $F^{(-)}(t)$ are the upper and lower bounds on an effective elastic property in which the phase elastic properties have been replaced by viscoelastic relaxation moduli or creep compliances; this relationship was illustrated by Schapery (1967) for the particular case of the axial shear modulus of a unidirectional, fibrous composite. Also, by recognizing that the approximation $\tilde{F}' \approx \tilde{F}$, with $\omega = s$, is often quite good for polymeric materials [e.g., compare Eqs. (89) and (90b)], Schapery (1967) indicated that the bounds on $\tilde{F}$ apply to F'. However, no method for bounding F'' was given.

When the phase loss tangents are small, we can deduce the bounds on F' directly from Eq. (125b):

$$F^{(-)}\big|_{p_j=p_j'} \le F' \le F^{(+)}\big|_{p_j=p_j'} \tag{139}$$

Thus, the upper and lower bounds on F' are the corresponding elastic bounds, in which phase elastic constants are replaced by the real part of the phase complex properties. The imaginary part F'' or loss tangent, however, depends on derivatives of the elastic solution, and is not, in general, bounded by the elastic solutions.

Bounds on the effective complex shear and bulk moduli (and compliances) for an isotropic composite consisting of isotropic, viscoelastic phases were derived by Roscoe (1969). No assumption was made regarding the magnitudes of the relative stiffnesses and loss tangents of the phases. His bounds reduce to the familiar Reuss and Voigt bounds when specialized to elastic behavior; these bounds are normally quite far apart unless the moduli of all phases are close to one another. Christensen (1969) has also derived bounds on the components of the complex moduli and compliances of isotropic composites. However, his are based on assumptions which are even more restrictive than those used to deduce Eqs. (137).

Bounds on the real part of the complex shear modulus and loss tangent of rubber-modified plastics have been calculated by Bucknall and Hall (1971). In this case, Hashin's (1962) theory of isotropic elastic moduli was used to predict the bounds; while the bounds on shear modulus may be at least approximately correct, those on $\tan \phi$ are suspect in view of our previous observations.

C. *Nonisothermal Properties*

All of the above considerations on the isothermal correspondence principle and the prediction of effective properties and their bounds are valid as long as temperature is timewise constant, independently of the type of temperature dependence possessed by the constituent properties. Use of these effective properties under transient temperatures is valid if the composite material is thermorheologically simple (TSM), but *not* if it is thermorheologically complex (TCM).

Let us first consider the former class of composites. The extension to transient temperatures is based on the use of constitutive Eqs. (63) or (64) when expressed in terms of effective properties. Suppose that the effective moduli or compliances and thermal coefficients are known at some reference temperature T_R in terms of the constituent properties. Further, suppose that only one phase is viscoelastic (over the temperature range of interest) and it is a TSM with shift factor a_T. Then obviously the composite itself is a TSM since we obtain its effective properties from isothermal, micromechanics solutions at T_R by simply replacing real time t with reduced time ξ, Eq. (40).

In order to illustrate this procedure consider, for example, the simple case in which one (or more) of the effective properties obeys Eq. (130) under isothermal conditions. The effective property which is to be substituted into Eq. (63) or (64) is

$$F(\xi) = gF_1(\xi) \tag{140}$$

where F and F_1 in this equation are identical to those in Eq. (130), except the argument is now ξ.

Observe that if *all* properties satisfy Eq. (130), but each with its own g (g_{ijkl}, say), then the constitutive equations simplify greatly. For example, if the effective compliances are [summation convention used in Eqs. (142) and (145)]

$$S_{ijkl}(\xi) = g_{ijkl}D(\xi) \tag{141}$$

where $D(\xi)$ is the uniaxial creep compliance of the one viscoelastic phase, then

$$\int_0^t S_{ijkl}(\xi - \xi')\, \partial\sigma_{kl}/\partial t'\, dt' = g_{ijkl} \int_0^t D(\xi - \xi')\, \partial\sigma_{kl}/\partial t'\, dt' \tag{142}$$

If Eq. (141) is known to be true in one coordinate system (e.g., the principal material system), both Eqs. (141) and (142) remain valid in all other systems which are obtained by a rotation of the original one.

The above considerations carry over to the effective thermal properties of a TSM. Consider, for example, the quasi-elastic expression for axial expansion coefficient of a two-phase, unidirectional fibrous composite derived by Schapery (1968),

$$\alpha_A(\xi) = \frac{\alpha_1 v_1 E_1(\xi) + \alpha_2 v_2 E_2}{v_1 E_1(\xi) + v_2 E_2} \tag{143}$$

where phase 1 is viscoelastic with relaxation modulus $E_1(\xi)$ and phase 2 is elastic; the constituent thermal expansion coefficients, α_1 and α_2, are assumed constant. Now, suppose a principal coordinate system is used in writing Eq. (64), in which the x_1 axis is parallel to the fibers; then set $\alpha_{11} = \alpha_A$ and obtain

$$\int_0^t \alpha_{11}(\xi - \xi') \frac{\partial \Delta T}{\partial t'}\, dt' = \int_0^t \frac{\alpha_1 v_1 E_1(\xi - \xi') + \alpha_2 v_2 E_2}{v_1 E_1(\xi - \xi') + v_2 E_2} \frac{\partial \Delta T}{\partial t'}\, dt' \tag{144}$$

Turning now to the TCM, recall that two types (designated as TCM-1 and TCM-2) were discussed in Section II,D. The type TCM-1, by definition, is a composite material having two or more TSM phases with different a_T's which exhibit viscoelastic behavior over a common temperature range.

The behavior of composites of type TCM-1 have been studied under isothermal conditions at different temperatures. For example, Halpin (1969) investigated a composite consisting of two SBR elastomers with different glass transition temperatures (−29 and −75°C). Fesko and Tscheogl (1971), among others, have studied the behavior of two-phase block copolymers at various constant temperatures, which are also of the type TCM-1. However, the author is not aware of any investigations using transient temperatures. This is a particularly complicated situation as a correspondence principle cannot be used; prediction of composite mechanical response in terms of constituent properties may, of course, be accomplished by direct numerical solution of the field and boundary equations for each temperature history of interest.

If one (or more) phase is of the type TCM-2, which has been discussed in detail in Section II,D, a correspondence principle does not generally exist under transient temperatures. Moreover, the composite itself usually will have thermorheological behavior which is of a more complex type than TCM-2. There is, however, an important special case for which *non-isothermal* composite behavior can be predicted in terms of *isothermal* constituent properties. This situation exists for those effective properties which satisfy Eq. (130) under isothermal conditions and for the axial modulus, Eq. (133). One can show the constitutive relations are obtained by replacing the integrals in Eqs. (63) and (64) [where Eqs. (130) and (133) are applicable] by integrals of the form in Eq. (50) or in (56) and (57). Further simplification results if *all* effective properties satisfy Eq. (130); for example, the integral expression Eq. (142) becomes

$$g_{ijkl}\left[D_{\mathrm{I}}\sigma_{kl} + \int_0^t \Delta D(\xi - \xi')\,\frac{\partial}{\partial t'}\left(\frac{\sigma_{kl}}{a_{\mathrm{G}}}\right) dt'\right] \tag{145}$$

V. Analysis of Viscoelastic Structures

A. Quasi-Static Problems

Prediction of stresses and deflections in viscoelastic structures is made using methods which are quite analogous to those for effective properties described in Section IV.

For a *stable* structure satisfying the requirements of one of the correspondence principles given in Section III, the analysis can be accomplished in a straightforward manner if the elasticity solution is available. With such problems, the most convenient approach normally consists of first obtaining the quasi-elastic solution for the indicial admittance, and then using superposition if the applied loading varies in time. Accuracy of this method is greatest if creep compliances are used in the elasticity solution when the prescribed quantities are surface and/or body forces, and if relaxation moduli are used when displacements are prescribed. However, even if this latter procedure is not followed (e.g., relaxation moduli used when forces are prescribed) and approximation Eq. (95) is used, the error usually is still small, especially when the viscoelastic phase(s) is a hard polymer (McCammond, 1967; Sims, 1972); for highly viscoelastic materials it may be desirable to use exact Eq. (93) and LT inversion when compliances (moduli) must be calculated from moduli (compliances).

Sims (1972) has predicted the quasi-static response of unidirectional laminae and laminated plates of fiber- (glass, boron, and graphite) reinforced epoxy by means of the quasi-elastic method; this study includes results on the effect of coupling between bending and extension in unbalanced laminates and on buckling of laminates. The stresses in a viscoelastic matrix material containing rigid inclusions and voids have been analyzed by Hackett using a Voigt model representation (1968) and using actual relaxation properties of an epoxy (1971); in the latter study, the collocation method of LT inversion was applied to associated elastic solutions obtained by the finite-element method. Still other analyses of viscoelastic composites have been made by Huang and Lee (1969) and Schapery (1967); in the former paper, the response of an internally pressurized, wire-reinforced cylinder was calculated by using a finite-difference integration method, while in the latter study the approximate methods outlined in Section III were applied to an orthotropic, internally pressurized cylinder and to the thermoviscoelastic analysis of shrinkage stress in a fiber-reinforced plastic.

Turning now to the problem of viscoelastic *buckling*, we first note that a correspondence principle can be used under isothermal (Biot, 1965) and nonisothermal (Schapery, 1974) conditions if (i) the in-plane (buckling) loads are independent of time, and (ii) the conditions of the correspondence principles in Section III are met; in-plane loads are applied at $t = 0$.

An introduction to this area is given by Kempner (1962), and results for bars characterized by simple spring and dashpot models are described. Buckling of layered viscoelastic media has been studied extensively by Biot (1965). The quasi-elastic method was used to predict buckling of a rubber column by Halpin and Meinecke (1969) and buckling of plastic struts by Salchev and Williams (1969); in both investigations quite good agreement between theoretical and experimental buckling times was achieved. Sims (1972) has employed this method to calculate buckling times for laminated plates of epoxy reinforced with boron and graphite fibers. A typical set of results is shown in Fig. 11. The percentage values indicate the relative difference between uniaxial loads N_{cr} which cause buckling at $t = 0$ and $t = 50$ hr; h is the total laminate thickness.

Although quasi-elastic buckling times agree reasonably well with experimental data on the rubber and plastic columns, as noted above, the prediction of a finite buckling time is inconsistent with the exact linear theory (which predicts exponential divergence). In order to clarify this matter, let us consider, for simplicity, the behavior of a pinned-end column; also we assume power law creep compliance, Eq. (76), is applicable. Euler's formula for the critical load P_{cr} of a column with length L and area moment of inertia I is

$$P_{cr} = \pi^2 EI/L^2 \tag{146}$$

Following the quasi-elastic method, we replace E by $D(t)^{-1}$ and then solve the resulting equation for buckling time t_q:

$$t_q = \left[\left(\frac{\pi^2 I}{P_{cr} L^2 D_I} - 1\right)\frac{D_I}{D_1}\right]^{1/n}, \qquad P_{cr} < \frac{\pi^2 I}{L^2 D_I} \tag{147}$$

where instantaneous buckling occurs if the far right inequality is not satisfied. In contrast, exact LT inversion yields a continuously increasing transverse deflection y_T, which is approximated quite well by (Schapery, 1974)

$$y_T \approx A e^{t/t_c} \tag{148}$$

when $t \gtrsim t_c$, where t_c and A are independent of time. [If desired, the modified direct method of LT inversion developed by Schapery (1974) can be used to predict the entire deflection history.] Also,

$$t_c = t_q[\Gamma(1 + n)]^{-1/n}, \qquad 0 < n \leq 1 \tag{149}$$

where $\Gamma = \Gamma(1 + n)$ is the gamma function with argument $(1 + n)$. For the limit of elastic behavior, $n \to 0$, we find $t_c = t_q e^c \approx 1.8 t_q$, where c is Euler's constant; the rule $t_c \approx 1.7 t_q$ is within 5% of Eq. (149) whenever $0 < n \lesssim 0.10$. Furthermore, the ratio t_c/t_q decreases monotonically with increasing n until it is unity at $n = 1$.

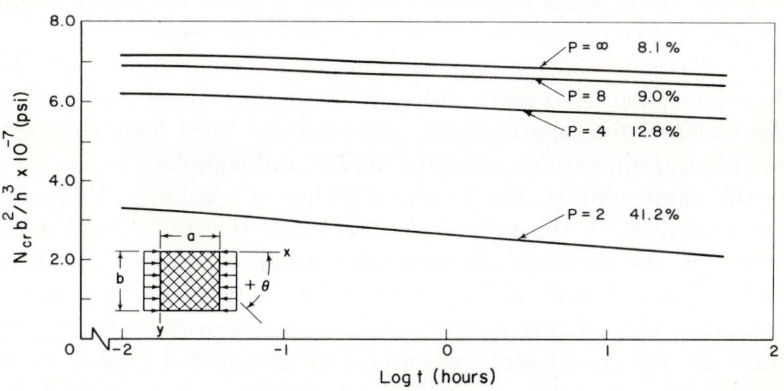

FIG. 11. Dependence of uniaxial load on buckling time for a boron–epoxy angle-ply plate, temperature is 73°F, a/b is an integer, P is the number of layers, and $\theta = 45°$ (after Sims, 1972).

The obvious significance of t_c is that it defines the time beyond which deflection grows rapidly; thus for $t > t_c$, the deflection quickly reaches the point at which the bar breaks or yields. According to Eq. (149), $t_q < t_c$, and therefore the quasi-elastic method tends to provide a conservative estimate of buckling time; this observation is not limited to materials with power law compliances (Schapery, 1974).

The small curvature condition upon which the quasi-elastic method is based is not satisfied by the deflection at times near to and beyond t_c (Schapery, 1974). This accounts for the fact that this approximate method yields a result which is qualitatively different from the exact LT inversion. However, considering the fact that the quasi-elastic prediction of buckling time is on the conservative side, and that the scatter in experimental values may be quite large (Salchev and Williams, 1969), this simple technique should be quite adequate for many engineering applications.

B. Dynamic Problems

Dynamic analysis of viscoelastic structures by means of the correspondence principle (CP) for steady-state vibrational loading and for transient loading will be examined here. In order to be able to make direct use of elastic solutions we must assume that the structures are nonaging and have timewise constant temperatures, or, at least, undergo transient changes in material properties which are relatively very small during each cycle of vibrational loading or over the time of application of transient loads. Recall that an additional requirement is that the boundaries cannot move, apart from small displacements, and a force boundary condition cannot change to a displacement condition and vice versa.

These CP's are valid for composite materials, regardless of whether or not the microstructure is taken into account. As stated previously, if wavelengths which define the dynamic response are very long compared to microstructural dimensions, effective moduli and compliances can be used; within this same context, the density ρ is that of a volume element which is large compared to the microscale (viz., ρ is the "effective" density). The bulk of the existing viscoelasticity (and elasticity) solutions for *bounded* media is based on the effective property theory. On the other hand, most of the existing results which explicitly account for the microstructure are for steady-state vibrations of unbounded media. As noted above, the dynamic CP's are valid for both cases, and we shall discuss both; with microstructural effects one merely uses phase properties instead of effective properties when converting from elastic to viscoelastic solutions.

1. Steady-State Vibration of Bounded Media

First, the simple example of a composite bar undergoing torsional vibrations will be discussed in order to illustrate the analysis method and the influence of viscoelastic damping on dynamic response. The effect of damping in more general situations is then examined.

The example is taken from a paper by Hashin (1970b), and is that of a unidirectional, fiber-reinforced, circular cylinder of length l (5.0 ft), diameter d (4.0 in.), density ρ (sp. gr. = 3.0). The fibers were assumed to be perfectly rigid and parallel to the composite cylinder's axis; therefore, supposing that the effective modulus theory was valid, the complex shear modulus components were calculated from Eq. (127), in which the fiber volume fraction was v_2 (0.6). For the matrix (phase no. 1), Hashin assumed a constant loss tangent,

$$\tan \phi_1 = 0.10 \tag{150}$$

which, in this problem, is also the effective loss tangent $\tan \phi$. The complex modulus of the matrix G_1^* was assumed to have the real part

$$G_1' = 0.5 \times 10^6[1 + 0.25 \log \omega] \text{ psi} \qquad \text{where} \quad 1 < \omega < 10^4 \quad \text{sec}^{-1} \tag{151}$$

It should be noted in passing that Eqs. (150) and (151) are mathematically inconsistent, in that the real part G_1' is for a complex modulus whose loss tangent is *not* constant; for the real part and loss tangent are interrelated (e.g., Ferry, 1970) and, strictly speaking, one should not use independent representations; however, we shall continue to use Eqs. (150) and (151) as none of the following manipulations and analytical results are directly dependent on this interrelationship.

With the cylinder built in at one end, and subjected to the forcing torque $T_0 e^{i\omega t}$ at the other end (where T_0 is a real constant), the complex viscoelastic solution for angle of twist at the driven end is

$$\theta = \theta_A e^{i\omega t} \tag{152}$$

where

$$\theta_A \equiv (32 l T_0 / \pi d^4 G_A^* \sqrt{\gamma})(\sin \sqrt{\gamma} / \cos \sqrt{\gamma}) \tag{153a}$$

and

$$\gamma \equiv \rho \omega^2 l^2 / G_A^* = \rho \omega^2 l^2 J_A^* \tag{153b}$$

in which $J_A^* = (G_A^*)^{-1}$ is the complex shear compliance (introduced for later convenience). This result was obtained directly from an elastic solution by substituting the effective complex shear modulus G_A^* for its elastic counterpart.

The coefficient θ_A is complex. It may be written in polar form

$$\theta_A = |\theta_A| e^{i\beta} \tag{154}$$

The physical angle of twist is found by combining Eqs. (152) and (154),

$$\theta = |\theta_A| e^{i(\omega t+\beta)} \tag{155}$$

and then taking the real part,

$$\text{Re}\{\theta\} = |\theta_A| \cos(\omega t + \beta) \tag{156}$$

which is the response to the real torque $T_0 \cos \omega t$; alternatively, the imaginary part is the physical response to torque $T_0 \sin \omega t$. [The task of evaluating $|\theta_A|$ and β from the complex functions in Eqs. (153), or performing a similar task in more involved problems, can be easily accomplished by using a digital computer together with built-in complex functions.] The frequency dependence of the amplitude of angle of twist $|\theta_A|$ is drawn in Fig. 12 for a frequency range which covers the first two resonances; the modulus $G_1 = 0.5 \times 10^6$ psi was used in obtaining the elastic solution indicated.

There are two features of the dynamic response curve which are of particularly practical interest; one is amplitude of angle of twist at each resonance, and the second is the width $\Delta\omega$ of the response curve. The amplitude at each resonance is readily calculated from Eq. (153) by using the fact that the loss tangent is relatively small. Because of this smallness, the maxima are very close to frequencies at which Eq. (153a) would be

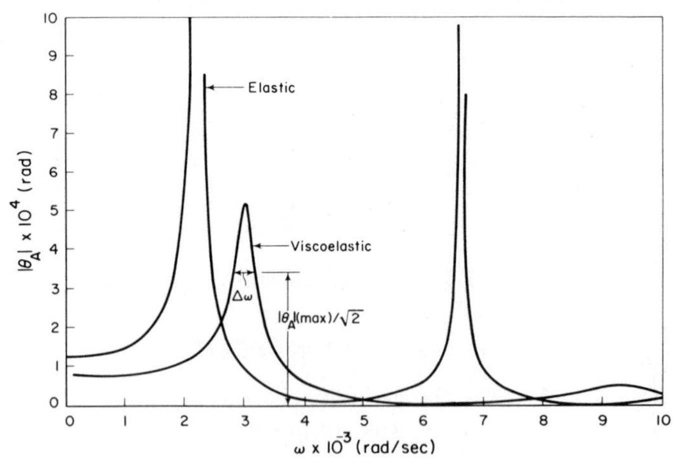

FIG. 12. Amplitude of angle of twist for composite cylinder (after Hashin, 1970b).

infinite if the material were elastic with compliance J_A' (which is easily proved by differentiation); denote these frequencies by ω_n, which correspond to the values $(\gamma_n)^{1/2} = n\pi/2$ with $n = 1, 3, \ldots$. Thus, from Eq. (153b),

$$\omega_n = (\rho J'_{An})^{-1/2} n\pi/2l \tag{157}$$

with $J'_{An} \equiv J_A'(\omega_n)$. Now, expand $\cos\sqrt{\gamma}$ with respect to J_A^* in a complex Taylor series about the point J'_{An} and neglect second- and higher-order terms in $\tan\phi$; there results, after using Eq. (157),

$$\cos\sqrt{\gamma} \approx (in\pi/4)\tan\phi \tag{158}$$

Substituting approximation Eq. (158) into Eq. (153a) and neglecting $\tan\phi$ compared to unity in all terms yields

$$|\theta_A|_{max} \approx \frac{256}{\pi^3}\frac{T_0 l}{d^4 n^2 G'_{An}\tan\phi} \tag{159}$$

and in terms of the angle of twist at zero frequency $|\theta_A|_0$,

$$|\theta_A|_{max}/|\theta_A|_0 \approx (8G_A/G'_{An})/n^2\pi^2\tan\phi, \qquad n = 1, 3, \ldots \tag{160}$$

where $G_A \equiv G'_A(0)$. Recognizing that $G'_{A1} \approx G'_{A3}$ [see Eq. (151)], we see that the second resonance peak is predicted to be approximately $\frac{1}{9}$ the value of the first one; this result is closely satisfied by the plot in Fig. 12. In contrast, as observed by Hashin (1970b), the resonance peaks in *transverse* vibrations decrease more slowly; see also Snowdon (1968).

The width of the resonance peak $\Delta\omega$ at the amplitude $|\theta_A| = |\theta_A|_{max}/\sqrt{2}$ is well known to be directly related to the loss tangent for small *viscous* damping, in that $\Delta\omega/\omega_1 = \tan\phi$; this relation provides a simple means for evaluating the loss tangent from dynamic response data. We shall prove that it holds (approximately) at each resonance for a relatively general viscoelastic representation, as defined by a frequency-dependent complex compliance, almost regardless of the type of structure involved; the only assumptions which will be made are (i) $\tan\phi$ is small compared to unity (but not necessarily constant); and (ii) the structure's response is dependent on a single complex compliance S^*, say. (The author is not aware of such a proof being in the literature; additionally, the analysis will bring out a means for evaluating dynamic viscoelastic response in terms of *numerical* elastic solutions.) Denote a steady-state sinusoidal input (e.g., displacement or force) by $I_A e^{i\omega t}$ and a response (e.g., displacement or acceleration at a particular point) by $R_A e^{i\omega t}$. The ratio is written in the form

$$R_A/I_A = 1/f \tag{161}$$

where $I_A f^{-1}$ is the *elastic* solution, except S^* $(= S' - iS'')$ appears in place of the elastic compliance. For the time being, assume R_A and I_A are the same type of physical quantities (e.g., both are accelerations or displacements). Then, by means of dimensional analysis,

$$f = f(\lambda) \tag{162a}$$

where the dimensionless parameter

$$\lambda \equiv \rho\omega^2 L^2 S^* \tag{162b}$$

has been defined; also ρ and L are density (at any point) and a length, respectively. The function f may, in addition, depend on other dimensionless parameters involving densities and lengths, but not on frequency or compliance; these additional parameters have no affect on the present considerations and, therefore, will not be explicitly shown. For a small loss tangent, $\tan\phi \equiv S''/S'$, the function f can be approximated by the first two terms of a complex Taylor series (assuming, of course, that f is analytic in the region of interest):

$$f \approx f_0 + (\lambda - \lambda')\, df_0/d\lambda \tag{163a}$$

where

$$f_0 \equiv f(\lambda'), \qquad df_0/d\lambda \equiv df/d\lambda\,|_{\lambda=\lambda'} \tag{163b}$$

and

$$\lambda' \equiv \rho\omega^2 L^2 S' \tag{163c}$$

Thus,

$$f \approx f_0 - i\rho\omega^2 L^2 S' \tan\phi\, df_0/d\lambda \tag{163d}$$

Denote the frequencies at which $f_0 = 0$ by ω_n; these are the resonant frequencies of an *elastic* structure with compliance S'. The corresponding values of λ' are

$$\lambda_n' \equiv \rho\omega_n^2 L^2 S_n' \tag{164}$$

where $S_n' \equiv S'(\omega_n)$. Evaluate f, Eq. (163d), in the neighborhood of any one ω_n by expanding f_0 about this point and dropping second- and higher-order terms; there results, finally,

$$f \approx \rho\omega_n^2 L^2 S_n' \frac{df_{0n}}{d\lambda}\left[(2 + k_n)\left(\frac{\omega - \omega_n}{\omega_n}\right) - i\tan\phi_n\right] \tag{165a}$$

where

$$\tan\phi_n \equiv \tan\phi(\omega_n) \tag{165b}$$

$$k_n \equiv d\log S'/d\log\omega\,|_{\omega=\omega_n} \tag{165c}$$

$$df_{0n}/d\lambda \equiv df/d\lambda\,|_{\lambda=\lambda_n'} \tag{165d}$$

Substitute expansion Eq. (165a) into Eq. (161) and obtain

$$R_A/I_A = |R_A/I_A|\, e^{i\beta} \tag{166}$$

where in the neighborhood of each resonance the amplitude is

$$\left|\frac{R_A}{I_A}\right| \approx \left|\rho\omega_n{}^2 L^2 S_n{}' \frac{df_{0n}}{d\lambda}\right|^{-1} \left\{(2+k_n)^2\left(\frac{\omega-\omega_n}{\omega_n}\right)^2 + \tan^2\phi_n\right\}^{-1/2} \tag{167a}$$

and the phase angle between input and response is

$$\beta = \tan^{-1}\{[\omega_n \tan\phi_n]/[(\omega-\omega_n)(2+k_n)]\} \tag{167b}$$

At resonance, $\omega = \omega_n$, we obtain $\beta = 90°$ and

$$|R_A/I_A|_{\max} \approx |\rho\omega_n{}^2 L^2 S_n{}' (df_{0n}/d\lambda)\tan\phi_n|^{-1} \tag{168}$$

Now, use Eqs. (167a) and (168) to find the ratio $\Delta\omega/\omega_n$ [with $\Delta\omega \equiv 2(\omega - \omega_n)$] for which $|R_A/I_A|_{\max} = \sqrt{2}\,|R_A/I_A|$; thus

$$\Delta\omega/\omega_n = [1 + (k_n/2)]^{-1}\tan\phi_n \tag{169}$$

If we can neglect k_n, this equation becomes identical to the well-known result for a simple damped system; however, it should be emphasized that Eq. (169), as well as Eqs. (165)–(168), are valid near and at all resonances of multidegrees of freedom and continuous systems (within the context of the previously stated assumptions, of course). The size of k_n can be *estimated* by referring to the power law compliance, Eqs. (90); for example, if $S_I \ll S'$, we find $-k_n \approx n'$, where n' is now the log–log slope of S' vs ω; since the loss tangent has been assumed small, n' must necessarily be small according to Eq. (90e). It can also be shown that the error due to neglecting k_n is of the same order of magnitude as that due to our neglecting changes in the imaginary part of Eq. (163d) when in the neighborhood of a resonance.

The input and response were initially assumed to represent the same type of physical quantities. This restriction is readily removed by again using dimensional analysis, and absorbing the additional parameters on the left side of Eq. (161). For example, let the response be an acceleration and the input a force. Then if I_A is replaced by $I_A S'\omega^2 L^{-1}$, all results, Eqs. (165)–(169), will be applicable.

A further generalization can be easily made in order to analyze structural response which is dependent on more than one modulus or compliance (if the associated loss tangents are small). In particular, assume the elastic response depends on the compliances S_j ($j = 1, 2, \ldots, N$). Then, on the basis of dimensional analysis, and for the viscoelastic structure, no generality is lost by writing f as a function of N independent dimensionless parameters involving all S_j^*. For convenience, we select these parameters as

$$\lambda_j \equiv \rho\omega^2 L^2 S_j^* \tag{170}$$

Then with notation analogous to that used above, and using a multiple Taylor series expansion, we obtain a generalization of Eq. (163d):

$$f \approx f_0 - i\rho\omega^2 L^2 \sum_{j=1}^{N} S_j' \tan\phi_j \frac{\partial f_0}{\partial \lambda_j} \tag{171}$$

which can be substituted into Eq. (161) to calculate dynamic response as a function of frequency. The response amplitude at each resonant frequency ω_n is, therefore,

$$\left|\frac{R_A}{I_A}\right|_{\max} \approx \left|\rho\omega_n^2 L^2 \sum_{j=1}^{N} S'_{jn} \tan\phi_{jn} \frac{\partial f_{0n}}{\partial \lambda_j}\right|^{-1} \tag{172}$$

The above analysis shows that a *numerical* (or analytical) elasticity solution can be used to predict the dynamic response of an arbitrary linear viscoelastic structure if the loss tangents are small. According to Eqs. (163d) and (171), one needs the reciprocal of the elastic solution f and the first derivative of the reciprocal $df/d\lambda$ (or $\partial f/\partial\lambda_j$ when more than one compliance enters) in which elastic compliance(s) is replaced by the real part of the corresponding complex compliance(s). This result is similar to that discussed previously in Section IV for the prediction of effective complex properties.†

Let us now turn to a review of some experimental results and their theoretical prediction. Schultz and Tsai studied the vibration response of cantilever beams of glass fiber-reinforced epoxy unidirectional composites (1968) and laminates (1969). Free and forced steady-state vibrations covering a frequency range of 5 to 10,000 Hz in the former investigation, and forced vibrations over the range of 30 to 9400 Hz in the latter one were used to obtain the real part of the beams' uniaxial modulus (storage modulus) and damping ratio; this damping ratio γ', say, is defined in the case of forced vibrations as $\Delta\omega/2\omega_n$ (which is approximately $\frac{1}{2}\tan\phi$) at each resonance, in accordance with Fig. 12. The unidirectional composites were found to have moduli which varied only slightly with frequency, while

† After writing the report (Schapery, 1972) upon which this chapter is based, the author received two reports from Hashin (1972a, b) which treat the vibrational analysis of composites with small loss tangents. The method of approximation and principal conclusions are essentially the same as given in this section. However, the elasticity solution, rather than its reciprocal, is used in the expansion (Hashin, 1972b), which precludes the prediction of response at resonant frequencies.

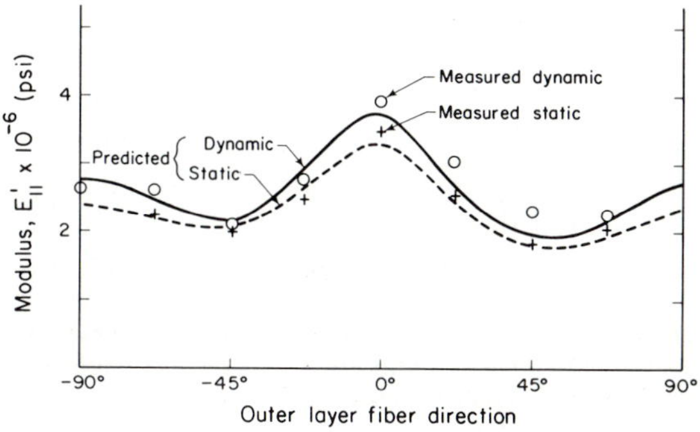

FIG. 13. Real part of complex uniaxial modulus (storage modulus) for $\pi/4$ laminates (after Schultz and Tsai, 1969).

the damping ratios (typically of magnitude 10^{-2}) increased noticeably with frequency. This increase may have been due, at least in part, to shear deformations in view of the high modes which were induced; the shear effect is reflected in the preceding theoretical considerations as an increase in $df_{0n}/d\lambda$ with frequency [see Eq. (168)]. Transformation relations for the uniaxial storage modulus and the damping ratio of beams with different fiber orientations were found to be quite well satisfied, although the agreement for the damping ratio was not as good as that for modulus.

The unidirectional properties were used by Schultz and Tsai (1969) to predict the uniaxial storage modulus and damping ratio of laminated beams having six plies with relative fiber directions of 0, $-\pi/3$, $\pi/3$, $\pi/3$, $-\pi/3$, 0 and having eight plies with fiber directions of 0, $\pi/2$, $\pi/4$, $-\pi/4$, $-\pi/4$, $\pi/4$, $\pi/2$, 0; both systems were quasi-isotropic for in-plane stresses, and had their plies arranged symmetrically with respect to the middle surface. Figures 13 and 14 show the theoretical and experimental results for

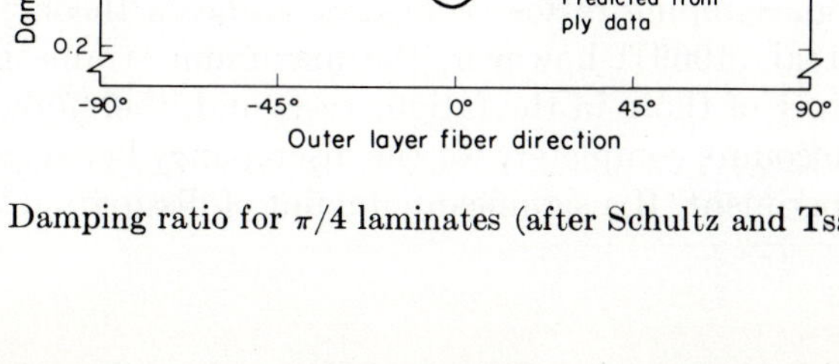

FIG. 14. Damping ratio for $\pi/4$ laminates (after Schultz and Tsai, 1969).

uniaxial storage modulus E'_{11} and damping ratio γ' of the second system (designated as $\pi/4$ laminates) as functions of angle between the outer-ply fibers and the beam's axis. The measured dynamic moduli were obtained from the first mode resonance, while the measured damping ratios ($\Delta\omega/2\omega_n$) correspond to the *smallest* of the values observed for all frequencies; at some frequencies the measured damping ratios were much higher than those in Fig. 14. As a summary of the results, Schultz and Tsai indicate that:

1. The storage moduli of the laminated beams at low frequency are larger than those measured statically. The moduli tend to increase slightly with increasing frequency and then in some cases to decrease.
2. The loss moduli are relatively large at low frequency, decreasing and then increasing with increasing frequency.
3. Both static and dynamic values of the storage moduli of the laminated beams can be predicted from unidirectionally reinforced ply data. They exhibit the transformation properties derived using elementary anisotropic beam bending theory.
4. The loss moduli are usually larger than those predicted from ply properties. They exhibit qualitative, but not quantitative, agreement with the predicted transformation properties (1969, p. 443).

Using laminated beam specimens and experimental techniques similar to those in the preceding study, Schultz and Warwick (1971) showed that an increase in crack damage could be correlated (qualitatively) with a decrease in resonant frequency, decrease in gage strain level (on the surface near the clamp), and increase in damping ratio. Measurements of these quantities were made under low-excitation accelerations ($\approx 5g$) before and after fatiguing the specimens at higher g-levels. Typical results for damping ratio changes are shown in Fig. 15 for four different specimens in which the outer-ply fibers were at 90° to the longitudinal beam axis; these ratios were found using $\gamma' = \Delta\omega/2\omega$ at the second resonant frequency. The percent change in damping ratio is approximately equal to the ordinate value times 100, as the initial damping ratios were on the order of 10^{-2}; e.g., specimens F and G exhibited a maximum change of approximately 40%. On the basis of the large change in damping ratio (or overall loss tangent) with damage, and its ease of measurement, the authors suggested that it could be used to nondestructively monitor the integrity of fibrous composite structures. The influence of damage on damping may partly or entirely account for the increase in damping ratio with strain reported by Schultz and Tsai (1968), and for the high damping ratios (compared to beam theory) observed by Schultz and Tsai (1969); however, the maximum strains in the (1969) study were $\frac{1}{2}$ to $\frac{1}{5}$ of those in the fatigue tests and, therefore, it is unlikely that damage accounts completely for the discrepancy between simple beam theory and experiment. If a significant amount of dissipation is truly due to nonlinear mechanisms, such as crack growth and/or interfacial rubbing, it probably will be revealed through superposition techniques; e.g., analyze response to transient inputs by using superposition of harmonic responses.

In a somewhat related investigation, Dudek (1970) reported on experimental and analytical results concerning limitations of the composite (two-layer) viscoelastic beam test, and identified sample geometry requirements which must be satisfied to minimize the effects of theory limitations and experimental errors on material property determinations. Nielsen and Lee (1972) suggested that a discrepancy between theoretical and experimental values of loss tangent for particulate-filled polystyrene composites was due to friction between particles in agglomerates, friction between matrix and particles, and friction due to rubbing between crack surfaces within the polymer; the influence of a polymer-rich skin on the complex modulus determined from the torsional or flexural tests was also studied, and a simple method for correcting modulus data is given.

Finally, Biot (1972) has developed new procedures for the dynamic analysis of multilayered, orthotropic, viscoelastic plates. They take into account high stress gradients which may exist near the surface of anisotropic materials (the "skin effect") *and* microstructural effects through the use of stress couples. Also analyzed are the damping effectiveness of each constituent and the relative importance of shearing compared to longitudinal damping on overall plate response under the assumption of small loss tangents. Further, it is shown that solutions for the dynamic response of

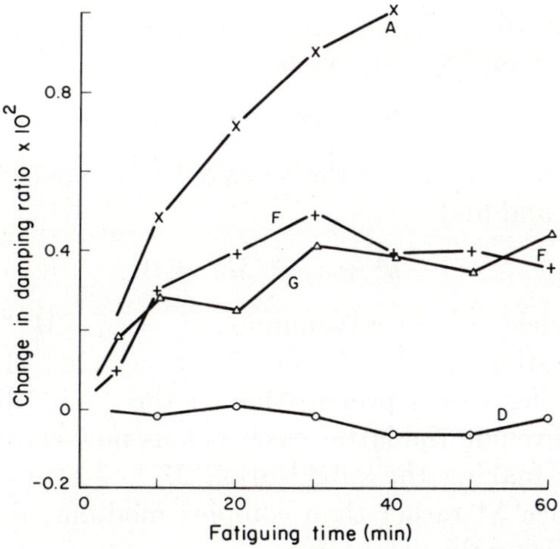

FIG. 15. Change in damping ratio with fatiguing time for 90° $\pi/4$ beams. Fatigue excitation level, g's: A = 50, D = 20, F and G = 40 (after Schultz and Warwick, 1971).

multilayered plates with rectangular, triangular, and circular plan forms can be deduced directly from a plane strain analysis.

2. Steady-State Vibration of Unbounded Media

We consider in this subsection the steady-state response of viscoelastic media in which the influence of external boundaries is neglected. Such a simplification is especially useful in the study of composite materials, as it enables the primary effects of inhomogeneity to be studied with a minimum of mathematics. The correspondence principle still can be used to convert elasticity solutions, with or without microstructure, to viscoelasticity. Moreover, as in the previous subsection, considerable simplification results if the constituents have small loss tangents. Apart from a brief review of the composites literature, our main concern here is with the implications of the small damping assumption.

In order to clarify the analysis procedure, let us first consider one-dimensional planar motion of an infinite, *homogeneous* body. Assume that the displacement in the x direction is $u = u(x, t)$, corresponding to extensional and compressional deformations. The governing equation for an elastic material is the wave equation (Fung, 1965):

$$\frac{\partial^2 u}{\partial x^2} - \frac{\rho}{M}\frac{\partial^2 u}{\partial t^2} = 0 \tag{173}$$

in which M is an elastic modulus ($= \lambda_e + 2\mu_e$, where λ_e and μ_e are the Lamé elastic constants).

For steady-state vibrations of frequency ω, the solution to this equation may be written in complex notation as

$$u = Ae^{-ikx}e^{i\omega t} = Ae^{i(\omega t - kx)} \tag{174}$$

where A is a constant and k is the so-called *wavenumber*. Substitute Eq. (174) into (173), and find

$$-k^2 + (\rho\omega^2/M) = 0 \tag{175}$$

Equation (175) yields a *real* wavenumber, $k = \omega(\rho/M)^{1/2}$, corresponding to waves propagating in the $(+x)$ direction, plus the solution $k = -\omega(\rho/M)^{1/2}$ defining waves propagating in the $(-x)$ direction; for our purposes we can exclude the latter case. Let us now convert the result to viscoelasticity by making the substitution $M^{-1} \rightarrow S^*$ $(\equiv S' - iS'')$; the complex compliance S^* rather than complex modulus, is introduced only for convenience. Using the polar form

$$S^* = |S| e^{-i\phi} \tag{176a}$$

where

$$|S| = S'(1 + \tan^2 \phi)^{1/2} \tag{176b}$$

we find

$$k = \omega(\rho\,|S|)^{1/2} e^{-i\phi/2} \tag{177}$$

Divide k into a real part k' and an imaginary part k'' according to the equation

$$k = k' - ik'' \tag{178a}$$

Then,

$$k' = \omega(\rho\,|S|)^{1/2} \cos(\phi/2) \tag{178b}$$

$$k'' = \omega(\rho\,|S|)^{1/2} \sin(\phi/2) \tag{178c}$$

Substitute Eq. (178a) into the displacement Eq. (174),

$$u = Ae^{-k''x} e^{i(\omega t - k'x)} \tag{179}$$

In as much as k'' and k' are both positive, this equation defines an exponentially damped *wavetrain* traveling in the $(+x)$ direction, having wavelength $2\pi(k')^{-1}$ and phase velocity $c' = \omega/k'$; from Eq. (178b),

$$c' = [(\rho\,|S|)^{1/2} \cos(\phi/2)]^{-1} \tag{180}$$

The constant A can be interpreted by supposing the body is semiinfinite. Assume the plane $x = 0$ is the surface of the body; then Eq. (174) yields

$$u(0, t) = Ae^{i\omega t} \tag{181}$$

Hence, A is the amplitude of the displacement imposed on the surface; if in real notation $u(0, t) = A \cos \omega t$, then the physical solution is the real part of Eq. (179),

$$u = Ae^{-k''x} \cos(\omega t - k'x) \tag{182}$$

Phase velocity, Eq. (180), depends on frequency through $|S|$ and ϕ; equivalently, we can say that phase velocity is a function of wavenumber (or wavelength) since k' and frequency are directly related according to Eq. (178b). Waves for which this dependency exists are said to be *dispersive*. In contrast, waves in the homogeneous elastic medium are nondispersive since Eq. (180) reduces to c' = constant.

With composite media the equation governing k is far more involved than Eq. (175), and an explicit, analytical solution cannot normally be obtained. However, the assumption of small damping greatly simplifies the analysis of composites. In order to illustrate the analysis method for this situation, we shall rederive the above solution for k by linearizing the equation with respect to tan ϕ. For the viscoelasticity problem, first write Eq. (175) in the form

$$f = f(k, S^*) = 0 \tag{183}$$

where, for simplicity, we have not shown the variables ρ and ω. Now, expand f in a multiple Taylor series about the point $k = k'$ and $S^* = S'$, and drop second- and higher-order terms in k'' and tan ϕ; there results

$$f_0 \equiv f(k', S') = 0 \tag{184a}$$

from the real part of $f = 0$, and

$$-(\partial f_0/\partial k)\, k'' - (\partial f_0/\partial S)\, S' \tan \phi = 0 \tag{184b}$$

from the imaginary part of $f = 0$, where f_0 is the function for an elastic material with modulus $M = (S')^{-1}$; also $\partial f_0/\partial k$ and $\partial f_0/\partial S$ are derivatives of the elasticity function $f(k, S)$ evaluated at $k = k'$ and $S = S'$. Equation (184b) yields

$$k'' = -\left(\frac{\partial f_0/\partial S}{\partial f_0/\partial k}\right) S' \tan \phi \tag{185}$$

Thus, with ϕ sufficiently small,

$$k' = \omega(\rho S')^{1/2} \tag{186a}$$

$$k'' = \omega(\rho S')^{1/2}(\phi/2) \tag{186b}$$

These are the same expressions that are found from Eqs. (178) by using the small damping approximations $\cos \phi \approx 1$, $\sin \phi \approx \phi$, and $|S| \approx S'$.

Let us now indicate how the above approximate analysis can be extended to existing solutions for elastic composite media. For the sake of argument we shall refer here to the "effective stiffness" theory of Herrmann and Achenbach (1970), in which an approximate energy method was used to derive differential equations which account for structural details in layered and fibrous composites. Consider, in particular, a unidirectional, two-phase composite in which transverse waves are propagating in the direction of the fibers (the x direction, say). Solutions of the type

$$A e^{ik(x-ct)} \tag{187a}$$

were assumed, and a so-called "dispersion relation," which replaces Eq. (175) for homogeneous media, was derived; see Eq. (63) in the paper by Herrmann and Achenbach (1970). This relation is of the form

$$g(k, c, E_{\mathrm{f}}, \mu_{\mathrm{f}}, \mu_{\mathrm{m}}) = 0 \tag{187b}$$

where only the variables which are needed for the study at hand have been explicitly shown; also, E_f and μ_f are the uniaxial and shear moduli, respectively, of the fibers, and μ_m is the matrix shear modulus. Even without introducing viscoelasticity, Eq. (187b) defines dispersive waves in that it yields a phase velocity c that depends on wavenumber k. Conversion to viscoelasticity is accomplished, of course, by replacing the moduli with corresponding complex moduli or compliances. One additional substitution, in which c is replaced by ω/k, simplifies the task of linearizing the dispersion relation with respect to damping parameters; this simplification results from the fact that ω is real and therefore does not explicitly enter the linearization process. Thus, if all moduli are viscoelastic, the dispersion relation becomes

$$f(k, \omega, E_f^*, \mu_f^*, \mu_m^*) = 0 \tag{188}$$

As before, we linearize Eq. (188) about the real values k', E_f', μ_f', μ_m', in terms of the complex variables $k = k' - ik''$, $E_f^* = E_f' + iE_f''$, etc. The result of this process is the set of two equations

$$f(k', \omega, E_f', \mu_f', \mu_m') = 0 \tag{189a}$$

$$-\frac{\partial f_0}{\partial k} k'' + \frac{\partial f_0}{\partial E_f} E_f'' + \frac{\partial f_0}{\partial \mu_f} \mu_f'' + \frac{\partial f_0}{\partial \mu_m} \mu_m'' = 0 \tag{189b}$$

The first equation, which defines k' (in terms of moduli and ω), is identical to the dispersion relation for an *elastic* material except elastic moduli have been replaced by the real part of complex moduli. The second equation can be solved explicitly for k'' in terms of partial derivatives of the elastic function f; by definition, the partial derivatives $\partial f_0/\partial k$, etc., are equal to the derivatives of f, except k', E_f', etc. appear in place of the elasticity variables.

When the actual expression for f is substituted into Eq. (189b), it turns out that k'' is predicted to be positive, just as in the previous example. However, when this positive k'' is substituted into response Eq. (187a), an exponentially diverging (rather than decaying) wave results. {Had the argument of the assumed response function been selected with the opposite sign, as it was in the first example [see Eq. (174)], the wave would have been correctly predicted to decay; on the other hand, this problem does not arise with elastic media as the nature of the solution is the same for Eqs. (174) and (187a).} All one has to do here is change the sign of k'', without making any other changes in the analysis, to obtain the correct result.

The above method of analysis is not, of course, limited to the Herrmann and Achenbach dynamic theory. Moreover, it is easily generalized to include additional constituent properties. Dimensional analysis can be employed to simplify the form of f, even when a specific representation is not used, by expressing it in terms of independent dimensionless parameters, just as was done in the preceding subsection.

Some viscoelastic solutions for time harmonic wavetrains in composites already exist in the literature; however, to the author's knowledge, the type of linearization process described above has not been used to simplify the analyses. Bedford and Stern (1971) and Stern *et al.* (1971) have developed and applied approximate theories for laminated and fibrous composites. Shaw and Bugl (1969) studied two-dimensional propagation of harmonic plane waves through a layered viscoelastic medium; the viscoelastic response was obtained through the use of an exact elasticity solution and the correspondence principle in terms of complex moduli. Finally, we note that the Herrmann and Achenbach (e.g., 1970) approximate elasticity theory has been extended by Grot and Achenbach (1970) to viscoelastic laminated composites with anisotropic layers and with small temperature changes, in which the effect of temperature variation on viscoelastic properties is neglected.

3. *Transient Dynamic Response*

Included in this problem area is the analysis of damped free vibrations of specimens which are used in the experimental determination of effective complex moduli or compliances, and the analysis of waves in composite media subjected to transient pulses.

The first catagory has been carefully investigated theoretically by Struik (1966). He derived easily applied, approximate equations for calculating the complex modulus from free vibration data on essentially arbitrary, linearly viscoelastic specimens; additionally, a method for estimating the error was given. One of the important findings concerns the accuracy of the equations commonly used for calculating complex moduli; these equations are based on the use of an elementary differential equation of motion for *free* vibrations which is identical to that for an elastic material except complex modulus appears in place of the elastic constant. Although this equation is not exact for most cases, Struik found that for viscoelastic materials with small loss tangents, such as amorphous polymers below T_g, this elementary theory can be utilized to calculate complex moduli which agree closely with the actual properties.

Many examples can be found in the literature on the use of free vibrations and the elementary theory for determining complex properties of monolithic and composite materials. For example, Schrager and Carey (1970) employed a torsion pendulum in a study of the temperature dependence of boron fiber–epoxy composites; Sierakowski *et al.* (1971) used free and forced vibrations of cantilever beams in order to determine the influence of particle size and percent filler on aluminum particle-filled epoxy composites.

Analysis of transient wave propagation in viscoelastic, composite media is, at this time, rather limited. Tauchert (1971) used Fourier synthesis (in which harmonic wavetrains are superposed) to predict velocity and attenuation of stress pulses propagating in rods of woven fabric–phenolic composites; theoretical predictions, based on the use of effective complex moduli obtained from forced vibration tests, were found to agree favorably with experimental results.

The simplicity and accuracy of the Fourier superposition technique has been noted by others for wave propagation in monolithic polymeric materials. For example, Knauss (1968) analyzed transient wave propagation in an amorphous polymer in the strongly viscoelastic glass-to-rubber transition region; Mao and Rader (1970) used this method to predict stress pulses in a bar of polymethyl methacrylate (which had a small, constant loss tangent) with varying cross-sectional area. However, the author is not aware of anyone having used Fourier synthesis to study the influence of microstructure on transient wave propagation in composites; the approximate wavetrain solutions discussed in the preceding subsection could be used directly in the analysis. Broad spectrum representation of viscoelastic constituent properties appears to be easily included in this method. Although numerical problems may be encountered with pulses having high timewise curvature, viscoelastic damping normally precludes this behavior, except possibly near the point of load application. Use of the *fast Fourier transform method* (Robinson, 1967) should greatly facilitate the analysis.

Finally, we refer to a recent, extensive study by Hegemier (1972) on steady-state and transient wave propagation in laminates and fibrous composites. Emphasis was placed on the analysis of dispersion phenomena in elastic composites. However, one solution was given for transient wave propagation normal to the layers in a viscoelastic laminate; it was derived by means of the correspondence principle and evaluation of the Laplace inversion integral. Figure 16 shows the stress time history at the midplane of the 13th layer from the point of load application; the input is a unit step function and, therefore, the wave would have been a unit step function if the body had been elastic and homogeneous. The quartz was assumed elastic, and the viscoelastic phenolic resin was assumed to be represented by the relaxation modulus $E(t) = E_\infty + E_1 \exp(-t/\rho_1)$ corresponding to a three-element spring–dashpot model. As should be expected, the viscoelasticity is seen to reduce the curvature present in the elastic solution.

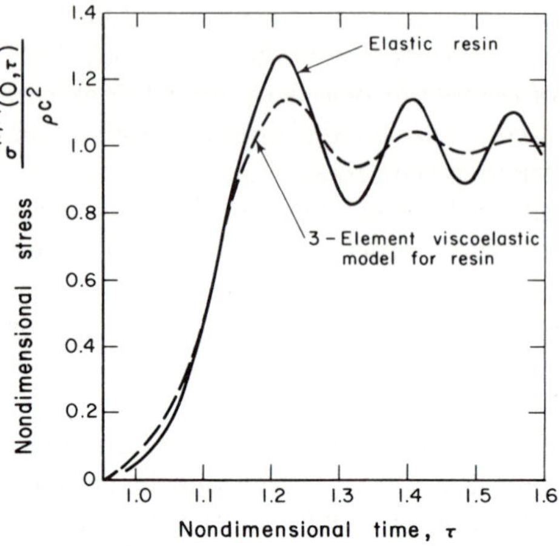

FIG. 16. Transient pulse traveling normal to laminates in a quartz–phenolic composite (after Hegemier, 1972).

VI. Nonlinear Behavior

A material or structure is *nonlinear* whenever one or both of the conditions of linearity [viz., proportionality, Eq. (2), and superposition, Eq. (3)] is not met. In this section we shall review the general nature and sources of nonlinear, viscoelastic behavior of polymeric composites, and then comment on techniques which have been proposed to characterize analytically mechanical nonlinearities. Some final comments pertain to nonlinear structural analysis.

A. Sources and Manifestations of Nonlinearity

Although the primary concern in all of the previous sections was linear behavior and analysis, nonlinear response of fibrous composites under cyclic loading was acknowledged in Section V,B,1. In this case, damping ratios were observed to be high compared to linear theory, and to change with time if the strains were high enough. The existence and growth of cracks was indicated as the probable source of much or all of this nonlinear behavior.

The general phenomenon of what we shall call *microstructural damage*, with crack formation and growth being a special case, is often the most important source of nonlinear behavior for particulate and fibrous composites. Moreover, as a result of high stress concentrations which exist near to and at particles or fibers in many composites, this damage and consequent nonlinearity may be quite significant at relatively small overall strains (or stresses) compared to ultimate values.

Probably the first well-documented nonlinearity due to microstructural damage is the so-called "Mullins effect." This effect was first studied in detail by Holt (1932). He showed that if a vulcanized rubber which contains carbon black is stretched and then released, it will follow a lower stress–strain curve when it is again stretched to the same elongation. He also found that additional prestretches to the first elongation softened the rubber further, but to a lesser degree than was observed on the first stretch; furthermore, the rubber was observed to regain a portion of its stiffness when allowed to rest in an unloaded state. The Mullins effect has since been observed in many other filled-rubber systems. Figure 17 illustrates the Mullins effect for very low rate (equilibrium) straining; it is to be noted that the upper most, nearly linear, stress–strain curve defines the behavior for a single stretch to the highest strain level shown. For a review of the early investigations in this area and a study of the underlying molecular mechanisms, see the work of Bueche (1960); generally speaking, Bueche attributed the Mullins effect to internal polymer chain failure and to separation of chains from filler surface. The "time-dependent Mullins effect" has been described and modeled in a recent paper by Farris (1971b); we shall return to this theory in Section VI,B.

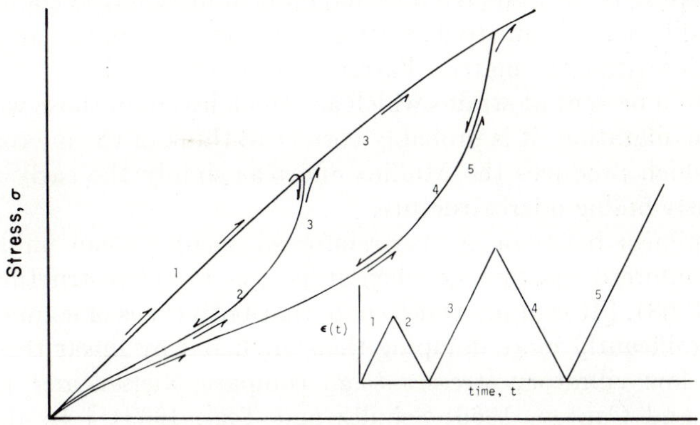

FIG. 17. Typical equilibrium behavior of a highly filled elastomer when the direction of straining is reversed (after Farris, 1971b).

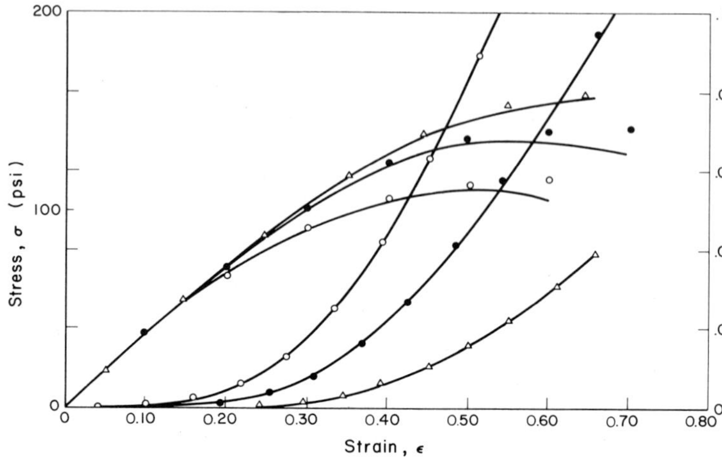

FIG. 18. Variation of stress and dilatation with strain for a highly filled elastomer (polyurethane) at a series of pressures: ○, 15 psi; ●, 50 psi; △, 165 psi; —, theoretical. The test temperature was 77°F; the strain rate was 0.66 min^{-1}; and the filler content was 63.5 vol % (after Farris, 1968b).

Dilatation is another manifestation of microstructural damage, especially for composites consisting of a rubbery matrix and a large volume fraction of hard filler particles (e.g., Farris, 1968a, b; Oberth, 1967). Typical behavior for a solid propellant under uniaxial stress σ plus pressure is shown in Fig. 18; dilatation, stress, and strain refer to increases from a state of hydrostatic pressure. Except for a very small amount of dilatation equal to $\sigma/3K$, where K is the effective bulk modulus in undamaged composite, the indicated dilatation is due to dewetting (separation of matrix and particle) and vacuoles within the matrix. Farris (1971b) observed that the Mullins effect is often present at strains which are much less than those which cause observable dilatation; it is probably correct to think of the microstructural damage which produces the Mullins effect as simply the early state of a continuously failing microstructure.

The nonlinear behavior of fiber-reinforced plastics is not unlike that of particle-reinforced elastomers when it is due to microstructural failure (Halpin, 1968). [It is of interest to note that both types of composites may exhibit significantly more damping than predicted by linear theory under relatively low vibratory stresses (e.g., compare Nielsen and Lee, 1972; Schapery and Cantey, 1966; Schultz and Tsai, 1969).] In the case of plastics much of the damage probably is in the form of relatively well-defined cracks. However, there is considerably less quantitative information currently available on relations between microstructural mechanisms and overall time-dependent behavior of fiber-reinforced plastics than of particulate composites.

In addition to the sources of nonlinearity discussed above, there are other ones which fall under the general heading of *reversible nonlinearities.* By this term, we mean behavior which can be reproduced on a given specimen after it is allowed to rest for an extended period of time; the effects of previous inputs are therefore assumed to fade away. This type of nonlinear behavior in reinforced plastics is due, in large part, to a highly stress-dependent viscosity; it shows up in a_T, which decreases rapidly at high stresses (Lou and Schapery, 1971; Schapery, 1969b). On the other hand, reversible nonlinearity in many elastomeric composites is a direct result of high strain levels which the polymer is capable of withstanding before failing.

Still another reversible nonlinearity may exist under fatigue loading conditions, assuming chemical changes do not occur. Temperature rise due to the conversion of mechanical energy into thermal energy can be very significant, especially in large specimens or structures; this rise may soften the material to the point that the structural response is altered, which, in turn, affects the temperature change (e.g., Schapery and Cantey, 1966).

B. Analytical Characterization Methods

The development of nonlinear viscoelastic constitutive equations for polymers and their experimental verification have received intensive study in the past five years. However, most of the theories are limited to reversible nonlinearities and have been applied almost entirely to monolithic materials; the term "fading memory" is commonly used instead of the analogous "reversible nonlinearities" when referring to these theories. Included in this catagory are most of the single- and multiple-integral representations; for recent reviews of these theories see, for example, the work of Christensen (1971), Oden (1972), Schapery (1969b), and Soussou and Moavenzadeh (1970); for studies on the application of these theories to amorphous and semicrystalline polymers see the above references and, for example, Morgan and Ward (1971), Pipkin and Rogers (1968), Stafford (1969), and Ward (1971).

Lou and Schapery (1971) used a single-integral, fading memory representation to characterize glass fiber-reinforced epoxy. However, it was necessary to employ mechanically conditioned specimens in order to achieve good agreement between theory and experiment. The conditioning consisted of loading and unloading each specimen ten times, using the

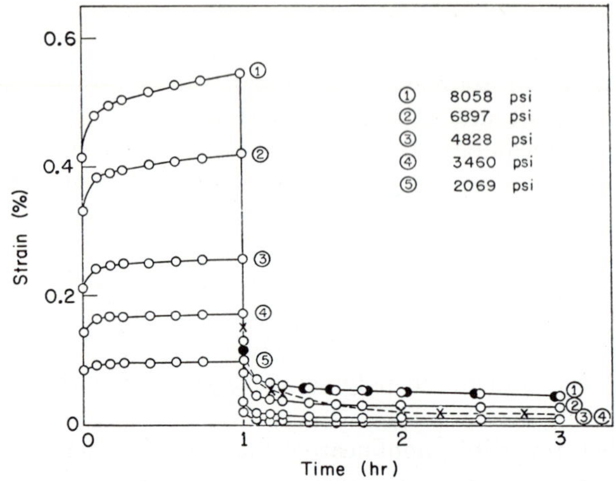

FIG. 19. Creep–recovery curves for a unidirectional, glass fiber-reinforced epoxy composite with $\theta = 30°$, temperature $164 \pm 2°F$, and relative humidity $21 \pm 1\%$ (after Lou and Schapery, 1971). Predictions for case (1) are shown for both modified superposition theory (×) and nonlinear theory (●).

highest stress level of interest; this cycling produced material having a nearly constant amount of damage when tested a short time later. However, rehealing apparently occurred when the specimens were allowed to rest without load for several days, as reflected by a change in mechanical response. Typical creep and recovery curves for a mechanically conditioned specimen are shown in Fig. 19. The prediction shown by the broken line was made using superposition Eq. (3); viz., the strain during recovery was assumed to be equal to the sum of the strain due to a stress applied at $t = 0$ and the strain due to an equal, but opposite, stress applied at $t = 1$ hr. The fact that strain measured during recovery was greater than predicted by superposition is quite typical of reinforced and unreinforced glassy polymers in their nonlinear range of behavior. In fact, it was found that this violation of superposition occurred at lower stresses than those which caused the creep compliance to change, and was greatest in unconditioned specimens; it should be noted that the variation of compliance with stress represents a violation of Eqs. (2) and (3).

The problem of characterizing composites with time-dependent microstructural damage would seem to be much more difficult than with reversible nonlinearities. Nevertheless, a realistic theory does exist. It was proposed by Fitzgerald and Farris (1970) for filled elastomers, and then successfully applied by Farris under isothermal (1971b) and nonisothermal (1971a) conditions; the main restriction is the strains must be small enough that dilatation due to dewetting and other internal vacuoles is negligible (e.g., $\epsilon < 10\%$ in Fig. 18). An important parameter in this theory is the so-called pth order Lebesgue norm of strain $\|\epsilon\|_p$:

$$\|\epsilon\|_p \equiv \left[\int_0^t |\epsilon(t')|^p\, dt'\right]^{1/p} \tag{190}$$

where $|\epsilon|$ is the absolute value of strain. Its introduction was based on a linear cumulative damage model for the failing microstructure (Farris, 1971b). A typical term in the theory involving Lebesgue norms is

$$\sigma = A[\|\epsilon\|_q/\|\epsilon\|_p]^m \epsilon \tag{191}$$

where A, m, p, and q are positive constants. Even this simple expression predicts a time-dependent Mullins effect when p and q are sufficiently large.

An interesting feature of this equation is that it obeys the proportionality condition, Eq. (2) (which the reader can readily verify by substituting $\epsilon' = c\epsilon$, with c constant), but *not* superposition; recall that the fiber-reinforced plastic just discussed behaved similarly in terms of stress, rather than strain. Responses which satisfy Eq. (2), but not necessarily Eq. (3), are said to be *homogeneous to degree one*. Farris (1971b) has shown that the stress predicted by Eq. (191) may be considerably different from linear theory, especially if the strain rate is *not* constant. This point is illustrated in Fig. 20 for a sand–asphalt concrete; similar graphs can be found in Farris' (1971b) paper for this type of strain history as well as more involved ones. The nonlinear theory used in predicting the result in Fig. 20 contained terms similar to Eq. (191); it should be added that the *same* relaxation modulus was used with both the linear and nonlinear theories.

A multiaxial theory for particulate composites was also proposed by Farris (1971b), and it has received a limited amount of experimental verification (Farris, 1971c). In addition, Schapery (1971, 1973) has used nonequilibrium thermodynamics and fracture mechanics to obtain an alternative multiaxial representation which includes both reversible non-linearities and microstructural damage. It should be pointed out, however, that the latter theory with both types of nonlinearity, with or without the presence of vacuole dilatation, is largely unverified, and has not yet been reduced to practice. Furthermore, the validity of an analogous theory (Schapery *et al.*, 1973) for fiber-reinforced plastics is not known; well-designed uniaxial and multiaxial test programs are needed at this time in order to assess existing theory.

The Farris theory for particulate composites has been incorporated into a two-dimensional finite element analysis (Farris, 1971c). However,

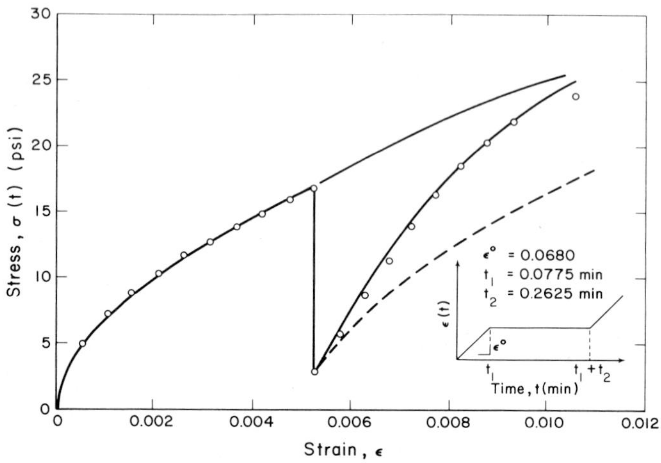

FIG. 20. Comparison of calculated and observed behavior of a sand–asphalt concrete subjected to an interrupted ramp strain input (after Vakili and Fitzgerald, 1972). The open circles represent calculated points for nonlinear viscoelasticity; the dashed curve shows the predicted curve from linear theory; and the solid curve is the experimental values.

analyses which account for nonlinear viscoelastic behavior of fibrous composites apparently do not exist. Some numerical solutions for these latter composites have been obtained by Adams (1970) by using the classical (metal-like) equations of rate-independent plasticity to describe nonlinear matrix behavior; some justification for this representation was provided through comparison with a very limited amount of experimental data on boron fiber-reinforced epoxy and aluminum composites under monotonically increasing transverse loading. It is believed that generalization of numerical techniques to include viscoelastic behavior under more complex loading conditions will be facilitated by recognizing the similarity between the constitutive equations for metals and polymers (Schapery 1969b, 1970); for example, the classical equations for metals with rate-dependent and rate-independent behavior are contained in the equations for thermorheologically simple materials if a_T is assumed to depend on octahedral stress, strain, and strain rate, as well as temperature.

Acknowledgment

This work was sponsored, in part, by the Air Force Materials Laboratory, Air Force Systems Command, under Contract number F33615-70-C-1260.

References

Adams, D. F. (1970). *J. Compos. Mater.* **4,** 310.

Anonymous (1955). Investigations of Creep Characteristics of Concrete for Canyon Ferry Dam Missouri River Basin Project. U.S. Bur. of Reclamation, Concrete Lab. Rep. C-789, Denver, Colorado.

Arenz, R. J. (1964). *J. Appl. Mech.* **31,** 17.

Arenz, R. J. (1965). *J. Appl. Mech.* **32,** 303.

Ashton, J. E., Halpin, J. C., and Petit, P. H. (1969). Primer on Composite Materials: Analysis. Technomic, Westport, Connecticut.

Barlow, A. J., Ergindav, A., and Lamb, J. (1969). *Proc. Roy. Soc. London* **A309,** 473.

Beckwith, S. W., Schapery, R. A., and Webb, L. D. (1970). Literature Survey of the Field of Fiber-Reinforced Plastic Composites: Part I. Rep. CMR-1, Texas A&M Univ.

Beckwith, S. W., Schapery, R. A., and Webb, L. D. (1971). Literature Survey of the Field of Fiber-Reinforced Plastic Composites: Part II. Rep. CMR-2, Texas A&M Univ.

Bedford, A., and Stern, M. (1971). *J. Appl. Mech.* **38,** 8.

Biot, M. A. (1954). *J. Appl. Phys.* **25,** 1385.

Biot, M. A. (1958). *Proc. U.S. Nat. Congr. Appl. Mech. ASME, 3rd* 1.

Biot, M. A. (1965). "Mechanics of Incremental Deformations." Wiley, New York.

Biot, M. A. (1972). *Int. J. Solids Struct.* **8,** 491.

Bucknall, C. B., and Hall, M. M. (1971). *J. Mater. Sci.* **6,** 95.

Bueche, F. (1960). *J. Appl. Polym. Sci.* **4,** 107.

Christensen, R. M. (1969). *J. Mech. Phys. Solids* **17,** 23.

Christensen, R. M. (1971). "Theory of Viscoelasticity." Academic Press, New York.

Cost, T. L. (1964). *AIAA J.* **2,** 2157.

Cost, T. L., and Becker, E. B. (1970). *Int. J. Num. Methods Eng.* **2,** 207.

Darlington, M. W., and Saunders, D. W. (1970). *J. Phys. D Appl. Phys.* **3,** 535.

Darlington, M. W., and Saunders, D. W. (1971). *J. Macromol. Sci.-Phys.* **B5(2)**, 207.

DeGroot, S. R., and Mazur, P. (1962). "Non-Equilibrium Thermodynamics." North-Holland Publ., Amsterdam.

Dudek, T. J. (1970). *J. Compos. Mater.* **4,** 74.

Farris, R. J. (1968a). *Trans. Soc. Rheol.* **12:2,** 303.

Farris, R. J. (1968b). *Trans. Soc. Rheol.* **12:2,** 315.

Farris, R. J. (1971a). Applications of Non-Linear Viscoelasticity and Cumulative Damage (A Realistic Evaluation of Real Propellant Behavior). Final Rep., NOSC Contract No. N00017-70-C-4441, Aerojet Solid Propulsion Co.

Farris, R. J. (1971b). *In* "Polymer Networks: Structural and Mechanical Properties" (A. J. Chompff and S. Newman, eds.), pp. 341–394. Plenum Press, New York.

Farris, R. J. (1971c). Unpublished research.

Ferry, J. D. (1970). "Viscoelastic Properties of Polymers," 2nd ed. Wiley, New York.

Fesko, D. G., and Tscheogl, N. W. (1971). *J. Polym. Sci. Part C* **35,** 51.

Fitzgerald, J. E., and Farris, R. J. (1970). Deficiencies of Viscoelastic Theories as Applied to Solid Propellants. Rep. UTEC TH 70-204, Univ. of Utah.

Flügge, W. (1967). "Viscoelasticity." Ginn (Blaisdell), Boston, Massachusetts.

Fried, N. (1970). *In* "Mechanics of Composite Materials" (F. W. Wendt, H. Liebowitz, and N. Perrone, eds.), pp. 813–837. Pergamon, Oxford.

Fung, Y. C. (1965). "Foundations of Solid Mechanics." Prentice-Hall, Englewood Cliffs, New Jersey.

Gradowczyk, M. H., and Moavenzadeh, F. (1969). *Trans. Soc. Rheol.* **13:2,** 173.
Grot, R. A., and Achenbach, J. D. (1970). *Acta Mech.* **9,** 245.
Hackett, R. M. (1968). *AIAA J.* **6,** 2442.
Hackett, R. M. (1971). *Polym. Eng. Sci.* **11,** 220.
Halpin, J. C. (1968). *In* "Composite Materials Workshop" (S. W. Tsai, J. C. Halpin, and N. J. Pagano, eds.) pp. 87–152. Technomic, Westport, Connecticut.
Halpin, J. C. (1969). Characterization of Orthotropic (Fiber-Reinforced) Polymeric Solids. Doctoral Dissertation, Univ. of Akron, Ohio.
Halpin, J. C., and Meinecke, E. A. (1969). Delayed Instability (Buckling) of Viscoelastic Materials. Air Force Mat. Lab. T.R. AFML-TR-68-331.
Halpin, J. C., and Pagano, N. J. (1968). *J. Compos. Mater.* **II,** 68.
Hashin, Z. (1962). *J. Appl. Mech.* **29,** 143.
Hashin, Z. (1965). *J. Appl. Mech.* **32,** 630.
Hashin, Z. (1966). *AIAA J.* **4,** 1411.
Hashin, Z. (1970a). *Int. J. Solids Struct.* **6,** 539.
Hashin, Z. (1970b). *Int. J. Solids Struct.* **6,** 797.
Hashin, Z. (1970c). Theory of Fiber Reinforced Materials. Univ. Pennsylvania, Towne School of Civil and Mech. Eng. NASA Contract NASI-8818, Langley Res. Center.
Hashin, Z. (1972a). Damping Characteristics of Fiber Composites. MED Rep. 36. Technion–Israel Inst. of Tech.
Hashin, Z. (1972b). Vibrations of Viscoelastic Bodies with Small Loss Tangents: Correspondence Principle and Application to Composites. MED Rep. 37. Technion–Israel Inst. of Tech.
Hayes, M. A., and Morland, L. W. (1969). *Trans. Soc. Rheol.* **13:2,** 231.
Hegemier, G. A. (1972). *In* "Dynamics of Composite Materials" (E. H. Lee, ed.), pp. 70–121. ASME, New York.
Herrmann, G., and Achenbach, J. D. (1970). *In* "Mechanics of Composite Materials" (F. W. Wendt, H. Liebowitz, and N. Perrone, eds.), pp. 337–360. Pergamon, Oxford.
Holt, W. L. (1932). *Rubber Chem. Technol.* **5,** 79.
Huang, N. C., and Lee, E. H. (1969). *J. Appl. Mech.* **36,** 573.
Huseby, T. W., and Matsuoka, S. (1967). *Mater. Sci. Eng.* **1,** 321.
Johnson, T. A., Fowlkes, C. W., and Dill, E. H. (1970). *Proc. Int. Congr. Rheol., 4th* **3,** 349.
Kaye, A., and Saunders, D. W. (1964). *Brit. J. Appl. Phys.* **15,** 1103.
Kempner, J. (1962). *In* "Handbook of Engineering Mechanics" (W. Flügge, ed.), Chapter 54. McGraw-Hill, New York.
Knauss, W. G. (1968). *J. Appl. Mech.* **35,** 449.
Krokosky, E. (1967). *In* "Modern Composite Materials" (L. J. Broutman and R. H. Krock, eds.), pp. 120–145. Addison-Wesley, Reading, Massachusetts.
Leeming, H. (1972). Solid Propellant Structural Test Vehicle Program. Final Rep. AFRPL-TR-72-29, Lockheed Propulsion Co.
Lou, Y. C., and Schapery, R. A. (1971). *J. Compos. Mater.* **5,** 208.
Lubliner, J., and Sackman, J. L. (1966). *J. Mech. Phys. Solids* **14,** 25.
Mao, M., and Rader, D. (1970). *Int. J. Solids Struct.* **6,** 519.
McCammond, D. (1967). *Trans. J. Plast. Inst.* **35,** 409.
McCrum, N. G., and Pogany, G. A. (1970). *J. Macromol. Sci.-Phys.* **B4(1),** 109.
McHenry, D. (1943). *Proc. ASTM* **40,** 1069.
Moehlenpah, A. E., Ishai, O., and Dibenedetto, A. T. (1969). *J. Appl. Polym. Sci.* **13,** 1231.

Moehlenpah, A. E., Ishai, O., and Dibenedetto, A. T. (1971). *Polym. Eng. Sci.* **11,** 129.
Morgan, C. J., and Ward, I. M. (1971). *J. Mech. Phys. Solids* **19,** 165.
Morland, L. W., and Lee, E. H. (1960). *Trans. Soc. Rheol.* IV, 233.
Nagamatsu, K., Takemura, T., Yoshitomi, T., and Takemoto, T. (1958). *J. Polym. Sci.* **33,** 515.
Nielsen, L. E., and Lee, B. L. (1972). *J. Compos. Mater.* **6,** 136.
Oberth, A. E. (1967). *Rubber Chem. Technol.* **40,** 1337.
Oden, J. T. (1972). "Finite Elements of Nonlinear Continua." McGraw-Hill, New York.
Pipkin, A. C. (1972). "Lectures on Viscoelasticity Theory." Springer-Verlag, Berlin and New York.
Pipkin, A. C., and Rogers, T. G. (1968). *J. Mech. Phys. Solids* **16,** 59.
Robinson, E. A. (1967). "Multichannel Time Series Analysis with Digital Computer Programs." Holden-Day, San Francisco, California.
Rogers, T. G., and Pipkin, A. C. (1963). *ZAMP* **14,** 334.
Roscoe, R. (1969). *J. Mech. Phys. Solids* **17,** 17.
Salchev, L. Z., and Williams, J. G. (1969). *Plast. Polym.* **37,** 159.
Schapery, R. A. (1961). A Simple Collocation Method for Fitting Viscoelastic Models to Experimental Data. Rep. GALCIT SM 61-23A, Calif. Inst. Tech.
Schapery, R. A. (1962). *Proc. U.S. Nat. Congr. Appl. Mech. ASME, 4th* 1075.
Schapery, R. A. (1964). *J. Appl. Phys.* **35,** 1451.
Schapery, R. A. (1965). *J. Franklin Inst.* **279,** 268.
Schapery, R. A. (1967). *J. Compos. Mater.* **1,** 228.
Schapery, R. A. (1968). *J. Compos. Mater.* **2,** 380.
Schapery, R. A. (1969a). Further Development of a Thermodynamic Constitutive Theory: Stress Formulation. AA&ES Rep. 69-2. Purdue Univ.
Schapery, R. A. (1969b). *J. Polym. Eng. Sci.* **9,** 295.
Schapery, R. A. (1970). *In* "Thermoinelasticity" (B. A. Boley, ed.), pp. 259–285. Springer-Verlag, Berlin and New York.
Schapery, R. A. (1971). Recent Developments in the Nonlinear, Viscoelastic Characterization of Solid Propellant. Lead Article, Solid Rocket Structural Integrity Abstr., Univ. of Utah (CPIA Pub. No. 208).
Schapery, R. A. (1972). Viscoelastic Behavior and Analysis of Composite Materials. Mech. Mat. Res. Center Rep. MM 72-3. Texas A&M Univ.
Schapery, R. A. (1973). Studies on the Nonlinear Viscoelastic Behavior of Solid Propellant. Mech. Mat. Res. Center Rep. MM 2803-73-2. Texas A&M Univ.
Schapery, R. A. (1974). An Approximate Method of Viscoelastic Buckling Analysis. Mech. Mat. Res. Center Rep. 74-5. Texas A&M Univ.
Schapery, R. A., and Cantey, D. E. (1966). *AIAA J.* **4,** 255.
Schapery, R. A., and Martin, R. E. (1972). On the Thermal Diffusivity of Filled Elastomers. Mech. Mat. Res. Center Rep. 72-2. Texas A&M Univ.
Schapery, R. A., Beckwith, S. W., and Conrad, N. (1973). Studies on the Viscoelastic Behavior of Fiber-Reinforced Plastic. Mech. Mat. Res. Center Rep. MM 2702-73-3 (AFML-TR-73-179). Texas A&M Univ.
Schrager, M., and Carey, J. (1970). *Polym. Eng. Sci.* **10,** 369.
Schultz, A. B., and Tsai, S. W. (1968). *J. Compos. Mater.* **2,** 368.
Schultz, A. B., and Tsai, S. W. (1969). *J. Compos. Mater.* **3,** 434.
Schultz, A. B., and Warwick, D. N. (1971). *J. Compos. Mater.* **5,** 394.
Schwarzl, F., and Staverman, A. J. (1952). *J. Appl. Phys.* **23,** 838.
Shaw, R. P., and Bugl, P. (1969). *J. Acoust. Soc. Amer.* **46,** 649.

Sierakowski, R., Hemp, G., and Hokstad, A. (1971). *J. Compos. Mater.* **5,** 417.
Sims, D. F. (1972). Viscoelastic Creep and Relaxation Behavior of Laminated Composite Plates. Doctoral Dissertation, Dept. Mech. Eng., Southern Methodist Univ.
Snowdon, J. C. (1968). "Vibration and Shock in Damped Mechanical Systems." Wiley, New York.
Sokolnikoff, I. S. (1956). "Mathematical Theory of Elasticity." McGraw-Hill, New York.
Soussou, J. E., and Moavenzadeh, F. (1970). Classical and Statistical Theories for the Determination of Constitutive Equations. Dept. of Civil Eng., Rep. R70-33. M.I.T.
Stafford, R. O. (1969). *J. Mech. Phys. Solids* **17,** 339.
Steel, D. J. (1965). *Trans. J. Plast. Inst.* **33,** 161.
Stern, M., Bedford, A., and Yew, C. H. (1971). *J. Appl. Mech.* **38,** 448.
Struik, L. C. E. (1966). *Rheol. Acta* **6,** 119.
Tauchert, T. R. (1971). *J. Compos. Mater.* **5,** 456.
Theocaris, P. S. (1964). *J. Mech. Phys. Solids* **12,** 125.
Theocaris, P. S., and Hadjijoseph, C. (1965). *Kolloid-Z.* **202,** 133.
Tobolsky, A. V., and Catsiff, E. (1956). *J. Polym. Sci.* **19,** 111.
Tsai, S. W. (1970). *In* "Mechanics of Composite Materials" (F. W. Wendt, H. Liebowitz, and N. Perrone, eds.), pp. 749–767. Pergamon, Oxford.
Vakili, J., and Fitzgerald, J. E. (1972). Nonlinear Characterization of Sand Asphalt Concrete by Means of Permanent Memory Norms. Dept. of Civil Eng., Rep. UTEC CE 72-079, Univ. of Utah.
Van der Pol, B., and Bremmer, H. (1955). "Operational Calculus." Cambridge Univ. Press, London and New York.
von Karman, T., and Biot, M. A. (1940). "Mathematical Methods in Engineering." McGraw-Hill, New York.
Wada, Y., Hirose, H., Asano, T., and Fukutomi, S. (1959). *J. Phys. Soc. Japan* **14,** 1066.
Ward, I. M. (1971). "Mechanical Properties of Solid Polymers." Wiley (Interscience), New York.
Watkins, L. A. (1973). Creep of an Epoxy Resin under Transient Temperatures. M.S. Thesis, Civil Eng., Texas A&M Univ.
Williams, M. L. (1964). *AIAA J.* **2,** 785.

5

Elastoplastic Behavior of Composites

DONALD F. ADAMS

Department of Mechanical Engineering
University of Wyoming
Laramie, Wyoming

General working theories governing elastoplastic material behavior are a relatively recent development. Thus, specific applications to composite material micromechanical behavior are as of yet somewhat limited. Nevertheless, it has been recognized that, because of the high stress and strain concentrations which exist in local regions between filaments, the elastic limit of the matrix material may be exceeded long before the gross behavior of the composite exhibits an obvious degree of nonlinearity. This localized elastoplastic behavior has a significant influence on the redistribution of stresses within the composite, and hence the initiation of composite failure. The importance of this behavior in engineering applications is reviewed, and a discussion of possible approaches to the problem is given. The elastoplastic constitutive relations are summarized and the several commonly utilized yield criteria compared. Existing applications of these methods of analysis are categorized as being either simple models or rigorous analyses. They are discussed in terms of their range of applicability, accuracy of representation, and assumptions incorporated. Particular attention is given to the rigorous analyses, this being the direction future investigations are expected to follow.

I. Introduction

A. Definition of Elastoplastic Behavior

Most studies of the micromechanical behavior of composite materials which have been conducted to date have assumed, either explicitly or implicitly, that the individual constituent materials remain linearly elastic. However, many materials, and particularly those which are typically selected for use as matrices in composite systems, do not remain linearly elastic when subjected to an applied loading.

For some materials this nonlinear response may be due, at least in part, to viscoelastic, i.e., time-dependent, effects as discussed in Chapter 4. On the other hand, most materials exhibit a nonlinear, time-independent, deformational response when the applied loading exceeds a limiting value, viz., the yield point of the material. This latter type of nonlinear response, exhibited beyond the elastic range, is termed *plasticity*. Hence the terminology "elastoplastic behavior" is commonly used to denote the full range of loading.

Typical uniaxial tension stress–strain curves for two commonly used matrix materials are shown in Fig. 1. As indicated, both the high polymer (Epon 828/1031 epoxy) and the metal (2024 aluminum alloy) exhibit nonlinear behavior, this being particularly pronounced in the case of the

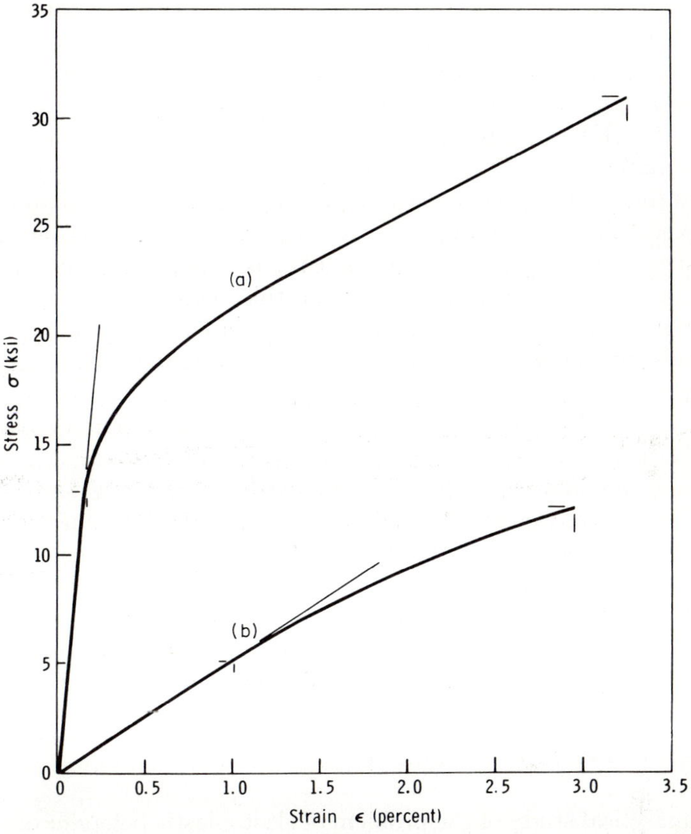

FIG. 1. Matrix tensile stress–strain curves (Adams, 1970b). Curve (a) is for 2024 aluminum alloy with a 2-hr anneal at 900°F; initial modules is 8.1×10^6 lb/in.^{2} and Poisson's ratio is 0.32 (assumed). Curve (b) is for Epon 828/1031 epoxy resion with an initial modules of 0.52×10^6 lb/in.^{2} and a Poisson's ratio of 0.35.

metal. It is obvious that an elastic analysis is applicable only in the initial portion of the stress–strain response.

B. Significance in Composite Applications

The reinforcing filaments in a composite material are typically selected on the basis of their high strength and stiffness characteristics. A primary role of the matrix material is to bind the filaments together into a solid mass, while permitting ease of fabrication into a desired configuration.

Implicit in this function of the matrix is its ability to transfer loads from filament to filament. Typically the fabricability requirement results in the selection of a matrix material having a low stiffness (relative to the filament reinforcement), and correspondingly a low strength. The resulting mismatch of stiffness properties between the filaments and the surrounding matrix material results in the development of high local strain concentrations in the matrix when loads are applied, as discussed in Chapter 3. Thus, even when the average applied stress on the composite is relatively low, local stresses and strains within the composite material may have already exceeded the elastic limit values. For this reason, the consideration of elastoplastic micromechanical behavior is important even for composite materials subjected to applied stresses which are intended to remain below the apparent elastic limit of the composite. For structures designed to an ultimate strength criterion, the need for consideration of elastoplastic material behavior is even more apparent.

To accommodate high local strains without inducing local failures, a matrix is normally selected which has a high strain to failure, as indicated in Fig. 1. This permits local plastic flow to occur in regions of high strain concentration, causing a redistribution of stresses (stress relief). This ability to accommodate large local strains is also important in regions of load path discontinuity, such as around the ends of noncontinuous reinforcements, e.g., chopped fibers, whiskers, and particulates.

C. *Approaches to the Problem*

An analytical study of the problem of elastoplastic behavior of composite materials can be approached from any one of a number of different levels of modeling accuracy. Historically, as is the case for most analyses, the initial analytical investigations of elastoplastic material behavior utilized very simple models. These models were known to be very crude representations of the actual material system. However, they were relatively easy to formulate, while providing at least qualitative indications of theretofore unknown behavior. Subsequent analyses have utilized somewhat more accurate models, having a broader range of applicability and providing numerical results of a greater degree of reliability.

These simple model analyses and the more rigorous analyses will both be discussed, in Sections III and IV, respectively. Before doing so, however, it will be informative to discuss the basic principles of the general theory of plasticity, in order to indicate the characteristics and basic assumptions which distinguish elastoplastic behavior from the types of material behavior discussed in Chapters 3 and 4 of this book.

II. Elastoplastic Constitutive Relations

A. The General Concepts of Plasticity

When a material is loaded beyond its elastic limit, the state of stress at any time is dependent upon the prior loading history. This is unlike elastic behavior, for which the instantaneous state of stress is dependent only upon the current state of strain. Thus, it is necessary to first establish a criterion for yielding, i.e., to define the limits of applicability of elastic (Hookean) behavior, and then to establish a suitable stress–strain relationship for material behavior beyond the elastic limit.

Since, up to the elastic limit, strains can be expressed directly in terms of stresses, it is obvious that a yield criterion can be expressed in terms of stresses only. Likewise, if the material is assumed to be isotropic, the functional relationship between stress components which expresses this criterion must be independent of the coordinate system selected, i.e., the functional relationship will be an invariant of the stress tensor. Thus, the yield criterion is governed by the various laws of elastic behavior.

Once the material has been stressed beyond its elastic limit, behavior during unloading and subsequent reloading is not necessarily governed by the theory of elasticity since the strain field is no longer uniquely determined by the state of stress. However, experiments verify that for most materials it is a reasonable approximation to assume that unloading and subsequent reloading to the previous state of stress does occur elastically, and that hysteresis effects are negligible.

Experiments also verify that a hydrostatic pressure does not produce appreciable plastic deformation. Thus, it can be assumed that the plastic component of strain is a function only of the deviatoric component of the stress tensor, being independent of the dilatational component.

These then are the basic distinguishing characteristics and assumptions of the general theory of plasticity. For a detailed presentation, the interested reader is referred to a text such as that by Hill (1950) or Prager and Hodge (1951), among others.

B. Comparisons of Various Yield Criteria

Various yield criteria have been proposed through the years, but most of them violate one or more of the basic experimental observations mentioned previously. In particular, most predict that the dilatational component of the stress tensor influences yielding and plastic flow. Two theories which do not are those attributed to Tresca and von Mises. Both are in common use, as much because of their relative simplicity as their demonstrated accuracy.

The Tresca yield criterion is the more complicated of the two. It assumes that yielding occurs when the maximum shear stress reaches a given value. The complication of this theory arises from the fact that, for a given problem, it is not known beforehand which of the three principal shear stresses will be the maximum. Thus it becomes necessary to use the more general form of the yield criterion, i.e.,

$$[(s_2 - s_3)^2 - 4k^2][(s_3 - s_1)^2 - 4k^2][(s_1 - s_2)^2 - 4k^2] = 0 \qquad (1)$$

where $s_1 = \sigma_1 - \frac{1}{3}(\sigma_1 + \sigma_2 + \sigma_3)$, etc., represent the principal components of the deviatoric stress tensor, σ_1, σ_2, σ_3 are the principal components of stress, and $2k$ is the yield value of the maximum shear stress. Expressed in terms of invariants of the deviatoric stress tensor, Eq. (1) becomes

$$4J_2^3 - 27J_3^2 - 36k^2J_2^2 + 96k^4J_2 - 64k^6 = 0 \qquad (2)$$

The Tresca yield criterion is therefore somewhat difficult to implement, in either of the forms represented by Eqs. (1) or (2).

The von Mises yield criterion is considerably simpler, being expressed as

$$J_2 - k^2 = 0 \qquad (3)$$

where here k is the yield limit of a simple shear test. In fact, von Mises originally proposed Eq. (3) as a simple approximation to the Tresca yield criterion, which was in common use at the time. The two criteria differ by only a relatively small amount (see, for example, Hill, 1950, p. 21). However, subsequent experimental work demonstrated that the von Mises yield criterion is actually in closer agreement with experimentally observed material behavior than is the Tresca criterion.

In most current practical applications of plasticity theory, including those related to composite materials, the von Mises yield criterion or one of its variations is being used. This has stemmed from the fact that it is both simple to use and correlates well with experimental observations of material behavior. Nevertheless, it is of interest to attempt to attach some physical significance to it. For example, Eq. (3) can be rewritten in terms of principal components of the deviatoric stress tensor as

$$s_1^2 + s_2^2 + s_3^2 = 2k^2 \qquad (4)$$

As Nadai (1937) first proposed, Eq. (4) implies that yielding begins when the shear stress acting on the octahedral planes reaches a limiting value; the octahedral shear stress has the value $(2k/3)^{1/2}$ at the yield point. A more detailed discussion of this and other physical interpretations of Eq. (3) is given by Prager and Hodge (1951, p. 26). This discussion of physical significance is introduced here primarily because an octahedral shear stress criterion will be utilized subsequently in deriving the required flow rule.

Yield criteria other than those of Tresca and von Mises have also been established, some of which provide even better agreement with experimental data. However, they are generally too complex for practical applications to most plasticity problems.

C. The Prandtl–Reuss Flow Rule

Having established a yield criterion to indicate the onset of plastic flow, it is then necessary to establish appropriate stress–strain relationships which govern plastic flow. The Prandtl–Reuss flow rule is most commonly used, the key assumption of which is that the rate of change of the plastic strain is at any instant proportional to the deviatoric stress, i.e.,

$$\dot{\epsilon}_{ij}^{(p)} = \dot{\lambda} s_{ij} \tag{5}$$

where $\dot{\lambda}$ is a positive scalar function, in general a function of both time and the spatial coordinates, and

$$s_{ij} = \sigma_{ij} - \tfrac{1}{3}\sigma_{kk}\delta_{ij} \tag{6}$$

is the deviatoric component of the stress tensor σ_{ij}, where δ_{ij} is the Kronecker delta. Latin indices have the range 1, 2, 3 and Greek 1, 2. The usual conventions of indicial notation apply.

In order to implement Eq. (5), the rate of change of the total strain can be separated into an elastic (recoverable) and a plastic (irrecoverable) component, i.e.,

$$\dot{\epsilon}_{ij} = \dot{\epsilon}_{ij}^{(e)} + \dot{\epsilon}_{ij}^{(p)} \tag{7}$$

The elastic component follows the same generalized Hooke's law as for completely elastic material behavior, i.e.,

$$\dot{\epsilon}_{ij}^{(e)} = \frac{1 - 2\nu}{3E}\,\dot{\sigma}_{kk}\delta_{ij} + \frac{1 + \nu}{E}\,\dot{s}_{ij} \tag{8}$$

where E and ν are the material Young's modulus and Poisson's ratio, respectively, and $\dot{s}_{ij}$ bears the same relationship to $\dot{\sigma}_{ij}$ as s_{ij} to σ_{ij}, which is indicated in Eq. (6).

In order for Eq. (5) to be useful, it is necessary to define $\dot{\lambda}$ in terms of known quantities. One method is as follows (Swedlow and Yang, 1966):

Equation (5) is multiplied by itself and solved for $\dot{\lambda}$ to give

$$\dot{\lambda} = (\dot{\epsilon}_{ij}^{(p)}\dot{\epsilon}_{ij}^{(p)})^{1/2}/(s_{kl}s_{kl})^{1/2} \tag{9}$$

Defining the octahedral plastic shear strain rate as

$$\dot{\epsilon}_0^{(p)} = (\tfrac{1}{3}\dot{\epsilon}_{ij}^{(p)}\dot{\epsilon}_{ij}^{(p)})^{1/2} \tag{10}$$

and the octahedral shear stress as

$$\tau_0 = (\tfrac{1}{3}s_{ij}s_{ij})^{1/2} \tag{11}$$

Eq. (9) can be rewritten as

$$\dot{\lambda} = \dot{\epsilon}_0^{(p)}/\tau_0 \tag{12}$$

thus defining the positive scalar function $\dot{\lambda}$ in Eq. (5), which becomes

$$\dot{\epsilon}_{ij}^{(p)} = (\dot{\epsilon}_0^{(p)}/\tau_0)s_{ij} \tag{13}$$

It remains to establish a relationship between $\dot{\epsilon}_o^{(p)}$ and τ_0. This can be done by applying Eq. (13) to a problem for which both $\dot{\epsilon}_o^{(p)}$ and τ_0 are known. For example, the results of a simple uniaxial test can be employed, for which

$$\sigma_{22} = \sigma_{33} = \sigma_{12} = \sigma_{13} = \sigma_{23} = 0, \qquad \sigma_{11} \neq 0 \tag{14}$$

and

$$\dot{\epsilon}_{22}^{(p)} = \dot{\epsilon}_{33}^{(p)} = -\tfrac{1}{2}\dot{\epsilon}_{11}^{(p)}, \qquad \dot{\epsilon}_{11}^{(p)} \neq 0 \tag{15}$$

$$\dot{\epsilon}_{12}^{(p)} = \dot{\epsilon}_{13}^{(p)} = \dot{\epsilon}_{23}^{(p)} = 0$$

Substituting Eqs. (15) into Eq. (10) and solving for $\dot{\epsilon}_{11}^{(p)}$ gives

$$\dot{\epsilon}_{11}^{(p)} = \sqrt{2}\dot{\epsilon}_0^{(p)} \tag{16}$$

Since σ_{11} and $\dot{\epsilon}_{11}^{(p)}$ are uniform throughout the test specimen, they are functions of t only, and Eq. (16) can be rewritten as

$$\dot{\epsilon}_{11}^{(p)} = \sqrt{2}\, d\epsilon_0^{(p)}/dt = \sqrt{2}\, d\epsilon_0^{(p)}/d\tau_0\, d\tau_0/dt = \sqrt{2}\dot{\tau}_0/2M_T \tag{17}$$

where, by definition,

$$2M_T = d\tau_0/d\epsilon_0^{(p)}$$

is the tangent modulus of the octahedral shear stress–octahedral plastic shear strain curve for the elastoplastic material, and

$$\dot{\tau}_0 = s_{ij}\dot{s}_{ij}/3\tau_0 \tag{18}$$

is obtained by differentiating Eq. (11). Substituting the rate of plastic strain values defined by Eqs. (15) and (17) into Eq. (10) we obtain

$$\dot{\epsilon}_0^{(p)} = \dot{\tau}_0/2M_T \tag{19}$$

Substituting Eq. (19) into Eq. (13) gives

$$\dot{\epsilon}_{ij}^{(p)} = \dot{\tau}_0 s_{ij}/2\tau_0 M_T \tag{20}$$

This is the form of the Prandtl–Reuss flow rule desired.

Substituting Eqs. (8) and (20) into Eq. (7), the constitutive relations can be expressed as

$$\dot{\epsilon}_{ij} = \frac{1-2\nu}{3E}\dot{\sigma}_{kk}\delta_{ij} + \frac{1+\nu}{E}\dot{s}_{ij} + \frac{\dot{\tau}_0 s_{ij}}{2\tau_0 M_T} \tag{21}$$

Substituting Eqs. (6) and (18) into Eq. (21) and rearranging terms gives

$$\dot{\epsilon}_{ij} = \frac{1+\nu}{E}\dot{\sigma}_{ij} - \frac{\nu}{E}\dot{\sigma}_{kk}\delta_{ij} + \frac{s_{ij}s_{kl}\dot{\sigma}_{kl}}{6\tau_0^2 M_T} \tag{22}$$

In obtaining the last term, use has been made of the fact that no plastic work is done by the hydrostatic component of the applied stress, i.e.,

$$s_{kl}\dot{s}_{kl} = s_{kl}(\dot{\sigma}_{kl} - \tfrac{1}{3}\dot{\sigma}_{mm}\delta_{kl}) = s_{kl}\dot{\sigma}_{kl}$$

since

$$s_{kl}\dot{\sigma}_{mm} = 0$$

Equation (22) is the general constitutive relation governing elastoplastic material behavior, relating the total strain and stress rates at any instant in terms of the octahedral shear stress τ_0.

Often this general constitutive relation is alternately expressed in terms of a quantity $\bar{\sigma}$ (which is variously known as the "effective stress" or "equivalent stress") rather than in terms of the octahedral shear stress τ_0 defined in Eq. (11). This effective stress is defined as

$$\bar{\sigma} = (\tfrac{3}{2}s_{ij}s_{ij})^{1/2} \tag{23}$$

and thus differs from τ_0 only in the magnitude of the constant coefficient.

Various other expressions of an elastoplastic constitutive relation are also available, which incorporate a yield criterion other than that of von Mises (e.g., that of Tresca) and/or a flow rule other than that of Prandtl–Reuss. Few of them are actually being used at present, however, primarily because of the complexities they introduce into the general analysis.

D. Perfectly Plastic Material Behavior

In the previous paragraphs of this section, a general theory of plasticity was discussed, which implies the consideration of a material stress–strain response of a general nature. That is, beyond the elastic limit, both an

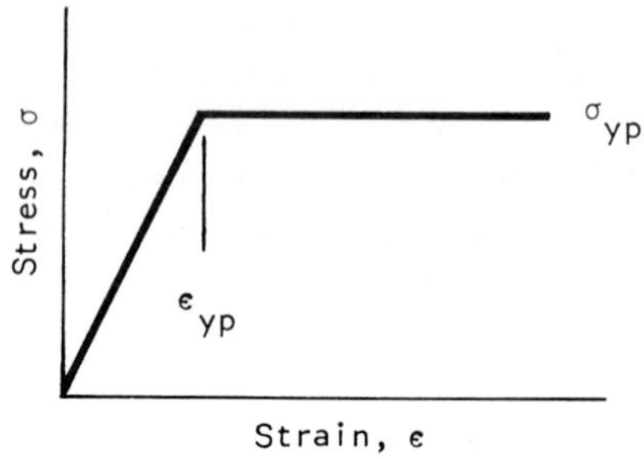

FIG. 2. Elastic–perfectly plastic material behavior.

elastic and a plastic component of strain increment are present, as indicated in Eq. (7). Since by definition the elastic component of the strain increment is associated with a proportional increment of stress [Eq. (8)], the resulting total stress–strain behavior [as represented, for example, by Eq. (22)] will exhibit a variation of stress with increasing strain beyond the elastic limit (as indicated in Fig. 1). This is frequently referred to as *strain hardening behavior.*

Because of the complexity associated with the introduction of a general flow rule, such as represented by Eq. (5), a simpler idealization was introduced early in the development of the plasticity theory, viz., perfectly plastic material behavior. The simplification is that beyond the elastic limit the elastic component of the strain is assumed to be zero. Thus, the stress–strain curve as represented by a simple uniaxial loading is bilinear, with the second segment of the curve being parallel to the strain axis, as indicated in Fig. 2. This eliminates the need for a flow rule: at strain levels below the elastic limit or yield strain ϵ_{yp}, the stress is directly proportional to the strain, at strain levels greater than ϵ_{yp}, the stress remains constant.

The stress–strain behavior of most engineering materials, including those of interest for use in composites (see, for example, Fig. 1) is not accurately represented by the simple bilinear stress–strain relation of Fig. 2. This is particularly true when the material is uniformly stressed, so that all regions reach the yield value simultaneously. The idealization would predict unrestricted plastic flow, i.e., unlimited strain with no increase in stress. However, when a loading creates stress gradients within the material, the highest stressed regions will yield first. However, this plastic flow is restricted since the surrounding material has not yet yielded. This is termed "contained plastic flow" and is typical of composites since high stress gradients usually exist due to the difference in stiffness properties of the constituents. Thus, if the actual stress–strain curves of the constituents (as represented by Fig. 1) are crudely approximated by a bilinear curve of the form of Fig. 2, an elastic–perfectly plastic analysis can be performed.

Perfectly plastic material behavior has been introduced here because such an assumption has frequently been used in analyses of composite material behavior, particularly in conjunction with the simple model analyses to be discussed in the next section.

III. Simple Model Analyses

A. Definition of a Simple Model

A simple model will be defined here as one which either utilizes gross approximations to physical reality, or restricts the applicability of the analysis to specific, relatively simple, physical configurations.

The former type implies an applicability to general problems, but offers no indication of the accuracy achieved. Thus, unless a rigorous analysis is also available to compare results with, this type of simple model provides, at most, a general, qualitative indication of material behavior. Quantitative results are available, but are of little significance.

On the other hand, the latter type of simple model can provide reliable quantitative results. Unfortunately, the range of practical application to physically realistic problems is very limited.

Examples of both types of simple model analyses will be presented here. Those included are considered as being representative of those most frequently referred to in the literature and/or of the greatest significance.

Simple model analyses of the elastoplastic behavior of composite materials have been developed for the same reasons as in any other area of investigation—the physical phenomenon was well recognized long before the basic mathematical descriptions and computational schemes required to develop a rigorous method of analysis were available. During the mid-1960s, when rigorous elastic analyses were being published, the incremental loading procedures and large capacity digital computers required to extend these analyses to elastoplastic material behavior were not yet well established. Thus, many investigators reverted to the prior simple model elastic analyses, extending them to consider elastoplastic material behavior also. Since rigorous elastic analyses were already available, it was well known that such efforts were, at best, an interim step, and of only limited quantitative value. Nevertheless, they can often provide some physical insight to actual material behavior.

In addition to the two basic types of simple model analyses which were defined above, and which will be discussed in greater detail in subsequent paragraphs, another approach is also available for predicting the stress–strain response of a composite material in terms of the properties of its constituents. This is essentially a semiempirical method. Various equation forms, particularly power laws, containing undefined constants are fitted to known experimental results to obtain empirical relationships. Such formulas typically relate stress to strain in terms of physical parameters such as the filament volume content and the material properties of the constituents. These stresses and strains might be selected local values, but most often they are average or composite values. In either case, such analyses are not truly micromechanical analyses since they do not provide a description of the local stress and strain gradients within the composite. Their utility is primarily as a simple-to-evaluate engineering estimate of the basic composite stress–strain response, which is information required as input to most macromechanical or laminate analyses.

Bert (1972) has recently presented one such semiempirical analysis. His work is cited here particularly since it contains an extensive and up to date list of references to similar other works of this type, providing the interested reader with an excellent summary and source of information. Like all studies of this type, Bert has chosen to make several very severe assumptions in order to keep the working formulas relatively simple. Thus, it is not surprising that the correlation with experimental data other than those used to evaluate the empirical constants is frequently not very good. However, until the rigorous analyses to be discussed in Section IV are adequately developed so that they can be readily used for routine engineering computations, results such as those presented by Bert serve a useful function.

B. Gross Approximations to Physical Reality

1. Series–Parallel Models

Since composite materials consist of distinct constituent material phases bonded together, the concept of a model consisting of elements connected together in series and parallel arrays has long been popular. Perhaps the first attempt to apply a series–parallel model to elastoplastic material behavior was that by Shaffer (1968). He assumed linearly elastic–perfectly plastic material behavior in analyzing transverse normal loading. This was an extension of his prior elastic analysis (Shaffer, 1964). A typical repeating unit of the composite cross section is isolated, as indicated by the dashed lines in Fig. 3. This unit is then divided into elemental strips of differential width and constant length p as indicated by the dotted lines in Fig. 3. For a filament volume less than 68%, some elemental strips of matrix material will exist which contain no filament material, as is obvious in Fig. 3. In this case the model consists of elemental strips of matrix material arranged in parallel with strips consisting of a series array of matrix and filament regions. For filament volume contents greater than 68%, the continuous strips of matrix material vanish and the model reduces to a parallel array of series elements, the geometry of which becomes more complicated.

Several gross approximations are made in this analysis, two of the most severe of which are the complete neglect of interactions between the surfaces of adjacent elemental strips and the fact that the stresses and strains are uniform within each elemental strip of length p. The first assumption reduces the problem to a one-dimensional stress analysis, i.e., σ_x is the only nonzero stress component. This is not physically realistic and reduces the yield criterion to the trivial case whereby yielding is predicted to occur when σ_x attains the yield value as determined from a simple uniaxial test, i.e., when $\sigma_x = \sigma_{\text{yp}}$. In actual composites, significant biaxial and triaxial stress states occur. The first approximation combined with the second also results in the prediction that the stress in any elemental strip (all of which are of length p, the filament center-to-center spacing) is constant. Thus, even for elemental strips which extend across both filament and matrix

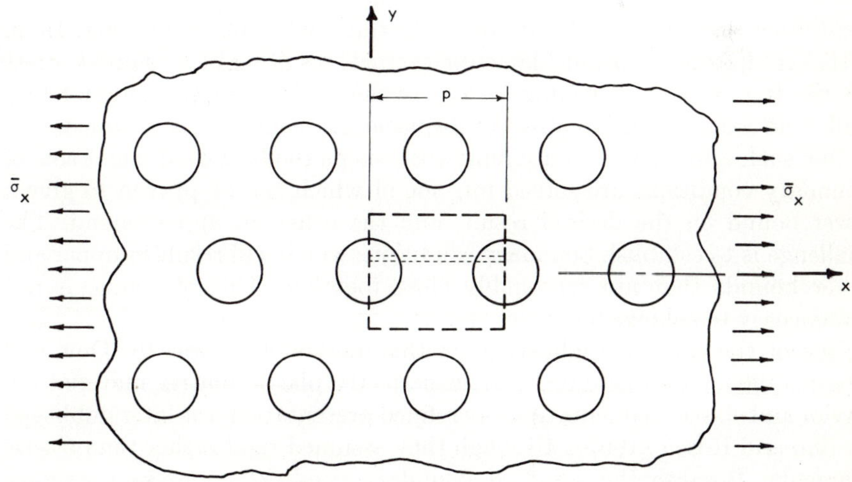

FIG. 3. Composite cross section indicating typical repeated unit (dashed lines) and an elemental strip (dotted lines).

material regions, the stress does not vary within the element. That is, the model does not predict gradients in the x direction.

In spite of these gross approximations, certain numerical results obtained from the original elastic analysis agreed quite well with the more rigorous analyses of other investigators (see Shaffer, 1964). It is possible that similar correlations may be achieved when results of rigorous elastoplastic analyses become more readily available. The problem is that a rigorous analysis (or very extensive experimental data) must be available for comparison and verification purposes before the simple model can be used with any degree of confidence. Hence, its principal value is, and will continue to be, as an easily used engineering tool, to be utilized only within the range of parametric variations for which it has previously been verified. Thus, it is in a sense comparable in its applicability to the semiempirical models such as those discussed by Bert (1972), even though it is not in itself a semiempirical model.

2. Concentric Cylinder Models

Like the series–parallel models, the concentric cylinder models represent a gross approximation to actual composite behavior and yet are very frequently used. This is because they permit relatively simple analyses and lead to closed form solutions. A typical model consists of a single filament, usually of circular cross section, imbedded in a concentric cylinder of matrix material. The gross approximation is associated with the assumed boundary conditions (explicit or implicit) on the surface of the outer cylinder. In an actual composite, adjacent filaments interact, resulting in a complex stress distribution in the intervening matrix. In the model, a simple, circumferentially and axially uniform stress or displacement condition is assumed.

One such approach is to use limit analysis methods. Two distinct sets of boundary conditions are solved for, one of which can be proven to give a lower bound on the desired result, and the other an upper bound. The challenge is to establish boundary conditions which will result in upper and lower bounds that are reasonably close together. This of course is not always easy to achieve.

One of the earliest applications of this methodology was by Dow and Rosen (1965), who assumed elastic–perfectly plastic matrix material behavior and elastic filaments. A more refined presentation was later published by Shu and Rosen (1967), although they assumed rigid rather than elastic filaments. Because the assumed boundary conditions represent average rather than local values, such analyses are typically used to estimate gross or composite properties rather than local stresses and strains. When so used, they cannot truly be considered to be micromechanical analyses since they do not predict local behavior.

Another analysis using a concentric cylinder model is that of Ebert and Gadd (1965). They performed a reasonably straightforward continuum analysis for two concentric circular cylinders of infinite length subjected to an axial load. Elastic–perfectly plastic material behavior was assumed, their particular interest being the magnitude of the transverse stresses induced. A minor extension of the original analytical development was later presented, along with experimental results, by Ebert *et al.* (1968).

The gross approximation which makes this analysis of little more than academic interest is that the surface of the outer cylinder is assumed to be stress free, i.e., interactions with adjacent filaments which would be present in a real composite are not modeled. Thus, the analysis is a reasonable approximation only for composites of very low (less than 10%) filament volume, for which the filaments are actually widely spaced.

Hill (1964) has also considered the concentric cylinder model, for an assumed plane strain condition and uniform lateral pressure loading. His general methodology is of interest even though he emphasized the estimation of composite stress–strain response rather than local stress distributions.

C. Analyses of Limited Applicability

1. Single Filament in an Infinite Matrix

The gross approximation to real composites of assuming a single filament in an infinite matrix is obvious. Actually, the model of Ebert and Gadd (1965) discussed in the previous paragraphs approaches this case as the diameter of the outer cylinder is increased relative to that of the inner one.

The single filament approximation is often assumed primarily because it permits the use of any one of several well-defined analytical methodologies. An example is the analysis presented by Mendelson (1968), who assumed a circular filament, equal biaxial tension at infinity, and a total deformation theory. Another presentation has been given by Sendeckyj (1969).

In addition to analytical methods, numerical techniques have also been used. For example, Owen *et al.* (1969) have used a concentric cylinder, finite-element formulation to study the stress distributions in an elastic–perfectly plastic matrix in the gap region between two colinear filaments subjected to an axial loading. The matrix region is actually of finite extent, but can be assumed to be large relative to the filament diameter. Thus, this model approximates a single (in this case discontinuous) filament in an infinite matrix adequately well for most purposes.

2. Idealized Boundary Condition Formulations

A major problem in correctly modeling composite material behavior is adequately representing the complex boundary conditions which are introduced when a local material region is isolated for analysis. For example, a single filament and the matrix material immediately surrounding it might be isolated from the composite. Stress and/or displacement conditions must then be established at the isolating surfaces which correctly represent the actual physical conditions. However, the stress and displacement distributions throughout the composite are very complicated due to the multiple interactions between filaments. Thus, the stresses and displacements along the isolating cutting planes, which become the boundary conditions of the analysis, are generally unknown distributions.

Necessary boundary conditions can often be established however, either by idealizing the material behavior, as will be discussed here, or by limiting the filament packing geometry to special configurations, as will be discussed in Section IV. Actually, the concentric cylinder models as previously discussed, when they are used in an attempt to represent a filament array, also utilize idealized (uniform) boundary conditions.

An example of an idealized boundary condition formulation is the analysis by Owen *et al.* (1969), which is an extension, presented in the same paper, of their previously cited work. Elastic–perfectly plastic material behavior was assumed in modeling the axial loading of a composite containing a discontinuous filament adjacent to continuous filaments (see Fig. 4). A finite element numerical technique was used, the region indicated in Fig. 4b being divided into a network of constant strain triangular elements. The particularly limiting boundary condition idealization is the assumption of plane strain. That is, the material is assumed to be uniform and of infinite extent in the x direction. Thus, the model is actually that of a discontinuous flat plate sandwiched between two continuous plates. This is not a very satisfactory approximation to the actual geometry of a filament-reinforced composite. A three-dimensional analysis would be required to model correctly the actual composite in this case.

Another example of an idealized boundary condition formulation is that of Huang (1971), although it is not a complete micromechanics analysis in that he only attempts to predict the composite stress–strain response and not local stress and strain distributions. Huang incorporates two very limiting boundary condition idealizations in analyzing the transverse normal loading problem. He assumed the filaments to be rigid so that a simple boundary condition, viz., zero normal displacement and strain at the filament–matrix interface, is obtained. The second idealization is that the filaments are uniaxially oriented but randomly distributed within the matrix so that macroscopically the composite is transversely isotropic. He then utilized the so-called "self-consistent" approach, the basic assumption of which is that an isolated filament can be regarded as being embedded in an "equivalent" homogeneous medium whose properties are the unknowns to be determined. This procedure averages out the effect of interactions between individual filaments, leading to average results rather than local values. This becomes a particularly limiting assumption for high filament volumes, when filaments are closely spaced and hence interact strongly.

IV. Rigorous Analyses

A. Definition of a Rigorous Analysis

A rigorous analysis will be defined here as one which satisfies the mathematical laws of analytical mechanics and at the same time accurately

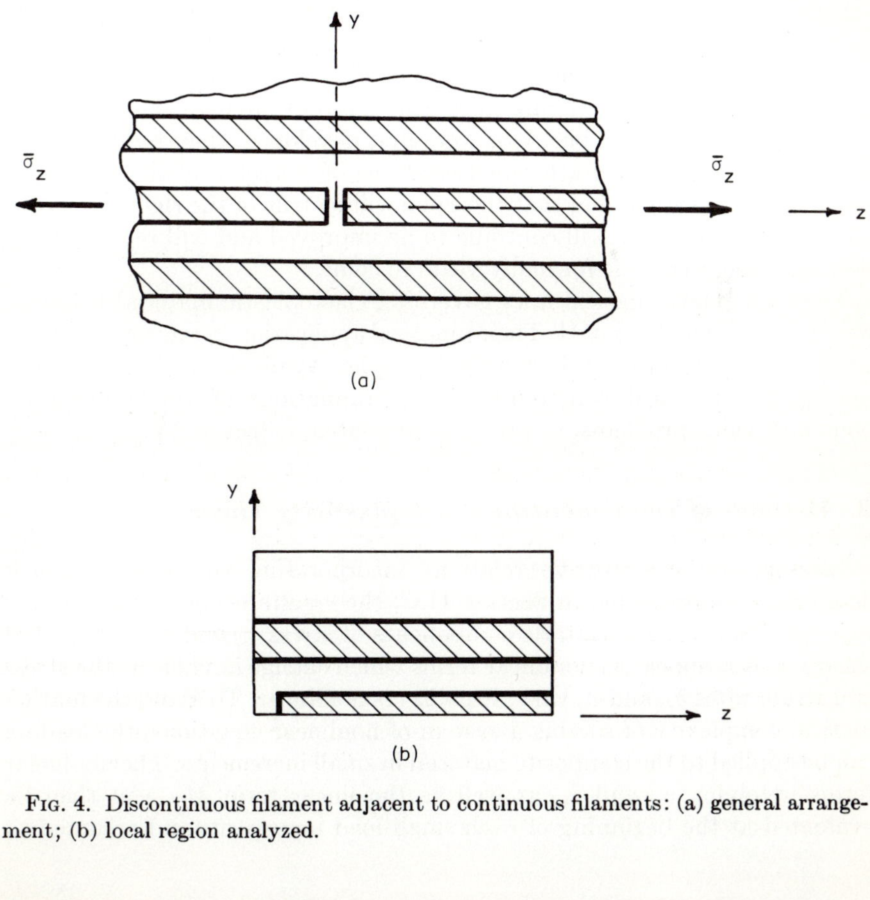

FIG. 4. Discontinuous filament adjacent to continuous filaments: (a) general arrangement; (b) local region analyzed.

models the known physical constraints. These physical constrains are associated with the material characteristics of the individual constituents and their geometric arrangement within the composite.

In Section III, simple model analyses were defined and discussed. In all cases, the true representation of actual physical behavior was severely compromised in order to simplify the analysis. Within the past several years, somewhat more rigorous representations have begun to be developed. Because of the complexities contained in these formulations, numerical solution techniques are required, typically involving the related development of digital computer programs. These programs tend to become very large as the accuracy of the model representation is increased, requiring the use of large capacity computers and large amounts of computer running time. Thus, the programs are time-consuming to develop and can be costly to use.

Nevertheless, at the present time this appears to be the only way of accurately representing the actual elastoplastic behavior of composite materials. It should be noted that the same progression of analysis methods, from simple models to rigorous analyses, was also followed in past years in developing analyses of linearly elastic behavior of composite materials. Even today, some 6–8 years since the rigorous numerical analyses of linearly elastic material behavior were first developed, they remain as the only complete and accurate representations of micromechanical behavior available. It is anticipated that the same will be true of the rigorous elastoplastic analyses. They will continue to be improved and will represent the best approach available for many years to come.

The basic mathematical laws governing elastoplastic material behavior were presented in Section II. Their practical application to rigorous analyses will be discussed in the following paragraphs. Available models for representing the physical constraints, i.e., formulations of the appropriate boundary value problems, will then be presented in Section IV,C.

B. Methods of Implementing Elastoplasticity Theory

Elastoplastic constitutive relations incorporating the Prandtl–Reuss flow rule were presented in Section II,C, the results being represented by Eq. (22). Since the deviatoric components of stress s_{ij} and the octahedral shear stress τ_0 appear as nonlinear terms which change in value as the stress and strain rates $\dot{\sigma}_{ij}$ and $\dot{\epsilon}_{ij}$ vary, Eq. (22) is nonlinear. To avoid the mathematical complexity of solving a system of nonlinear equations, the loading can be applied to the composite material in small increments. The nonlinear terms involving s_{ij} and τ_0, as well as the linear term M_{T}, can then be evaluated at the beginning of each small load increment, and assumed to remain constant during the load increment. Thus, by utilizing an incremental loading procedure, the nonlinear problem is reduced to a series of linear problems. While this does represent an approximation to the basic mathematical laws, its extent can be governed by controlling the size of the load increments permitted.

Two basic methods have evolved for incorporating this incremental loading procedure. They are generally referred to as the "method of initial strains" and the "tangent modulus method."

1. *Method of Initial Strains*

The method of initial strains (Mendelson and Manson, 1959) is the older and perhaps more commonly used of the two methods available for analyzing incremental loading problems. The elastic and plastic components of the total strain increment as given by Eq. (22) are kept separate when constructing the set of matrix equations so that the stiffness matrix includes only the elastic components of the strain increment, i.e., it contains only the elastic material constants E and ν. Since these do not change from one load increment to the next, this stiffness matrix need only be inverted once. The increments of plastic strain, represented by the last term of Eq. (22), are treated as unknown constants.

An iterative solution technique is utilized within each load increment. The plastic strain increment at each material point is first estimated and treated as an initial condition. A linearly elastic analysis is performed, subject to this initial condition, to obtain the elastic strain increment field. The total strain increments (the sum of the initial plastic strain increments and the computed elastic strain increments) are then substituted into the inverse of Eq. (22) to obtain the total stress increments $\dot{\sigma}_{ij}$. A revised estimate of the plastic strain increment field is then obtained by using those values of $\dot{\sigma}_{ij}$ to evaluate the plastic strain increment components, i.e., the last term of Eq. (22). These new values of $\dot{\epsilon}_{ij}^{(p)}$ then become the initial conditions for the next iteration cycle. This iterative procedure is repeated until some prescribed degree of convergence is attained.

There are actually two options available for determining the revised estimate of $\dot{\epsilon}_{ij}^{(p)}$ within each iteration, as discussed by Lansing *et al.* (1966). The computationally quicker "constant stress method" has been found to be subject to "sudden and catastrophic divergence," the onset of which is associated with the size of the load increment assumed. The "constant strain method," although converging much less rapidly, is much more stable and is therefore generally used.

An early use of the method of initial strains was that by Padlog *et al.* (1960). However, one of the first applications specifically to composite materials was that by Foye and Baker (1971). Foye actually outlined the method of application in unpublished reports as early as 1966.

2. Tangent Modulus Method

The tangent modulus method (Marcal and Turner, 1963) permits the use of the same basic procedures as earlier developed for linearly elastic analyses. The elastoplastic constitutive relation represented by Eq. (22) is utilized in place of the generalized Hooke's law [Eq. (8)], the complete loading history being obtained as the sum of individual linear (but not elastic) solutions. The quantities s_{ij}, τ_0, and M_T in Eq. (22) are reevaluated at the beginning of each load increment and treated as constants, resulting in a linear relation between the variables $\dot{\epsilon}_{ij}$ and $\dot{\sigma}_{ij}$, the total strain and stress components at each point within the material, respectively. Thus, a direct solution can be obtained for each load increment, with no iteration being required.

As will be described in Section IV,D, numerical rather than closed form analytical solutions are necessary for all of the rigorous analyses to be discussed here. This usually requires that a set of linear algebraic equations be solved for in order to obtain values of $\dot{\sigma}_{ij}$ or $\dot{\epsilon}_{ij}$ at each material point. In terms of matrix algebra, a stiffness matrix must be constructed and inverted. The tangent modulus method requires that this stiffness matrix be constructed and inverted for each applied load increment since new values of s_{ij}, τ_0, and M_T must be introduced each time.

3. Comparisons and Relative Merits

Marcal (1968), in making a comparative study, presented a derivation of the equations governing the method of initial strains directly from those of the tangent modulus method. He demonstrated that the method of initial strains does not converge and therefore is not applicable to the special case of an elastic–perfectly plastic material. That convergence becomes very slow as elastic–perfectly plastic material behavior is approached, i.e., for materials exhibiting a very small change in stress with increasing strain for strains beyond the elastic limit, has been confirmed by Foye and Baker (1971). On the other hand, Adams (1970a, b) has found that the tangent modulus method provides good results in this case.

Marcal (1968) indicated that for one specific numerical example which he considered, the computing time was somewhat less when using the tangent modulus method. However, such a comparison is undoubtedly a function of the problem geometry and the assumed material stress–strain curve. A valid comparison for composite material micromechanical analysis applications would be to compare the tangent modulus method as utilized by Adams (1970) and the method of initial strains as utilized by Foye and Baker (1971). Insufficient information concerning accuracy achieved and computing time requirements is given in the published reports to make such a comparison possible, however.

C. Formulation of the Boundary Value Problem

Having selected a method of implementing the general theory governing elastoplastic material behavior, whether it be the tangent modulus method or the method of initial strains, there remains to formulate a boundary value problem which is representative of the physical geometry to be modeled by the analysis. The problem is similar to that encountered for linearly elastic analyses such as discussed in Chapter 3. A representative local material region must first be isolated from the bulk composite. Advantage is usually taken of some assumed filament packing symmetry in order to establish displacement and shear stress boundary conditions. That such conditions necessarily exist can be argued from simple physical considerations. The assumed symmetry is usually with respect to the filament cross-sectional shape and the packing geometry. For example, it is often assumed that the filament shapes are symmetric about orthogonal axes which are also both axes of filament packing symmetry and principal load axes. This is not as restrictive as it may first appear. For example, the rectangular and diamond packing arrays shown in Figs. 5b and c both satisfy these conditions, symmetry being maintained about both the x and the y axes. The filaments need not be circular in cross section; any shape which is symmetrical about both the x and y axes is admissible. A complete discussion of boundary conditions associated with Fig. 5 can be found in the works of Adams *et al.* (1967) and Foye (1968).

Ideally, a full three-dimensional analysis of the boundary value problem would be preferred. This would permit the consideration of stress and strain gradients in the direction of filament reinforcement (the 3 direction indicated in Fig. 5a) as well as in the transverse 1, 2 plane. For example, this would permit the analysis of the behavior of discontinuous reinforcements such as whiskers and chopped fibers (e.g., the transfer of stresses at the fiber ends and into adjacent fibers), and particulate-reinforced composites. It would also facilitate the modeling of voids, local debonding, and similar manufacturing defects. The analytical formulation of a truly three-dimensional model is straightforward. However, the complexity of the boundary value problem makes a numerical solution technique mandatory. Large amounts of digital computer storage are required and the computer running times are relatively long. Most versions of the present generation of digital computers, e.g., the IBM Model 360 and its counterparts, do not

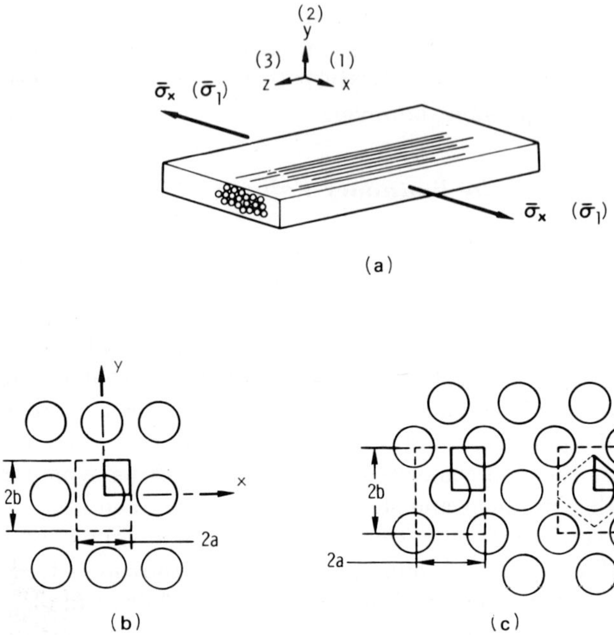

FIG. 5. Filament packing geometries: (a) transverse normal loading of a unidirectionally reinforced composite; (b) rectangular array; (c) diamond array.

have sufficient storage capacity and computational speed to make the consideration of practical three-dimensional analyses feasible. Just a few years ago, when computers such as the IBM Model 7094 and its counterparts were typically being used, the same type of comments then applied to two-dimensional analyses—it would have been difficult to overcome storage limitations, and computer running times required would have been prohibitive in most cases. Thus, in the future, solutions of full three-dimensional problems will undoubtedly be routinely performed. At the present time, however, it is necessary to make certain idealizations in order to make the problem amenable to practical solution.

One way of achieving this is to reduce the problem to a two-dimensional formulation. For example, for continuous-filament-reinforced composites, it is reasonable to assume that stress and strain gradients in the axial direction (the 3 direction in Fig. 5a) are negligible relative to those in the transverse plane. This is the classical plane stress or plane strain condition. A plane stress condition assumes that the stress in the direction normal to the plane of interest (the σ_{33} component of stress, normal to the 1, 2 plane in Fig. 5a) is zero. This is usually associated with the analysis of thin plates (thin in the 3 direction) subjected to in-plane loadings. However, for a continuous-filament-reinforced lamina, as indicated in Fig. 5a, the dimension in the direction of reinforcement (the 3 direction) is typically very large. This corresponds more closely to the plane strain condition, i.e., zero displacement in the 3 direction. Since there are assumed to be no gradients in the 3 direction, the strains ϵ_{i3} and correspondingly the increments of strain $\dot{\epsilon}_{i3}$ are equal to zero, i.e.,

$$\dot{\epsilon}_{13} = \frac{\partial \dot{u}_1}{\partial x_3} + \frac{\partial \dot{u}_3}{\partial x_1} = 0, \qquad \dot{\epsilon}_{23} = \frac{\partial \dot{u}_2}{\partial x_3} + \frac{\partial \dot{u}_3}{\partial x_2} = 0, \qquad \dot{\epsilon}_{33} = \frac{\partial \dot{u}_3}{\partial x_3} = 0 \quad (24)$$

where $\dot{u}_i$ and x_i represent the displacement increment components and spatial coordinates, respectively.

For plane strain,

$$\dot{\sigma}_{13} = \dot{\sigma}_{23} = 0 \quad (25)$$

as can be verified by substitution into Eq. (22), making use of Eqs. (24) also.

Although, by definition, the axial strain increment $\dot{\epsilon}_{33}$ is zero everywhere, the axial stress increment $\dot{\sigma}_{33}$ is not. It is not an independent quantity, however, and can be expressed in terms of the other nonzero stress increment components as follows. Setting $i = j = 3$ in Eq. (22) we obtain

$$(1 + \nu)\dot{\sigma}_{33} - \nu\dot{\sigma}_{kk} + (E/A)s_{33}s_{kl}\dot{\sigma}_{kl} = 0 \quad (26)$$

where $A = 6\tau_0{}^2 M_{\text{T}}$. Solving Eq. (26) for $\dot{\sigma}_{33}$

$$\dot{\sigma}_{33} = (1/D)(A\nu\dot{\sigma}_{\gamma\gamma} - Es_{33}s_{\gamma\delta}\dot{\sigma}_{\gamma\delta}) \quad (27)$$

where $D = A + Es_{33}^2$, and, as previously noted, the Greek indices have the range 1, 2.

Utilizing Eq. (25), Eq. (27) can be substituted into Eq. (22) to give, after considerable simplification,

$$\dot{\epsilon}_{\alpha\beta} = \frac{1 + \nu}{E}(\dot{\sigma}_{\alpha\beta} - \nu\dot{\sigma}_{\gamma\gamma}\delta_{\alpha\beta}) + \frac{1}{D}t_{\alpha\beta}t_{\gamma\delta}\dot{\sigma}_{\gamma\delta} \quad (28)$$

where

$$t_{\alpha\beta} = s_{\alpha\beta} + \nu s_{33}\delta_{\alpha\beta}$$

For a detailed presentation of this derivation, see the work of Adams (1973) or Swedlow (1969).

Because displacement-type boundary conditions are typically encountered in composite material micromechanics problems, it is often convenient to formulate the numerical analysis in terms of displacement components. Thus, the inverse of Eq. (27) is needed so that stress increments can be readily expressed in terms of strain components, which in turn are simple functions of the displacements.

The inverse of Eq. (28) can be obtained by first expanding it into the three equations it represents, i.e., $\dot{\epsilon}_{11}$, $\dot{\epsilon}_{22}$, and $\dot{\epsilon}_{12}$, and then solving these equations for $\dot{\sigma}_{11}$, $\dot{\sigma}_{22}$, and $\dot{\sigma}_{12}$ by successive elimination. However, the process is lengthy (see Adams, 1973). The result is

$$\dot{\sigma}_{\alpha\beta} = \frac{E}{1+\nu}\left[\dot{\epsilon}_{\alpha\beta} + \frac{\nu}{1-2\nu}\dot{\epsilon}_{\gamma\gamma}\delta_{\alpha\beta} - \frac{s_{\alpha\beta}s_{\gamma\delta}\dot{\epsilon}_{\gamma\delta}}{3\tau_0{}^2 + (1+\nu)(A/E)}\right] \tag{29}$$

For use in numerical analyses, it is convenient to expand Eq. (29) into the three equations it represents, i.e., $\dot{\sigma}_{11}$, $\dot{\sigma}_{22}$, and $\dot{\sigma}_{12}$. The index notation can also be changed to conventional engineering notation. Thus, $\dot{\sigma}_{11}$, $\dot{\sigma}_{22}$, $\dot{\sigma}_{12}$, $\dot{\epsilon}_{11}$, $\dot{\epsilon}_{22}$, $\dot{\epsilon}_{12}$ will be replaced by $\dot{\sigma}_x$, $\dot{\sigma}_y$, $\dot{\tau}_{xy}$, $\dot{\epsilon}_x$, $\dot{\epsilon}_y$, $\frac{1}{2}\dot{\gamma}_{xy}$, respectively. This notation is more consistent with that used in previously published micromechanics analyses. Note in particular that $\dot{\epsilon}_{12}$ has been replaced by $\frac{1}{2}\dot{\gamma}_{xy}$, to convert from tensoral shear strain to engineering shear strain. Engineering shear strain has typically been used in these numerical analyses.

The resulting expansion of Eq. (29) can be expressed in matrix notation as

$$[\dot{\sigma}] = [H][\dot{\epsilon}] \tag{30}$$

where

$$[\dot{\sigma}] = \begin{bmatrix} \dot{\sigma}_x \\ \dot{\sigma}_y \\ \dot{\tau}_{xy} \end{bmatrix}, \qquad [\dot{\epsilon}] = \begin{bmatrix} \dot{\epsilon}_x \\ \dot{\epsilon}_y \\ \dot{\gamma}_{xy} \end{bmatrix}$$

$$[H] = \frac{E}{1+\nu}\begin{bmatrix} \left(\dfrac{1-\nu}{1-2\nu} - \dfrac{s_{11}^2}{B}\right) & \left(\dfrac{\nu}{1-2\nu} - \dfrac{s_{11}s_{22}}{B}\right) & \left(-\dfrac{s_{11}s_{12}}{B}\right) \\ \left(\dfrac{\nu}{1-2\nu} - \dfrac{s_{11}s_{22}}{B}\right) & \left(\dfrac{1-\nu}{1-2\nu} - \dfrac{s_{22}^2}{B}\right) & \left(-\dfrac{s_{22}s_{12}}{B}\right) \\ \left(-\dfrac{s_{11}s_{12}}{B}\right) & \left(-\dfrac{s_{22}s_{12}}{B}\right) & \left(\dfrac{1}{2} - \dfrac{s_{12}^2}{B}\right) \end{bmatrix}$$

and

$$B = 3\tau_0{}^2 + (1+\nu)(A/E)$$

Note that $[H]$ is a symmetric matrix.

Equation (30) represents the elastoplastic constitutive relation for a plane strain condition. This can also be readily modified to consider a generalized plane strain condition, i.e., $\dot{\epsilon}_z$ = constant rather than $\dot{\epsilon}_z = 0$, or to include the longitudinal shear loadings, i.e., $\dot{\tau}_{zx}$ and $\dot{\tau}_{zy}$ (see, for example, Foye and Baker, 1971).

Having defined the material region to be analyzed, the loading conditions, the boundary conditions, and the governing constitutive relations, there remains to select a specific solution technique.

D. Solution Techniques

As discussed in Section IV,A, the rigorous analyses typically require a numerical solution technique of some type. Early works, not necessarily related directly to composite materials analyses, utilized finite difference methods extensively, while current emphasis is clearly toward the use of finite element techniques. In addition, there have been occasional attempts to adopt stress function approaches directly.

1. Finite Difference Methods

Finite difference methods were first applied to plane elastoplastic (more specifically, elastic–perfectly plastic) problems by Allen and Southwell (1950), at which time hand computations using relaxation methods were being used to obtain numerical results. Relaxation methods were described by Southwell (1940) and in more detail later by Allen (1954). Spencer (1962) subsequently extended the work of Allen and Southwell (1950) to the treatment of multiply connected plane bodies, a key step toward subsequent applications to composite material problems. His analysis was also restricted to the consideration of elastic–perfectly plastic material behavior, no unloading, and a statically determined problem (this latter condition being associated with the assumption of elastic–perfectly plastic material behavior).

Havner (1966) later presented an alternate approach which is applicable to general elastoplastic material behavior. He obtained a finite difference discretization through a minimization of the total potential energy function, which leads to a positive definite symmetric matrix. He did not present numerical results, however, and did not consider the two-phase material problem which is representative of a composite system.

In fact, no applications of the finite difference methods specifically to elastoplastic analyses of composite material problems have been published. Adams *et al.* (1967), who utilized the finite difference method in developing elastic analyses of composite material behavior, subsequently switched to a finite element technique when considering elastoplastic material behavior (Adams, 1970a, b).

The principal reason for the lack of applications of finite difference methods to elastoplastic material behavior of composites is not associated with any unique mechanical properties of the constituents, however. Rather, it is a geometry problem, the same difficulty that is encountered in any application of the finite difference method to problems involving curved boundaries, viz., constraining the finite difference node points to lie on the curved boundaries. Even the use of irregular grid spacings does not completely eliminate this problem, as discussed by Adams *et al.* (1967). The use of triangular finite element grids completely avoids this major problem, making the finite element method a much more versatile approach.

2. Finite Element Techniques

A significant number of applications of the finite element method to elastoplastic material behavior have been published during the past few years. In fact, much of the work relating to the development of practical methods of implementing plasticity theory, in particular the method of initial strains and the tangent modulus method discussed in Section IV,B, has been conducted using finite element analyses. Several of these applications are directly associated with composite materials investigations, while others could be readily adapted to this purpose.

Finite element analyses had their beginning in the development by Turner *et al.* (1956) of what they termed the "method of direct stiffness calculation." This in turn was an extension of the computation of influence coefficients by Levy (1953). In the direct stiffness method, the displacements of specific points within the body are considered as the unknowns. An infinite number of systems of mutually compatible deformations are possible; the correct displacement field is that for which the governing equations of equilibrium are satisfied. Gallagher *et al.* (1962) first applied the direct stiffness method to elastoplastic material behavior, using the method of initial strains. For a chronology of subsequent applications of the direct stiffness method, using both the method of initial strains and the tangent modulus method, see, for example, the work of Marcal (1969). A number of these studies were of stress distributions around holes, notches, and similar discontinuities in flat plates subject to in-plane loadings. By assuming the discontinuity to be an inclusion, e.g., a filament, rather than a cutoff of some type, a direct extension to composite materials is achieved. The finite element method, in its current state of development, is described in detail in a number of texts, including that by Zienkiewicz (1971).

The first rigorous application to elastoplastic composite material behavior of the finite element method using the method of initial strains was by Foye (1968), and subsequently in greater detail by Foye and Baker (1971). The first rigorous application using the tangent modulus method was by Adams (1970a, b), and subsequently in greater detail by Adams (1973).

(a) METHOD OF INITIAL STRAINS. The basic procedure outlined in Section IV,B,1 for implementing the method of initial strains is directly applicable to the finite element technique. Details of the computational procedure and an actual computer program are given by Foye and Baker (1971). Constant strain triangular elements were used, the analysis being a direct extension of a prior elastic analysis (Foye, 1968). Generalized plane strain was assumed, the analysis being readily modified to handle the assumption of plane stress also, if desired. The generalized plane strain condition permits combined axial and transverse normal loadings. In addition, longitudinal shear loading is included. This is readily achieved since the generalized plane strain solution involves only the in-plane (x, y) components of the node point displacements while the shear loading involves only axial (z) components. Thus, a complete combined loading problem can be analyzed.

One potential limitation of this particular computer program is that it permits the use of a maximum of only 49 node points, which limits the degree of grid refinement possible. The manner in which the program is written would make it somewhat difficult to enlarge this capacity.

Foye and Baker (1971) presented numerical results for a square array of circular filaments, for filament volumes of 40, 50, and 60%. Both single load component and combined loadings were assumed, mostly for material properties representative of a boron–epoxy composite, but also for glass–epoxy, graphite–epoxy, and boron–aluminum. While these results are not numerous enough to establish detailed parametric variations, they do provide a good indication of the development pattern of zones of elastoplastic material behavior.

(b) TANGENT MODULUS METHOD. The basic procedure outlined in Section IV,B,2 for implementing the method of initial strains is directly applicable to the finite element technique. A detailed presentation of the general method of analysis and a description of the required computational procedure is given by Adams (1973). He analyzed the problem of transverse normal loading of a regular array of filaments, using constant strain triangular elements and assuming a plane strain condition. Combined loading was not considered. A number of numerical results for both square and rectangular arrays, at filament volumes of 34 and 70%, for boron–aluminum, boron–epoxy, and glass–epoxy composites were presented in earlier publications by Adams (1970a, b). However, these publications contain only a very brief discussion of the method of analysis itself.

The computer program developed as part of this investigation (Repnau and Adams, 1973) readily permits the consideration of large finite element grid arrays. The program is presently dimensioned to handle a maximum of 200 node points and 350 elements. This has been found to be quite adequate for this type of problem. Even so, the program could be readily modified to permit larger arrays, being limited only by the capacity of the particular computer to be used and the cost of computer time required to solve such large problems.

By converting this plane strain analysis to generalized plane strain, i.e., by permitting an increment of average applied strain in the axial direction $\Delta\bar{\epsilon}_z$ for each applied load increment rather than requiring that $\Delta\bar{\epsilon}_z = 0$, the problem of axial loading can also be considered. This is achieved by specifying that $\Delta\bar{\sigma}_x = \Delta\bar{\sigma}_y = 0$ and $\Delta\bar{\sigma}_z \neq 0$ for each increment. This has been done recently by Lin *et al.* (1972), using an analysis which incorporates linear strain rather than constant strain triangular elements. Axial loading results are presented for both a boron–epoxy and a boron–aluminum composite of 50% filament volume, assuming a square array.

3. Stress Function Methods

Relatively little work has been done to date using stress functions to analyze elastoplastic material behavior. The combination of a complex geometry boundary value problem and the complex constitutive relations governing elastoplastic material behavior make most stress function approaches unattractive.

One analysis which can be cited, however, is that by Rybicki (1968). He applied an incremental complementary energy approach to both plane stress and generalized plane strain problems. The method bears some similarity and hence has some of the desirable characteristics of the finite element method in the sense that the structure or local material region to be analyzed is divided into an array of discrete elements. Rybicki considered two types of discrete elements, viz., a rectangular membrane element in representing plane stress and an axisymmetric cylindrical or ring element in representing generalized plane strain. A second-order Hermite interpolation polynomial was used to represent the stress function within each element. In this way material nonlinearities can be included in the formulation as desired. Solutions were obtained using a minimization algorithm.

Although Rybicki actually considered only a relatively simple composite material problem, i.e., a two-material concentric cylinder model subject to generalized plane strain, his work does represent an approach that could, at least in principle, be extended to the more complex boundary value problems typical of rigorous composite material analyses.

While it is not clear that this or any other stress function approach offers any significant advantages over the more conventional finite element methods as described in the previous subsection, it does provide an interesting alternative for consideration.

E. Typical Numerical Results

As discussed in the previous subsection, only a very limited number of numerical results of the rigorous analyses are available as of yet, and most

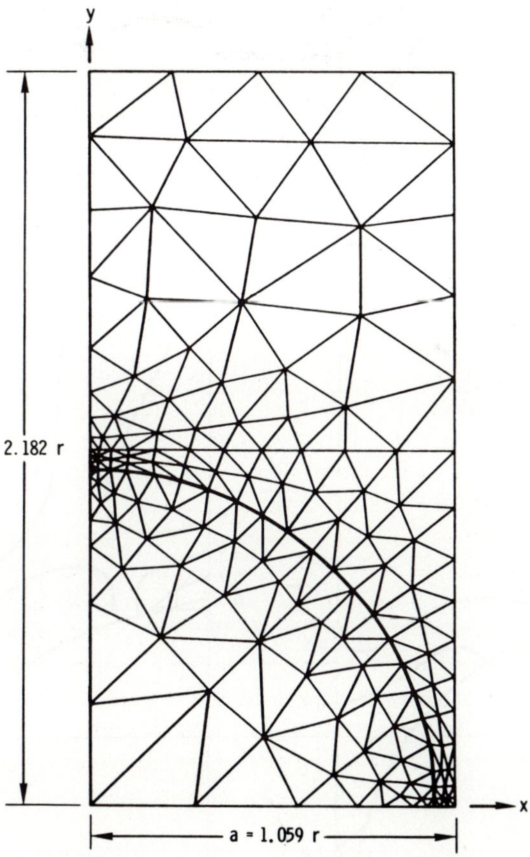

FIG. 6. Triangular finite element grid, rectangular filament array, $v_f = 34\%$, 272 elements, 161 nodes (Adams, 1970b).

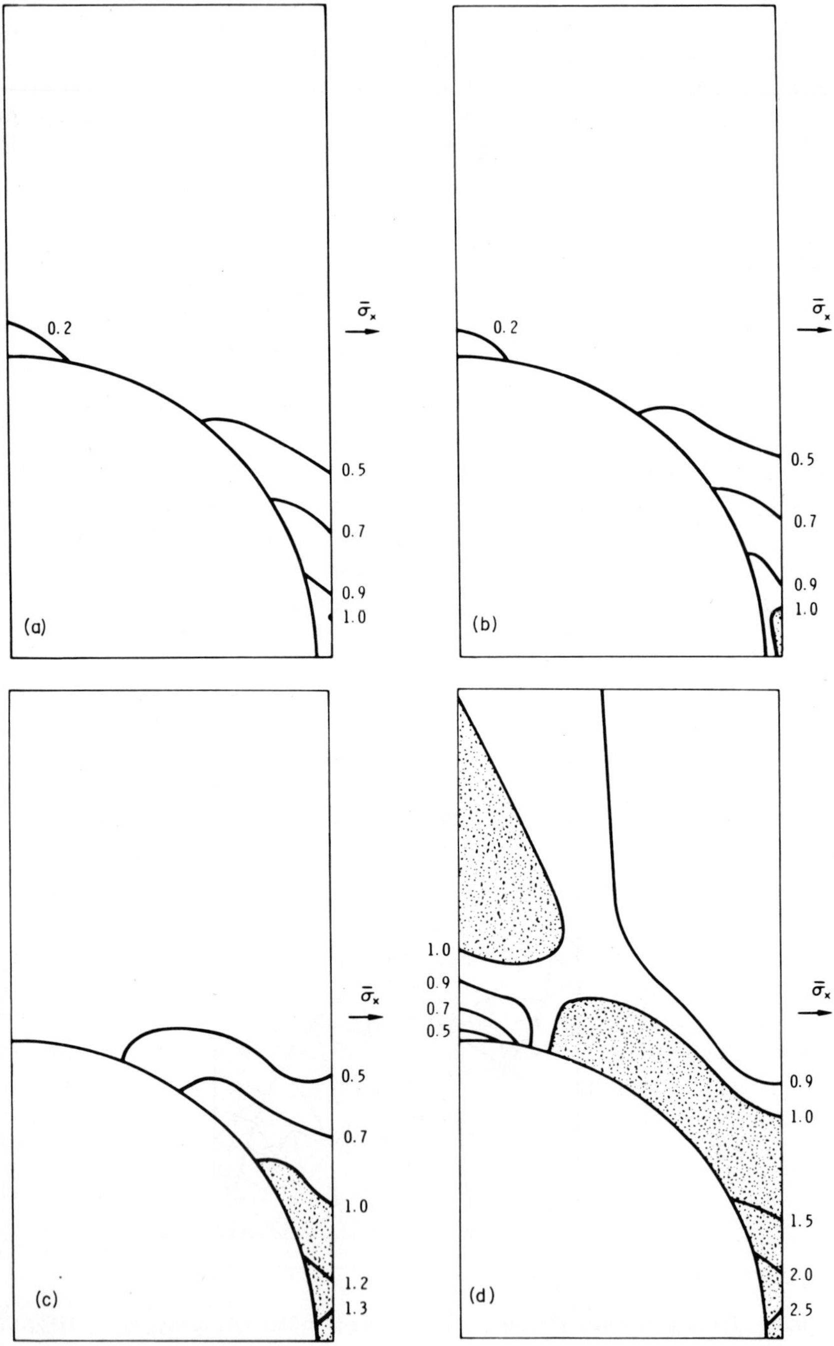
0.2
$\bar{\sigma}_x$
0.5
0.7
0.9
1.0
(a)
0.2
$\bar{\sigma}_x$
0.5
0.7
0.9
1.0
(b)
$\bar{\sigma}_x$
0.5
0.7
1.0
1.2
1.3
(c)
1.0
0.9
0.7
0.5
$\bar{\sigma}_x$
0.9
1.0
1.5
2.0
2.5
(d)

of these have been obtained using finite element analyses. The publications by Adams (1970a, b) and Foye and Baker (1971) contain numerical results of particular interest. Examples selected from the work of Adams (1970b) will be included here to indicate the type of information which can be obtained.

A condition of plane strain is assumed, and a loading in one transverse direction on a unidirectional composite containing a rectangular array of circular filaments has been selected for illustration. Both the epoxy and the aluminum matrix materials indicated in Fig. 1 will be utilized.

A typical finite element grid is indicated in Fig. 6. This models a composite containing 34 vol% of circular filaments of radius r packed in a rectangular array. Because of the assumed symmetry, as discussed in Section IV,C, it is only necessary to analyze the first quadrant of the repeating unit, as shown in Fig. 5b. The first quadrant represented in Fig. 6 indicates the assumed filament spacings; the filaments are closely spaced in the x direction and widely spaced in the y direction. The 272 triangular elements are sized to provide small elements in regions of expected high stress and strain gradients, and larger elements elsewhere.

Using this finite element grid, the aluminum matrix of Fig 1, and assuming boron filaments, an incremental loading in the x direction provides the following results.

Octahedral shear stress contours in the aluminum matrix for applied stress increment nos 1, 2, 5, and 10 are shown in Fig. 7. These contours, and all other octahedral shear stress contours presented here, have been normalized by dividing by the value of octahedral shear stress at which yielding of the material occurs ($\tau_{0(\text{yp})} = 6128$ lb/in.^{2} for the aluminum matrix of Fig. 1). Thus, regions for which $\tau_0/\tau_{0(\text{yp})} \geq 1$ (shown shaded in Fig. 7 and subsequent contour plots) represent material stressed beyond the elastic limit.

It should be noted that the shaded regions do not represent an abrupt change in material behavior (as would be the case for an elastic–perfectly plastic material) since the matrix stress–strain curve (see Fig. 1) makes a smooth transition from linear elastic to nonlinear response. In fact, the elastic limit (defined in plasticity theory as being the proportional limit) of the matrix is in itself not clearly defined from the experimental stress–strain curve and must be estimated. The shading is used primarily to aid the

---

FIG. 7. Octahedral shear stress contours (normalized), boron–aluminum composite, $v_f = 34\%$, rectangular array (Adams, 1970b). (a) Applied stress increment no. 1 with $\bar{\sigma}_x = 8098$ lb/in.$^2 = 1.00\bar{\sigma}_{x(\text{yp})}$; (b) increment no. 2 with $\bar{\sigma}_x = 8597$ lb/in.$^2 = 1.06\,\bar{\sigma}_{x(\text{yp})}$; (c) increment no. 5 with $\bar{\sigma}_x = 12{,}470$ lb/in.$^2 = 1.54\,\bar{\sigma}_{x(\text{yp})}$; (d) increment no. 10 with $\bar{\sigma}_x = 24{,}600$ lb/in.$^2 = 3.04\,\bar{\sigma}_{x(\text{yp})}$.

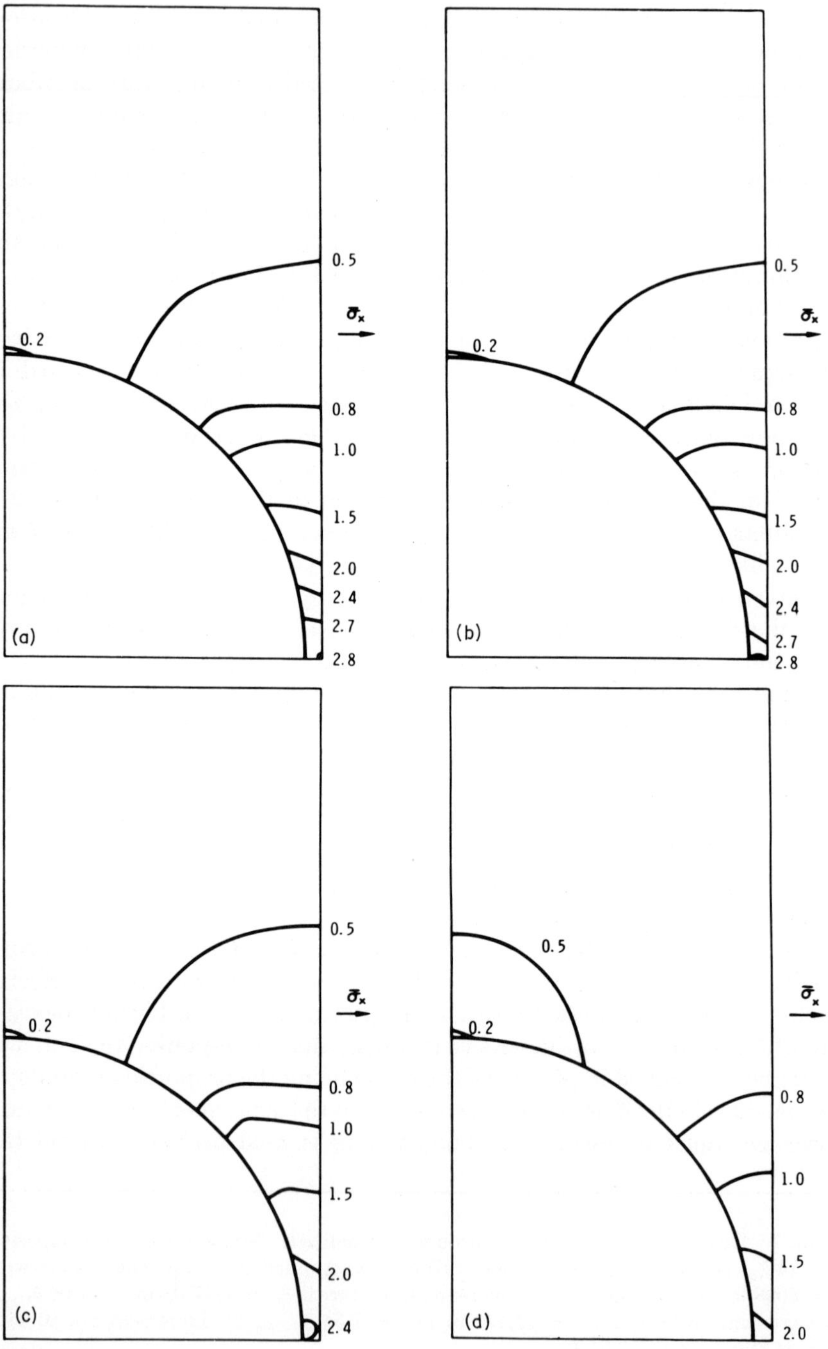
0.5
$\bar{\sigma}_x$
0.2
0.8
1.0
1.5
2.0
2.4
2.7
(a)
2.8
0.5
$\bar{\sigma}_x$
0.2
0.8
1.0
1.5
2.0
2.4
(b)
2.7
2.8
0.5
$\bar{\sigma}_x$
0.2
0.8
1.0
1.5
2.0
(c)
2.4
0.5
$\bar{\sigma}_x$
0.2
0.8
1.0
1.5
(d)
2.0

reader in observing the extent of nonlinear response for a given loading condition.

As in all examples to be presented here, the first applied stress increment ($\bar{\sigma}_x = 8098$ lb/in.^{2} in Fig. 7) represents the elastic limit loading of the composite. First yielding occurs in the matrix on the midline between filaments, indicated by the contour (a point) marked 1.0. As the applied stress is incrementally increased, the region of yielding spreads to the filament–matrix interface. As indicated for increment no. 10 [part (d)], an additional region appears and spreads through the matrix material between the vertically spaced filaments at the higher applied stress levels.

First failure occurs, i.e., the octahedral shear stress in the most stressed element exceeds the ultimate strength value dictated by the material stress–strain behavior of Fig. 1, in the local region between closely spaced filaments. The numerical solution technique of Adams (1970a, b) terminates at this point since it does not contain a provision for the introduction and propagation of a crack (a local material failure). The presentation of a procedure for predicting crack initiation, propagation, and total failure, i.e., composite ultimate strength, is contained in a recent publication by Adams (1973).

Contour plots of maximum principal stress for the same four increments of applied stress (increment nos. 1, 2, 5, and 10), are given in Fig. 8. These maximum principal stress contours have been normalized by dividing by the accumulated applied stress $\bar{\sigma}_x$ on the composite for that increment. Thus, the contour values shown indicate the level of stress concentration at a given applied stress. Note that the highest value of maximum principal stress (on the x axis midway between filaments) does not occur at the same point as the maximum octahedral shear stress (since the minimum principal stress, which also contributes to the computation of the latter value, is larger on the midline away from the x axis, while the maximum principal stress decreases only slightly). This is just one example of how the predicted location of first yielding can be dependent upon the particular yield criterion used. As discussed earlier, the analysis of Adams (1970a, b) utilizes an octahedral shear stress criterion.

As also indicated in Fig. 8, the normalized maximum principal stress contours away from the local region of high stress concentration change very little from increment to increment. However, due to the local yielding,

---

FIG. 8. Maximum principal stress contours (normalized), boron–aluminum composite, $v_f = 34\%$, rectangular array (Adams, 1970b). (a) Applied stress increment no. 1 with $\bar{\sigma}_x = 8098$ lb/in.^{2} $= 1.00\ \bar{\sigma}_{x(\text{yp})}$; (b) increment no. 2 with $\bar{\sigma}_x = 8597$ lb/in.^{2} $= 1.06\ \bar{\sigma}_{x(\text{yp})}$; (c) increment no. 5 with $\bar{\sigma}_x = 12{,}470$ lb/in.^{2} $= 1.54\ \bar{\sigma}_{x(\text{yp})}$; (d) increment no. 10 with $\bar{\sigma}_x = 24{,}600$ lb/in.^{2} $= 3.04\ \bar{\sigma}_{x(\text{yp})}$.

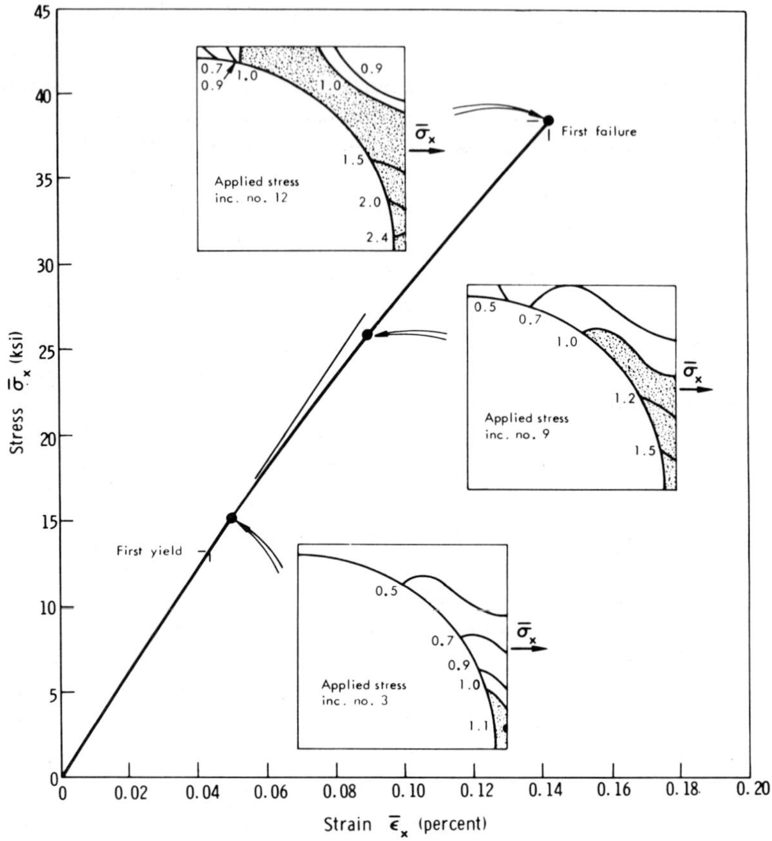

FIG. 9. Boron–aluminum composite, predicted stress–strain curve, and corresponding octahedral shear stress contours, $v_f = 70\%$, square array (Adams, 1970b).

the highest normalized maximum principal stress decreases from a value of 2.8 at the elastic limit [part (a)] to less than 2.1 at first failure [part (d)].

It should be kept in mind when comparing trends between Figs. 7 and 8 that the octahedral shear stress contours have been normalized by dividing by the constant value $\tau_{0(\mathrm{yp})}$, the octahedral shear stress at yield of the matrix, while the maximum principal stress contours have been normalized by dividing by $\bar{\sigma}_x$, the incrementally increasing applied stress. The finite element analysis also provides a complete description of the filament behavior, and contour plots are similarly available.

It is interesting to consider the behavior of the same boron filament and aluminum matrix constituents in a square packing array such that the spacing between filaments in both coordinate directions is the same as that in the x direction for the rectangular array of Figs. 6–8. This spacing corresponds to a square array filament volume of 70%. Selected results are presented in Fig. 9. First yielding is predicted to occur at an applied stress level of 13,230 lb/in.^{2} (as indicated by the tick marks in Fig. 9). Because of the very nearly linear behavior, however, a deviation from linearity is not apparent in the plotted curve until a somewhat higher stress level, at about the third applied stress increment, corresponding to 15,200 lb/in.^{2}. As indicated in the corresponding octahedral shear stress contour plot in Fig. 9, considerable local yielding has occurred by this time. Thus, if this were an experimental data curve, the elastic limit which would probably be assumed would actually correspond to a condition of local yielding, i.e., permanent deformation.

All results presented to this point have been for a specific pair of boron filament and aluminum matrix constituents. Results for a boron–epoxy composite, assuming a 70% filament volume square array, are presented in Fig. 10. The stress–strain curve for the epoxy matrix is given in Fig. 1. The same boron filament properties as previously used were assumed here. As in the case of the 70% boron–aluminum composite (Fig. 9), the apparent composite elastic limit (about 6100 lb/in.^{2}) is considerably higher than the predicted value of 4149 lb/in.^{2}. The extent of yielding at first failure (end of the curve) is much less than for the boron–aluminum system, however. This is due to the fact that the elastic limit strain for the epoxy (1.0%) is much greater than that for the aluminum (0.15%), even though the ultimate strains are similar (see Fig. 1).

Results for an S glass–epoxy composite (having a filament modulus of 12.4×10^6 lb/in.^{2} as compared to 55×10^6 lb/in.^{2} for the boron with all other conditions identical) are very similar to those presented in Fig. 10 for boron–epoxy and therefore are not presented. Specifically, similarities include the magnitude of applied stress at first failure (between 12,500 and 13,000 lb/in.^{2} in both cases), and the extent of local yielding at any given level of applied stress. The yield stress of the S glass–epoxy composite (4681 lb/in.^{2}) is higher than that for the boron–epoxy composite (4149 lb/in.^{2}) because of the lower stress concentration induced in the former.

It is not the purpose of the present discussion to attempt an exhaustive presentation of numerical results. However, some of the important parametric studies which can be performed now that rigorous analyses are available should be mentioned.

The importance of filament packing geometry, particularly the spacing between filaments, is apparent even from the selected results presented here. The close spacing in the direction of applied stress in the case of the rectangular array results in a severe local stress concentration. However,

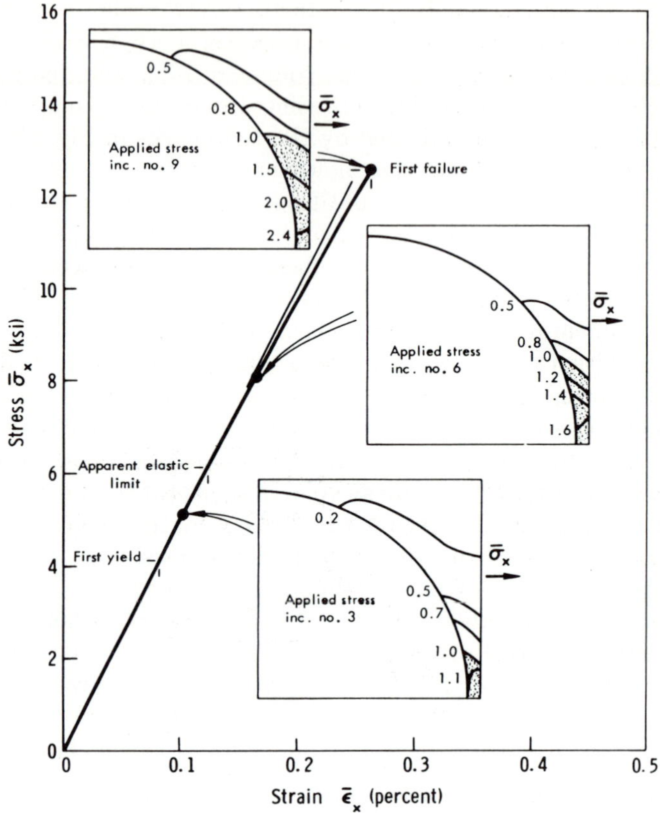

FIG. 10. Boron–epoxy composite, predicted stress–strain curve, and corresponding octahedral shear stress contours, $v_f = 70\%$, square array (Adams, 1970b).

had the stress been applied in the direction of wide spacing of the rectangular array, significantly different results would have been obtained. A complete parametric study of the influence of filament spacing, for different combinations of filament and matrix materials, will be very useful.

These analyses can also be used to determine the proper combination of variables to achieve a specific characteristic of the resulting composite. For example, this characteristic may be high ultimate strength, large strain at failure, or good biaxial transverse loading characteristics. The prediction of an optimum stress–strain curve for both the filament and the matrix to achieve a given composite property is a relatively unexplored area. This type of information will be very useful to those investigators who are actually developing new constituent materials.

The influence of filament shape on local material behavior, and hence on composite properties, is becoming an important consideration now that graphite filaments having very irregular cross sections are being utilized. The question of whether there are advantages in using the more nearly circular graphite filaments which are also being produced remains to be explored.

The introduction of a third material, for example in the form of a coating on each filament, can also be analyzed. In addition to providing a means of controlling the minimum spacing between filaments, a material with properties different from either the filaments or the matrix can provide a beneficial transition zone for transferring stresses between them.

The influence of voids in the filaments or the matrix material on the local state of stress is another problem of significant practical importance which can be analyzed.

These are just a few of the detailed parametric studies which remain to be explored. All can be performed using these rigorous analyses. In many cases it would be nearly impossible to isolate and to evaluate independently these many variables experimentally.

V. Needs for Additional Research

The basic methodology required to perform meaningful elastoplastic analyses of composite material behavior is now available, as discussed in the previous sections of this chapter. Simple model analyses have already been developed to about the fullest extent practicable. They cannot be expected to provide the accurate, detailed information required in future composite materials developments. On the other hand, the rigorous micromechanical analyses, which can provide this information, have not yet been adequately exercised, although the basic methodology is now in hand and the required computational schemes and computer capacity are available.

There are several reasons for this lack of widespread use of the rigorous methods up to the present time. They have not been available for use very long, for only the past two or three years at most. They require the use of large-scale, complex computer programs, which require some effort to become familiar with, even when the programs are thoroughly documented, which not all are. Most of these computer programs require a relatively large computer facility and significant amounts of computer time. Thus, they can be expensive to utilize extensively. The significant effort required to develop such a computer program, or to even modify an existing one, has limited the number of investigators active in this area of research.

On the other hand, the most difficult part of the basic development work has now been completed, and it has been demonstrated that these rigorous analyses can be successfully utilized to study elastoplastic micromechanical behavior of composites. At the present time, the finite element technique is the most highly developed method, with two equally serviceable options being available for modeling the particular geometries and boundary conditions characteristic of composite materials. These are the method of initial strains as adapted by Foye and Baker (1971) and the tangent modulus method as incorporated by Adams (1970a, b and 1973). Both approaches are fully described in these references and the associated computer programs are available and adequately documented.

These analyses need to be exercised by utilizing them in a variety of actual applications by various investigators. In this way, they will be further refined and expanded in capability.

List of Symbols

A	$6\tau_0{}^2M$
a, b	Dimensions of finite element grid
B	$3^2\tau_0 + (1 + \nu)(A/E)$
D	$A + Es^2{}_{33}$
E	Young's modulus
$[H]$	Plane strain elastoplastic stiffness matrix
J_2, J_3	Invariants of the deviatoric stress tensor
k	Yield criterion constant
$2M_{\mathrm{T}}$	Tangent modulus of the $\tau_0 - \epsilon_0{}^{(\mathrm{p})}$ curve
p	Filament spacing in Shaffer's model
r	Filament radius
s_1, s_2, s_3	Principal components of the deviatoric stress tensor
s_{ij}	Deviatoric stress components
$t_{\alpha\beta}$	$s_{\alpha\beta} + \nu s_{33}\delta_{\alpha\beta}$
$\gamma_{xy}, \gamma_{zx}, \gamma_{zy}$	Engineering shear strain components
δ_{ij}	Kronecker delta
$\epsilon_{ij}^{(\mathrm{e})}$	Elastic strain components
$\epsilon_{ij}^{(\mathrm{p})}$	Plastic strain components
$\epsilon_0{}^{(\mathrm{p})}$	Octahedral plastic shear strain
$\epsilon_x, \epsilon_y, \epsilon_z$	Normal strain components
ϵ_{yp}	Yield point strain
λ	Positive scalar function
ν	Poisson's ratio
$\bar{\sigma}$	Effective stress
$\sigma_1, \sigma_2, \sigma_3$	Principal stress components
σ_{ij}	Total stress components
$\sigma_x, \sigma_y, \sigma_z$	Normal stress components

$\bar{\sigma}_x$, $\bar{\sigma}_y$, $\bar{\sigma}_z$	Components of average applied stress
$\Delta\bar{\sigma}_x$, $\Delta\bar{\sigma}_y$, $\Delta\bar{\sigma}_z$	Increments of applied stress
σ_{yp}	Yield point stress
τ_0	Octahedral shear stress
$\tau_{0(\text{yp})}$	Octahedral shear stress at yield point

References

Adams, D. F. (1970a). *J. Compos. Mater.* **4,** 310.

Adams, D. F. (1970b). Inelastic Analysis of a Unidirectional Composite Subjected to Transverse Normal Loading. Rand Corp. Rep. RM-6245-PR.

Adams, D. F. (1973). An Elastoplastic Analysis of Crack Propagation in a Unidirectional Composite. Rand Corp. Rep., R-1070-PR.

Adams, D. F., Doner, D. R., and Thomas, R. L. (1967). Mechanical Behavior of Fiber-Reinforced Composite Materials. Contractor Rep. AFML-TR-67-96.

Allen, D. N. de G. (1954). "Relaxation Methods." Oxford Univ. Press, London and New York.

Allen, D. N. de G., and Southwell, R. V. (1950). *Phil. Trans. Roy. Soc.* **A242,** 379.

Bert, C. W. (1972). Plasticity and Creep Analysis of Filmentary Metal-Matrix Composites. Sandia Corp. Rep. SC-DR-720055.

Dow, N. F., and Rosen, B. W. (1965). Evaluation of Filament-Reinforced Composites for Aerospace Structural Applications. NASA Contractor Rep. CR-207.

Ebert, L. J., and Gadd, J. D. (1965). *In* "Fiber Composite Materials," Chapter 5. Amer. Soc. Met., Metals Park, Ohio.

Ebert, L. J., Hecker, S. S., and Hamilton, C. H. (1968). *J. Compos. Mater.* **2,** 458.

Foye, R. L. (1968). Advanced Design Concepts for Advanced Composite Airframes. Contractor Rep. AFML-TR-68-91, Vol. 1.

Foye, R. L., and Baker, D. J. (1971). Design/Analysis Methods for Advanced Composite Structures. AFML-TR-70-299, Vol. I, Analysis; Vol. II, Computer Programs.

Gallagher, R. H., Padlog, J., and Bijlaard, P. P. (1962). *ARS J.* **32,** 700.

Havner, K. S. (1966). *Quart. Appl. Math.* **23,** 323.

Hill, R. (1950). "The Mathematical Theory of Plasticity." Oxford Univ. Press, London and New York.

Hill, R. (1964). *J. Mech. Phys. Solids* **12,** 199.

Huang, W. (1971). *J. Compos. Mater.* **5,** 320.

Lansing, W., Jenson, W. R., and Falby, W. (1966). *In* "Matrix Methods in Structural Mechanics" (J. S. Przemieniecki *et al.*, eds.), AFFDL-TR-66-80, pp. 605–633.

Levy, S. (1953). *J. Aerosp. Sci.* **20,** 449.

Lin, T. H., Salinas, D., and Ito, Y. M. (1972). *J. Compos. Mater.* **6,** 48.

Marcal, P. V. (1968). *AIAA J.* **6,** 157.

Marcal, P. V. (1969). Finite Element Analysis with Material Nonlinearities—Theory and Practice. Brown Univ.

Marcal, P. V., and Turner, C. E. (1963). *J. Mech. Eng. Sci.* **5,** 3.

Mendelson, A. (1968). Elastoplastic Analysis of a Circular Cylindrical Inclusion in a Uniformly Stressed Infinite Homogeneous Matrix. NASA TN D-4350.

Mendelson, A., and Manson, S. S. (1959). Practical Solution of Plastic Deformation Problems in the Elastic-Plastic Range. NASA TR R28.

Nadai, A. (1937). *J. Appl. Phys.* **8,** 205.

Owen, D. R. J., Holbeche, J., and Zienkiewicz, O. C. (1969). *Fibre Sci. Technol.* **1,** 185.

Padlog, J., Huff, R. D., and Holloway, F. G. (1960). Unelastic Behavior of Structures Subjected to Cyclic, Thermal and Mechanical Stressing Conditions. WADD TR-60-271.

Prager, W., and Hodge, P. G., Jr. (1951). "Theory of Perfectly Plastic Solids." Wiley, New York.

Repnau, T., and Adams, D. F. (1973). A Finite Element Computer Program for the Elastoplastic Analysis of Crack Propagation in a Unidirectional Composite. Rand Corp. Rep., R-1392-PR.

Rybicki, E. F. (1968). An Incremental Complementary Energy Method of Nonlinear Stress Analysis. Case Western Res. Univ., Solid Mech., Struct. and Mech. Design Div., Rep. No. 17.

Sendeckyj, G. P. (1969). Integrated Research on Carbon Composite Materials. AFML-TR-66-310, Part III, p. 259.

Shaffer, B. W. (1964). *AIAA J.* **2,** 348.

Shaffer, B. W. (1968). *AIAA J.* **6,** 2316.

Shu, L. S., and Rosen, B. W. (1967). *J. Compos. Mater.* **1,** 366.

Southwell, R. V. (1940). "Relaxation Methods in Engineering Science." Oxford Univ. Press, London and New York.

Spencer, A. J. M. (1962). *Appl. Sci. Res.* **A12,** 391.

Swedlow, J. L. (1969). *Int. J. Fracture Mech.* **5,** 33.

Swedlow, J. L., and Yang, W. H. (1966). Stiffness Analysis of Elasto-Plastic Plates, California Inst. Technol. Rep. GALCIT SM 65-10.

Turner, M. J., Clough, R. W., Martin, H. C., and Topp, L. J. (1956). *J. Aerosp. Sci.* **23,** 805.

Zienkiewicz, O. C. (1971). "The Finite Element Method in Engineering Science." McGraw-Hill, New York.

6

Application of Statistical Theories for the Determination of Thermal, Electrical, and Magnetic Properties of Heterogeneous Materials

MARK J. BERAN

Towne School of Civil and Mechanical Engineering
University of Pennsylvania
Philadelphia, Pennsylvania

I. Introduction

The purpose of this chapter is to discuss the application of explicitly statistical theories for the determination of the properties of heterogeneous materials. For the most part we shall be concerned with variables governed by equations of the form

$$\frac{\partial}{\partial x_i}[\epsilon_{ij}(\mathbf{x})E_j(\mathbf{x})] = \rho(\mathbf{x}) \tag{1a}$$

$$\delta_{ijk}\frac{\partial}{\partial x_j}E_k(\mathbf{x}) = 0 \tag{1b}$$

where δ_{ijk} is the antisymmetric tensor of Levi–Civita, and ϵ_{ij}, E_j, and ρ may represent thermal, electrical, or magnetic properties. For example, we may choose the following interpretations: $\epsilon_{ij}(\mathbf{x})$—conductivity tensor, $E_j(\mathbf{x})$—temperature gradient, $\rho(\mathbf{x})$—heat source density. The same techniques (albeit with more complex equations) may also be applied to elastic properties (Beran, 1971).

We consider here problems in which $\epsilon_{ij}(\mathbf{x})$ may be most conveniently treated from a statistical point of view. For convenience we shall assume that $\rho(\mathbf{x})$ is deterministic, but if this function is also treated statistically, no real additional difficulties are introduced. We suppose $\phi(\mathbf{x})$ [where $E_j(\mathbf{x}) = \partial\phi(\mathbf{x})/\partial x_j$] or $n_i\epsilon_{ij}E_j$ is given on some surface S, and we wish to study the material properties in the interior region V. We choose both the surface geometry and the boundary conditions to be deterministic. A statistical variation in ϕ or $\epsilon_{ij}E_j$ can be included in the formulation, but introducing a stochastic variation in the geometry of S is very difficult and is a subject that has received very little attention. (See, however, Lomakin, 1968, for a perturbation approach to this problem.)

In many cases Eqs. (1) have been treated from a deterministic point of view, assuming $\epsilon_{ij}(\mathbf{x})$ has some arbitrary variation with position. Sometimes this approach has been fruitful, while at other times it has served to conceal vital aspects of the problem. We shall not discuss these approaches here but will, when applicable, compare results obtained by both methods. We shall, however, point out problems in which the physical case can only be interpreted properly from a statistical point of view.

The statistical theory we present here is similar to the statistical theory of turbulence. Because Eqs. (1) are linear many more theoretical results have been obtained, but much of the underlying statistical formulation is formulated along parallel lines. We also encounter here the infinite hierarchy of equations that must be terminated. Unfortunately, the experimental results (in which statistical quantities are determined from measurement) lag far behind the many significant experiments performed in turbulence theory.

For people with a nonmathematical orientation, the developments in the statistical theory of heterogeneous materials may seem unduly cumbersome. The results, however, in many cases are very tractable. In this chapter we shall try to present the basic statistical formulation in enough detail so that it may be understood by those interested in further developing this area. At the same time we shall omit subsequent detailed mathematical derivations and concentrate instead on results which may be readily applied.

In Section II we shall discuss the basic deterministic equations given in Eqs. (1). We consider the type of analytic solutions that may be obtained and state the variational principles associated with these equations. Some remarks will be made about the determination of bounds for effective constants.

An explicitly statistical formulation will be given in Section III. The concept of an ensemble average will be introduced, and its relation to volume averages will be considered. The infinite hierarchy of statistical equations will be discussed, and it will be pointed out that the general functional formulation allows the only complete solution to the problem. Some remarks will be made about numerical solutions.

The mean field $\{E_i(\mathbf{x})\}$ may be shown to satisfy an integrodifferential equation which is called a *Dyson equation.* This equation is presented in Section IV. The kernel of the integral portion of the equation depends on all order statistical information associated with the function $\epsilon_{ij}(\mathbf{x})$, and it is usually not possible to determine this function except in the limit of small perturbations. The equation, however, provides us with a phenomenological form that is more general than the differential equations that hitherto have been postulated to govern $\{E(\mathbf{x})\}$. The integrodifferential form is absolutely necessary in the neighborhood of rapidly varying functions $\rho(\mathbf{x})$ or near the boundaries of the medium. It is shown under what circumstances the integrodifferential equation reduces to a differential equation.

Effective constants have received the most attention in the literature of heterogeneous materials. In Section V we shall present bounds for the effective constants in terms of statistical information. The final expressions will be simple algebraic ones. (Most of this section may be read without reading the previous ones.) While the derivation of bounds is a theoretical problem of much interest, bounding expressions which contain volume fractions, geometric factors, and packing factors are a necessity to the design engineer working in composite materials. The innumerable number of possible combinations makes measurement of initial designs impossible, and bounds are needed to provide guides to allow an engineer to design a composite material to meet a particular need.

In Section VI we shall discuss the comparison of theory and experiment. This discussion will be confined to effective constants. We shall be particularly concerned with experiments that allow the possibility of determining statistical information about $\epsilon_{ij}(\mathbf{x})$. Most experiments have been limited to measurements of only volume fractions, but some recent work shows how statistical information may be measured.

The principal application of the statistical approach is in the design of composite materials with desired effective constants. We shall discuss the problem of increasing the effective thermal conductivity of thermal storage materials in Section VII. We also wish to point out in this section that the bounding techniques that have been used to gain information about effective constants yielded statistical parameters that may be of use in other areas of application.

We conclude this chapter with a summary in Section VIII followed by a reference list and bibliography.

II. Basic Equations

The basic equations which govern the problems we wish to treat are given in Eqs. (1a) and (1b). Equation (1b) is satisfied if we choose

$$E_i = \partial\phi/\partial x_i \tag{2}$$

Equation (1a) then becomes

$$\partial/\partial x_i[\epsilon_{ij}(\mathbf{x})\ \partial\phi(\mathbf{x})/\partial x_j] = \rho(\mathbf{x}) \tag{3}$$

The boundary conditions associated with this equation are the specification of ϕ or $\epsilon_{ij}\ \partial\phi/\partial x_j\ n_i$ on some surface S enclosing a volume V.

The problem of interest to us here is the case in which $\epsilon_{ij}(\mathbf{x})$ may be conveniently described from a statistical point of view. Thus, for example, we shall not consider here functions $\epsilon_{ij}(\mathbf{x})$ that may be represented by simple polynomials or a single sine function. In an effort to simplify the mathematical development we shall restrict our attention to the case where

$$\epsilon_{ij}(\mathbf{x}) = \epsilon(\mathbf{x})\delta_{ij} \tag{4}$$

All the developments we present may be generalized to the arbitrary tensor function.

For the most part we shall consider $\epsilon(\mathbf{x})$ functions that are statistically homogeneous and isotropic. We shall explain this concept more fully in the next section, but we note here that in this case we may write

$$\epsilon(\mathbf{x}) = \bar{\epsilon} + \epsilon'(\mathbf{x}) \tag{5}$$

where $\bar{\epsilon}$ is independent of $\mathbf{x}$ and $\epsilon'(\mathbf{x})$ is a fluctuating function whose mean value is zero. Using Eqs. (4) and (5), Eq. (3) may be written as

$$\bar{\epsilon}\nabla^2\phi(\mathbf{x}) = -\frac{\partial}{\partial x_i}\left[\epsilon'(\mathbf{x})\frac{\partial\phi(\mathbf{x})}{\partial x_i}\right] + \rho(\mathbf{x}) \tag{6}$$

If $\epsilon'(\mathbf{x})$ is a random function, we are, in general, unable to solve Eq. (6) by the analytic methods we would use if $\epsilon'(\mathbf{x})$ was a known deterministic function. We are forced to use special techniques that rely on the fact that we are most interested in the average value of $\phi(\mathbf{x})$ [which we denote as $\{\phi(\mathbf{x})\}$] rather than a detailed specification of $\phi(\mathbf{x})$. We should note immediately, however, that we cannot find $\{\phi(\mathbf{x})\}$ simply by averaging Eq. (6). If we average Eq. (6), we find

$$\bar{\epsilon}\nabla^2\{\phi(\mathbf{x})\} = -\frac{\partial}{\partial x_i}\left\{\epsilon'(\mathbf{x})\frac{\partial\phi(\mathbf{x})}{\partial x_i}\right\} + \rho(\mathbf{x}) \tag{7}$$

In general, the term $\{\epsilon'(\mathbf{x})\ \partial\phi(\mathbf{x})/\partial x_i\}$ is not simply related to $\{\phi(\mathbf{x})\}$, and thus Eq. (7) is not of use for finding $\{\phi(\mathbf{x})\}$.

In a sense the basic problem we seek to solve in a statistical formulation is to find a simple relation between $\{\phi(\mathbf{x})\}$ and $\{\epsilon'(\mathbf{x})\ \partial\phi(\mathbf{x})/\partial x_i\}$. We shall show what has been accomplished in this direction in Section IV.

Equation (6) can, in principle, be solved numerically if $\epsilon'(\mathbf{x})$ is specified. If the computer capacity is available, it does not matter how complex the function $\epsilon'(\mathbf{x})$ that we consider. In numerical studies we are thus faced with two problems: (1) the specification of $\epsilon'(\mathbf{x})$, and (2) the estimation of computer capacities sufficient to calculate $\phi(\mathbf{x})$. The specification of $\epsilon'(\mathbf{x})$ is clearly a nontrivial problem if we wish to describe a real physical material. To understand the full implications of what need be done, we must first introduce an explicitly statistical formulation. We shall, therefore, defer discussion of this point, and return to this question again in Section III.

In an effort to circumvent much of the difficulties inherent in studying heterogeneous materials, workers have sought to replace Eq. (7) by the equation

$$\epsilon^*\nabla^2\{\phi(\mathbf{x})\} = \rho(\mathbf{x}) \tag{8}$$

Here ϵ^* is an effective constant, and from an examination of Eq. (7) we see that if Eq. (8) is applicable, then ϵ^* is determined from the function $\partial/\partial x_i\{\epsilon'(\mathbf{x})\ \partial\phi(\mathbf{x})/\partial x_i\}$.

Equation (8) is applicable if the function $\rho(\mathbf{x})$ varies very slowly in a characteristic distance l_c over which $\epsilon'(\mathbf{x})$ varies rapidly. (In statistical language l_c would be a characteristic correlation length.) We denote L_ρ to be a characteristic length associated with $\rho(\mathbf{x})$. In addition, we would require that l_c shall be compared to a characteristic length of the material sample L_V. If these conditions are met, it is meaningful to speak of an effective constant. The problem then is to determine ϵ^* from a knowledge of the statistical properties of $\epsilon'(\mathbf{x})$.

As we shall see in the next section, a complete statistical specification of the $\epsilon'(\mathbf{x})$ field is necessary if ϵ^* is to be determined without any approximation. If we only have a few statistical properties of the $\epsilon'(\mathbf{x})$ field, then we can only bound ϵ^* between two limits, ϵ_{L} and ϵ_{B}. That is, we can only find

$$\epsilon_{\mathrm{L}} \leq \epsilon^* \leq \epsilon_{\mathrm{B}} \tag{9}$$

For example, if we have a two-phase medium and the only information that we have available is the volume fraction of the phases, we can only determine bounds for ϵ^*. When the ratio of the properties of the two phases is large, the ratio $\epsilon_{\mathrm{B}}/\epsilon_{\mathrm{L}}$ may be large, and little can be said about the true value of ϵ^*. In order to find more restrictive bounds we must include more statistical information that gives shape and packing information.

To find bounds for ϵ^* we make use of two standard variational principles. The principle of minimum potential energy states that the integral

$$U = (1/2V) \int_V \epsilon E_i E_i \, d\mathbf{x} \tag{10}$$

subject to the subsidiary condition

$$\mathrm{curl}(\mathbf{E}) = 0 \qquad (E_i = \partial\phi/\partial x_i)$$

is stationary and a minimum for

$$\frac{\partial}{\partial x_i} \epsilon \frac{\partial \phi}{\partial x_i} = 0 \tag{11}$$

where V is the volume of the medium, and we denote S to be the surface enclosing V. As a boundary condition we require that the variation of ϕ, $\delta\phi$, is zero on the surface S.

When ϕ is the solution to the problem, then U is the energy density of the field U_{E}. Any other choice of ϕ satisfying the boundary condition and the subsidiary condition yields a value of U such that

$$U_{\mathrm{E}} \leq U \tag{12}$$

When the condition $l_c \ll L_V$ (for $\rho = 0$) is satisfied, we may write

$$U_E = \tfrac{1}{2}\epsilon^* \{\partial\phi/\partial x_i\} \{\partial\phi/\partial x_i\} \le U \tag{13}$$

where $\{\partial\phi/\partial x_i\}$ is the ensemble average value of the field in V.

If, for example, the trial solution is taken to be the average field $\{\partial\phi/\partial x_i\}$, then we find from Eqs. (9) and (12) the simple result

$$\epsilon^* \le \bar{\epsilon} \tag{14}$$

Using a better choice of the $\phi(\mathbf{x})$ field we may find a more restrictive upper bound for ϵ^*.

The lower bound for ϵ^* is found from the complementary variational principle. This principle states that the integral

$$U = (1/2V) \int_V D_i D_i/\epsilon \, d\mathbf{x} \tag{15}$$

subject to the subsidiary condition

$$\partial D_i/\partial x_i = 0 \tag{16}$$

is stationary and a minimum for

$$\mathrm{curl}[\mathbf{D}/\epsilon] = 0 \tag{17}$$

Equation (16) implies that

$$\mathbf{D} = \mathrm{curl}\ \mathbf{A} \tag{18}$$

where $\mathbf{A}$ is a vector field. The boundary condition for the variation of A_i, δA_i, is zero on the surface S.

If we assume as a trial function that D_i is the average value of D_i in the volume V, we find the lower bound $1/(\overline{1/\epsilon})$, where $(\overline{1/\epsilon})$ is the average value of $1/\epsilon$ in V. Using this result and Eq. (14) we find

$$1/(\overline{1/\epsilon}) \le \epsilon^* \le \bar{\epsilon} \tag{19}$$

The bounds given in Eq. (19) are not very restrictive. They contain only information about the probability of ϵ having a particular value. (For a two-phase medium this is simply the volume fraction.) One of the aims of a statistical theory is to introduce further information into the bounds. This is done by choosing better trial functions for ϕ and D_i. As we shall see in Section V, the trial functions generated from a statistical perturbation analysis are suitable for introducing shape and packing information into the bounds.

III. Statistical Formulation

A. Concept of an Ensemble, Probability Density Functions, and Correlation Functions

1. Material Properties

In order to properly study heterogeneous materials from a statistical point of view, it is necessary to introduce the concept of an ensemble. The formulation is similar to that used in turbulence and classical statistical mechanics. In an ensemble approach we do not treat a single material sample but rather a collection of material samples that are all prepared in the same macroscopic manner. By this we mean that each sample is prepared by the same manufacturing process, has the same fraction of components, has the same geometrical configuration, and in any overall sense is indistinguishable from the other samples of the collection. The differences in the samples can only be ascertained on a submacroscopic level. For example, suppose we form a material mixture by combining two different powdered metals. We manufacture the samples by choosing for each sample the same volume fractions, we compress the powders in the same shaped mold, and use the same compression apparatus and process control. The samples we obtain are macroscopically identical, but under close examination we see that a plot of material versus position would be quite different in each sample. The collection of samples so constructed is called an *ensemble.*

For an ensemble constructed in a manner similar to the above example we can only describe the material properties in a statistical sense. The function $\epsilon(\mathbf{x})$ is complicated in each sample, and if we focus our attention on a point $\mathbf{x}_1$, the only way we can describe $\epsilon(\mathbf{x}_1)$ is by giving the probability density function

$$P_1\left(\epsilon(\mathbf{x}_1)\right) d\epsilon(\mathbf{x}_1)$$

This function gives the probability that $\epsilon(\mathbf{x}_1)$ lies between $\epsilon(\mathbf{x}_1)$ and $\epsilon(\mathbf{x}_1) + d\epsilon(\mathbf{x}_1)$. We may determine this function experimentally by measuring $\epsilon(\mathbf{x}_1)$ in a large number of samples and determining the percentage of samples in which $\epsilon(\mathbf{x}_1)$ lies between ϵ_a and $\epsilon_a + \Delta\epsilon_a$. In the two-phase mixtures described above $\epsilon(\mathbf{x}_1)$ can only take on one of two values, ϵ_1 or ϵ_2, and

$$P_1\left(\epsilon(\mathbf{x}_1)\right) = v_1\delta\left(\epsilon(\mathbf{x}_1) - \epsilon_1\right) + v_2\delta\left(\epsilon(\mathbf{x}_1) - \epsilon_2\right) \qquad (20)$$

where v_1 and $v_2 = 1 - v_1$ are the volume fractions of the materials (assuming they are prepared in a homogeneous manner). In general, however, we may consider samples in which $\epsilon(\mathbf{x})$ may take on a continuum of values.

From a statistical point of view P_1 gives only limited information since it tells us nothing about how $\epsilon(\mathbf{x})$ varies from point to point. For example, in the above two-phase case it tells us nothing about the shape of the powders or how they are packed together. To gain some of this additional information we must have knowledge of the probability density function

$$P_2\left(\epsilon(\mathbf{x}_1), \epsilon(\mathbf{x}_2)\right) d\epsilon(\mathbf{x}_1)\, d\epsilon(\mathbf{x}_2)$$

This function gives the probability that $\epsilon(\mathbf{x}_1)$ lies between $\epsilon(\mathbf{x}_1)$ and $\epsilon(\mathbf{x}_1) + d\epsilon(\mathbf{x}_1)$ and $\epsilon(\mathbf{x}_2)$ lies between $\epsilon(\mathbf{x}_2)$ and $\epsilon(\mathbf{x}_2) + d\epsilon(\mathbf{x}_2)$. [We note that $P_1\left(\epsilon(\mathbf{x}_1)\right) = \int P_2\left(\epsilon(\mathbf{x}_1), \epsilon(\mathbf{x}_2)\right) d\epsilon(\mathbf{x}_2)$.] In other words, it tells how the values of $\epsilon(\mathbf{x})$ at $\mathbf{x}_1$ and $\mathbf{x}_2$ are related. This relationship is best given by the correlation function $\{\epsilon(\mathbf{x}_1)\epsilon(\mathbf{x}_2)\}$ defined as

$$\{\epsilon(\mathbf{x}_1)\epsilon(\mathbf{x}_2)\} = \int_0^\infty \int_0^\infty \epsilon(\mathbf{x}_1)\epsilon(\mathbf{x}_2) P_2\left(\epsilon(\mathbf{x}_1), \epsilon(\mathbf{x}_2)\right) d\epsilon(\mathbf{x}_1)\, d\epsilon(\mathbf{x}_2) \tag{21}$$

When $\mathbf{x}_1 = \mathbf{x}_2$, then the correlation function reduces to the mean square fluctuation in $\epsilon(\mathbf{x})$ at $\mathbf{x} = \mathbf{x}_1$. When $|\mathbf{x}_1 - \mathbf{x}_2| \to \infty$, the values of $\epsilon(\mathbf{x})$ at $\mathbf{x}_1$ and $\mathbf{x}_2$ are assumed to be independent and

$$P_2\left(\epsilon(\mathbf{x}_1), \epsilon(\mathbf{x}_2)\right) = P_1\left(\epsilon(\mathbf{x}_1)\right) P_2\left(\epsilon(\mathbf{x}_2)\right) \tag{22}$$

Then

$$\{\epsilon(\mathbf{x}_1)\epsilon(\mathbf{x}_2)\} = \{\epsilon(\mathbf{x}_1)\}\{\epsilon(\mathbf{x}_2)\} \tag{23}$$

where $\{\epsilon(\mathbf{x}_i)\}$ is the average value of $\epsilon(\mathbf{x}_i)$.

It is convenient to work with the function

$$\epsilon'(\mathbf{x}) = \epsilon(\mathbf{x}) - \{\epsilon(\mathbf{x})\} \tag{24}$$

where

$$\{\epsilon'(\mathbf{x})\} = 0$$

The correlation function

$$R'_{2\epsilon}(\mathbf{x}_1, \mathbf{x}_2) = \{\epsilon'(\mathbf{x}_1)\epsilon'(\mathbf{x}_2)\} \tag{25}$$

has a maximum value $\{\epsilon'^2(\mathbf{x}_1)\}$ when $\mathbf{x}_1 = \mathbf{x}_2$ and decays to zero as $|\mathbf{x}_1 - \mathbf{x}_2| \to \infty$.

The distance in which $R'_{2\epsilon}(\mathbf{x}_1, \mathbf{x}_2)$ becomes effectively zero will be denoted by l_c. This distance is called a *characteristic correlation length* and plays a central role in all statistical theories. As we pointed out above, the relationship between l_c and L_ρ or L_V is crucial in understanding the behavior of the material.

Sometimes $R'_{2\epsilon}(\mathbf{x}_1, \mathbf{x}_2)$ may be predicted analytically if something is known about the manufacturing process. It is not difficult to do this in a two-phase mixture if the concentration of inclusions in a matrix is low. For high concentrations it is very difficult to do unless the manufacturing process may be shown to conform to a simple theoretical model. Frisch (1965), Gilbert (1962), Weissberg and Prager (1962), Beran (1968), and Corson (1971) give a discussion of correlation functions for various types of materials.

Up to this point we have not assumed that the statistical properties of the material sample need be the same in different parts of the sample. That is, $\{\epsilon(\mathbf{x}_1)\}$, for example, could depend upon $\mathbf{x}_1$. In this case the only way to measure $R'_{2\epsilon}(\mathbf{x}_1, \mathbf{x}_2)$ would be to perform measurements on a large number of material samples. In many practical applications it is, however, possible to assume that the samples are statistically homogeneous. That is, we assume that $\{\epsilon(\mathbf{x}_1)\}$ is independent of $\mathbf{x}_1$, $R'_{2\epsilon}(\mathbf{x}_1, \mathbf{x}_2)$ depends only on $|\mathbf{x}_1 - \mathbf{x}_2|$, and in general, all statistical functions are independent of absolute position.

When the samples are statistically homogeneous, we usually invoke an ergodic hypothesis and assume that spatial averages are identical to ensemble averages. The spatial average, denoted by angular brackets $\langle\ \rangle$ is defined as

$$\langle\ \rangle = \lim_{V\to\infty} (1/V) \int_V (\ )\, d\mathbf{x} \tag{26}$$

Since V is never actually infinite, we require in application that $L_V \gg l_c$. When the identification $\{\ \} = \langle\ \rangle$ is possible, all statistical information may be determined from a single sample and measurement is attendantly greatly simplified.

Just as $P_2(\epsilon(\mathbf{x}_1), \epsilon(\mathbf{x}_2))$ contains far more information than $P_1(\epsilon(\mathbf{x}_1))$, so the function $P_3(\epsilon(\mathbf{x}_1), \epsilon(\mathbf{x}_2), \epsilon(\mathbf{x}_3))$ contains information not contained in P_2 and P_1. In particular the three-point correlation defined as

$$\begin{aligned} R'_{3\epsilon}(\mathbf{x}_1, \mathbf{x}_2, \mathbf{x}_3) &= \{\epsilon'(\mathbf{x}_1)\epsilon'(\mathbf{x}_2)\epsilon'(\mathbf{x}_3)\} \\ &= \int_0^\infty \int_0^\infty \int_0^\infty \epsilon'(\mathbf{x}_1)\epsilon'(\mathbf{x}_2)\epsilon'(\mathbf{x}_3) \\ &\quad \times P_3(\epsilon(\mathbf{x}_1), \epsilon(\mathbf{x}_2), \epsilon(\mathbf{x}_3))\, d\epsilon(\mathbf{x}_1)\, d\epsilon(\mathbf{x}_2)\, d\epsilon(\mathbf{x}_3) \end{aligned} \tag{27}$$

contains significantly more information than $R'_{2\epsilon}(\mathbf{x}_1, \mathbf{x}_2)$. Often $R'_{3\epsilon}(\mathbf{x}_1, \mathbf{x}_2, \mathbf{x}_3)$ contains shape information not contained in $R'_{2\epsilon}$. If we have a two-phase material, $R'_{3\epsilon}$ can give good information about the shape of the inclusion, distinguishing between spherical and needlelike shapes. As we shall see subsequently, $R'_{3\epsilon}$ appears naturally when determining bounds for the effective constant ϵ^*.

In general, we require the function

$$P_n\left(\epsilon(\mathbf{x}_1), \epsilon(\mathbf{x}_2), \cdots, \epsilon(\mathbf{x}_n)\right)$$

in the limit $n \to \infty$ to describe completely the statistical properties of the ensemble. The amount of information contained in P_n for n large is enormous and of little practical value [unless P_n is simply related to P_m $(m < n)$ as in the case of Gaussian statistics]. From a conceptual point of view, however, it is essential to realize that if only P_2 or P_3 is specified, we have only partial statistical information and any subsequent solutions we obtain for $\{\phi(\mathbf{x})\}$ or ϵ^* can only be approximate.

2. Statistical Field Properties

Just as we required statistical quantities like $\{\epsilon(\mathbf{x}_1)\}$, $R'_{2\epsilon}(\mathbf{x}_1, \mathbf{x}_2)$, $R'_{3\epsilon}(\mathbf{x}_1, \mathbf{x}_2, \mathbf{x}_3)$, so we require analogous quantities to describe the statistical properties of ϕ or E_i. We require quantities of the form

$$\begin{aligned} &\{\phi(\mathbf{x})\} \\ R'_{2\phi}(\mathbf{x}_1, \mathbf{x}_2) &= \{\phi'(\mathbf{x}_1)\phi'(\mathbf{x}_2)\} \\ R'_{3\phi}(\mathbf{x}_1, \mathbf{x}_2) &= \{\phi'(\mathbf{x}_1)\phi'(\mathbf{x}_2)\phi'(\mathbf{x}_3)\} \\ &\vdots \end{aligned}$$

We note that $\{E_i(\mathbf{x})\} = \partial\{\phi(\mathbf{x})\}/\partial x_i$ and that, in general, R'_{nE_i} can be obtained from $R'_{n\phi}$ by differentiation. For complete generality we need the probability density function

$$P_n\left(\phi(\mathbf{x}_1), \phi(\mathbf{x}_2), \cdots, \phi(\mathbf{x}_n)\right)$$

in the limit as $n \to \infty$.

In most practical problems we usually consider only $\{\phi(\mathbf{x})\}$ and $R'_{2\phi}(\mathbf{x}_1, \mathbf{x}_2)$. The basic difficulty in the statistical theory of heterogeneous materials is that we wish to determine *exactly* these quantities in terms of the statistical properties of the $\epsilon(\mathbf{x})$ field, we require not only $\{\epsilon(\mathbf{x})\}$ and $R'_{2\epsilon}(\mathbf{x}_1, \mathbf{x}_2)$, but all the information contained in $P_n\left(\epsilon(\mathbf{x}_1), \ldots, \epsilon(\mathbf{x}_n)\right)$ in the limit as $n \to \infty$. This is in contrast to the problem described by the equation

$$(d^2x/dt^2) + kx = F(t) \tag{28}$$

where $F(t)$ is a random forcing function. Here, if $\{F(t)\}$ is known, $\{x(t)\}$ may be found; if $\{F(t_1)F(t_2)\}$ is known, $\{x(t_1)x(t_2)\}$ may be found, and so on. Our problem is similar to the equation

$$(d^2x/dt^2) + k(t)x = F(t) \tag{29}$$

where both $F(t)$ *and* $k(t)$ are random functions. In this case if we wish to find $\{x(t)\}$ or $\{x(t_1)x(t_2)\}$ *exactly* we must know the probability density function

$$P_n(F(t_1), \ldots, F(t_n); k(t_1), \ldots, k(t_n))$$

in the limit $n \to \infty$.

In practice we cannot, of course, deal with the function P_n (as $n \to \infty$) except in very special circumstances. We are thus forced to find approximate solutions for $\{\phi(\mathbf{x})\}$ and $R'_{2\phi}(\mathbf{x}_1, \mathbf{x}_2)$ in terms of low-order statistical information like $\{\epsilon(\mathbf{x})\}$, $R'_{2\epsilon}(\mathbf{x}_1, \mathbf{x}_2)$. Unfortunately, it is not always possible to do this since in many cases the function $\{\phi(\mathbf{x})\}$ critically depends on higher-order information. We shall discuss this case explicitly in the next section. We do note here that one case where it is possible to find $\{\phi(\mathbf{x})\}$ and $R'_{2\phi}(\mathbf{x}_1, \mathbf{x}_2)$ in terms of $\{\epsilon(\mathbf{x})\}$ and $R'_{2\epsilon}(\mathbf{x}_1, \mathbf{x}_2)$ is when the fluctuations in $\epsilon(\mathbf{x})$ about the mean value $\{\epsilon(\mathbf{x})\}$ are small.

From a conceptual point of view it is worth pointing out that a functional equation can be derived which governs the function

$$P_n(\phi(\mathbf{x}_1), \ldots, \phi(\mathbf{x}_n); \epsilon(\mathbf{x}_1), \ldots, \epsilon(\mathbf{x}_n))$$

in the limit $n \to \infty$. (Actually, we usually treat the characteristic functional associated with this functional.) The boundary condition for this functional equation is

$$P_n(\epsilon(\mathbf{x}_1), \ldots, \epsilon(\mathbf{x}_n))$$

For this equation $\phi(\mathbf{x})$ satisfies the usual condition on the enclosing surface S. Little has been accomplished in the solution of this functional equation in the heterogeneous materials problem, but in turbulence theory this has been an active area of investigation that began with Hopf's early work (Hopf, 1952). Beran (1968) gives a number of references, and we also refer the reader to some recent work by Rosen (1971). At this stage it would appear useful to begin research from a functional point of view on the heterogeneous material problem.

B. Statistical Hierarchy of Equations

We have stated above, in general terms, the difficulties of finding the solution for $\{\phi(\mathbf{x}_1)\}$, $R_{2\phi}(\mathbf{x}_1, \mathbf{x}_2)$, and other statistical quantities associated with ϕ. Here we show explicitly why exact solutions cannot be obtained with limited information about the statistical properties of $\epsilon(\mathbf{x})$. In the course of this discussion we shall derive the first two equations of the infinite hierarchy of equations that govern the statistical properties of $\phi(\mathbf{x})$.

We write

$$\phi(\mathbf{x}) = \{\phi(\mathbf{x})\} + \phi'(\mathbf{x})$$

and thus Eq. (6) becomes

$$\bar{\epsilon}\nabla^2\{\phi(\mathbf{x})\} + \frac{\partial}{\partial x_i}\left[\epsilon'(\mathbf{x})\frac{\partial}{\partial x_i}\{\phi(\mathbf{x})\}\right]$$

$$= -\bar{\epsilon}\nabla^2\phi'(\mathbf{x}) - \frac{\partial}{\partial x_i}\left[\epsilon'(\mathbf{x})\frac{\partial\phi'(\mathbf{x})}{\partial x_i}\right] + \rho(\mathbf{x}) \tag{30}$$

We have assumed for convenience that $\bar{\epsilon}$ is independent of $\mathbf{x}$.

We next take the ensemble average of Eq. (3) and find

$$\bar{\epsilon}\nabla^2\{\phi(\mathbf{x})\} = -\frac{\partial}{\partial x_i}\left\{\epsilon'(\mathbf{x})\frac{\partial\phi'(\mathbf{x})}{\partial x_i}\right\} + \rho(\mathbf{x}) \tag{31}$$

If $\epsilon'(\mathbf{x}) \equiv 0$ (the nonstatistical case), we find

$$\bar{\epsilon}\nabla^2\{\phi(\mathbf{x})\} = \rho(\mathbf{x}) \tag{32}$$

and we can easily solve for $\{\phi(\mathbf{x})\}$. When $\epsilon'(\mathbf{x})$ is not identically zero, we find that the quantity

$$\{\epsilon'(\mathbf{x})\ \partial\phi'(\mathbf{x})/\partial x_i\}$$

appears in the equation governing $\{\phi(\mathbf{x})\}$. Nothing is known about this quantity and if we wish to solve for $\{\phi(\mathbf{x})\}$, we must find an additional equation which governs this quantity. Such an equation is found from Eq. (30). We multiply both sides of Eq. (30) by the quantity $\epsilon'(\mathbf{x}_1)$. This yields

$$\epsilon'(\mathbf{x}_1)\bar{\epsilon}\nabla^2\{\phi(\mathbf{x})\} + \frac{\partial}{\partial x_i}\left[\epsilon'(\mathbf{x}_1)\epsilon'(\mathbf{x})\frac{\partial}{\partial x_i}\{\phi(\mathbf{x})\}\right]$$

$$= -\bar{\epsilon}\frac{\partial}{\partial x_i}\left[\epsilon'(\mathbf{x}_1)\frac{\partial\phi'(\mathbf{x})}{\partial x_i}\right] - \frac{\partial}{\partial x_i}\left[\epsilon'(\mathbf{x}_1)\epsilon'(\mathbf{x})\frac{\partial\phi'(\mathbf{x})}{\partial x_i}\right] + \epsilon'(\mathbf{x}_1)\rho(\mathbf{x}) \tag{33}$$

Averaging Eq. (33) then gives

$$\frac{\partial}{\partial x_i}\left[R'_{2\epsilon}(\mathbf{x}_1, \mathbf{x})\frac{\partial}{\partial x_i}\{\phi(\mathbf{x})\}\right]$$

$$= -\bar{\epsilon}\frac{\partial}{\partial x_i}\left\{\epsilon'(\mathbf{x}_1)\frac{\partial\phi'(\mathbf{x})}{\partial x_i}\right\} - \frac{\partial}{\partial x_i}\left\{\epsilon'(\mathbf{x})\epsilon'(\mathbf{x})\frac{\partial\phi'(\mathbf{x})}{\partial x_i}\right\} \tag{34}$$

Equations (31) and (34) are two equations governing the functions

$$\{\phi(\mathbf{x})\}, \qquad \{\epsilon'(\mathbf{x}_1)\ \partial\phi'(\mathbf{x})/\partial x_i\}$$

but now Eq. (34) contains the additional unknown quantity $\{\epsilon'(\mathbf{x}_1)\epsilon'(\mathbf{x})\cdot \partial\phi'(\mathbf{x})/\partial x_i\}$.

If we continue in this manner, we will derive a sequence of N equations governing the N quantities

$$\{\phi(\mathbf{x})\},\quad \{\epsilon'(\mathbf{x}_1)\ \partial\phi'(\mathbf{x})/\partial x_i\},\quad \ldots,\quad \{\epsilon'(\mathbf{x}_{N-1})\ \cdots\ \epsilon'(\mathbf{x}_1)\ \partial\phi'(\mathbf{x})/\partial x_i\}$$

and the additional function

$$\{\epsilon'(\mathbf{x}_N)\ \cdots\ \epsilon'(\mathbf{x}_1)\ \partial\phi'(\mathbf{x})/\partial x_i\}$$

Only in the limit $N \to \infty$ may the sequence of equations be considered in any sense determinate.

The sequence of equations, Eq. (30), Eq. (34), . . ., is termed an *infinite statistical hierarchy of equations*, and in order to obtain a solution for $\{\phi(\mathbf{x})\}$ some approximation must be made to terminate the hierarchy. Perhaps the simplest assumption is that made possible when the fluctuations in $\epsilon(\mathbf{x})$, about $\bar{\epsilon}$, are small. In this case we may assume that in Eq. (34) the three-point correlation term is negligible. We find

$$\frac{\partial}{\partial x_i}\left[R'_{2\epsilon}(\mathbf{x}_1, \mathbf{x})\ \frac{\partial\{\phi(\mathbf{x})\}}{\partial x_i}\right] = -\bar{\epsilon}\,\frac{\partial}{\partial x_i}\left\{\epsilon'(\mathbf{x}_1)\ \frac{\partial\phi'(\mathbf{x})}{\partial x_i}\right\} \tag{35}$$

Equations (31) and (35) form a determinate set of equations to find

$$\{\phi(\mathbf{x})\} \qquad \text{and} \qquad \{\epsilon'(\mathbf{x}_1)\ \partial\phi'(\mathbf{x})/\partial x_i\}$$

When the fluctuations in $\epsilon(\mathbf{x})$ are not small, it is very difficult to find a suitable approximation to terminate the hierarchy. In the statistical theory of heterogeneous material very little work has been done to find suitable approximations to terminate the hierarchy for $\{\phi(\mathbf{x})\}$ other than the small perturbation approximation. Rather, attention in this field has been directed toward terminating the hierarchy governing the field quantity

$$\{\epsilon'(\mathbf{x}_1)\ \partial\phi'(\mathbf{x})/\partial x_i\}$$

when $\{\phi\} = 0$. The reason for this is that this quantity determines the effective constant ϵ^* when $l_c/L_V \to 0$. We shall report on this work in Section V. There has been, however, a formal solution to the hierarchy governing $\{\phi(\mathbf{x})\}$ [or more conveniently $\{E_i(\mathbf{x})\}$]. This leads to an integrodifferential equation with a kernel function that depends on all statistical information contained in the $\epsilon'(\mathbf{x})$ field. We shall discuss in Section IV how this approach may be of use.

We want to stress that the infinite hierarchy of statistical equations that resulted from multiplying Eq. (30) by 1, $\epsilon'(\mathbf{x}_1)$, . . ., $\epsilon'(\mathbf{x}_1)\epsilon'(\mathbf{x}_2)$ . . ., and averaging is not a result of this particular procedure. Multiplying by any set of functions and subsequently averaging will lead to an indeterminate set. This may be seen by deriving the functional equation governing the functional associated with $P_n\,(\phi(\mathbf{x}_1), \ldots, \phi(\mathbf{x}_n)\,; \epsilon(\mathbf{x}_1), \ldots, \epsilon(\mathbf{x}_n)\,)\ (n \to \infty)$. If we expand the functional in a Taylor series, we find the hierarchy given above. If a different expansion is used, a different hierarchy results. There is, however, no way in general to find a finite set without some approximation.

C. Numerical Solutions

In any real ensemble of material samples statistical equations are essentially solved by nature in any experiment, and it is only necessary to average the results to find the statistical quantities we desire. The question always then arises as to how difficult it would be to solve numerically Eq. (6) for several representations of the $\epsilon'(\mathbf{x})$ field and then average the resulting solutions.

The numerical solution of Eq. (6) has two aspects. First, we must construct an $\epsilon'(\mathbf{x})$ field with specified statistical properties. Secondly, for these fields we have to determine if computer capacity is available to solve the resultant equation. Until now, both aspects have presented formidable obstacles to numerical solutions, but in the future the specification of statistical functions will be the most difficult problem to overcome. Computer capacity increases yearly, but very little work has been done on constructing statistical fields that have certain statistical properties.

We shall confine our remarks here to the two-dimensional problem. This restriction is often appropriate for studying fiber-reinforced materials and is thus not simply of academic interest. The full three-dimensional problem requires considerably more computer capacity than the two-dimensional case, and more attention has been directed toward the simpler geometry.

We also restrict our attention to statistically homogeneous $\epsilon'(\mathbf{x})$ fields. This means that if $\epsilon'(\mathbf{x})$ is specified numerically for one material sample, we may (invoking an ergodic hypothesis) calculate the statistical functions $R'_{n\epsilon}$ by using volume averages.

From a statistical point of view we wish to determine the statistical properties of the $\phi(\mathbf{x})$ field from a knowledge of the statistical properties of the $\epsilon'(\mathbf{x})$ field. There are two ways to specify the statistical properties of the $\epsilon'(\mathbf{x})$ field. In the first method we construct numerical representations of $\epsilon'(\mathbf{x})$ such that the associated statistical functions $R_{n\epsilon}$ are specified functions. In the second method we mathematically construct a material model according to certain general statistical rules and then find $R'_{n\epsilon}$, either analytically or numerically. We discuss below some of the problems that occur in both procedures.

It is very difficult to construct an $\epsilon'(\mathbf{x})$ field for which all the $R'_{n\epsilon}$ are specified. At present our experience is very much limited to specification of $R'_{2\epsilon}$. This may be done by Fourier analysis in which a complete set of harmonic functions are superimposed so that the square of the amplitudes are proportional to the spatial Fourier transform of $R'_{2\epsilon}$, which we denote by $\Phi'_{2\epsilon}$. Here

$$\Phi'_{2\epsilon}(\mathbf{k}) = \int_{-\infty}^{\infty} \int e^{-i\mathbf{k}\cdot\mathbf{r}} R'_{2\epsilon}(\mathbf{r})\, d\mathbf{r}, \qquad \mathbf{r} = \mathbf{x}_2 - \mathbf{x}_1 \tag{36}$$

where $R'_{2\epsilon}$ may also be specified optically by noting that there is a Fourier transform relation between an incoherent source and the coherence function of the far-field radiation field.

If the $\epsilon(\mathbf{x})$ field that we wish to specify is Gaussian, then all higher-order functions $R'_{n\epsilon}$ may be determined from $R'_{2\epsilon}$, and we have only to superimpose our sine functions in such a way (independent phases) so that the field is Gaussian. Unfortunately, since $0 \leq \epsilon(\mathbf{x})$, the field cannot be taken to be Gaussian except in the special case of small perturbations where analytic solutions are available. For cases of numerical interest the $\epsilon(\mathbf{x})$ field cannot be considered to be Gaussian, and the three-point moment ($R'_{3\epsilon} = 0$ for Gaussian statistics) is of central importance in our study.

Specification of $R'_{3\epsilon}$ creates a far more difficult problem than the specification of $R'_{2\epsilon}$ since nothing as simple as harmonic wave superposition with only amplitude specification is available to us. We know of no way to construct analytically a random field with a prespecified function $R'_{3\epsilon}$ beyond attempting to construct laboriously a harmonic field that yields the appropriate Fourier transform of $R'_{3\epsilon}$. The Fourier transform of $R'_{3\epsilon}$ is given by the expression

$$\Phi'_{3\epsilon}(\mathbf{k}_1, \mathbf{k}_2) = \iint_{-\infty}^{\infty} \iint \exp(-i\mathbf{k}_1\cdot\mathbf{r}_1 - i\mathbf{k}_2\cdot\mathbf{r}_2) R'_{3\epsilon}(\mathbf{r}_1, \mathbf{r}_2)\, d\mathbf{r}_1\, d\mathbf{r}_2 \tag{37}$$

and we see that since $\Phi'_{3\epsilon}(\mathbf{k}_1, \mathbf{k}_2)$ depends on $\mathbf{k}_1$ and $\mathbf{k}_2$, it is a very complex problem to superimpose harmonic waves in such a way that this function is reproduced. For $R'_{n\epsilon}$, $n > 3$, the problem is even more difficult.

In a sense, the procedure of constructing fields with given correlation functions $R'_{n\epsilon}$ is an ideal since if it were possible to do this, we could consider any materials we construct as belonging to a particular class of statistical functions. Since this approach is so intractable, however, attention is inevitably directed toward analyzing particular models where a material construction procedure is specified, and the statistical functions associated with the $\epsilon'(\mathbf{x})$ field are calculated numerically or sometimes analytically.

In this latter approach we may, for example, construct a two-phase Poisson material. This is done by choosing N points per unit volume at random in a plane surface. We term these points, *Poisson points*. Next, again at random, choose $v_1 = M/N$ of the Poisson points to belong to one material and $1 - v_1$ points to belong to another material. Finally, we decide which material may be associated with all other points by stating that the material at a point is the same as the material at the nearest Poisson point. The entire procedure may be carried out on a computer, and we would have constructed a random $\epsilon'(\mathbf{x})$ field from which all the functions $R'_{n\epsilon}$ may be calculated.

Another model would be to state that all Poisson points and some surrounding area belong to one material. At each point a circle of radius r_0 might encompass this material. All other area in the plane would be composed on the second material. If $r_0 \ll l_p$, where $l_p = [V/N]^{1/3}$ is the mean distance between Poisson points, then we have the case of a dilute concentration of circles in a matrix. When $r_0 \approx O(l_p)$, however, the above model would contain many overlapping circles, and it would not appear as a material with circular inclusions. If we still wished to have a model with circular inclusions, we should have to specify a rule whereby whenever circles from neighboring Poisson points overlap, they are to be displaced so that they do not overlap. This, of course, is what would happen if we were constructing a fiber-reinforced material with circular fibers. The shape of the inclusions need not be circles, but could be any shape. The entire procedure could again be carried out on a computer, and we shall have generated a random $\epsilon'(\mathbf{x})$ field.

Numerical work has proceeded on periodic arrays of inclusions. This problem is much simpler than the random problem since the symmetry can be used to reduce greatly the calculations. Some of this work is reviewed by Chamis and Sendeckyj (1968). Adams and Tsai (1969) attempted to extend the notion of a periodic array to the random packing problem by placing inclusions only at random points in the periodic array. Using a number of assumptions to simplify the calculation, they numerically found information about effective elastic constants. They demonstrated that the random model may yield results that differ considerably from strictly periodic arrays.

There are many other models that can be constructed, and it seems likely that as the computer capacity grows, more investigators will turn toward numerical solutions. Finite element and similar techniques (Zienkiewicz, 1971) seem to be particularly appropriate for this type of study. Numerical studies are also being considered in other statistical problems (for example, scattering from a rough surface) and it is a fruitful area for future research.

IV. Equations Governing the Mean Field

In Section III,B we pointed out that in order to determine $\{\phi(\mathbf{x})\}$ we must in principle solve an infinite hierarchy of equations that contain all the statistical information in the $\epsilon'(\mathbf{x})$ field. In this section we wish to show that $\{E_i(\mathbf{x})\} = \{\partial\phi(\mathbf{x})/\partial x_i\}$ formally satisfies an integrodifferential equation usually termed a *Dyson equation*. The kernel of the integral portion of the equation is a function of all statistical information in the $\epsilon'(\mathbf{x})$ field. Since all the statistical information is contained in a single kernel function, this equation may be used phenomenologically in the absence of detailed information about the $\epsilon'(\mathbf{x})$ field. The character of the integrodifferential equation is quite different from that of an ordinary differential equation, and we shall point out the need for such an equation near rapidly varying source functions or boundaries.

To derive the Dyson equation for $\{E_i(\mathbf{x})\}$ it is convenient to write Eq. (6) in operator form. [The details are given by Beran and McCoy (1970). See also Bourret (1962), Frisch (1968), and Brown (1955).]

We write the definitions

$$\{L\} \equiv \bar{\epsilon}\, \partial^2/\partial x_i\, \partial x_i, \qquad L' \equiv \partial/\partial x_i[\epsilon'(\mathbf{x})\, \partial/\partial x_i] \tag{38}$$

and use the operators

$$IA = A \qquad \text{(the identity operator)}$$

$$PA = \{A\} \qquad \text{(the averaging operator)}$$

Equation (6) becomes

$$[\{L\} + L']\phi = \rho \tag{39}$$

Averaging Eq. (39) yields

$$\{L\}\{\phi\} + \{L'\phi'\} = \rho \tag{40}$$

Equation (40) is Eq. (39) in operator form.

We next subtract Eq. (40) from Eq. (39). We find

$$[\{L\} + (I - P)L']\phi' = -L'\{\phi\} \tag{41}$$

Up to this point we have simply performed some straightforward manipulations in operator notation. We now proceed formally and solve Eq. (41) by writing

$$\phi' = [\{L\} + (I - P)L']^{-1}[-L'\{\phi\}] \tag{42}$$

where $[Q]^{-1}$ is the inverse of the operator $[Q]$. In Eq. (42) the inverse operator is given by its binomial expansion as

$$[\{L\} + (I - P)L']^{-1} = \sum_{n=0}^{\infty} (-1)^n [\{L\}^{-1}(I - P)L']^n \{L\}^{-1} \tag{43}$$

where

$$\{L\}^{-1}u(\mathbf{x}) = (1/\bar{\epsilon}) \int G(\mathbf{x}, \mathbf{x}')u(\mathbf{x}')\, d\mathbf{x}' \tag{44}$$

and u is any function of position. Then $G(\mathbf{x}, \mathbf{x}')$ is the Green's function associated with the Laplace operator $\{L\}$ and appropriate boundary conditions.

To find an equation governing $\{\phi\}$ we operate on both sides of Eq. (42) by L', average this equation and substitute $\{L'\phi'\}$ in Eq. (40). We find finally

$$\{L\}\{\phi\} = \rho + [\{L'[\{L\} + (I - P)L']^{-1}L'\}]\{\phi\} \tag{45}$$

In terms of the operator definitions Eq. (45) may be written

$$\bar{\epsilon} \frac{\partial}{\partial x_i} \{E_i\} + \frac{\partial}{\partial x_i} [K_{ij}(\mathbf{x}, \mathbf{x}')\{E_j(\mathbf{x}')\}] = \rho(\mathbf{x}) \tag{46}$$

where K_{ij} is represented by the infinite summation

$$\begin{aligned} K_{ij}(\mathbf{x}, \mathbf{x}') &= \{\epsilon'(\mathbf{x})A_{ij}(\mathbf{x}, \mathbf{x}')\} + \{\epsilon'(\mathbf{x})(I - P)A_{ik}(\mathbf{x}, \mathbf{x}_1)A_{kj}(\mathbf{x}_1, \mathbf{x}')\} \\ &\quad + \{\epsilon'(\mathbf{x})(I - P)A_{ik}(\mathbf{x}, \mathbf{x}_1)(I - P)A_{kl}(\mathbf{x}_1, \mathbf{x}_2)A_{lj}(\mathbf{x}_2, \mathbf{x}')\} + \cdots \end{aligned} \tag{47}$$

Here

$$A_{ij}(\mathbf{x}, \mathbf{x}')u_j(\mathbf{x}') = \frac{1}{\bar{\epsilon}} \int \frac{\partial G}{\partial x_i}(\mathbf{x}, \mathbf{x}') \frac{\partial}{\partial x_{j'}} [\epsilon'(\mathbf{x}')u_j(\mathbf{x}')]\, d\mathbf{x}' \tag{48}$$

Equation (46) is an integrodifferential equation where the kernel $K_{ij}(\mathbf{x}, \mathbf{x}')$ is given by an infinite number of integrodifferential operators. In general, $K_{ij}(\mathbf{x}, \mathbf{x}')$ is very complex for arbitrary geometry even if Eq. (47) is truncated at the first term. We shall thus in the remainder of this section restrict ourselves to the case of an infinite medium. We point out that this solution is valid except in a boundary layer near a boundary.

If we consider an infinite medium, then

$$G(\mathbf{x}, \mathbf{x}') = \frac{1}{4\pi} \frac{1}{r}; \qquad r = |\mathbf{x} - \mathbf{x}'| \tag{49}$$

After some manipulation in which the various terms in Eq. (47) are integrated by parts and orders of integration are interchanged, $K_{ij}(\mathbf{x}, \mathbf{x}')$ may be put in the form of an algebraic operator. It remains an infinite series dependent on $G(\mathbf{x}, \mathbf{x}')$ and all the functions $R'_{n\epsilon}$. We find finally

$$\epsilon^* \frac{\partial}{\partial x_i} \{E_i(\mathbf{x})\} + \frac{\partial}{\partial x_i} \int M_{ij}(\mathbf{x} - \mathbf{x}')\{E_j(\mathbf{x}')\}\, d\mathbf{x}' = \rho(\mathbf{x}) \tag{50}$$

where $M_{ij}(\mathbf{x} - \mathbf{x}')$ is an algebraic operator that depends only on the vector $\mathbf{x} - \mathbf{x}'$ and ϵ^* is a constant given by the expression

$$\epsilon^* = \bar{\epsilon} - \tfrac{1}{3}(C_\Sigma/\bar{\epsilon}) + \eta \tag{51}$$

Here η is the sum of an infinite number of integrals over the functions $R'_{n\epsilon}(\mathbf{x}_1, \ldots, \mathbf{x}_n)$, C_Σ is an infinite sum containing only the moments $\{\epsilon'^n\}$, and ϵ^* and $M_{ij}(\mathbf{x} - \mathbf{x}')$ have been defined so that $\int M_{ij}(\mathbf{x} - \mathbf{x}')\, d\mathbf{x}' = 0$. Subsequently ϵ^* will be identified with the effective constant of the medium.

If the constant ϵ^* and the function $M_{ij}(\mathbf{x} - \mathbf{x}')$ are known, then Eq. (50) may be used to solve for $\{E_i(\mathbf{x})\}$. For the average field this equation replaces the equation

$$\epsilon \frac{\partial}{\partial x_i} E_i(\mathbf{x}) = \rho(\mathbf{x}) \tag{52}$$

which governs $E_i(\mathbf{x})$ in the absence of fluctuations in ϵ. From a theoretical point of view it is, however, virtually impossible to calculate ϵ^* and $M_{ij}(\mathbf{x} - \mathbf{x}')$ from a knowledge of the statistical properties of the $\epsilon'(\mathbf{x})$ field except in the limit of small fluctuations. Both ϵ^* and $M_{ij}(\mathbf{x} - \mathbf{x}')$ are represented by an infinite series of complex integrals. Rather, the more practical approach appears to be to assume Eq. (50) is the correct form of the equation governing $\{E_i(\mathbf{x})\}$ and ϵ^* and $M_{ij}(\mathbf{x} - \mathbf{x}')$ are quantities that must be determined experimentally or chosen phenomenologically to agree with experiment.

From the earliest work on heterogeneous materials it was often assumed that Eq. (52) could be replaced by the analogous equation

$$\epsilon^* \frac{\partial}{\partial x_i} \{E_i(\mathbf{x})\} = \rho(\mathbf{x}) \tag{53}$$

where ϵ^* was a constant to be determined from experiment in the same way that ϵ was determined from experiment. The reasoning was simply that to define ϵ we averaged over atomic dimensions and to define ϵ^* we will average over a dimension large compared to a scale, say l_c, over which ϵ^* varies. This reasoning is sound provided only that $\{E_i(\mathbf{x})\}$ does not vary appreciably in distances of the order of l_c. If this is true and we assume that

$M_{ij}(\mathbf{x} - \mathbf{x}') \to 0$ in distances of order l_c, then the integral term in Eq. (50) becomes

$$\frac{\partial}{\partial x_i}\{E_j(\mathbf{x})\}\left[\int M_{ij}(\mathbf{x} - \mathbf{x}')\,d\mathbf{x}'\right]$$

The integral over $M_{ij}(\mathbf{x} - \mathbf{x}')$ is equal to zero by construction and thus Eq. (50) reduces to Eq. (53).

The assumption that $\{E_i(\mathbf{x})\}$ is essentially constant in distances of order l_c depends on two conditions. First, the material sample must have a characteristic dimension L_V such that $L_V \gg l_c$. We have in our development beyond Eq. (48) assumed this is true since we have taken the material to be infinite. Second, we require $L_\rho \gg l_c$, where L_ρ is a characteristic distance over which the function ρ varies appreciably. This latter assumption is not made in our derivation of Eq. (50), and in fact, the principal purpose of presenting Eq. (50) is to provide an equation which is valid when $l_c \leq L_\rho$, e.g., when the characteristic size of a heat source is comparable to the characteristic size of the variation in conductivity.

In those cases where $l_c \ll L_V$, $l_c \ll L_\rho$, then Eq. (53) is applicable and ϵ^* may be identified with the effective material constant. The only other condition we have required is that $M_{ij}(\mathbf{x} - \mathbf{x}') \to 0$ in distances of order l_c. Since M_{ij} depends on a sum of all order correlation functions, we cannot prove this to be so, but it is considered a reasonable assumption if we require that $R'_{n\epsilon}$ approach a constant value in distances of the order of l_c.† We shall present in the next section bounds for ϵ^* and also point out an assumption that allows ϵ^* to be calculated explicitly. In general, however, we may always, if necessary, measure ϵ^* in exactly the same way that ϵ is measured when it is a constant parameter. We perform our experiment on a large sample in the absence of a source $\rho(\mathbf{x})$.

While it is relatively easy to measure ϵ^* and use this value in the solution of Eq. (50), it is very difficult to measure $M_{ij}(\mathbf{x} - \mathbf{x}')$. To our knowledge no measurements of this function exist at this time. In principle, it may be determined from measurement of $\phi(\mathbf{x})$ for a known field $\rho(\mathbf{x})$. This is so since the integral term in Eq. (50) is in the form of a convolution integral. For simplicity consider the case of isotropic statistics. Writing

$$M_j(\mathbf{x} - \mathbf{x}') \equiv \frac{\partial}{\partial x_i} M_{ij}(\mathbf{x} - \mathbf{x}')$$

and taking the Fourier transform of both sides of Eq. (50) yields after some

† *Note added in proof:* Recent work indicates that if the ratio of the maximum and minimum values of $\epsilon(\mathbf{x})$ is large then the characteristic decay distance associated with M_{ij} is also dependent on this ratio.

manipulation

$$\{\phi(\mathbf{k})\} = -\hat{\rho}(\mathbf{k})/k^2[\epsilon^* + \hat{M}(k)] \tag{54}$$

Here the caret indicates a Fourier transform, and since M_{ij} is an isotropic tensor in either space

$$\hat{M}_j(\mathbf{k}) = k_j\hat{M}(k) \tag{55}$$

If $\hat{\rho}(\mathbf{x})$ is known and $\{\phi(\mathbf{x})\}$ is measured, then $\hat{\rho}(\mathbf{k})$ and $\{\phi(\mathbf{k})\}$ may be calculated and $\hat{M}(k)$ determined from Eq. (54). Then $M_j(\mathbf{x} - \mathbf{x}')$ is obtained by Fourier inversion.

The procedure is very tedious, but possible if $\{\phi(\mathbf{x})\}$ may be measured. This means, for example, that in the temperature case, $T(\mathbf{x})$ must be determined in the interior of a material sample. To circumvent the difficulty of measuring $M_{ij}(\mathbf{x} - \mathbf{x}')$ the most practical procedure would be to try to establish the general properties of M_{ij} and then assume a reasonable form for this function. In the isotropic case this would mean guessing the form of the single function $\hat{M}(k)$. At this point the explicit perturbation solution, obtained by keeping only the first term in Eq. (47), is of use. For this case we can find the nature of $\hat{M}(k)$ as $k \to 0$ and $k \to \infty$ and the general form at intermediate points. From these results we can infer something about $\hat{M}(k)$ when the fluctuations are not small.

It may be shown (Beran and McCoy, 1970) that

$$\hat{M}(k)\bigg|_{k\to 0} \to -\frac{2}{15}\frac{(kl_c)^2}{\bar{\epsilon}}\int_0^\infty \left(\frac{r}{l_c}\right) R'_{2\epsilon}\left(\frac{r}{l_c}\right) d\left(\frac{r}{l_c}\right) \tag{56}$$

$$\hat{M}(k)\bigg|_{k\to\infty} \to -\frac{2}{3}\frac{R'_{2\epsilon}(0)}{\bar{\epsilon}} = \frac{1}{\{1/\epsilon\}} - \epsilon^* = \tilde{\epsilon} \tag{57}$$

Thus $\hat{M}(k)$ varies as ak^2 for small k and asymptotically $(k \to \infty)$ approaches the constant value $\tilde{\epsilon}$. The characteristic distance over which $\hat{M}(k)$ varies from zero to its asymptotic value is $k_c = 1/l_c$.

Assuming these characteristics apply in general, it is easy to construct representations of $\hat{M}(k)$ that may be used for comparison with experiment. For example, a first guess might be

$$\hat{M}(k) = \tilde{\epsilon}[1 - \exp[l_c^2k^2/\tilde{\epsilon}]] \tag{58}$$

(Note $\tilde{\epsilon}$ is negative.)

In the theory of elasticity there was great interest in determining the first nonzero effects resulting from the finite value of l_c. There is a whole literature (see Mindlin and Eshel, 1968) on strain gradient and similar theories. There has not been such an active development of theories for the case we consider here, but it is of interest to present the equation which to lowest order takes into account the finite effect of l_c. This equation may be obtained by expanding $\{E_j(\mathbf{x}')\}$ under the integral sign in a Taylor series. Equivalently, we may replace $\hat{M}(k)$ in Eq. (54) by ak^2. In either case we find the following equation for $\{\phi(\mathbf{x})\}$:

$$\epsilon^* \nabla^2\{\phi(\mathbf{x})\} - a^2\nabla^4\{\phi(\mathbf{x})\} = \rho(\mathbf{x}) \tag{59}$$

Equation (59) is not valid if l_c is of order L_ρ, but only if $l_c \ll L_\rho$. If $L_\rho \leq l_c$, Eq. (50) must be used. Equation (59) is also not valid near boundaries. By choosing to consider an infinite medium we, of course, chose to ignore boundaries in our development beyond Eq. (47). More importantly, however, an integrodifferential equation of the type we consider here can never be replaced by a differential equation with a finite number of terms in the neighborhood of wall. Near the wall or near the maximum value of $\rho(\mathbf{x})$ when $L_\rho \leq l_c$, a differential equation gives erroneous results. It can be shown, however, that the effect of the wall is confined to a boundary layer of order l_c and beyond this layer approximations can be made.

As we stated above, the main interest of research workers has been the study of ϵ^* rather than $\hat{M}(k)$. We shall now in the next section turn our attention to the determination of this constant using statistical techniques.

V. Effective Constants

In Section IV we pointed out that if the characteristic correlation length l_c is small compared to both L_V, the characteristic size of the material sample, and L_ρ, the characteristic distance over which the source $\rho(\mathbf{x})$ varies significantly, then the equation governing the average field $\{E_i(\mathbf{x})\}$ may be written [see Eq. (53)] as

$$\epsilon^* \frac{\partial}{\partial x_i} \{E_i(\mathbf{x})\} = \rho(\mathbf{x}) \tag{60}$$

Here ϵ^* is an effective constant which may be determined by experiment as the constant ϵ is, when there are no statistical fluctuations. We also derived a formal expression for ϵ^* given by Eq. (51). Unfortunately, the η portion of Eq. (51) depends on all n-point correlation functions of the medium and cannot be readily evaluated unless we assume small perturbations [see Eq. (56)]. Equation (51) can be evaluated if some assumptions are made about η (C_Σ can be explicitly evaluated for a two-phase material) but it is difficult to relate the mathematical assumptions to physical materials. Some interesting work along these lines has been done by Kroner (1967) and Bolotin and Moskalenko (1968). These works are related to the self-consistent approximation in which it is essentially assumed that any inclusion in a matrix behaves as though it were in a matrix with some constant effective property. [See also Kerner (1956) and Hashin (1968).]

Exact solutions of ϵ^* of the above kind are valuable but restrictive. This is so because as we know from bounds we shall present below that it is impossible to specify ϵ^* from just a knowledge of low-order statistical information. Using only volume fractions Hashin and Shtrikman (1962) have derived the best bounds for a two-phase material. This clearly shows that additional geometrical information is necessary if ϵ^* is to be specified for a particular material sample. Solutions obtained from approaches like the self-consistent method apply to only a particular class of geometries. Moreover, since experiments show that all values of ϵ^* between the Hashin–Shtrikman bounds may be obtained, these solutions are only of practical value if the geometrical structure to which they apply is known.

This situation, rather than being a limitation, may be used to an engineer's advantage. He may strive to construct materials that have a particular value of ϵ^* (lying between the bounds for fixed volume fractions) that is of particular interest to him. To do this he must have some knowledge of how shape and packing information change the value of ϵ^*. In the remainder of this section we shall show how this geometrical information may be included in the bounds for ϵ^*.

In Section II we stated two variational principles [see Eqs. (10)–(19)] which allow us to bound ϵ^*. These principles were formulated in terms of volume averages, but they may be reformulated in terms of ensemble averages since we assume an ergodic hypothesis holds when ϵ^* is defined (Beran, 1968). Moreover, the definition of ϵ^* given in Eq. (13) is in terms of energy

$$U = \epsilon^* \{E_i\}\{E_i\} \tag{61}$$

while the definition given in the previous section is essentially in terms of field quantities

$$\{D_i\} = \epsilon^* \{E_i\} \tag{62}$$

The definitions may be shown to be equivalent, however, since $\{D_i' E_i'\} = 0$.

In terms of ensemble averages the variational principle which leads to the upper bound yields

$$\epsilon^* \{E_i\}\{E_i\} \leq \{\epsilon E_{Ai} E_{Ai}\} \tag{63}$$

where E_{Ai} is a trial solution such that

$$\{E_{Ai}\} = \{E_i\} \tag{64}$$

The simplest trial solution is perhaps

$$\{E_{Ai}\} = \{E_i\} \tag{65}$$

and this yields [see Eq. (14)]

$$\epsilon^* \leq \bar{\epsilon} \tag{66}$$

Similar manipulation on the corresponding variational principle yields finally

$$1/\{1/\epsilon\} \leq \epsilon^* \leq \bar{\epsilon} \tag{67}$$

Equation (67) bounds ϵ^* in terms of only one-point statistical information and contains no information about the geometry of the medium. To include geometrical information in the bounds we must use a better trial function than that given in Eq. (65). Before doing this, however, we wish to set down the Hashin–Shtrikman bounds for a two-phase material.

Hashin and Shtrikman (1962) used a generalized variational principle to find bounds for ϵ^* that are more restrictive than those given in Eq. (67), in the special case when $\epsilon(\mathbf{x})$ lies between two nonzero finite values ϵ_1 and ϵ_2. In particular, the bounds for a two-phase material are given by the following expression:

$$\epsilon_1 + \frac{v_2}{[1/(\epsilon_2 - \epsilon_1)] + (v_1/3\epsilon_1)} \leq \epsilon^* \leq \epsilon_2 + \frac{v_1}{[1/(\epsilon_1 - \epsilon_2)] + (v_2/3\epsilon_2)} \tag{68}$$

They further showed that the bounds are the best bounds if only the volume fractions, v_1 and v_2, are specified.

For a two-phase material we find from Eq. (67) that

$$\epsilon_1\epsilon_2/(v_1\epsilon_2 + v_2\epsilon_1) \leq \epsilon^* \leq v_1\epsilon_1 + v_2\epsilon_2 \tag{69}$$

It can be proven directly that the bounds given in Eq. (68) always lie within the bounds given in Eq. (69). Moreover, when $\alpha \equiv \epsilon_2/\epsilon_1$ is large, Eq. (68) gives significantly better results.

Returning to Eq. (63) we now choose a trial function E_{Ai} that allows us to include statistical information in the bounds (Beran, 1965). We choose as our trial function a weighted sum of perturbation solutions of Eq. (30) $(E_i = \partial\phi/\partial x_i)$. We write

$${}_N E_{iA} = \{E_3\}\delta_{i3} + \lambda_1 E_i^{(1)}(\mathbf{x}) + \lambda_2 E_i^{(2)}(\mathbf{x}) + \cdots \tag{70}$$

Here $\{E_3\}$ is the average field (chosen in the "3" or z direction) and $E_i^{(j)}(\mathbf{x})$ is the jth order contribution to the perturbation solution. The perturbation is in terms of $\epsilon'/\bar{\epsilon}$, where the statistical fluctuation in ϵ' is assumed to be small compared to $\bar{\epsilon}$.

To obtain bounds that contain principally shape information we consider only the first two terms (i.e., we set $\lambda_j = 0; j \geq 2$). We substitute Eq. (70) into Eq. (63) and choose λ_1 so that the expression is a minimum. After considerable manipulation we find finally

$$\epsilon^* \leq \bar{\epsilon} - \frac{(\frac{1}{3}\{\epsilon'^2\}/\bar{\epsilon})}{[1 + 3\bar{\epsilon}I/\{\epsilon'^2\}]} \tag{71}$$

where

$$I = \frac{1}{16\pi^2\bar{\epsilon}^2} \int_{V'} \int_{V''} \frac{\partial^2}{\partial r_3 \, \partial s_3} \{\epsilon'(0)\epsilon'(\mathbf{r})\epsilon'(\mathbf{s})\} \frac{r_i s_i}{r^3 s^3} \, d\mathbf{r} \, d\mathbf{s} \tag{72}$$

The bound in Eq. (71) thus depends on an integral over the three-point correlation function:

$$R'_{3_\epsilon}(\mathbf{r}, \mathbf{s}) = \{\epsilon'(0)\epsilon'(\mathbf{r})\epsilon'(\mathbf{s})\}$$

The integral I contains shape information not contained in either Eqs. (67) or (68). Proceeding similarly with the other variational principle we find a lower bound for ϵ^*. We have for the complete set

$$\left[\left\{\frac{1}{\epsilon}\right\} - \frac{[\frac{4}{3}\{(\epsilon'/\epsilon)\}]^2[1/4\bar{\epsilon}^2]}{[(1/3\bar{\epsilon}^2)\{\epsilon'^2/\epsilon\}] + K}\right]^{-1} \leq \epsilon^* \leq \bar{\epsilon} - \frac{(\frac{1}{3}\{\epsilon'^2\}/\bar{\epsilon})}{[1 + 3\bar{\epsilon}I/\{\epsilon'^2\}]} \tag{73}$$

where

$$K = \frac{1}{16\pi^2\bar{\epsilon}^2} \int_{V'} \int_{V''} \frac{\partial^2}{\partial r_3 \, \partial s_3} \left\{\frac{\epsilon'(\mathbf{r})\epsilon'(\mathbf{s})}{\epsilon(0)}\right\} \frac{r_i s_i}{r^3 s^3} \, d\mathbf{r} \, d\mathbf{s} \tag{74}$$

The integrals I and K are difficult to evaluate since they require knowledge of two three-point correlation functions:

$$\{\epsilon'(0)\epsilon'(\mathbf{r})\epsilon'(\mathbf{s})\}, \qquad \{\epsilon'(\mathbf{r})\epsilon'(\mathbf{s})/\epsilon(0)\}$$

For any real materials these functions are generally very complex. In order for the bounds to be useful in practical applications some model must be used to simplify I and K. Miller (1969a) developed such a model, and we shall summarize the results he obtained. Before doing this, however, it is important to point out that in order to gain a deep understanding of real materials it is necessary to perform scientific studies in which the three-point correlation functions are measured and I and K evaluated. By doing this we can check the validity of the various models used and generate new models by developing approximate expressions for the correlation functions. Corson (1971) measured both three-point correlation functions for samples compressed from powdered mixtures of Pb–Al and Pb–Fe. Working with the elastic properties rather than thermal or electric properties, he then evaluated the elastic analog of the integrals I and K (I and K can be reduced to three-fold integrals and evaluated numerically). He showed that the measured effective constant for shear lay within the bounds given by the elastic analog of Eq. (73). The bounds lay within the Hashin–Shtrikman bounds and reduced their width by about one-half. In the course of his work he developed approximate analytic expressions for the three-point correlation function.

One of the simplest material models is perhaps the two-phase symmetric material (Brown, 1965; Beran and Molyneux, 1966). Here $v_1 = v_2 = \frac{1}{2}$, and it is assumed that on the basis of geometry alone the two-phases are indistinguishable. In this case it may be shown that

$$\{\epsilon'(0)\epsilon'(\mathbf{r})\epsilon'(\mathbf{s})\} = 0$$
$$\{\epsilon'(\mathbf{r})\epsilon'(\mathbf{s})/\epsilon'(0)\} = (\bar{\epsilon}/\epsilon_1\epsilon_2)\{\epsilon'(\mathbf{r})\epsilon'(\mathbf{s})\} \qquad (75)$$

Using Eq. (75), Eq. (73) may be evaluated explicitly, and we find

$$\left(\frac{6\alpha^2 + 6\alpha}{\alpha^2 + 10\alpha + 1}\right)\frac{1}{\alpha^{1/2}} \le \frac{\epsilon^*}{(\epsilon_1\epsilon_2)^{1/2}} \le \left(\frac{2\alpha^2 + 8\alpha + 2}{6 + 6\alpha}\right)\frac{1}{\alpha^{1/2}} \qquad (76)$$

where $\alpha = \epsilon_2/\epsilon_1$, $\epsilon_2 > \epsilon_1$. Equation (76) lies within the Hashin-Shtrikman bounds and reduces the width by about half. The reason, of course, for the reduction in bound width is that in addition to volume fraction information, Eq. (76) includes some information about the geometry.

The symmetric medium described above is limited to $v_1 = v_2 = \frac{1}{2}$. Partially in an effort to extend the concept of symmetry to other volume fractions, Miller (1969a) developed the cell model for two-phase materials. He assumed that the material volume may be covered by a complete set of nonoverlapping continuous closed cells. Within each cell the material property is constant. Adjacent cells may or may not be composed of the same material. He assumed that the cell distribution was statistically homogeneous and isotropic. The key assumption in his model is that the material property (ϵ_1 or ϵ_2) of any cell is independent of the properties of adjacent cells. When it is assumed that the cell geometry is the same for either phase, the material may be considered to be a generalization of the symmetric material given above. The model, however, may also be considered for materials in which the phase geometries are different.† Generally speaking,

† *Note added in proof:* W. F. Brown (in press) has pointed out that for asymmetric materials the cell model leads to an inconsistency. He provides, however, a consistent generalization which leads to the upper bound in Eq. (78). The G's again describe the geometry of the cells, but the independence assumption is replaced by other restrictions imposed on the two- and three-point probability functions.

the model is most easily understood when applying it to materials composed of inclusions in a matrix. The cells corresponding to the inclusion phase are given the same geometrical shape as the inclusions, while the cells corresponding to the matrix geometry are assumed to be arbitrary in shape.

Miller shows how the three-point correlation functions may be expressed in terms of the functions $g_i(0, \mathbf{r}, \mathbf{s})$, defined as the probability of a triangle $(0, \mathbf{r}, \mathbf{s})$ having all three vertices in a single cell of property ϵ_i, given that one of the points is in the cell. He shows for example that

$$\{\epsilon'(0)\epsilon'(\mathbf{r})\epsilon'(\mathbf{s})\} = \epsilon_2'^3\left[v_2 g_2(0, \mathbf{r}, \mathbf{s}) - \frac{v_2^3}{(1-v_2)^2} g_1(0, \mathbf{r}, \mathbf{s})\right] \tag{77}$$

Using Eq. (77) and a similar expression for $\{\epsilon'(\mathbf{r})\epsilon'(\mathbf{s})/\epsilon(0)\}$, he finds the following bounds for ϵ^*:

$$\frac{\epsilon^*}{(\epsilon_1\epsilon_2)^{1/2}} \le \frac{1}{\alpha^{1/2}}\left[1 + v_2(\alpha - 1) - \frac{(\frac{1}{3})v_2(1-v_2)(\alpha-1)^2}{1 + (\alpha - 1)\left(v_2 + 3[(1-v_2)^2 G_2 - v_2^2 G_1]\right)}\right]$$

$$\frac{\epsilon^*}{(\epsilon_1\epsilon_2)^{1/2}} \ge \alpha^{1/2}\left[[\alpha - v_2(\alpha - 1)] - \frac{\frac{4}{3}(1-\alpha)^2(1-v_2)v_2}{1 + \alpha + 3(\alpha - 1)[v_2^2 G_1 - (1-v_2)^2 G_2]}\right]^{-1} \tag{78}$$

where

$$G_i = \frac{1}{16\pi^2}\int_V\int_{V'} \frac{\partial^2 g_i(0, \mathbf{r}, \mathbf{s})}{\partial r_3\,\partial s_3}\frac{r_j s_j}{r^3 s^3}\,d\mathbf{r}\,d\mathbf{s}$$

Here G_i are numbers that describe the shape of the cells. Miller has shown that $\frac{1}{9} \le G \le \frac{1}{3}$ and that $G = \frac{1}{9}$ corresponds to a spherical shape, $G = \frac{1}{6}$ to a needle shape, and $G = \frac{1}{3}$ to a disk or platelike shape. The value of G can be assigned exactly to any ellipsoidal shape. If v_2 is the volume fraction of the inclusion phase of a material, the bounds in Eq. (78) coincide in the limit $v_2 \to 0$ for either $G = \frac{1}{9}$ or $G = \frac{1}{3}$. Thus, in these cases very close bounds may be achieved when v_2 is small even if α is large.

The number G in Eq. (78) describes only the shape characteristic of the cells and does not characterize how they may be packed together. By packing we mean here both the placement of the cells relative to each other and their size distribution. Thus G is independent of the size of the cells. In order to find some packing information we must consider the λ_2 term in Eq. (70). ElSayed (1972) has treated this problem and found that in this case the bounds depend upon two additional numbers which also can be bounded. Physically, these numbers give information about the probable positions of two inclusions in the limit $v_2 \to 0$. Work on the precise physical meaning of the additional two numbers is currently in progress, but preliminary results show that in certain cases the assignment of even one of these numbers gives exceptionally close bounds.

The bounds in Eq. (78) are for the three-dimensional problem. In fiber-reinforced materials a two-dimensional representation is often appropriate. In this case we find for the bounds the following expressions (Beran and Silnutzer, 1971):

$$\frac{\epsilon_T^*}{(\epsilon_1\epsilon_2)^{1/2}} \leq \frac{1}{\alpha^{1/2}}\left[1 + v_2(\alpha - 1) - \frac{\frac{1}{2}v_2(1 - v_2)(\alpha - 1)^2}{1 + (\alpha - 1)(v_2 + 2[(1 - v_2)^2G_2 - v_2^2G_1])}\right]$$

$$\frac{\epsilon_T^*}{(\epsilon_1\epsilon_2)^{1/2}} \geq \alpha^{1/2}\left[\alpha - v_2(\alpha - 1) - \frac{\frac{1}{2}(1 - \alpha)^2(1 - v_2)v_2}{\alpha + (1 - \alpha)v_2 + 2(\alpha - 1)[v_2^2G_1 - (1 - v_2)^2G_2]}\right]^{-1} \tag{79}$$

where $\frac{1}{4} \leq G_i < \frac{1}{2}$. Here $G = \frac{1}{4}$ corresponds to a circular shape and $G = \frac{1}{2}$ corresponds to the shape of parallel lamellae (two-dimensional needles) and ϵ_T^* is the transverse effective constant. Along the fiber direction the effective constant, ϵ_L^* is equal to $\bar{\epsilon}$.

The bounds presented in Eqs. (78) and (79) allow an engineer to take into account the effect of shape in a two-phase composite material. In addition generalized bounds to take into account packing information will soon be available. For fiber-reinforced materials, for example, it will thus be possible to change ϵ^* in a known manner by changing the shape of the fibers and the way they are packed. Circular fibers and ribbonlike fibers behave very differently, and for a fixed shape the manner of packing further changes ϵ^*. For large α these changes are very dramatic, and the designer should no longer be content with dealing with only one shape of fiber packed in an unspecified manner.

VI. Comparison of Theory and Experiment

This section and the next on applications should comprise a substantial portion of this chapter. Unfortunately, statistical ideas are only just beginning to be considered in experimental work and in applications. For example, there is considerable amount of data on the measurement of effective constants, but very little has been done with statistical ideas clearly in mind. In this section we only consider data applicable to the statistical theories we have treated in this chapter.

In Section IV we considered the equations governing the mean field. We pointed out that $\{E_i(\mathbf{x})\}$ satisfied the equation

$$\epsilon^* \frac{\partial}{\partial x_i} \{E_i(\mathbf{x})\} + \int M_j(\mathbf{x} - \mathbf{x}') \{E_j(\mathbf{x}')\} \, d\mathbf{x}' = \rho(\mathbf{x}) \qquad (80)$$

in an infinite medium, where ϵ^* is the effective constant and $M_j(\mathbf{x} - \mathbf{x}')$ is an unknown function that is a function of all the statistical information in the $\epsilon'(\mathbf{x})$ field. Since it is virtually impossible to determine $M_j(\mathbf{x} - \mathbf{x}')$ theoretically except in the limit of small perturbations, it was suggested that Eq. (80) could be considered as a phenomonological equation and $M_j(\mathbf{x} - \mathbf{x}')$ could be determined experimentally. To our knowledge no experiments of this type have been performed.

The effective constant ϵ^* is also determined by all the statistical information in the $\epsilon'(\mathbf{x})$ field. Again, it is unlikely that we should be able to calculate ϵ^* except in the small fluctuation limit or other very special cases, but here we are fortunate in having theoretical expressions that bound the possible values of ϵ^*. These expressions bound ϵ^* in terms of low-order statistical information of the $\epsilon'(\mathbf{x})$ field.

The lowest-order statistical information is $P(\epsilon(\mathbf{x}_1))\, d\epsilon(\mathbf{x}_1)$. This function is, roughly, the probability of the material having a particular value of $\epsilon(\mathbf{x}_1)$. The bounds in terms of this information alone are given by Eq. (67). These bounds are very unrestrictive, and essentially all available data lies within these bounds. When a material has only two-phases, better bounds than Eq. (67) are available. They are given in Eq. (68) and have been shown to be the best bounds possible if only v_1 and v_2 are known. By and large these bounds have been shown to agree with experiment. This is true for both ϵ^* and the analogous bounds derived for elastic effective constants. We refer the reader to the work of Landauer (1952), DeLoor (1956), Nishimatsu and Gurland (1960), and Hasselman and Fulrath (1968). We note, however, that not all experimental data is in agreement [see Krock (1963)].

The above bounds on ϵ^* had been developed without the statistical theory. From the statistical point of view treated in this chapter we are more concerned with the confirmation of the bounds given in Eqs. (73), (78), and (79). The former bounds depend explicitly on three-point statistical information, and the later bounds depend on numbers developed from a statistical theory.

For purposes of a statistical theory there should be available measurements of the correlation functions. To obtain this data is a tedious but straightforward process. For the case of statistically homogeneous and isotropic materials it is possible to obtain the correlation functions from a single material sample. Corson (1971) obtained two- and three-point correlation functions for two-phase mixtures of powdered metals. He polished a plane section of the composite material and then photographed the two phases using a microscope. In Fig. 1 we present a photograph of a 20% Fe–80% Pb mixture magnified 200×. Corson obtained the correlation functions by analyzing microphotographs of this type. In the course of his analysis Corson obtained the functions $P_{11}(r)$ and $P_{22}(r)$, respectively, the probability that both points are in the iron phase or in the lead phase. We present these graphs in Figs. 2 and 3. We note that $R_{2\epsilon}$ is a linear

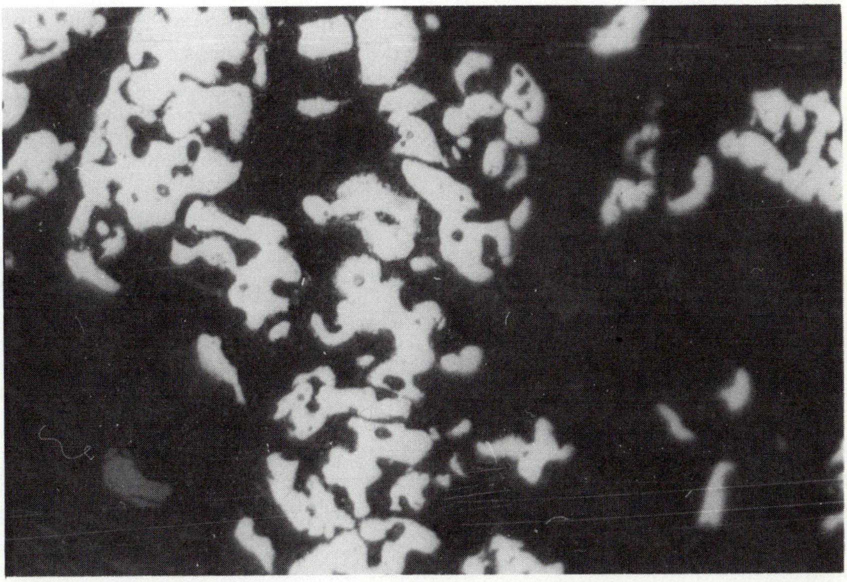

FIG. 1. Microphotograph of 20% Fe–80% Pb system, approximately 200× magnification (Corson, 1971).

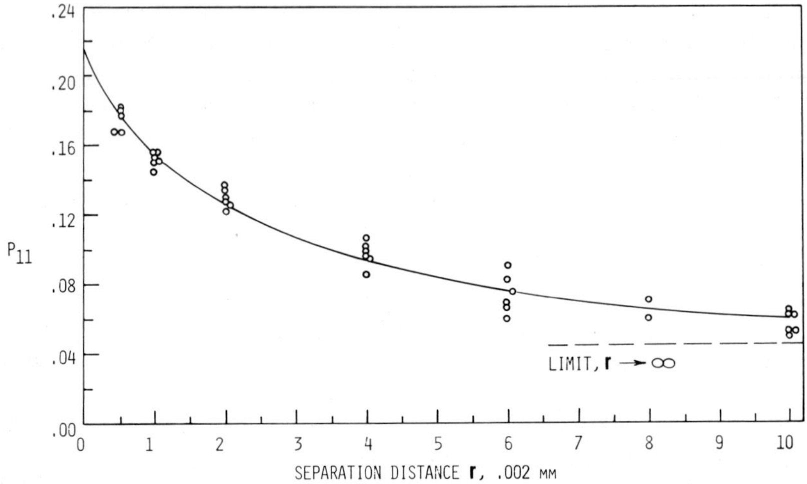

FIG. 2. P_{11} versus separation distance in 20% Fe–80% Pb system: — is constructed function, ○ is data. (From Corson, 1971.)

combination of P_{11} and P_{22}. In Figs. 4 and 5 two typical three-point correlation functions are shown. For completeness the functions must be obtained for all possible triads of three points. Each triad may be specified by two distances r_1 and r_2 and the angle θ between the vectors $\mathbf{r}_1$ and $\mathbf{r}_2$. The functions shown in Figs. 4 and 5 are for fixed θ and six choices of r_1.

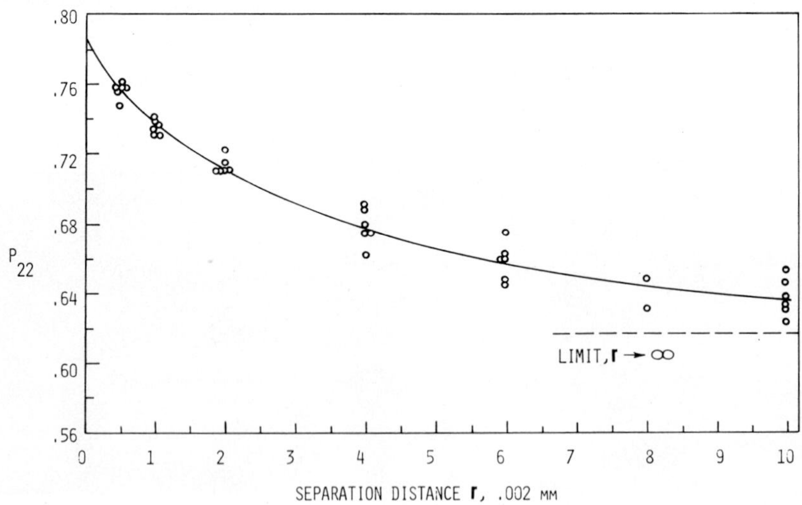

FIG. 3. P_{22} versus separation distance in 20% Fe–80% Pb system: — is constructed function; ○ is data. (From Corson, 1971.)

To our knowledge Corson's experiments represent the only thorough experimental determination of correlation functions for statistical heterogeneous materials. It is hoped that other investigators will perform similar experiments so that an experimental store of information will become available. Once a reasonable amount of data has been accumulated, it should not be difficult to formulate approximate analytical expressions to represent the correlation functions. Corson's data, for example, may be approximated by exponential type expressions.

Expressions like those given by Corson may be used in Eq. (73) to yield closer bounds than the bounds given in Eq. (68). As we stated previously, Corson evaluated the analogous elasticity bounds using his measured correlation functions and obtained more restrictive bounds. The effective constants he measured lay within these more restrictive bounds.

While the experimental verification of the bounds given in Eq. (73) requires very thorough experimental work, the bounds given by Eqs. (78) or (79) are much easier to check. It is only necessary to assign values of G_1 and G_2 to the two phases. For an inclusion-matrix material structure G_1

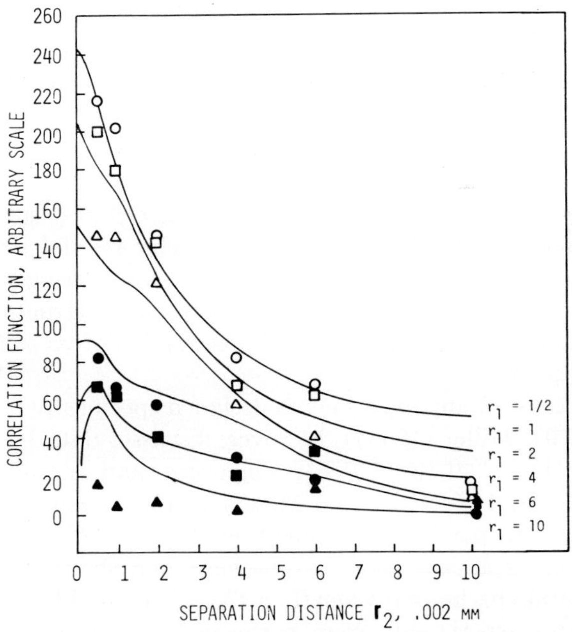

FIG. 4. $R'_{3\epsilon}(r_1, r_2, \theta)$ versus r_2, $\theta = 30°$ in 20% Fe–80% Pb system: — is constructed function; ○ is $r_1 = \frac{1}{2}$; □ is $r_1 = 1$; △ is $r_1 = 2$; ● is $r_1 = 4$; ■ is $r_1 = 6$; ▲ is $r_1 = 10$. (From Corson, 1971.)

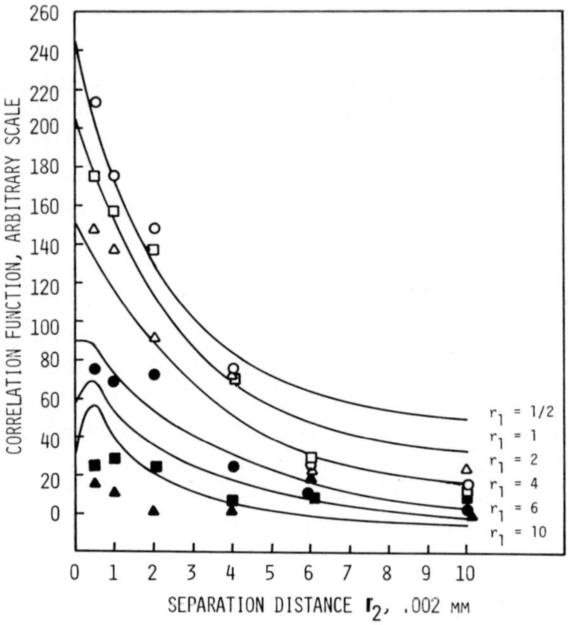

FIG. 5. $R'_{3_\epsilon}(r_1, r_2, \theta)$ versus r_2, $\theta = 60°$ in 20% Fe–80% Pb system: — is constructed function; ○ is $r_1 = \frac{1}{2}$; □ is $r_1 = 1$; △ is $r_1 = 2$; ● is $r_1 = 4$; ■ is $r_1 = 6$; ▲ is $r_1 = 10$. (From Corson, 1971.)

is determined by the inclusion geometry (i.e., $G_1 = \frac{1}{9}$ spherical shape, $G_2 = \frac{1}{3}$ platelike shape). Often G_2 is hard to specify and if desired, it may be left arbitrary by choosing values that give the least stringent bounds. In materials where there is no evident inclusion-matrix structure it is more difficult to choose G_1 and G_2, but visual observation often suffices to make reasonable estimates.

We do not know of any experiments designed specifically to check either Eq. (78) or (79). Miller (1969a), however, discusses data by a number of investigators, i.e., DeLoor (1956), Sugawara and Yoshizawa (1962), Herring (1960), and Woodside and Messmer (1961), and shows agreement with his theory. In particular, he mentions the data of Landauer (1952) where both phases appear spherical in shape and shows that the results obtained lie within his bounds when $G_1 = G_2 = \frac{1}{9}$. It would be most desirable if in the future experiments were performed specifically to check Eqs. (78) and (79) (and their subsequent generalizations to include packing information) since it is important to establish the validity of the Miller cell model for real materials.

VII. Applications

The principal application of statistical ideas to problems of heterogeneous materials will probably be in the design of materials with desired effective constants. The simplest way to illustrate this point is to consider the design of a two-phase composite material that has a maximum effective thermal conductivity. For example, high thermal storage–low thermal conductivity materials are sometimes seeded with high-conductivity inclusions to increase their effective conductivity. Since the inclusion material is usually expensive relative to the matrix material, it is important that inclusion shape and packing be optimally chosen.

If the inclusions are placed at random in the matrix we may assume that ϵ^* is bounded by the expressions in Eq. (78), G_1 and G_2 are numbers bounded by $\frac{1}{9}$ and $\frac{1}{3}$, where $\frac{1}{9}$ corresponds to a spherical shape, $\frac{1}{3}$ to a platelike shape, and $\frac{1}{6}$ to a needlelike shape. As an example, let us suppose that the ratio of inclusion conductivity matrix to matrix conductivity α is 100, and that because of the cost we are interested in inclusion concentrations of less than 10% (i.e., $v_2 < 0.1$). For these values $\alpha - 1 \approx \alpha$ and $v_2{}^2G_1$ may be neglected in comparison to $(1 - v_2)^2G_2$. The latter approximation indicates that the matrix geometry is unimportant and only the inclusion shape is important. In Table I we present values of the bounds for $v_2 = 0.05$ and 0.1 and $G_2 = \frac{1}{9}, \frac{1}{6}$, and $\frac{1}{3}$. With $\epsilon_L{}^*/\epsilon_1$ the lower bound of ϵ^* divided by the matrix conductivity and $\epsilon_U{}^*/\epsilon_1$ the upper bound of ϵ^* divided by the matrix conductivity.

We see immediately from Table I that if we wish to maximize the effective thermal conductivity for a fixed volume fraction, then spherical

TABLE I

BOUNDS ON EFFEECIVE THERMAL CONDUCTIVITY FOR VARIOUS VALUES OF VOLUME FRACTION AND *G* NUMBER

$\epsilon_L{}^*/\epsilon_1$	$\epsilon_U{}^*/\epsilon_1$	v_2	G_2
1.3	3.1	0.1	$\frac{1}{9}$
1.4	6.0	0.1	$\frac{1}{6}$
2.6	7.7	0.1	$\frac{1}{3}$
1.2	1.6	0.05	$\frac{1}{9}$
1.2	2.9	0.05	$\frac{1}{6}$
2.4	3.3	0.05	$\frac{1}{3}$

inclusions ($G = \frac{1}{9}$) are not as effective as platelike or needle inclusions. *However*, and this is a crucial point in any design procedure, there are packing distributions of spherical inclusions leading to $\epsilon_U^*/\epsilon_1 = 3.1$ ($v_2 = 0.1$), while on the other hand, there are packing distributions of platelike inclusions that lead to $\epsilon_L^*/\epsilon_1 = 2.6$ ($v_2 = 0.1$). Thus, if one is not careful about how the inclusions are placed within the matrix and what relative size distributions of inclusions are chosen, then it is quite possible to have a less effective material with platelike inclusions than with spherical inclusions. Beyond this, of course, one must determine the best packing so that if we use platelike inclusions, we come close to $\epsilon_U^*/\epsilon_1 = 7.7$ rather than $\epsilon_U^*/\epsilon_1 = 2.6$.

Unfortunately, at this time, no one has published specifications that will tell the engineer how to place and size-distribute the inclusions in order to obtain the maximum value of ϵ_U^*/ϵ_1. As we stated in Section IV, ElSayed has extended Miller's bounds to include packing information. For the problem discussed here two new numbers are required to specify the packing. Preliminary results show that often specifying one of these numbers enables the engineer to approach close to the maximum upper limit $\epsilon_U^*/\epsilon_1 = 7.7$. Currently an attempt is made to identify this number with a particular mode of packing. In addition to this theoretical work, it is important that experimental work proceed at the same time, but we do not know of any work proceeding along these lines.

If $v_1 = 0.05$ rather than $v_1 = 0.1$, the bounds for ϵ_L^*/ϵ_1 and ϵ_U^*/ϵ_1 are much closer together and packing information is less important. Here if plate-like inclusions are used, the results will always be better than if spherical inclusions are used. The ratio 3.3/2.4 is significantly less than 7.7/2.6 and packing is less crucial in design. If $\alpha = 10$ rather than $\alpha = 100$, packing information need not be considered at all.

The reader should not draw the conclusion from this discussion that to enhance conductivity one must not use spherical inclusions but only platelike inclusions. It may, for example, turn out that platelike inclusions are too expensive to manufacture or that it is too difficult to handle them. The purpose of this section is only to point out that the gain in conductivity obtained by proper shape and packing of inclusions is significant enough to make them a major factor in the design of composite materials.

The same design problems enter in the manufacture of fiber-reinforced materials used in structural materials and bounds for the effective elastic constants have also been derived Miller (1969b). Here the problem is more difficult since more than one effective constant must be considered. Thermoelastic problems have also received some recent attention from a statistical point of view (Alvarez-Vara, 1972), but more work is required before guides to design are available.

In closing it should be pointed that the statistical techniques used in the study of heterogeneous materials also have application in areas not directly related to this physical problem. One such area is pattern recognition. At present, the two-point correlation function and its spatial Fourier transform are used to distinguish different types of objects and to characterize random patterns. For isotropic patterns, however, three-point correlation functions which lead to the G number given above are necessary to characterize the shape of objects. Four-point correlation functions are necessary to distinguish how objects cluster and their relative size distribution. As an example, the G number and the packing numbers that appear in El-Sayed's generalization may be used to characterize topographic maps. Two-phase maps could be formed by considering all land above some elevation z_c to be one phase and all land below to be the second phase. The G and other numbers associated with a few values of z_c could be used to characterize a particular terrain. Similar analyses could be used to characterize sea states and other types of maps.

VIII. Summary

In this chapter we discussed the application of explicitly statistical theories for the determination of properties of heterogeneous materials. Equations (1a) and (1b) are the governing equations of the materials we considered, and $\epsilon_{ij}(\mathbf{x})$ was taken to be a function that could best be described in statistical terms. The theory is applicable to thermal, electrical, and magnetic properties of heterogeneous materials, and has been extended to treat elastic and thermoeleastic properties.

The basic difficulty in treating heterogeneous materials from a statistical point of view is that a finite number of governing equations cannot be formulated for quantities like the mean $\{\phi(\mathbf{x})\}$ or the two-point correlation function $R_{2\phi}$. Rather, as is shown in Section III,B, an infinite hierarchy results. Only if some approximation is made, like the small perturbation approximation used in Eq. (35), is it possible to solve for $\{\phi\}$ or other statistical quantities. Little progress has been made in finding useful approximations, but it has been possible to iterate formally Eq. (30). It was shown that for an infinite statistically homogeneous material $\{E_i\}$ satisfies Eq. (50), which is an integrodifferential equation. Equation (50) may be viewed as a phenomenological equation in which the effective constant ϵ^* and kernel function $M_{ij}(\mathbf{x} - \mathbf{x}')$ are determined either experimentally or by general theoretical requirements. When $l_c \ll L_\rho$, Eq. (50) may be simplified, and it is found that $\{\phi\}$ satisfies Eq. (59).

Perhaps the most important property of a heterogeneous material is its effective constant ϵ^*. Like the quantities $\{\phi\}$ and $R_{n\phi}$, ϵ^* depends on all-order statistical information contained in the $\epsilon'(\mathbf{x})$ field. Fortunately, it is possible to use variational principles in order to bound ϵ^* using only low-order statistical information. For example, for a two-phase material the bounds in Eq. (68) are given in terms of volume fractions. Bounds that depend on higher-order statistical information are given in Eq. (73). For two-phase cell materials these latter bounds reduce to those given in Eq. (78) which depend on volume fractions and a shape factor G_i. Presently, bounds are being derived which depend on v, G, and numbers that provide packing information. These bounds are important to design engineers since it is impossible to measure ϵ^* for all the possible composite materials that may be constructed. The engineer must have some theoretical guidance in designing materials that are optimum for given specifications.

Very little experimentation has been performed with statistical parameters clearly in mind. This is in contrast to turbulence theory where many excellent experimental studies were carried out to measure statistical quantities. Similarly, while composite materials are assuming more and more importance in a wide variety of applications, most design is done on the basis of deterministic ideas. In Section VII it is pointed out how the bounds on ϵ^* that were derived using statistical concepts may aid the engineer in design of composite materials.

References

Adams, D., and Tsai, S. (1969). *J. Compos. Mater.* **3**, 368.

Alvarez-Vara, F. (1972). On Some Problems in the Statistical Theory of Heterogeneous Media. Ph.D. Thesis, Univ. of Pennsylvania, Philadelphia, Pennsylvania.

Beran, M. (1965). *Nuovo Cimento* **38**, 771.

Beran, M. (1968). "Statistical Continuum Theories." Wiley (Interscience), New York.

Beran, M. (1971). *Phys. Status Solidi* **6**, 365.

Beran, M., and McCoy, J. (1970). *Quart. Appl. Math.* **28**, 245; (1972) **29**, 586.

Beran, M., and Molyneux, J. (1966). *Quart. Appl. Math.* XXIV, No. 2, 107.

Beran, M., and Silnutzer, N. (1971). *J. Compos. Mater.* **5**, 246.

Bolotin, V., and Moskalenko, V. (1968). *Sov. Phys.-Dokl.* **13**, 73.

Bourret, R. (1962). *Nuovo Cimento* **26**, 1.

Brown, W. F., Jr. (1955). *J. Chem. Phys.* **23**, 1514.

Brown, W. F., Jr. (1965). *Trans. Soc. Rheol.* **9**, Pt. 1, 357.

Brown, W. F., Jr. (in press). *J. Math. Phys.*

Chamis, C., and Sendeckyj, G. (1968). *J. Compos. Mater.* **2**, 332.

Corson, P. (1971). An Experimental Study of the Elastic Properties of Two-Phase Solids. Ph.D. Thesis, Univ. of Pennsylvania, Philadelphia, Pennsylvania.

DeLoor, G. (1956). Dielectric Properties of Heterogeneous Mixtures. Thesis, Univ. of Leiden.

ElSayed, M. (1972). The Determination of Bounds for Thermal, Electrical and Magnetic Properties of Heterogeneous Materials Using Higher Order Statistical Information. Ph.D. Thesis, Univ. of Pennsylvania, Philadelphia, Pennsylvania.

Frisch, H. (1965). *Trans. Rheol. Soc.* Pt. 9, No. 1, 293.

Frisch, U. (1968). *In* "Probabilistic Methods in Applied Mathematics" (A. T. Bharucha-Reid, ed.). Academic Press, New York.

Gilbert, C. (1962). *Ann. Math. Stat.* **33,** 958.

Hashin, Z. (1968). *J. Compos. Mater.* **2,** 284.

Hashin, Z., and Shtrikman, S. (1962). *J. Appl. Phys.* **33,** 3125.

Hasselman, D., and Fulrath, R. (1968). "Mechanical Properties of Continuous Matrix, Dispersed Phase Ceramic Systems, Ceramic Microstructures," p. 343. Wiley, New York.

Herring, C. (1960). *J. Appl. Phys.* **31,** 1939.

Hopf, E. (1952). *J. Ratl. Mech. Anal.,* **1,** 87.

Kerner, E. (1956). *Proc. Phys. Soc. B* **69,** 802.

Krock, R. (1963). *Proc. ASTM* **63,** 605.

Kroner, E. (1967). *J. Mech. Phys. Solids* **15,** 319.

Landauer, R. (1952). *J. Appl. Phys.* **23,** 779.

Lomakin, V. (1968). *Prikl. Mekh.* **4,** 1.

Miller, M. (1969a). *J. Math. Phys.* **10,** 1988.

Miller, M. (1969b). *J. Math. Phys.* **10,** 2005.

Mindlin, R., and Eshel, N. (1968). *Int. J. Solids Struct.* **4,** 109.

Nishimatsu, C., and Gurland, J. (1960). *Trans. ASM* **52,** 469.

Rosen, G. (1971). *J. Math. Phys.* **12,** 812.

Sugawara, A., and Yoshizawa, Y. (1962). *J. Appl. Phys.* **33,** 3135.

Weissberg, W., and Prager, S. (1962). *Phys. Fluids* **5,** 1390.

Woodside, W., and Messmer, J. (1961). *J. Appl. Phys.* **32,** 1688.

Zienkiewicz, O. (1971). "The Finite Element Method in Engineering Science." McGraw-Hill, New York.

Bibliography

The following references were not explicitly used in this chapter, but may be of interest to the reader. They are listed alphabetically.

Beran, M. (1968). Bounds on Field Fluctuations in a Random Medium, *J. Appl. Phys.* **39,** 5712.

Beran, M., and McCoy, J. J. (1970). Mean Field Variations in a Statistical Sample of Heterogeneous Linearly Elastic Solids, *Int. J. Solids Struct.* **6,** 1035.

Beran, M., and McCoy, J. J. (1970). The Use of Strain Gradient Theory for Analysis of Random Media, *Int. J. Solids Struct.* **6,** 1267.

Bolotin, V. V. (1969). "Statistical Methods in Structural Mechanics." Holden-Day, San Francisco, California.

Bolotin, V., and Moskalenko, V. (1968). The Problem of Determining Elastic Constants in Micro-Inhomogeneous Media, *Prikl. Mekh. Tekh. Fiz.* No. 1, 66.

Bolotin, V., and Moskalenko, V. (1969). About Computation of Macroscopic Permanent Strongly Isotropic Composite Materials, *Solid State Mech.* No. 3, 106.

Budiansky, B. (1965). On the Elastic Moduli of Some Heterogeneous Materials, *J. Mech. Phys. Solids* **13,** 223.

Eshelby, J. D. (1957). The Determination of the Elastic Field of an Elipsoidal Inclusion, and Related Problems, *Proc. Roy. Soc. A* **241,** 376.

Eshelby, J. D. (1961). Elastic Inclusions and Inhomogeneities, *in* "Progress in Solid Mechanics," Vol. 2 (Sneddon and Hill, eds.). Wiley (Interscience), New York.

Fokin, A. (1971). Dielectric Constant of Mixture, *Sov. Phys.-Tech. Phys.* **16,** 849.

Fokin, A., and Shermerov, T. (1967). Elastic Moduli for Textured Materials, *Mekh. Tverd. Tela* No. 1, 129.

Frantsevich, I., and Karpinos, D. (eds.) (1970). Fibrous Composites. Translated from Russian by NASA 7T F-675 and NSF Israel Program for Scientific Translations, Jerusalem, 1972. (Available from U.S. Dept. of Commerce Natl. Tech. Inf. Service, Springfield, Virginia.)

Greszczuk, L. B. (1963). Effect of Foreign Inclusions on Properties of Solids, *J. Eng. Mech. Div. ASCE* **92,** No. EM6, Proc. Paper 5029, 63.

Hashin, Z. (1962). The Elastic Moduli of Heterogeneous Materials, *J. Appl. Mech. Trans. ASME* **29,** 143.

Hashin, Z. (1964). Theory of Mechanical Behavior of Heterogeneous Media, *Appl. Mech. Rev.* **17,** 1.

Hashin, Z. (1965). Elasticity of Random Media, *Trans. Soc. Rheol.* **9,** 1, 381.

Hashin, Z. (1970). Theory of Composite Materials, *Mech. Compos. Mater. Proc. 5th Symp. Naval Struct. Mech., Philadelphia, Pa., May 1967* Pergamon, Oxford.

Hashin, Z. (1970). Theory of Fiber Reinforced Materials. Contract NAS1-8818, NASA, Towne School, Univ. of Pennsylvania, Philadelphia, Pennsylvania.

Hashin, Z., and Shtrikman, S. (1961). Note on a Variational Approach to the Theory of Composite Elastic Materials, *J. Franklin Inst.* **271,** 336.

Hashin, Z., and Shtrikman, S. (1963). A Variational Approach to the Theory of the Elastic Behavior of Multiphase Materials, *J. Mech. Phys. Solids* **11,** 127.

Hill, R. (1965). A Self-Consistent Mechanics of Composite Materials, *J. Mech. Phys. Solids,* **13,** 213.

Khoroshun, L. P. (1966). On the Method of Determining Elastic Moduli of Reinforced Bodies, *Mekh. Polim.* **1,** 78.

Khoroshun, L. P. (1967). Theory of Isotropic Deformation of Elastic Bodies with Random Inhomogeneities, *Prikl. Mekh.* **3,** No. 9, 12.

Krivoglaz, M. A., and Cherevko, A. S. (1959). On the Elastic Moduli of a Two-Phase Solid, *Fiz. Met. Metall.* **8,** No. 2, 161.

Lifshits, M., and Rosentsveig, M. (1946). *Zh. Eksp. Teor. Fiz.* **16,** 967. Cited by Krivoglaz and Cherevko.

Lomakin, V. (1965). Deformation of Microscopically Nonhomogeneous Elastic Bodies, *Appl. Math. Mech.* **29,** 888.

Lomakin, V. (1966). On the Elasticity Theory of the Deformation of Microscopically Nonhomogeneous Materials and Its Relationship to the Couple Stress Theory of Elasticity, *PMM, Appl. Math. Mech.* **30,** 875.

Lomakin, V. (1967). Moment Equations for Some Statistical Problems in Elasticity, *Appl. Math. Mech.* **31,** 670.

Mandel, J., and Dantu, P. (1963). Contribution a l'etude theorique et experimentale du coefficient d'elasticite d'un milieu heterogene, mais statistiquement homogene, *Ann. Ponts Chaussees* **133,** 1, 115.

Masumoto, H., and Kobayashi, T. (1965). On the Thermal Expansion Coefficient and the Temperature Coefficient of Young's Modulus of Iron–Platinum Alloys, *Trans. Jap. Inst. Metals* **6,** 113.

Masumoto, H., Saito, H., and Kobayashi, T. (1963). The Thermal Expansion Coefficients and the Temperature Coefficients of Young's Modulus of the Alloys of Iron and Palladium, *Trans. Jap. Inst. Metals* **4,** 114.

McCoy, J. J. (1970). On the Displacement Field in an Elastic Medium with Random Variations of Material Properties, *Rec. Advan. Eng. Sci. Annu. Meeting Conf. Eng. Sci., 6th, November 1968, Princeton, New Jersey,* p. 235.

McDanels, D. L., Jech, R. W., and Weeton, J. W. (1965). Analysis of Stress-Strain Behavior of Tungsten-Fiber-Reinforced Copper Composites, *Trans. AIME* **233,** 636.

Molyneux, J. (1969). Statistical Properties of Polycrystalline Dielectrics, *J. Math. Phys.* **10,** 912.

Molyneux, J. (1970). Effective Permittivity of a Polycrystalline Dielectric, *J. Math. Phys.* **11,** 1172.

Molyneux, J., and Beran, M. (1965). Statistical Properties of the Stress and Strain Fields in a Medium with Small Random Variations in Elastic Coefficients, *J. Math. Mech.* **14,** 337.

Nemat-Nasser, S. (1966). On the Response of Continuous Media to Random Excitations, *Int. J. Solids Struct.* **2,** 371.

Paul, B. (1960). Prediction of Elastic Constants of Multiphase Materials, *Trans. AIME* **218,** 36.

Savin, G. N., and Khoroshun, L. P. (1967). Statistically Isotropic Deformation of Composite Materials, *Dokl. Akad. Nauk. Ukr. SSR, Ser. A* No. 9, 804.

Sendeckyj, G. P. (1969). Statistical Theory of Elastic Moduli of Particulate Composites, *J. Compos. Mater.* **3,** 177.

Somoroff, A. (1970). A Variational Approach to Statistically Nonhomogeneous Fields, *Quart. Appl. Math.* **28,** 219.

Stavrov, V. P., Dolgikh, V. Ya., and Volkov, S. D. (1967). On Elastic Constants of Stochastically Reinforced Plastics, *Mekh. Polim.* No. 2, 259.

Umekawa, S., Kotfila, R., and Sherby, O. D. (1965). Elastic Properties of a Tungsten–Silver Composite Above and Below the Melting Point of Silver, *J. Mech. Phys. Solids* **13,** 229.

Umekawa, S., and Sherby, O. D. (1966). Young's Modulus of Eutectic Alloy Systems with Special Emphasis on the Silver-Lead System, *J. Mech. Phys. Solids* **14,** 65.

Walpole, L. J. (1966). On Bounds for the Overall Elastic Moduli of Inhomogeneous Systems—I, *J. Mech. Phys. Solids* **14,** 151.

Walpole, L. J. (1966). On Bounds for the Overall Elastic Moduli of Inhomogeneous Systems—II, *J. Mech. Phys. Solids* **14,** 289.

Walpole, L. J. (1969). On the Overall Elastic Moduli of Composite Materials, *J. Mech. Phys. Solids* **17,** 235.

7

Finite Deformations of Ideal Fiber-Reinforced Composites

ALLEN C. PIPKIN

Division of Applied Mathematics
Brown University
Providence, Rhode Island

I. Introduction

In the present chapter we outline a theory of large deformations of materials composed of strong fibers embedded in a matrix of more compliant material, such as nylon cords in a rubber matrix or metal filaments bonded together with ductile aluminum. Our object is not to predict the mechanical properties of the composite material from the properties of its constituents, nor do we consider any of the important problems in which it is necessary to distinguish points in the matrix from points in the fibers. Rather, we seek to predict the gross mechanical behavior of the composite material, which is treated as a continuum with properties defined by its macroscopic mechanical response.

Because of the technological importance of rubber reinforced with strong cords, there is a moderately large body of theoretical work on large elastic deformations of materials reinforced with inextensible cords, beginning with the work of Adkins and Rivlin (1955). The basic work in this area is summarized in the book by Green and Adkins (1960). When rubber is reinforced with strong cords, the tensile modulus for extensions parallel to the cord direction is largely determined by the modulus of the cords, while the shear moduli of the material are of the order of magnitude of the shear modulus of the unreinforced rubber. The shearing compliances are thus very large in comparison to the extensibility of the material in the cord direction. In problems for which some shearing mode of deformation is admissible, it is an acceptable idealization to neglect the extensibility of the cords, and treat them as material curves that cannot change length in any deformation. Thus, a geometrical constraint condition replaces a more complicated relation between stress and deformation, and leads to a more tractible theory.

Fiber inextensibility is not the only geometrical constraint that can be used as a simplifying idealization. Because the bulk compressibility of rubber is very small in comparison to its shear compliance, it is often a good approximation to regard the compressibility as zero. The geometrical constraint of no change of volume of any material element replaces a more detailed and accurate but more complicated description of the bulk response of the material. The use of the assumption of incompressibility in making problems more tractible is well known from Rivlin's earliest work on finite elastic deformations (see, for example, Rivlin, 1948; 1949).

In most of the work on reinforced rubber, such as the basic work by Adkins (1956a, b), it is assumed that the cords lie in one or more discrete surfaces embedded in the material. The cords are treated as continuously distributed in the surface, so that on the surface there is no distinction between points in a cord and points in the matrix. In contrast, the materials that we wish to consider in the present review are essentially composed of the fibers themselves, rather than being merely reinforced by fibers in discrete layers. The theory treats the fibers as continuously distributed throughout the whole volume of the material. In the mathematical idealization, the fibers form a family of material curves, with one curve of the family through each point of the material. We implicitly restrict attention to a single family of fibers, by treating the fiber direction as a continuous function of position. For the case of elastic materials, a theory of this kind is described by Green and Adkins (1960), but the theory has only recently been exploited.

The three main idealizations that we have mentioned are that the fibers are *continuously distributed*, that they are *inextensible*, and that the composite is *incompressible* in bulk. Mulhern *et al.* (1967) used exactly these idealizations in their theory of fiber-reinforced plastic materials. We refer to the mathematical models based on these three assumptions as *ideal fiber-reinforced composites*, whether the stress response is elastic, plastic, viscoelastic, or any other response. Pipkin and Rogers (1971a) have shown that many features of the mechanical behavior of such materials are independent of the nature of the stress response. In the present review we concentrate on such general features.

We wish to make it clear that problems of finite deformation of ideal composites can be solved by very elementary methods. For this reason, we show the details of the analysis in a small number of problems, at the expense of a broader discussion of the general theory. A great variety of extensions and generalizations of the theory are discussed in a recent book by Spencer (1972b).

As an introduction to the theory in the simplest possible context, in Section II we briefly consider infinitesimal, plane deformations of ideal composites with elastic shearing response. The solutions of such problems can be compared with solutions obtained from the theory of infinitesimal elastic deformations of transversely isotropic materials with small but nonzero extensibility and compressibility. By doing so, we learn how to interpret the results of the idealized theory and how to judge its accuracy. In particular, we discuss the stress concentration layers that are an unusual feature of solutions within the idealized theory.

In Section III, which is by far the longest section of this chapter, we consider finite plane strain problems. Here we go into some detail about methods of solution. Displacement boundary value problems can be solved purely kinematically, without making use of detailed assumptions about stress response and even without use of the equilibrium equations. In traction boundary value problems, or in mixed problems, it is necessary to postulate some form of shearing stress response. For simplicity, we restrict attention to elastic shearing stress response, or to the quasi-elastic response of plastic or viscoelastic materials. The basic theory in Section III is from the paper by Pipkin and Rogers (1971a).

Finite plane strain superimposed on uniform stretching orthogonal to the plane of deformation is discussed in Section IV. This is not quite a trivial generalization of plane strain, because some work is involved in determining the so-called uniformly stretched state; stretching in the axial direction causes fibers in cross-sectional planes to curl up.

Axisymmetric deformations without twist are considered in Section V. Problems of this kind are more difficult to solve than plane problems, but we find that one of the most important results concerning plane problems is also true for axisymmetric deformations: for every kinematically admissible deformation there is a statically admissible stress field compatible with it.

The general theory for deformations with no special symmetry is discussed very briefly in Section VI. There we also mention previous work on transversely isotropic or orthotropic materials that is applicable to materials with extensible fibers.

Because we concentrate our attention on special classes of problems in which no elaborate general theory is required, the reader need not have any prior knowledge of the theory of finite elastic deformations. However, such previous knowledge is desirable, especially in Section VI, where we state many results without proof. The review articles by Rivlin (1960, 1970) cover all the general theory that we need, and much more.

The book by Green and Adkins (1960) is an important general reference that contains a great deal of material pertaining to fiber-reinforced and laminated materials; in particular, this book contains a general discussion of geometrical constraints and the stress reactions to such constraints, a chapter on problems involving transversely isotropic materials, orthotropic materials, and materials with curvilinear aelotropy, which applies to cases in which fibers are initially curved, and a chapter on reinforcement by inextensible fibers, with applications to cases in which the fibers lie in discrete surfaces.

The book by Spencer (1972b) covers much of the material that we discuss in this chapter and a great many related topics. In particular, this book contains extensive discussions of reinforcement by more than one family of inextensible fibers and by inextensible surfaces. Plastic stress response is also considered in detail.

II. Infinitesimal Plane Deformations

Because of the constraints of incompressibility and inextensibility, boundary value problems for ideal fiber-reinforced materials are not posed in exactly the same way as for unconstrained materials, and solutions exhibit certain unusual features. In order to examine these peculiarities in the simplest possible context, in the present section we examine infinitesimal plane deformations of materials reinforced with fibers that are initially straight and parallel. We find, among other things, that the stress field in an ideal fiber-reinforced material may involve Dirac deltas representing finite forces supported by single fibers or sheets. The physical interpretation of such singular stresses as thin layers of high stress concentration is discussed.

A. Displacement Boundary Value Problems

We consider plane deformations of a body in which the fibers are initially straight and parallel to the x direction. The cross section of the body is the same in each plane z = constant. The displacement field for an infinitesimal plane deformation has the form

$$u = u(x, y), \qquad v = v(x, y), \qquad w = 0 \tag{1}$$

Fibers are treated as continuously distributed, so that every material line initially parallel to the x direction is called a fiber. The fibers are treated as inextensible; no segment of any material line parallel to the x direction can change length in any kinematically admissible deformation. In terms of strain, inextensibility means that the strain component ϵ_{xx} $(=u_{,x})$ is zero at every particle. Consequently, the displacement component u parallel to a fiber must have the same value everywhere on that fiber: $u = u(y)$.

The material is taken to be incompressible in bulk. This means that in any kinematically admissible deformation, the volume dilation ϵ_{ii} is zero. Since ϵ_{zz} is zero in plane deformations and ϵ_{xx} is zero because of fiber inextensibility, the volume dilation is ϵ_{yy} $(=v_{,y})$. Hence, $v = v(x)$. Inextensibility and incompressibility, together, imply that the distance between two fibers cannot change in a plane deformation. The displacement v parallel to a normal line $x =$ constant is constant along each normal line.

From the constraint conditions alone, we have found that the displacement field for any kinematically admissible plane deformation must have the form

$$u = u(y), \qquad v = v(x), \qquad w = 0 \tag{2}$$

We observe that in boundary value problems with prescribed surface displacements, it is not possible to prescribe displacements arbitrarily around the entire perimeter of the plane region considered. The deformation is determined uniquely when the component u has been prescribed at one point on each fiber and v has been prescribed at one point on each normal line. (By a *normal line* we shall always mean a curve orthogonal to the fiber direction at each of its points.)

B. Mixed Boundary Value Problems

We have just seen that with sufficient displacement boundary data, the deformation can be determined without specifying anything about the shearing stress response of the material. However, most problems of practical interest are mixed boundary value problems, in which displacements are prescribed over a portion of the surface of the body and tractions are prescribed on the remainder of the surface. In order to deal with cases in which the deformation cannot be determined purely kinematically, it is necessary to state the relation between stress and deformation. For simplicity we consider the case of linear elastic shearing stress response, for which the shearing stress component σ_{xy} is equal to $2G\epsilon_{xy}$. Here G is the shear modulus for shearing parallel to the fiber direction. In view of the special form that the displacement field must take, we have

$$\sigma_{xy} = G[u'(y) + v'(x)] \tag{3}$$

As a specific example we consider the deformation of a cantilever under an end load (Rogers and Pipkin, 1971a). The end of the beam at $x = 0$ is bonded to a rigid support so that $u = v = 0$ there. On the end $x = L$ there is a tangential traction $\sigma_{xy} = -F/D$, $\sigma_{xx} = 0$. Here F is the total force on the end per unit of length in the z direction and D is the thickness of the beam. The faces of the beam, along $y = 0$ and $y = D$, are free from applied tractions (Fig. 1).

Because $u(y) = 0$ at the end $x = 0$ then u is equal to zero everywhere. Observe that the displacement boundary condition at the support is satisfied exactly, rather than approximately as in Saint Venant's classical treatment for elastic materials (see, for example, Lekhnitskii, 1963).

To determine $v(x)$ we use a method that seems to be rather widely applicable (see Sections II,C and III,G). Rather than considering the pointwise distribution of stress, at first we consider only the total force exerted across a normal line $x =$ constant, and in particular the shearing force parallel to this line. From (3), with $u = 0$, we find that the total shearing force on a normal section is $DGv'(x)$. Since this must be equal to the end load $-F$, we have a direct evaluation of $v'(x)$. With the displacement boundary condition $v(0) = 0$, we obtain

$$v(x) = -xF/GD \tag{4}$$

This is a simple shearing displacement field, rather than the flexural type of deformation postulated in the classical elastic approximation.

FIG. 1. Cantilever under end load.

With the deformation already fully determined, it is a simple matter to complete the computation of the stress field accompanying the deformation. The stress component $\sigma_{xx} = T$, say, is not related to the deformation by a constitutive equation. Rather, since this stress component does no work in any kinematically admissible deformation, it is a reaction to the constraint of fiber inextensibility. Similarly, the component $\sigma_{yy} = -P$, say, is the reaction to the constraint requiring that the distance between two fibers does not change. Both of these reactions take on whatever values may be necessary in order to enforce the constraints. These values are determined from the equilibrium equations and the traction boundary conditions.

For example, since the shearing stress is constant, the y component of the equilibrium equation yields

$$P_{,y} = 0 \tag{5}$$

The equation determines the variation of P along each normal line separately, in a trivial way in the present case. To complete the evaluation of P we need to know its value at one point on each normal line. The boundary condition $\sigma_{yy} = 0$ on $y = 0$ supplies the required information, and we obtain $P = 0$ everywhere. The condition $\sigma_{yy} = 0$ on $y = D$ is then satisfied automatically because the condition of zero resultant normal force was used implicitly in determining $v(x)$.

We should point out that the process of solution has involved no guess work or trial hypotheses. We emphasize this point because the solution appears, at first, to have an inconsistent feature. We have found that the shearing stress is equal to $-F/D$, but if no shearing traction is being applied to the upper and lower surfaces $y = D$ and $y = 0$, then $\sigma_{xy} = 0$ there. The explanation of the seeming discrepancy is that the value $\sigma_{xy} = -F/D$ is derived from the constitutive equation, which is valid at points in the material, while the boundary value $\sigma_{xy} = 0$ refers to an exterior value not connected to the deformation by a constitutive equation. The shearing stress is simply discontinuous at the boundary. The shearing stress is not constant, but takes the form

$$\sigma_{xy} = -(F/D)[H(y) - H(y - D)] \tag{6}$$

where H is Heaviside's unit step function.

It remains to be shown that a discontinuous shearing stress can satisfy the equilibrium equation. The x component of the equilibrium equation governs the tensile reaction $\sigma_{xx} = T$:

$$T_{,x} = -\sigma_{xy,y} \tag{7}$$

The equation determines the variation of T along each fiber separately, and the value of T at one point on each fiber will be required in order to complete the solution. The condition $\sigma_{xx} = 0$ on the end $x = L$ supplies this information. By using (6) in (7) and performing the integration, we obtain

$$T = (F/D)(L - x)[\delta(y - D) - \delta(y)] \tag{8}$$

Here $\delta(y) = H'(y)$ is the Dirac delta. [For a rigorous discussion of generalized functions see the work of Lighthill (1958).] Thus, the tension T is zero everywhere except in the two boundary fibers (i.e., sheets), where it is infinite in such a way as to represent finite forces supported by these fibers. The upper fiber carries a finite force $(F/D)(L - x)$ by itself, and the lower fiber carries an equal compressive load. Since the lower surface is not supported against buckling on its outside, we anticipate that the material may fail at this surface.

In the present example we have obtained a solution, given boundary data of exactly the same kind that would make the problem well set if we were dealing with classical elasticity theory. Although no general theorem concerning the existence and uniqueness of solutions of mixed boundary value problems for ideal composites has been proved, we can guess that boundary data yielding a well-set problem in elasticity theory will also give a well-set problem for ideal composites, provided that u is prescribed at no more than one point on each fiber and v is prescribed at no more than one point on each normal line.

C. Traction Boundary Value Problems

Problems in which tractions are prescribed at every boundary point are relatively difficult to solve, because purely kinematical conditions are less useful. Fortunately, pure traction boundary value problems are not encountered frequently in practice.

The prescribed tractions must satisfy the usual conditions of global equilibrium.

As an example we consider the same beam discussed in the preceding section, but now with tractions prescribed everywhere on its boundary. To solve the problem we consider the resultant force on finite sections of the body bounded by a fiber or a normal line, a method that will remain valuable in the case of finite deformations (see Section III,G). From the given tractions we first compute $F_y(x)$, the y component of the resultant traction on the part of the surface of the body to the right of the normal line x = constant. We also compute $F_x(y)$, the x component of the resultant traction on the surface above the fiber y = constant. Then by considering the global equilibrium of these parts of the body, and using expression (3) for the shearing stress on the normal line or fiber, respectively, we obtain

$$F_y(x) = G[u(D) - u(0) + Dv'(x)] \tag{9}$$

and

$$F_x(y) = G[v(L) - v(0) + Lu'(y)] \tag{10}$$

The first of these equations can be integrated to yield $v(x)$, apart from a constant of integration, in terms of the boundary tractions and the unknown displacement difference $u(D) - u(0)$. Similarly, the second gives u, up to a constant of integration, in terms of the boundary tractions and the unknown displacement difference $v(L) - v(0)$. It is then a matter of algebra to determine the displacement differences. With u and v then known completely the stress field can be computed in the same way as in the mixed problem previously considered. Now, however, there will in general be stress concentration layers not only in the two bounding fibers, but also in the two bounding normal lines $x = 0$ and $x = L$. In these normal lines, because of mismatch between the interior shearing stress and that prescribed on the exterior, the reaction $\sigma_{yy} = -P$ will involve Dirac deltas signifying finite forces carried by these lines.

For bodies of more general shape, the procedure outlined here leads to equations governing the difference between the displacements u at the two ends of each normal line and the difference between the values of v at the two ends of each fiber. In the present simple example there were only two distinct values to be determined, and this could be done by simple algebra. Some nontrivial problems in which the displacement differences cannot be determined purely algebraically have been solved by England (1972). The existence of solutions for bodies of rather arbitrary shape has been proved by a method that at the same time gives a prescription for the construction of approximate solutions (Pipkin and Sanchez, 1974). This partly verifies the hypothesis that data yielding well-set problems in elasticity theory also yield well-set problems in the theory of ideal composites.

D. *Physical Interpretation of Stress Concentration Layers*

We have found that solutions may involve shearing stress discontinuities across a fiber or a normal line. This occurs most commonly in cases in which the fiber or normal line is part of the boundary of the body, and prescribed shearing tractions do not match the interior shearing stress adjacent to the boundary. A line segment with different shearing forces on the two sides of it must have tensile forces on its ends that differ by a finite amount, in order to be in equilibrium. These finite forces carried by infinitesimally thin layers are represented in terms of stress as Dirac deltas.

In applying the theory to a real material with small but nonzero extensibility and compressibility, we interpret these stress singularities as thin layers in which the tensile stress along the direction of the layer is very large. The finite force predicted by the idealized theory is interpreted as the integral of the large but finite tensile stress across the small but nonzero thickness of the layer. Given an estimate of the thickness of the layer, the magnitude of the tensile stress in it can be estimated by dividing the force (predicted by the idealized theory) by the thickness.

To establish that these interpretations are correct in some sense, and to obtain estimates of the thicknesses of stress concentration layers, Everstine and Pipkin (1971) examined some exact solutions in the theory of elastic deformations of transversely isotropic materials. Young's modulus E for extensions parallel to the fiber direction was assumed to be very large in comparison to the shear modulus G. The Poisson's ratio ν for contraction perpendicular to the fiber direction due to tension in an orthogonal direction, also perpendicular to the fibers, was assumed to be close to unity. It was found that elasticity theory does, indeed, yield thin layers of high tensile stress at the places where the idealized theory predicts that they should be. For stress concentration layers along the fiber direction, the thickness was found to be of the order of $(G/E)^{1/2}L$, where L characterizes the length of the layer. Similarly, it was found that stress concentration layers along normal lines, whose existence is connected with the small compressibility of the material, have thicknesses of the order of $(1 - \nu)^{1/2}L$. In both cases it was found that the maximum tensile stress is predicted with acceptable accuracy by dividing the resultant, given by the idealized theory, by the appropriate thickness.

The fundamental parameters $(G/E)^{1/2}$ and $(1 - \nu)^{1/2}$ also play an important role in another kind of estimation. The idealized theory predicts that disturbances can propagate without attenuation infinitely far along a fiber or a normal line, in marked contrast with the assertion called "Saint Venant's principle." From the examination of exact elastic solutions it was found that this unattenuated propagation of disturbances should be interpreted as propagation over distances of the order of $L/(G/E)^{1/2}$ along fibers and $L/(1 - \nu)^{1/2}$ along normal lines. In this case L is a characteristic length perpendicular to the stress concentration layer.

Approximate solutions of elasticity problems for slightly extensible and compressible materials can be obtained by ordinary perturbation methods, except in stress concentration layers, where singular perturbation methods are required. An approximate solution of the cantilever problem in Section II,B for an elastic material has been obtained by standard methods of boundary layer analysis (Everstine, 1971; Everstine and Pipkin, 1973). This solution confirmed the predictions of the idealized theory, with one important exception. At the corners where the surfaces of the cantilever meet the supporting wall, there are weak stress singularities of the same kind that would be present in an isotropic material. In the idealized theory such singularities can escape notice when they are buried within extended stress concentration layers, as is the case in the cantilever problem.

Limits on the range of validity of the idealized theory can be derived from the preceding order of magnitude estimates of the thicknesses of stress concentration layers. The idealized theory is not likely to be accurate if the thickness is so large that the layer occupies an appreciable fraction of the total volume; extensional deformations in the stress concentration layers, although small, may allow the body to undergo flexural modes of deformation that the idealized theory does not allow (see Section III,H).

III. Finite Plane Deformations

In the case of finite deformations, as in the infinitesimal case, problems involving bodies with an axis of translational symmetry are exceptionally easy to solve. Even if there were no problem of practical importance involving conditions approximating plane strain, the insight into the mechanical behavior of fiber-reinforced materials that can be obtained by studying these explicit solutions would be sufficient reason for examining them.

Because we wish to emphasize that finite deformation problems can be solved by very elementary methods, and because we wish to put some emphasis on practical conclusions from the theory, the theory is not laid out in utmost generality at the outset. The reader who would prefer an outline of the general theory should read the sections in the following order: III, A, B, N, E, I. The results in these sections and in many others are from a paper by Pipkin and Rogers (1971a).

For the same pair of reasons, we consider plane strain problems here, rather than the plane stress of sheets reinforced with two families of inextensible fibers. Plane stress problems, which may be of more practical importance, can be solved by methods similar to those outlined here but are less amenable to explicit analytical solution. The plane stress theory is due to Rivlin (1955) and Adkins (1956b). A brief, interesting review of the theory has been given by Rivlin (1964). The plane stress theory is connected to the more general theory of elastic bodies reinforced with inextensible fibers, which is due to Adkins and Rivlin (1955) and Adkins (1956a).

To keep to the simplest possible illustrations of the theory, for the most part we discuss examples involving an initially rectangular body reinforced with fibers that are initially straight and parallel to one pair of boundaries. We refer to the body variously as a beam, cantilever, slab, column, or plate in order to suggest various applications. Only in Sections III, M and N do we consider the case of initially curved fibers quite explicitly; however, much of the general theory in the sections before these applies just as well to curved fibers as to straight fibers.

Problems can often be solved purely kinematically, without making use of any but the weakest hypotheses about the stress response of the material. For illustrations in cases that do require knowledge of the stress response, we suppose that this response is elastic or quasi-elastic (Section III,E). However, it should be noted that most of the theory is applicable equally well to cases in which the stress response is genuinely plastic or viscoelastic.

At appropriate places we mention the results in most of the previously published papers on plane deformations. The paper by Everstine and Rogers (1971) on the machining problem is an exception; while using just the methods described here, it is somewhat more complicated than the simple examples that we discuss. The paper by Spencer (1972a) on laminated plates shows how to extend the theory to cases in which the fibers are not all parallel to the plane of deformation. Many generalizations and extensions of the theory are discussed at length in the book by Spencer (1972b).

A. Fibers and Normal Lines

Throughout Section III we consider only plane deformations, in which the particle initially at X, Y, Z moves to a place x, y, z given by a mapping of the form

$$x = x(X, Y), \qquad y = y(X, Y), \qquad z = Z \tag{11}$$

In direct vector notation we write $\mathbf{x} = \mathbf{x}(\mathbf{X})$. Because of the one-to-one correspondence between the initial place $\mathbf{X}$ and the final place $\mathbf{x}$, physical quantities associated with a given particle can be treated as functions of either $\mathbf{X}$ or $\mathbf{x}$, as convenience dictates. We use the same symbol for a function under either interpretation and often change from one interpretation to the other with no warning. The gradient operator with respect to $\mathbf{X}$ is denoted ∇_0, and the gradient with respect to $\mathbf{x}$ is ∇.

The body is reinforced with a single family of fibers that lie in the planes Z = constant but need not be straight. We treat the fibers as continuously distributed. Through each point there is a plane curve that we call a *fiber*. The field of fiber directions in the undeformed body is specified by giving the angle $\theta_0(X, Y)$ between the fiber direction at the point X, Y and some fixed direction.

A fiber is a material curve, always composed of the same particles. After a plane deformation, the particles on a given fiber again form a plane curve. The field of fiber directions in the deformed body is denoted by $\theta(x, y)$. As the notation indicates, we suppose that fibers are arranged in the same way in each plane Z = constant, so that θ_0 and θ are independent of Z.

We define two fields of unit vectors, $\mathbf{a}$ and $\mathbf{n}$, by

$$\mathbf{a} = \mathbf{i} \cos\theta + \mathbf{j} \sin\theta \tag{12}$$

and

$$\mathbf{n} = -\mathbf{i} \sin\theta + \mathbf{j} \cos\theta \tag{13}$$

Here $\mathbf{i}$ and $\mathbf{j}$ are Cartesian unit vectors, with $\mathbf{i}$ along the direction $\theta = 0$. The unit vector $\mathbf{a}$ at any point is tangential to the fiber through that point, and $\mathbf{n}$ is perpendicular to the fiber. We use the notation $\mathbf{a}_0$ and $\mathbf{n}_0$ for the corresponding vector fields in the undeformed body, which are defined by replacing θ by θ_0 in (12) and (13). We observe that the derivatives of $\mathbf{a}$ and $\mathbf{n}$ with respect to θ are

$$\mathbf{a}'(\theta) = \mathbf{n}(\theta) \qquad \text{and} \qquad \mathbf{n}'(\theta) = -\mathbf{a}(\theta) \tag{14}$$

The curves with $\mathbf{n}_0$ or $\mathbf{n}$ as tangent vectors are called *normal lines*. Unlike fibers, normal lines are not material curves in general. A set of particles lying along a normal line in the undeformed state will ordinarily not lie along a normal line when the body is deformed. We note that the fiber angle θ is also the angle between a normal line and the fixed direction $\mathbf{j}$, and this is often the more convenient geometrical interpretation.

The curvatures of fibers and normal lines are the rates of changes of their angles with respect to position:

$$1/r_a = \mathbf{a}\cdot\nabla\theta, \qquad 1/r_n = \mathbf{n}\cdot\nabla\theta \tag{15}$$

The reciprocals of the curvatures are the radii of curvature, r_a and r_n. These radii can be positive, negative, or infinite. The initial radii of curvature are denoted r_{a0} and r_{n0}. From (14) and (15) we obtain the Serret–Frenet relations

$$(\mathbf{a}\cdot\nabla)\mathbf{a} = \mathbf{n}/r_a, \qquad (\mathbf{a}\cdot\nabla)\mathbf{n} = -\mathbf{a}/r_a$$

$$(\mathbf{n}\cdot\nabla)\mathbf{a} = \mathbf{n}/r_n, \qquad (\mathbf{n}\cdot\nabla)\mathbf{n} = -\mathbf{a}/r_n \tag{16}$$

The gradient operator is expressed in terms of directional derivatives by

$$\boldsymbol{\nabla} = \mathbf{a}(\mathbf{a}\cdot\boldsymbol{\nabla}) + \mathbf{n}(\mathbf{n}\cdot\boldsymbol{\nabla}) + \mathbf{k}(\mathbf{k}\cdot\boldsymbol{\nabla}) \tag{17}$$

Here $\mathbf{k}$ is the constant unit vector in the z direction. The divergences of the fields $\mathbf{a}$ and $\mathbf{n}$ are

$$\boldsymbol{\nabla}\cdot\mathbf{a} = \boldsymbol{\nabla}\theta\cdot\mathbf{a}'(\theta) = 1/r_n \tag{18}$$

and

$$\boldsymbol{\nabla}\cdot\mathbf{n} = \boldsymbol{\nabla}\theta\cdot\mathbf{n}'(\theta) = -1/r_a \tag{19}$$

B. Deformation Gradients

We first consider the homogeneous deformation of a body whose cross section in the plane of deformation is a rectangle bounded by the lines $X = 0$, $X = L$, $Y = 0$, and $Y = D$. The fibers are initially straight and parallel to the X direction, so that $\theta_0 = 0$ at each particle.

In a homogeneous deformation the rectangle deforms into a parallelogram. The base of the parallelogram must have the length L since fibers cannot change length. Furthermore, the constraint of incompressibility implies that areas are preserved in plane deformations, so the area of the parallelogram must be LD. Consequently, its altitude is D. Inextensibility and incompressibility, together, imply that the distance between two fibers cannot change.

The deformation is a simple shearing, apart from a rigid rotation taken into account by the directions of the vectors $\mathbf{a}$ and $\mathbf{n}$, and a rigid translation that we omit:

$$\mathbf{x} = (X + kY)\mathbf{a} + Y\mathbf{n} + Z\mathbf{k} \tag{20}$$

Here k is the *amount of shear*, and the vector $\mathbf{k}$ is a unit vector in the axial direction. The directional derivatives of $\mathbf{x}$ with respect to the initial coordinates are

$$(\mathbf{a}_0\cdot\boldsymbol{\nabla}_0)\mathbf{x} = \mathbf{a}, \qquad (\mathbf{n}_0\cdot\boldsymbol{\nabla}_0)\mathbf{x} = \mathbf{n} + k\mathbf{a}, \qquad (\mathbf{k}_0\cdot\boldsymbol{\nabla}_0)\mathbf{x} = \mathbf{k} \tag{21}$$

These are the *deformation gradients.*

In an inhomogeneous deformation, infinitesimal elements can still be regarded as homogeneously deformed, and the deformation gradients still must take the forms shown in (21). However, for an inhomogeneous deformation the amount of shear k and the fiber angle θ will vary from point to point. The vectors $\mathbf{a}$ and $\mathbf{n}$, which are functions of θ, will vary with position. With these provisions, the deformation gradients (21) are the most general that are kinematically admissible under the given constraints. The importance of the deformation gradients is that the local state of distortion and rotation of material elements is completely defined by these quantities.

C. Compatibility Conditions

The two functions k and θ specifying the amount of shear and the fiber angle are not arbitrary functions of $\mathbf{x}$ or $\mathbf{X}$. When these functions are given, the deformation gradients (21) are completely defined. The condition that these gradients actually be the derivatives of a single function $\mathbf{x}(\mathbf{X})$ imposes compatibility conditions on k and θ. The compatibility conditions are obtained by eliminating $\mathbf{x}$ from (21) by cross-differentiation. If fibers are initially straight and parallel, so that $\mathbf{a}_0$ is in the X direction and $\mathbf{n}_0$ is in the Y direction everywhere, from (21) we obtain

$$\partial \mathbf{a}/\partial Y = \partial^2 \mathbf{x}/\partial X\, \partial Y = \partial(\mathbf{n} + k\mathbf{a})/\partial X \qquad (22)$$

On carrying out the differentiation, with the aid of (14), and then separating the resulting expression into components in the $\mathbf{a}$ and $\mathbf{n}$ directions, we obtain

$$\partial(\theta - k)/\partial X = 0 \qquad \text{and} \qquad (\partial\theta/\partial Y) - (k\, \partial\theta/\partial X) = 0 \qquad (23)$$

The first compatibility condition immediately yields

$$k = \theta + f(Y) \qquad (24)$$

The amount of shear is equal to the fiber angle, plus an amount $f(Y)$ that is constant along each fiber Y = constant.

To interpret the second compatibility condition we first notice that

$$\mathbf{n} = (\partial \mathbf{x}/\partial Y) - (k\, \partial \mathbf{x}/\partial X) \qquad (25)$$

from which it follows that

$$\mathbf{n}\cdot\nabla = (\partial/\partial Y) - (k\, \partial/\partial X) \qquad (26)$$

Consequently, (23b) is simply

$$\mathbf{n}\cdot\nabla\theta = 0 \qquad (27)$$

Thus, the curvature of every normal line in the deformed body is zero. Normal lines are *straight* lines.

The fibers lie along curves that are the orthogonal trajectories of a family of straight lines. Consequently, the fibers lie along *parallel curves,* such as parallel straight lines or concentric circles. The distance between two curves in a parallel family, measured along a straight normal line, is the same for each normal line. Thus, fibers that are initially straight and parallel remain parallel, although not straight, in any kinematically admissible deformation. The distance between two fibers remains the same as it was initially.

In Section III,M we show that initially *parallel* fibers, whether initially straight or not, must remain parallel in any kinematically admissible plane deformation.

Because normal lines are straight, the fiber angle θ is constant along each normal line, and it is often convenient to identify a normal line by the value of θ on it. We speak of the angle θ of the normal line in such cases.

D. Example: Bending

By using the conservation of fiber length and fiber spacing, and the consequent straightness of normal lines, it is a simple matter to construct kinematically admissible deformations.

As a trivial example, we consider a slab with ends at $X = \pm L$ and sides along the lines $Y = 0$ and $Y = D$. The side $Y = D$ is forced into contact with a rigid cylinder of radius r_d and bonded to it, and the remaining surfaces of the slab are left free from traction.

Since the fiber $Y = D$ lies along a known curve after the deformation, the straight normal lines perpendicular to it can be constructed immediately. These are the radial lines $\theta =$ constant in an appropriate system of polar coordinates. The remaining fibers lie along the orthogonal trajectories of this family of normal lines, so the fibers lie along concentric circles. Since the distance between two fibers must be the same before and after the deformation, the fiber $Y =$ constant lies along the circle of radius

$$r(Y) = r_d + D - Y \tag{28}$$

Since the arclength $r\,d\theta$ of a fiber element must be the same as its initial length dX, then the particle initially at X, Y goes to r, θ, where θ is given by

$$\theta = A(Y) + [X/r(Y)] \tag{29}$$

Under mild assumptions about the nature of the shearing stress response of the material, the deformation will be symmetrical, with $A(Y) = 0$ (Fig. 2).

Since the fiber angle is constant along each straight normal line, in the present example it is convenient to let the fiber direction $\theta = 0$ be defined as the direction of the fibers on the polar axis $\theta = 0$, so that the fiber angle at any point is the same as the polar angle at that point. The fiber radius of curvature r_a is the same as the polar coordinate r.

To calculate the amount of shear we use (21b), which takes the form

$$\partial \mathbf{x}/\partial Y = \mathbf{n} + k\mathbf{a} \tag{30}$$

We observe that $\mathbf{x} = r(Y)\mathbf{i}_r[\theta(X, Y)]$, where $\mathbf{i}_r$ is a radial unit vector and

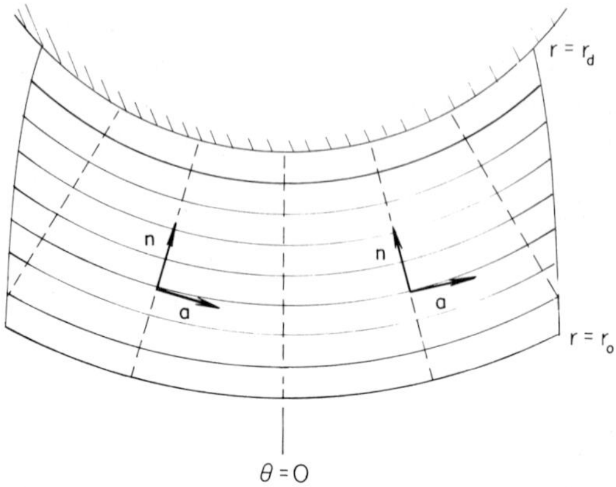

FIG. 2. Slab bonded to rigid cylinder.

the functions r and θ are given by (28) and (29). We also note that $\mathbf{n} = -\mathbf{i}_r$ and $\mathbf{a} = \mathbf{i}_\theta$. Then after carrying out the indicated differentiation we identify the coefficient of $\mathbf{i}_\theta$ as k:

$$k = \theta - A(Y) + r(Y)A'(Y) \tag{31}$$

As required by (24), $k - \theta$ is constant along each fiber $Y =$ constant.

In the present example we have implicitly assumed that the normal lines through the boundary fiber of given shape cover the whole region in which the body could lie, and that the directions of fibers and normal lines do not change discontinuously. Other deformations possible if one or the other of these assumptions is not satisfied are discussed in connection with similar examples by Pipkin and Rogers (1971a, b). The basic kinematic result concerning discontinuous changes of direction is that the line of discontinuity must bisect the angle between the fibers on either side of it, since the normal distance between two fibers must be the same on either side of the discontinuity.

E. Stress

The stress in an ideal fiber-reinforced material has the form

$$\boldsymbol{\sigma} = T\mathbf{aa} - P(\mathbf{I} - \mathbf{aa}) + \mathbf{S} \tag{32}$$

or, in terms of components with respect to an orthonormal system of base vectors,

$$\sigma_{ij} = Ta_ia_j - P(\delta_{ij} - a_ia_j) + S_{ij} \tag{33}$$

The term $T\mathbf{aa}$ represents a tensile stress of magnitude T in the fiber direction $\mathbf{a}$. This is a stress that can do no work in any deformation that preserves fiber lengths, and the most general form of stress with this property. It accordingly represents the reaction to the constraint of inextensibility, and is not connected to the deformation of the material by constitutive equations.

The isotropic pressure $-P\mathbf{I}$ is the reaction to the constraint of volume incompressibility. It is convenient to subtract from this stress its component in the fiber direction, so that T stands for the total tensile stress on surfaces normal to the $\mathbf{a}$ direction.

The remaining part of the stress $\mathbf{S}$ is to be specified by a constitutive equation in terms of the deformation or deformation history. Without yet restricting attention to elastic, plastic, viscoelastic, or other behavior, we can make some general observations about the nature of the extra stress $\mathbf{S}$ in a plane deformation.

First, we assume that the material has reflectional symmetry in planes z = constant. It follows that in plane deformations there can be no shearing stress on such planes. We also observe that if we define T and $-P$ to be the total normal stresses on surfaces perpendicular to the directions $\mathbf{a}$ and $\mathbf{n}$, respectively, then $\mathbf{S}$ can make no contribution to these stresses. It follows that $\mathbf{S}$ must have the form

$$\mathbf{S} = S(\mathbf{an} + \mathbf{na}) + S_3\mathbf{kk} \tag{34}$$

In terms of the total stress, S is the shearing stress

$$S = \mathbf{a} \cdot \boldsymbol{\sigma} \cdot \mathbf{n} \tag{35}$$

and S_3 is a normal stress difference:

$$S_3 = \sigma_{zz} - \sigma_{nn}, \qquad \sigma_{zz} = -P + S_3 \tag{36}$$

We assume that S and S_3 are determined by the deformation gradient at the particle considered, or by its history. For a plane deformation, the deformation with respect to the local axes $\mathbf{a}$, $\mathbf{n}$, and $\mathbf{k}$ is characterized completely by the amount of shear k. Thus, S and S_3 are determined by the value of k or its history.

We restrict our attention to elastic or quasi-elastic shearing response. For elastic response, S is a function $S(k)$ of the amount of shear. We assume that the material has sufficient symmetry to make S an odd function of k,

$$S(-k) = -S(k) \tag{37}$$

In particular, there is no stress if there is no shear. The strain energy density $W(k)$ is related to the stress by the work–energy relation $S\,dk =$

dW, i.e., $S = W'(k)$, or

$$W(k) = \int_0^k S(k')\,dk' \tag{38}$$

By quasi-elastic response we mean any kind of response in which, under special conditions, we can treat S as a function of the concurrent value of k. For example, if the shearing stress response is elastic–plastic, then S depends in general on the history of the shear k and not on its concurrent value alone. However, in cases that involve no unloading (meaning that k changes monotonically in the present context), it is well known that plastic response can be treated in the same way as nonlinear elastic response. For plastic response we use W as an abbreviation for the integral of S defined in (38); W is then not the stored energy.

For viscoelastic materials, S will generally depend on the history of k in a nontrivial way. However, the response is quasi-elastic if the deformation takes place all at one instant and there are no further changes of deformation. In such cases the stress depends on the fixed value of k and the amount of time elapsed since the change of shape. We can then regard S as a function of k, whose form depends parametrically on time.

For the solution of problems of plane deformation, nothing need be said about the form of the function S_3 except that it is assumed to be independent of z. The relation of S and S_3 to functions specifying the material response in more general forms of deformation is discussed further in Sections IV,A and VI,C.

F. Example: Simple Shearing

As an experiment to determine the form of the function $S(k)$, we consider the shearing of a block with boundaries $X = 0, L$ and $Y = 0, D$, which is bonded to rigid plates along the sides $Y = 0, D$. The fibers are initially parallel to the X direction. The upper plate is displaced parallel to the X direction while the lower plate is held fixed. The end surfaces $X = 0$ and $X = L$ are left free from traction.

In isotropic elasticity theory the solution of this problem is not a simple shearing deformation, because simple shearing requires tractions on the ends of the block as well as on its upper and lower surfaces. The problem, as posed, has not been solved exactly in isotropic elasticity theory. We shall show that for ideal fiber-reinforced materials, simple shearing is the exact solution.

Although we could postulate a simple shearing deformation and then verify that it is possible to construct a statically admissible stress field consistent with this deformation, let us proceed deductively in order to illustrate methods of solution that are of general use.

We know the shapes of two fibers in the deformed body, the straight fibers at the upper and lower surfaces. The straight normal lines orthogonal to these fibers are the lines x = constant. The remaining fibers, orthogonal to the normal lines, then lie along the straight lines y = constant. In view of the conservation of fiber spacing and fiber lengths, the deformation must have the form

$$x = X + u(Y), \qquad y = Y \tag{39}$$

The amount of shear is $k(y) = u'(y)$. The shearing stress $S(k)$ is then constant along each fiber y = constant. The total shearing force along the complete length of such a line is $LS(k)$, and for equilibrium this must be equal to F, the total shearing force on the top plate. Thus, if $S(k)$ is invertible, the shear k must be constant, and the deformation is

$$x = X + kY, \qquad y = Y \tag{40}$$

The relevant stress components are

$$\sigma_{xx} = T, \qquad \sigma_{xy} = S(k), \qquad \sigma_{yy} = -P \tag{41}$$

Since σ_{xy} is constant, the equilibrium equations yield

$$T = T(y) \qquad \text{and} \qquad P = P(x) \tag{42}$$

i.e., T is constant along each fiber and P is constant along each normal line.

A vector normal to the end $X = L$ after the deformation is given by $\mathbf{N} = \mathbf{a} - k\mathbf{n}$. Consequently, the condition of no traction there $\boldsymbol{\sigma}\cdot\mathbf{N} = \mathbf{0}$, yields

$$T = kS \qquad \text{and} \qquad P = -S/k \tag{43}$$

The same conditions must be satisfied on the end $X = 0$ as well. Indeed, whenever a boundary that was a normal line in the undeformed state is traction-free in the deformed state, the boundary conditions have the form (43).

The value of T is specified at two points on each fiber, but the values are consistent because the deformation was determined by using this consistency implicitly. We find that $T = kS(k)$ everywhere. The fact that the fiber ends are traction-free does *not* imply that T is zero.

In the region $kD < x < L$ covered by normal lines that do not cross the ends of the sheared block, P is indeterminate until we specify its value at one point on each normal line (see Fig. 3). If we regard the condition of no vertical displacement along the line $Y = 0$ as a boundary condition, then the condition of no vertical displacement along the line $Y = D$ is not an

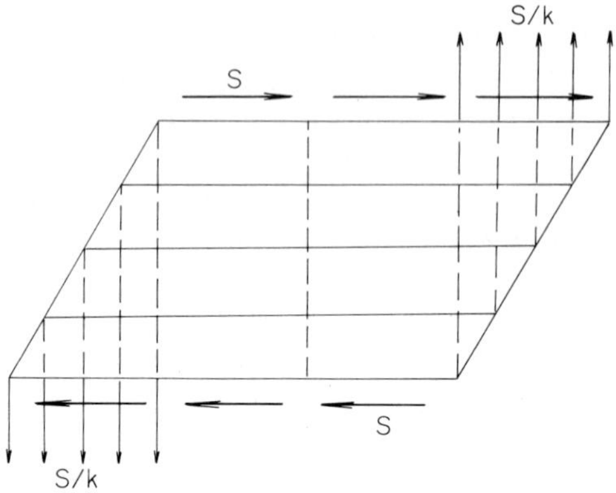

FIG. 3. Tractions supporting simple shearing.

independent boundary condition, but a conclusion from the equations. As an independent boundary condition we specify that $\sigma_{yy} = 0$ on $Y = D$, $kD < x < L$. Then P is zero for all y in this range of values of x.

In the triangular regions formed by the normal lines that cross the ends of the sheared block, i.e., for $0 \leq x \leq kD$ and $L \leq x \leq L + kD$, the boundary conditions (43) specify P at one point on each normal line. Since P is constant along normal lines in a simple shearing deformation, then $P = -S/k$ everywhere in these regions.

The deformation is supported in part by shearing tractions $S(k)$ on the upper and lower surfaces. However, these tractions have a resultant couple $DLS(k)$. Since there are no tractions on the ends to balance this couple, it is balanced by normal forces on the upper and lower surfaces. We have just found that on the upper surface, $\sigma_{yy} = S/k$ in the region $x \geq L$, and on the lower surface, $\sigma_{yy} = S/k$ for $x \leq kD$. Even for very small deformations, these normal stresses may be large enough to cause the bonds between plates and block to fail. In the limit as k approaches zero, the tension S/k approaches G, the shear modulus for infinitesimal deformations. In Section III,L we return to the problem of shearing and consider the deformation occurring when the corners of the block fail to adhere to the plates.

G. Equilibrium of Force Resultants

In the examples in Sections II,B, II,C, and III,F, it was necessary to use the equilibrium equations in order to determine the deformation. However, this did not entail the pointwise determination of the stress field. The required information could be obtained by considering only the resultant shearing force over the full length of each fiber or normal line. In the present section we use this technique to obtain some simple results concerning finite deformations.

We consider a column or plate of constant thickness D, with fibers initially parallel to the X direction. We have found (24) that in any deformed state, the amount of shear k is equal to the fiber angle θ plus an amount f that is constant along each fiber. We have also found (27) that the fiber angle is constant along each normal line, i.e., normal lines are straight. It follows that in elastic or quasi-elastic response, the resultant shearing force on a normal line that passes completely across the beam is

$$\int_0^D S[\theta + f(Y)]\, dY = S_f(\theta). \tag{44}$$

For a given function f, the integral depends only on θ. For example, if f is constant, then $S_f(\theta)$ is equal to $DS(\theta + f)$.

Let the body be divided into two parts by the normal line θ = constant. Let $\mathbf{R}(\theta)$ be the resultant surface traction on the part toward which $\mathbf{a}(\theta)$ points. Then equilibrium requires that

$$\int_0^D [T\mathbf{a}(\theta) + S\mathbf{n}(\theta)]\, dY = \mathbf{R}(\theta) \tag{45}$$

where the integral is taken along the normal line. Since $\mathbf{a}$ and $\mathbf{n}$ are constant in the integration, the $\mathbf{n}$ component of this equation is, with (44),

$$S_f(\theta) = \mathbf{n}(\theta) \cdot \mathbf{R}(\theta) \tag{46}$$

As an illustration of the use of this result, let us suppose that in the region covered by some family of normal lines, there is no surface traction on either side of the beam. Then in that region, $\mathbf{R}$ has the same value for all normal lines, and (46) becomes

$$S_f(\theta) = \mathbf{R} \cdot \mathbf{n}(\theta) \tag{47}$$

For example, if $f = 0$ and $S(k) = Gk$, the equation is explicitly

$$DG\theta = \mathbf{R} \cdot \mathbf{n}(\theta) \tag{48}$$

This equation, or (47), can be satisfied by only one or a few discrete values of θ. It follows that if θ does not change discontinuously, it must be constant throughout the region under consideration. Consequently, all fibers in the region are *straight*. The beam cannot be curved in any region where it has no surface tractions on either side.

There is an important exception to this statement. If (47) has two solutions, a *boundary* fiber may change direction discontinuously from one of these values to the other, forming an inside corner with a fan of straight normal lines emerging radially from the corner. The boundary fiber will in general be carrying a finite load. It is not possible to regard this load as being perpendicular to all of the normal lines that pass through the corner. Consequently, the total force tangential to such a normal line is not equal to $S_f(\theta)$, and the preceding proof that the beam must be straight is not valid. The beam may have straight sections with fiber angles satisfying (47), connected by corner fans.

Cases in which the function $f(Y)$ is constant, for which $S_f(\theta)$ is merely $DS(\theta + f)$, are relatively common. It is often known from the boundary data that some material line that was a normal line before the deformation is still a normal line after the deformation. The amount of shear on such a line is zero, so $-f$ is the value of θ on that line.

For example, if the end of the beam $X = 0$, say, is bonded to a rigid support, then on that end $(\mathbf{n}_0 \cdot \nabla_0)\mathbf{x} = \mathbf{n}_0$ in any deformation. Then $\mathbf{n} + k\mathbf{a}$ is equal to the unit vector $\mathbf{n}_0$ there, whence $k = 0$ and $\mathbf{n} = \mathbf{n}_0$. Hence f is constant (zero in this case).

H. Examples: Cantilevers and Columns

As an application of the preceding results we consider a plate of length L and width D, bonded at the end $X = 0$ to a rigid support. The sides $Y = 0$ and $Y = D$ are free from traction. On the end $X = L$ there is a dead load with resultant

$$\mathbf{R} = F(-\mathbf{i} \cos A + \mathbf{j} \sin A) \tag{49}$$

We will not go into any detail about the distribution of the load over the end, and will accordingly not be able to determine the deformation in the neighborhood of the loaded end.

Because the end $X = 0$ is bonded to a rigid support, the amount of shear k is equal to θ everywhere, and $S_f(\theta)$ in (47) is equal to $DS(\theta)$. Since the sides of the plate are not loaded, the deformed plate must be straight over most of its length, with a fiber angle θ_1 that is determined by using (49) in (47):

$$DS(\theta_1) = F \sin(\theta_1 + A) \tag{50}$$

Next to the support there is a fan region in which normal lines with angles in the range from 0 to θ_1 diverge radially from the corner $X = 0$, $Y = D$, and in this region the fibers lie on concentric circular arcs (Fig. 4).

To be concrete, let us suppose that the shearing stress response is linear, $S(k) = Gk$. [For rubber this is not a bad approximation even for k of the order of unity: see Treloar (1958).]

In the case $A = \pi/2$ the structure represents a cantilever under an end load F. In this case (50) gives

$$F/DG = \theta_1/\cos\theta_1 \tag{51}$$

Thus, there is a unique deflection angle θ_1 between 0 and $\pi/2$ for each positive value of F.

Both from (51) and the more general form (50), we see that the deflection angle does not depend on the length of the beam. The beam is in a state of simple shear over most of its length, and the deflection is determined by the shearing stress rather than by a relation between moment and curvature.

In the case $A = 0$ the structure represents a column supporting a dead weight F (regarding $-\mathbf{i}$ as the downward direction). In this case (50) is satisfied by $\theta_1 = 0$ for any value of F; the undeflected state is an equilibrium state. However, for a sufficiently large load this state may be unstable. Setting $S(k) = Gk$ for concreteness, we obtain

$$F/DG = \theta_1/\sin\theta_1 \tag{52}$$

if θ_1 is not zero. The smallest load at which a nonzero deflection is possible is $F_c = GD$. If $S(k)$ is not linear, so that $G = S'(0)$ is merely the shear modulus for infinitesimal deformations, then $F_c = GD$ is the smallest load at which small deflections are possible, but not necessarily the smallest load for which a finite deflection is possible.

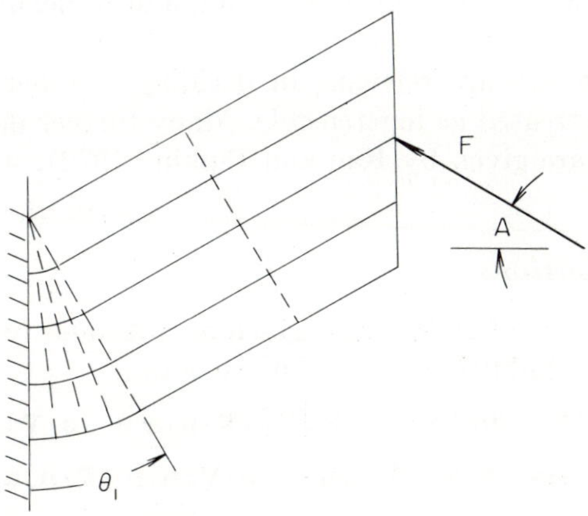

FIG. 4. Deformed cantilever or column.

For example, for elastic, perfectly plastic response with a yield stress S_0, in place of (52), we obtain

$$F/DS_0 = 1/\sin \theta_1 \tag{53}$$

for large deflections. Hence the column can remain in a completely collapsed state ($\theta_1 = \pi/2$) at a load DS_0 much smaller than DG, and in the presence of imperfections DS_0 is the maximum safe load.

The Euler buckling load, computed by assuming that the column deforms in flexure, is $F_e = \pi^2 ED^3/48L^2$ (per unit of length in the z direction). Here E is Young's modulus for extension in the fiber direction, which is treated as infinite in the idealized theory, and F_e depends on the length L of the column, of course, while the critical load $F_c = GD$ computed from the idealized theory does not. These two loads are equal when

$$D \approx 2(G/E)^{1/2}L \tag{54}$$

When D is smaller than this value, the Euler load is smaller than F_c and should be regarded as the maximum safe load. For D larger than this value, F_c is the maximum safe load.

As remarked in Section II,D, stress concentration layers have thicknesses of the order of $(G/E)^{1/2}L$, where L is the length of the layer. In the present problem there are stress concentration layers along the fibers $Y = 0$ and $Y = D$ since the exterior shearing traction on these boundaries is zero while the interior shearing stress is not. We see that F_c is the safe load if the total thickness of these two layers is smaller than the thickness of the column, while the Euler load is smaller when the so-called boundary layers are so thick as to cover the whole column. Flexural (extensional) deformation is confined to the stress concentration layers; the column deforms in shear when these layers are thin, and in flexure when they fill the column.

The same estimates are relevant in deciding whether or not a given cantilever can be treated as inextensible. Many further details concerning column buckling are given by Kao and Pipkin (1972), who obtained the preceding results.

I. Stress Distributions

When the expression for the stress given by (32) and (34) is used in the stress equation of equilibrium, $\nabla \cdot \boldsymbol{\sigma} = \mathbf{0}$, we obtain

$$(\nabla T \cdot \mathbf{a})\mathbf{a} - \nabla P \cdot (\mathbf{I} - \mathbf{aa}) + (T + P)[(\nabla \cdot \mathbf{a})\mathbf{a} + (\mathbf{a} \cdot \nabla)\mathbf{a}] \\ + \nabla S \cdot (\mathbf{an} + \mathbf{na}) + S[(\nabla \cdot \mathbf{a})\mathbf{n} + (\mathbf{a} \cdot \nabla)\mathbf{n} + (\nabla \cdot \mathbf{n})\mathbf{a} + (\mathbf{n} \cdot \nabla)\mathbf{a}] = \mathbf{0} \tag{55}$$

We have assumed that S_3 is independent of z, and it has accordingly dropped out of the problem. By using (16), (18), and (19) to simplify the equation, and then separating it into components in the fiber, normal, and z directions, we obtain

$$\mathbf{a}\cdot\nabla T + [(P + T)/r_n] = (2S/r_a) - \mathbf{n}\cdot\nabla S \tag{56}$$

$$\mathbf{n}\cdot\nabla P - [(P + T)/r_a] = (2S/r_n) + \mathbf{a}\cdot\nabla S \tag{57}$$

and $\partial P/\partial z = 0$. We are presently interested only in cases for which fibers are initially straight and parallel. In these cases, normal lines are straight in any state of deformation, so that $1/r_n = 0$. Hence, we obtain

$$\mathbf{a}\cdot\nabla T = (2S/r_a) - \mathbf{n}\cdot\nabla S \tag{58}$$

and

$$\mathbf{n}\cdot\nabla P - (P/r_a) = (T/r_a) + \mathbf{a}\cdot\nabla S \tag{59}$$

Given the deformation, so that S as well as the fields $\mathbf{a}$ and $\mathbf{n}$ are known, (58) specifies the variation of T along each fiber, and it can be integrated directly. It is necessary to specify the value of T at one point on each fiber, to determine the constant of integration. We note that T can change discontinuously from one fiber to the next; the equation does not in any way limit its variation in the normal direction.

When T has been determined, (59) is, for each normal line, a first-order ordinary differential equation for P. Boundary data specifying the value of P at one point on each normal line must be supplied.

We have seen that in problems with sufficient displacement boundary data, it is possible to determine the deformation purely kinematically, without any use of equilibrium conditions. The complement of this result is that, as we now see, for any kinematically admissible deformation it is possible to construct a statically admissible stress field compatible with it.

In mixed problems such as those in Section III,H, it is not possible to determine the deformation purely from kinematic considerations, but the equilibrium equations are used only as equations among force resultants. In these cases T may be specified at more than one point on each fiber or P may be specified at more than one point on each normal line, but the deformation is chosen to be such that the values are consistent with the limitations imposed by (58) and (59).

In pure traction boundary value problems, it may be possible to find the deformation by using the equilibrium of force resultants, but this is not certain; no problems of this kind have been solved except trivial ones. The explicit integrals of the equilibrium equations given later in this section and in Section III,J should be of use in solving traction boundary value problems.

In integrating (58) and (59) it is convenient to treat straight sections and curved sections of the body separately. In straight sections, where θ is constant, the fiber curvature $1/r_a$ is zero. We can install Cartesian coordinates x and y along the fiber and normal directions, respectively. Then (58) and (59) yield

$$T(x, y) = T_0(y) - \int_{x_0}^{x} \partial S/\partial y \, dx' \tag{60}$$

and

$$P(x, y) = P_0(x) + \int_{y_0}^{y} \partial S/\partial x \, dy'. \tag{61}$$

Here T_0 and P_0 are constants of integration.

In curved sections, where θ is not constant, we can number each normal line with the value of θ on it and number each fiber with a label Y such that Y measures distance along normal lines. If the fibers are initially straight and parallel to the X direction, the number Y of a fiber can be taken to be its initial Cartesian coordinate. Because fibers lie along parallel curves, the fiber radius of curvature varies linearly along each normal line:

$$r_a(\theta, Y) = r_0(\theta) - Y \tag{62}$$

The directional derivatives in the fiber and normal directions are then

$$\mathbf{a}\cdot\nabla = \partial/r_a\partial\theta \qquad \text{and} \qquad \mathbf{n}\cdot\nabla = \partial/\partial Y \tag{63}$$

Then (58) yields

$$T(\theta, Y) = T_0(Y) + \int_{\theta_0}^{\theta} (2S - r_a \, \partial S/\partial Y) \, d\theta' \tag{64}$$

From (59) we obtain

$$[Y - r_0(\theta)] \, \partial P/\partial Y + P = -T - \partial S/\partial\theta. \tag{65}$$

The solution of this equation is easily found to be

$$[r_0(\theta) - Y]P(\theta, Y) = r_0(\theta)P_0(\theta) + \int_{0}^{Y} (T + \partial S/\partial\theta) \, dY' \tag{66}$$

J. Stress in Quasi-Elastic Response of Plates

The general integrals of the equilibrium equations given in (60), (61), (64), and (66) are valid for bodies of any cross-sectional shape and any kind of shearing stress response. We now restrict our attention to homogeneous plates bounded initially by the fibers $Y = 0$ and $Y = D$, as in the explicit examples considered previously, and we suppose that the response of the material is elastic or quasi-elastic. As a further simplifying assumption, we suppose that the problem is such that in the expression $k = \theta + f$ for the amount of shear, the function f is constant, and we write $f = -\theta_1$.

With f constant, the amount of shear is constant along each normal line. Since the response is quasi-elastic, the shearing stress S is then also constant along each normal line, except at the boundaries $Y = 0$ and $Y = D$, where the applied shearing stress need not be equal to the interior shearing stress. Let $\bar{S}_0$ and $\bar{S}_d$ be the applied shearing stresses on the two boundaries. Then the derivative of the shearing stress in the normal direction is

$$\partial S/\partial Y = [S(\theta - \theta_1) - \bar{S}_0]\delta(Y) + [\bar{S}_d - S(\theta - \theta_1)]\delta(Y - D) \tag{67}$$

There is a confusion of notation because in the left-hand member, S is the shearing stress, while in the right-hand member, S stands for the function $S(k)$ that defines the response of the material.

By using (67) in (64) we find that the tension in the fiber direction is given in curved regions by

$$T = T_0(Y) + 2W(\theta - \theta_1) + F_0\delta(Y) + F_d\delta(Y - D) \tag{68}$$

Here W is the integral of S defined in (38). The two singular terms represent the finite forces F_0 and F_d carried by the two bounding fibers. These forces satisfy the equations

$$(\mathbf{a}\cdot\nabla)F_0 = -S(\theta - \theta_1) + \bar{S}_0 \qquad \text{and} \qquad (\mathbf{a}\cdot\nabla)F_d = -\bar{S}_d + S(\theta - \theta_1) \tag{69}$$

Equivalently, F_0 and F_d are integrals of the shearing stress discontinuities with respect to arclength along the boundaries.

At interior points, expression (68) is constant along fibers in straight regions where the fiber angle is constant, as required by the expression for the tension in straight regions, (60). Thus (68) is valid throughout the whole body, with the same integration function $T_0(Y)$, although it was first derived only for curved sections.

The tension T thus consists of (i) a tension $T_0(Y)$ that is constant along each fiber, and arbitrary until the tension at one point of the fiber has been specified, and (ii) an explicit term $2W(k)$ that depends on the amount of shear at the point considered. In addition, there are (iii) finite forces in the "strings" along the boundaries, which are found separately by integrating the shearing stress discontinuity along the string.

In straight regions the material is in a state of simple shear, since we have assumed that f is constant. Then P is constant along normal lines in such regions, as stated in (61). On each normal line, P has the same value that it had at the boundary. In curved regions, P varies along normal lines because the tension in a curved fiber exerts a net sideways force. By using (68) in (66), we obtain

$$[r_0(\theta) - Y]P(\theta, Y) = r_0(\theta)P_0(\theta) + \int_0^Y T_0(Y')\,dY' + Y[2W(\theta - \theta_1) + S'(\theta - \theta_1)] + F_0 H(Y) + F_d H(Y - D) \qquad (70)$$

We recall that $r_0 - Y$ is the fiber radius of curvature at the point considered and P_0 is the externally applied pressure on the fiber $Y = 0$. The pressure in the body directly inside this fiber is larger than P_0 by the amount F_0/r_0, just as the pressure inside a bubble is greater than that outside by an amount proportional to the surface tension. The pressure jump at $Y = D$ is F_d/r_d.

K. Example: Stress in Bent Slab

As an explicit example we consider the deformation of a slab described in Section III,D, with no surface tractions except along the rigid cylinder of radius r_d. The stress distribution for this particular problem was first worked out by Mulhern *et al.* (1967) for the case of plastic response (Fig. 2).

The ends $X = \pm L$ lie along the curves $r\theta = \pm L$ if the deformation is symmetrical [$A = 0$ in (29)]. The conditions of zero traction take the form (43) on these boundaries, since they were normal lines in the undeformed state. From (68) the tension T is then

$$T = (L/r)S(L/r) + 2W(\theta) - 2W(L/r) + F\delta(r - r_0) \qquad (71)$$

Here the fiber curvature r and fiber angle θ have the same meanings as ordinary polar coordinates, and $r_0 = r_d + D$. Since the result obtained by imposing the boundary condition at $X = L$ also satisfies the boundary condition at $X = -L$, the assumption that $A(Y) = 0$ (symmetrical deformation) is consistent. With no shearing traction applied to the surface $r = r_0$, the force F in the stress concentration layer along this surface is found from (69) to be

$$F = r_0W(L/r_0) - r_0W(\theta) \qquad (72)$$

We have arbitrarily taken the shearing stress applied to the bonded fiber $Y = D$ to be equal to that in the material, so that there is no stress concentration layer along this fiber.

The normal lines θ = constant that cross the end $X = L$ lie in the range $L/r_0 < \theta < L/r_d$. On these fibers the boundary value of P is specified by (43), and from (70), with (71) and (72), after a little manipulation we obtain

$$rP(r,\theta) = -(L/\theta^2)S(\theta) + [(L/\theta) - r][S'(\theta) + 2W(\theta)] + rW(L/r) - (L/\theta)W(\theta) - \int_r^{L/\theta} W(L/r')\,dr' \tag{73}$$

A similar expression is obtained for P in the region at the other end of the slab covered by normal lines that cross the end $X = -L$. The normal lines in the range $L/r_0 \leq \theta \leq L/r_0$, cross the surface $r = r_0$, where the applied pressure is zero. For this region, (70) and (71) yield

$$rP(r,\theta) = (r_0 - r)[S'(\theta) + 2W(\theta)] + r_0[W(L/r_0) - W(\theta)]H(r_0 - r) + rW(L/r) - r_0W(L/r_0) - \int_r^{r_0} W(L/r')\,dr' \tag{74}$$

While apologizing for the complicated appearance of this result, we should point out that it is the exact solution for all forms of quasi-elastic response (Pipkin and Rogers, 1971a).

The case of rigid-plastic response reveals a particularly interesting effect. Since the amount of shear is positive in the region $\theta > 0$ and negative in the region $\theta < 0$, then according to the rigid-plastic model the shearing stress changes discontinuously from $-S_0$ to S_0 across the normal line $\theta = 0$. This produces a stress concentration layer interior to the material rather than along a boundary. The singular part of P given by (74) is

$$P_s = 2S_0\delta(\theta)(r_0 - r)/r \tag{75}$$

For the more realistic elastic-plastic model, the Dirac delta would be replaced by a function sharply peaked at $\theta = 0$, but spread over the range of values $|\theta| \leq S_0/G$ ($\ll 1$) covering the small region in which the response is elastic. The slab will exert an intense pressure on the supporting cylinder in the neighborhood of $\theta = 0$.

L. *Example: Shearing with Loss of Contact*

In Section III,F we considered the simple shearing deformation of a slab bonded to two rigid plates. We found that high tensile stresses are produced in the bonds between the slab and the plates, near the acute corners of the sheared slab (Fig. 3). Let us now suppose that the bond cannot support a tensile stress.

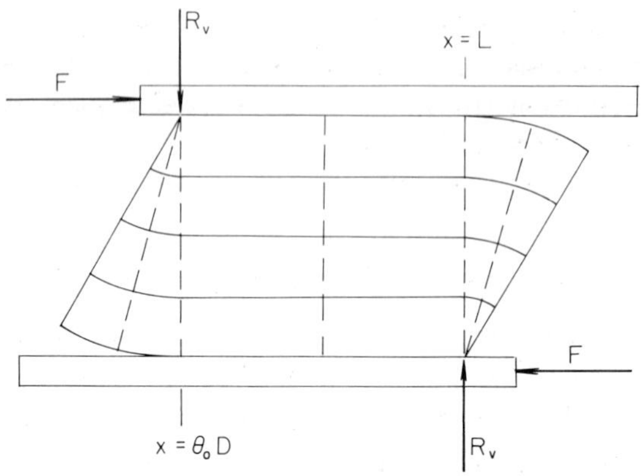

FIG. 5. Shearing with loss of contact.

In the previous solution, the parts of the boundary under a tensile stress were the segments of the fibers $y = 0$ and $y = D$ intersected by normal lines that also cross the ends of the sheared slab. It is natural to seek a solution in which these segments are no longer in contact with the plates. Then on each normal line through one of these segments, P will be prescribed ($P = 0$) at the intersection of the normal line with the free boundary. To avoid a possibly contradictory specification of P where the normal line crosses the end of the slab, we seek a configuration in which no normal line crosses the end. Then the ends of the deformed slab must be normal lines themselves, and the deformation shown in Fig. 5 is suggested. There are two fans of normal lines adjacent to the ends of the slab, running out radially from the corners $x = L$, $y = 0$ and $x = \theta_0 D$, $y = D$, with angles in the range $-\theta_0 \leq \theta \leq 0$ in both fans. The free boundaries lie along circular arcs, as do all of the fibers in the corner fans. The condition of no shearing traction on the ends is satisfied because the shear is zero on the ends, the ends being normal lines both before and after the deformation. The shear is given everywhere by

$$k = \theta + \theta_0 \tag{76}$$

In particular, $k = \theta_0$ in the simply sheared region $\theta_0 D \leq x \leq L$.

We have in effect simply guessed a deformation. The kinematic constraint conditions are so suggestive geometrically that it is usually possible to arrive at right answers by guessing, and this accounts for a good part of the power of the theory. We know that it will be possible to construct an equilibrium stress field compatible with this deformation (Section III,I), but we need to verify that a stress field satisfying the given traction boundary conditions can be obtained. The conditions are that there is no traction on any part of the boundary that is not in contact with one of the plates, and that any normal stress between the slab and the plates is nonpositive; tensile stress in the bond is not allowed.

By using the results in Section III,J we easily find that for quasi-elastic response, T is given by

$$T = 2W(\theta + \theta_0) - DW(\theta + \theta_0)[\delta(Y) + \delta(Y - D)] \tag{77}$$

In the simply sheared part of the slab, P is zero if we specify that P is zero on one of the boundaries ($y = 0$ or $y = D$) of this part. In the right-hand corner fan, at interior points,

$$P = S'(\theta + \theta_0)(D - Y)/Y + W(\theta + \theta_0)(D - 2Y)/Y \tag{78}$$

In this region Y is the same as the radius. A similar expression is valid in the left-hand fan region.

We note that in (78), $P = -W$ at $Y = D-$. The prescribed value $P = 0$ on $Y = D+$ differs from this by the boundary tension divided by the radius D, as required. The pressure is negative just inside the free boundary, and for large values of θ_0 it may be large enough to cause failure. The pressure increases as the radius decreases, and it is definitely positive (equal to S') half-way across the slab. It becomes infinite at the corner, where $Y = 0$.

The pressure singularity at the corner is balanced by a point force of reaction exerted by the plate on the slab at the corner. Let R_v be the vertical component of this force, at $x = L$, $y = 0$. There is an opposite force at the corner $x = \theta_0 D$, $y = D$. Then R_v is most easily computed by considering the equilibrium of the part of the slab to the right of any vertical normal line through the simply sheared region. Since the resultant downward shearing force on such a line is $DS(\theta_0)$, then

$$R_v = DS(\theta_0) \tag{79}$$

The two vertical reactions impose a couple $R_v(L - D\theta_0)$ on the slab. If F is the total shearing force exerted by the upper plate on the slab, then the shearing forces have a couple FD. Hence, the relation between F and θ_0 is

$$F = S(\theta_0)(L - D\theta_0) \tag{80}$$

We notice that $S(\theta_0)$ is not F/L, but F divided by the length of the contact region.

The preceding computation of F does not make use of any assumption about the distribution of the shearing traction in the contact region. The expression for the boundary fiber forces in (77) does make use of such an assumption, however; it is assumed that the applied shearing stress is the same as the interior shearing stress $S(\theta_0)$ at each point, or thus that the total shearing force F is distributed uniformly. This assumption is not necessary; we could specify a nonuniform distribution of shearing traction, and the deformation would be the same, but the tensile force in the boundary fiber would then vary instead of remaining constant at the value $-DW(\theta_0)$ in the contact region. Indeed, we could specify that part of the force F is contributed by a horizontal component of the point force of reaction at the corner.

The latter alternative might be considered seriously; for the tensile force $-DW(\theta + \theta_0)$ in the boundary fiber $y = 0$ was derived by starting from the value zero at the left-hand end of the fiber and it is not again zero at the right-hand end. We can interpret the result to mean that the force is $-DW(\theta_0)$ at $x = L-$, and that the force then changes discontinuously to zero at $x = L+$ as the fiber passes through the stress singularity at the corner. Alternatively, we could assume that the applied shearing traction is not $S(\theta_0)$, and choose it to be such that the tensile force in the boundary fiber reaches the value zero at $x = L-$. In this case the difference between the integral of the applied shearing stress and the total shearing force F would have to be compensated by a horizontal component of the point force of reaction at the corner.

We emphasize that in spite of this ambiguity, there is no ambiguity in the relation between the deformation and the total shearing force. However, the situation is disturbing. Since we treat the boundary fiber as inextensible, zero horizontal displacement on this fiber is not a boundary condition (except at one point), but a conclusion from the equations, and to pose the problem unambiguously we are required to specify the horizontal component of traction on the boundary (except at one point), rather than determine it from the theory. All specifications are equally acceptable in the idealized theory, but which one most closely resembles the solution for a real material, or for an ideal elastic material with slightly extensible fibers?

Questions of this kind arise more frequently than one might wish. Such questions can probably be answered by formulating a theory of slightly extensible materials that treats deformations as small perturbations on the deformation given by the idealized theory. This has been done by Everstine and Pipkin (1973) for infinitesimal elastic deformations, but the subject needs further study, even in this simplest special case.

M. Initially Curved, Parallel Fibers

If fibers initially lie along curves that are parallel but not straight, such as concentric circles, the theory of plane deformations is nearly the same as for initially straight, parallel fibers. The kinematical theorem of main importance is that if normal lines are straight in some kinematically admissible configuration of the body, they are straight in every kinematically admissible configuration. Equivalently, fibers always lie along parallel curves if they are parallel initially. To understand this without going into unnecessary detail, observe that in a deformation of a body with initially straight fibers, all kinematically admissible configurations have parallel fibers, and the mapping from any configuration to any other necessarily preserves this parallelism. We can treat any of these configurations as the initial state of a body with initially curved fibers. An analytical derivation is given in Section III,N.

The formula for the amount of shear is a slight generalization of (24). For a body with initially curved, parallel fibers, we can imagine a state of deformation in which all fibers are straightened out and all lie parallel to the direction $\theta = 0$. Let Y be a Lagrangian coordinate constant along each fiber; we can identify Y with a Cartesian coordinate of the particle in the straightened state. Then the amount of shear in the deformation from the initial state to the final state, with the straight state as an intermediate configuration, the shear is $-(\theta_0 + f_1)$, plus $(\theta + f_2)$. Writing $f(Y)$ for $f_2 - f_1$, we have

$$k(\mathbf{X}) = \theta(\mathbf{X}) - \theta_0(\mathbf{X}) + f(Y) \tag{81}$$

The shear is the *change* of fiber angle, plus an amount that is constant along any given fiber. An analytical derivation of this result in the special case of fibers initially arranged in concentric circles was given by Rogers and Pipkin (1971b). A general analytical derivation is given in Section III,N. The argument given here is due to T.G. Rogers.

The statements about the stress in Section III,E have nothing to do with the global shapes of fibers, of course. The stress equations of equilibrium retain the forms (58) and (59) used for initially straight fibers; in deriving these from (56) and (57) we used only the straightness of normal lines. The general integrals of the equilibrium equations, given in (60), (61), (64), and (66), remain valid.

As an example, let us consider a tube of internal radius R, reinforced by concentric circular fibers. Let $R + Y$ be the initial radius of a fiber, with $0 \leq Y \leq D$. Consider a deformation in which the normal line $\theta_0 = 0$ is still a normal line, with $\theta = 0$, after the deformation. Then $k = \theta - \theta_0$. Let $r(\theta)$ be the radius of curvature of the fiber $Y = 0$ after the deformation. Then the conditions of fiber inextensibility and constant fiber spacing yield

$$(R + Y)\theta_0 = \int_0^\theta [r(\theta') + Y]\, d\theta' \tag{82}$$

Consequently, the amount of shear is not constant along a normal line

$\theta =$ constant, but equal to

$$k = I(\theta)/(R+Y) \tag{83}$$

where

$$I(\theta) = R\theta - \int_0^\theta r(\theta')\, d\theta' \tag{84}$$

For quasi-elastic response, the resultant shearing stress along a normal line is not given by (44) but instead by

$$S_I(\theta) = \int_0^D S[I(\theta)/(R+Y)]\, dY \tag{85}$$

The dependence of this resultant on θ is determined by the form of I and thus, ultimately, by the form of the inner boundary as determined by its radius of curvature $r(\theta)$.

Rogers and Pipkin (1971b) considered the problem of a tube under internal pressure, mashed between rigid parallel plates. As we have just seen, the deformation is determined when the radius of curvature $r(\theta)$ at the inner boundary is known, or thus when $I(\theta)$ is known. By considering the equilibrium of stress resultants, a nonlinear integral equation for $I(\theta)$ was obtained. The nonlinearity is produced by the nonlinear dependence of $S_I(\theta)$ on $I(\theta)$. The equation was put into the form of an integral equation in order to facilitate the computation of approximate solutions by iterative methods. For the special case of linear shearing response, $S(k) = Gk$, the equation is linear and was solved explicitly. The equation can also be solved analytically in the case of rigid-plastic or elastic-plastic response, but these results were not published.

The problem is a simplified version of the problem of deformation of a radial tire under the weight of an automobile, if the linear shearing response, which is reasonably accurate for rubber, is used. For values of the parameters typical of a tire under normal inflation pressure, it was found that the exact solution does not differ much from that obtained by treating the tire as a membrane, even though the ratio of sidewall thickness to radius is not especially small. For lower pressures, corresponding to an underinflated tire, the sidewall is in compression and acts as a shear beam, much like a curved version of the cantilever discussed in Section III,H. Stress concentration layers occur at the inner and outer boundaries, suggesting that the innermost and outermost plies of a tire are subjected to the greatest loads.

N. Nonparallel Fibers

Materials reinforced with a single family of nonparallel fibers are of smaller practical interest because limitations on methods of fabrication tend to ensure parallelism. This is fortunate, because the analysis of deformations of such materials is more complicated than in the case of parallel fibers. There is no solution of any such problem in the literature. Here we merely outline the basic equations governing the plane deformations of such materials.

The basic geometric relations in Section III,A remain valid for any plane field of fiber directions, and the deformation gradients retain the form (21) in any case. However, the compatibility conditions among k, θ, and θ_0 are not the same as those in Section III,C.

To derive compatibility conditions, we wish to eliminate $\mathbf{x}$ from (21) by cross-differentiation. We observe that directional derivatives in the fiber and normal directions do not commute, in general. Rather, we find that

$$\begin{aligned}(\mathbf{n}\cdot\nabla)(\mathbf{a}\cdot\nabla) - (\mathbf{a}\cdot\nabla)(\mathbf{n}\cdot\nabla) &= [(\mathbf{n}\cdot\nabla)\mathbf{a} - (\mathbf{a}\cdot\nabla)\mathbf{n}]\cdot\nabla \\ &= [(\mathbf{n}/r_n) + (\mathbf{a}/r_a)]\cdot\nabla \qquad (86)\end{aligned}$$

where we have used (16). The same relation is valid for derivatives in the initial fiber and normal directions, of course. Derivatives with respect to initial position $\mathbf{X}$ are related to derivatives with respect to final position $\mathbf{x}$ through the deformation gradients (21):

$$\mathbf{a}_0\cdot\nabla_0 = \mathbf{a}\cdot\nabla, \qquad \mathbf{n}_0\cdot\nabla_0 = (\mathbf{n} + k\mathbf{a})\cdot\nabla \qquad (87)$$

On eliminating $\mathbf{x}$ between (21a) and (21b) by cross-differentiation, using the commutator (86) for the initial fields, we obtain

$$(\mathbf{n}_0\cdot\nabla_0)\mathbf{a} - (\mathbf{a}_0\cdot\nabla_0)(\mathbf{n} + k\mathbf{a}) = [(\mathbf{n} + k\mathbf{a})/r_{n0}] + [\mathbf{a}/r_{a0}] \qquad (88)$$

On using (15), (16), and (87) here, and then separating the resulting expression into components in the fiber and normal directions, we obtain

$$(\mathbf{a}\cdot\nabla)(\theta - \theta_0 - k) = k/r_{n0} \qquad \text{and} \qquad r_n = r_{n0} \qquad (89)$$

The second of these relations states that at a given particle, the radius of curvature of the normal line through it after the deformation is the same as the radius of curvature of the (different) normal line through it before the deformation. If fibers are initially parallel, so that $1/r_{n0} = 0$, then $1/r_n = 0$; normal lines are still straight and fibers are parallel, after the deformation. This is the analytical proof of one of the results in Section III,M.

In the special case of initially straight normal lines, for which $1/r_{n0} = 0$, (89a) states that $\theta - \theta_0 - k$ is constant along fibers. This is the result (81).

If the normal lines are not initially straight, the result (89b) does not in general allow us to state the shapes of the normal lines in the deformed state immediately. Instead, the problem of determining the shapes of the normal lines is coupled with that of determining the amount of shear and, possibly, the stress state (depending on precisely what problem is considered).

The array of mathematical methods known as the "method of characteristics" is relevant to the present problem. Descriptions of the use of such methods in problems closely similar to the present one are given by Adkins (1956a,b, 1958) and Rivlin (1955, 1959, 1964).

The stress equations of equilibrium now take the forms given in (56) and (57). These may be regarded as equations for T and P in characteristic form. The characteristics are the fibers and the normal lines, where T is the characteristic variable along fibers and P is the characteristic variable along normal lines. However, the two equations are now coupled, in general, and must be treated simultaneously. In traction boundary value problems it would even be necessary to determine the shapes of the characteristics from (89) simultaneously with the determination of T and P. This is numerically quite feasible, and it is possible that analytical results can be obtained in special cases, but the subject has not been investigated.

In closing this section on plane deformations, we remark that even in the case of initially parallel fibers, or even straight, parallel fibers, the theory is virtually undeveloped. For example, we have considered no problem involving a body whose boundaries are not composed of fibers and normal lines. No traction boundary value problems were solved. These and other areas are undoubtedly amenable to analysis, and some such problems are discussed in the book by Spencer (1972b), but little has been done as yet.

IV. Plane Deformations with Uniform Axial Stretching

The theory of plane deformations is more widely applicable if it is extended to include the possibility of a uniform stretching orthogonal to the plane of deformation. The final position $\mathbf{x}$ of a particle initially at $\mathbf{X}$ is then given by relations of the form

$$x = x(X, Y), \qquad y = y(X, Y), \qquad z = \lambda Z \tag{90}$$

The extension ratio λ is a constant, now different from unity in general.

It might be anticipated that such cases could be treated by considering the uniformly stretched state to be the "undeformed" state and then applying the theory outlined in Section III. In the case of initially straight, parallel fibers there is no difficulty in doing this. However, stretching in the Z direction necessarily entails contraction in the orthogonal plane if volume is preserved, and this contraction must be purely normal to the fibers since they cannot change length. Consequently, the cross section of the body in the XY plane must change shape. This distortion is equivalent to a mere change of scale in the normal direction if the fibers are initially straight and parallel, but with initially curved fibers the determination of the cross-sectional distortion due to axial stretching is a problem in itself.

In Section IV,A we derive expressions for the stress, generalizing those in Section III,E by allowing dependence on the stretching λ. In Section IV,B compatibility conditions slightly generalizing those in (89) are given. In Section IV,C we turn to the problem of determining the configuration of the body when it is in a state of pure axial tension, if there is one. We find that if fibers are initially parallel, although possibly curved, such a state exists and can be determined relatively easily. In such cases the pure tensile state can be treated as the initial state for the superimposed plane deformation, and all of the analysis in Section III is applicable (Section IV,D). We give a single, rather trivial example (Section IV,E). The results in this section are from a paper by Pipkin (1974).

A. Stress

We first consider the homogeneous deformation of a unit cube of material with fibers initially straight and parallel to the X direction. Stretching the block to λ times its initial length in the Z direction requires contraction to $1/\lambda$ times its initial length in the Y direction in order to preserve volume, so the particle initially at X_1, Y_1, Z_1 moves in a pure stretching deformation to X, Y, Z, where

$$X = X_1, \qquad Y = Y_1/\lambda, \qquad Z = \lambda Z_1 \tag{91}$$

If a homogeneous plane deformation is now superimposed on the stretched state, the final position of the particle is

$$x = X + kY, \qquad y = Y, \qquad z = Z \tag{92}$$

Notice that the amount of shear k is defined as the shearing with respect to the stretched state.

The stress still has the general form given by (32) and (34), given the same mild symmetry assumptions. Now, however, the functions S and S_3 depend on λ as well as k. We add the assumption that $S(k, \lambda)$ remains an odd function of k for each value of λ, so that $S(0, \lambda)$ is zero.

For elastic materials, the strain energy density W is also a function of k and λ in deformations of the kind considered here. The relation of the functions S and S_3 to W is found by considering the work done on a unit block in a change of extension $d\lambda$ and a change of shear dk. In such a deformation, the work done by tractions on opposite faces cancel, except for the normal stress S_3 on the surfaces z = constant and the shearing stress S on the surfaces y = constant. Now, λS is the total shearing force on such a surface, which initially had unit area, and S_3/λ is the tensile stress times the area of the face on which it acts. In a change of λ and k, the tensile force acts through a distance $d\lambda$, but the shearing force acts through a distance $d(k/\lambda)$. Then by setting the incremental work equal to the change of stored energy, we obtain

$$dW = (\lambda S)\,d(k/\lambda) + (S_3/\lambda)\,d\lambda \tag{93}$$

From this relation it follows that

$$S = \partial W/\partial k \tag{94}$$

just as in the special case $\lambda = 1$, and that

$$S_3 = \lambda\,\partial W/\partial\lambda + k\,\partial W/\partial k \tag{95}$$

In particular, the expression for S_3, evaluated at $\lambda = 1$, is the expression in terms of the energy density of the function $S_3(k)$ considered in Section III,E. The relation of $W(k, \lambda)$ to the energy density for a general kinematically admissible deformation is given in Section VI,C.

B. Inhomogeneous Deformations. Compatibility Conditions

Let $\mathbf{a}_1$, $\mathbf{n}_1$, and $\mathbf{k}_1$ be unit vectors in the fiber, normal, and axial directions before any deformation, and let $\mathbf{a}$, $\mathbf{n}$, and $\mathbf{k}$ be the corresponding vectors after the deformation. Then for the homogeneous deformation defined by (91) and (92), the deformation gradients are

$$(\mathbf{a}_1\cdot\nabla_1)\mathbf{x} = \mathbf{a}, \qquad (\mathbf{n}_1\cdot\nabla_1)\mathbf{x} = \lambda^{-1}(\mathbf{n} + k\mathbf{a}), \qquad (\mathbf{k}_1\cdot\nabla_1)\mathbf{x} = \lambda\mathbf{k} \tag{96}$$

For the general case of a plane deformation superimposed on a uniform axial stretching, with an arbitrary plane field of initial fiber directions $\theta_1(X, Y)$, we can still treat each infinitesimal material element as homogeneously deformed. Then the deformation gradients, which describe the deformation locally, must have the same forms (96). For an inhomogeneous deformation, λ, $\mathbf{k}$, and $\mathbf{k}_1$ are constant, but all other quantities are functions of position in the cross-sectional plane.

Compatibility relations among k, θ, and θ_1 are found in the same way that the relations for the special case $\lambda = 1$ were obtained in Section III,N. As generalizations of relations (89), we obtain

$$r_n = r_{n1} \tag{97}$$

and

$$(\mathbf{a}\cdot\boldsymbol{\nabla})(\theta - \lambda\theta_1 - k) = k/r_{n1} \tag{98}$$

If we can identify $\lambda\theta_1$ with the fiber angle $\theta_0(\mathbf{X})$ for some intermediate state of the body, then (97) and (98) become completely identical with the corresponding relations for the case $\lambda = 1$.

C. Simple Axial Tension

We wish to regard deformations as carried out in two steps; first, a pure axial stretching by the amount λ, and second, an ordinary plane deformation of the stretched body. If we use the simply stretched state as reference state, the whole theory of plane deformations previously outlined is applicable without any change.

We accordingly focus our attention on determining the deformation that we have called the "simply stretched state." Of course, stretching in the axial direction necessarily involves a change in the size and shape of the cross section. For the case of initially straight, parallel fibers, the mapping from the initial state to the simply stretched state is given by (91). In this case the cross-sectional deformation is a simple contraction in the normal direction. Since there is no shearing, the shearing stress S is zero, and the equilibrium equations are satisfied by $T = P = 0$. (The equilibrium equations have exactly the same forms as for purely plane deformations.) The only nonzero stress component is $S_3(0, \lambda)$, the tensile stress in the axial direction.

In more general cases, we seek the state of deformation produced by stretching the body to λ times its initial length and applying no tractions to the cross-sectional boundary. It will be particularly convenient if in this state, there is no stress at all except an axial tension. We call such a state a *simple tensile state*. In general there is no such state, as we shall see; axial stretching produces a stress field different from simple tension. However, in the most important special case, the case of initially parallel fibers, axial stretching does produce a simple tensile state.

In a simple tensile state the shearing stress S is zero, and thus the amount of shear k is zero as well, in any case of practical importance. Thus, a simple tensile deformation must involve no shear. Conversely, with $k = 0$ then $S = 0$, and the equilibrium equations are satisfied by $T = P = 0$. With k and λ both constant, then S_3 is constant, so the state of stress is a uniform axial tension.

As an example we consider a body with fibers initially along concentric circular arcs, with polar coordinates R_1, θ_1 in the range $R_a \leq R_1 \leq R_b$, $0 \leq \theta_1 \leq A$. In the simple tensile state, the particle initially at R_1, θ_1, Z_1 has the coordinates

$$R = R_1/\lambda, \qquad \theta = \lambda\theta_1, \qquad Z = \lambda Z_1 \tag{99}$$

We observe that the radial distance between any two fibers contracts in the ratio $1/\lambda$. Fiber lengths are preserved since $R\theta = R_1\theta_1$. To verify that there is no shear, we notice that normal lines are radial both before and after the deformation, and that since $\theta = \lambda\theta_1$, normal lines are material lines in this deformation. It therefore follows from (96b) that $k = 0$. The cross section does not have the same shape after the deformation, of course. Axial stretching causes all of the fibers to curl tighter.

For the general case of initially curved, parallel fibers, the deformation can be visualized by regarding the body as divided into infinitesimal wedges bounded by the normal lines θ_1 and $\theta_1 + d\theta_1$, say. According to (96b), with $k = 0$, the material curves formed by these lines are still normal lines after the deformation, and according to (97) they are still straight. From (98), with $k = 0$, we see that $\theta = \lambda\theta_1$ (plus an arbitrary constant), and that the radii of the fibers crossing the wedge decrease in the proportion $1/\lambda$. Thus, each infinitesimal wedge deforms as the finite wedge in the preceding example did.

For the less important case of fibers that are initially straight but not parallel, such as radial fibers like spokes on a wheel, the compatibility condition (98) implies that if $k = 0$, then θ is constant along each fiber, so the fibers remain straight. The normal distance between two fibers must decrease in the ratio $1/\lambda$, so $r_n\, d\theta = r_{n1}\, d\theta_1/\lambda$. Since the normal curvature is conserved according to (97), then $\theta = \theta_1/\lambda$ (plus an arbitrary constant signifying a rigid rotation of the whole body). In this case, $\lambda\theta_1$ is not the fiber angle in the simple tensile state.

The preceding class of examples is an exception to the general rule that if fibers are *not* initially parallel, there is no stretched state with $k = 0$ everywhere. This means that in the stretched state with no tractions on the cross-sectional boundary, shearing stresses and nonzero values of T and P to balance them will nevertheless be present. Since such cases are less important and more difficult, we do not consider them further.

D. *Parallel Fibers*

We have found that if fibers are initially parallel, there is a simple tensile state with no stress except the axial stress $S_3(0, \lambda)$. The fiber angle θ_0 at a given particle in the simple tensile state is related to the initial angle θ_1 by

$\theta_0 = \lambda\theta_1$. With $\lambda\theta_1$ replaced by θ_0, the compatibility conditions (97) and (98) have exactly the same form as for a plane deformation without axial stretching. The theory of plane deformations superimposed on the simple tensile state is then identical with that for the case of no axial stretching, except that S and S_3 depend parametrically on λ.

E. *Example: Tube Forming*

As an example of an unusual kind we consider a body in the form of a segment of a tube wall, with fibers along concentric circular arcs, as described in Section IV,C. The planes $\theta_1 = 0$ and $\theta_1 = A$ are brought into contact and bonded together, so that a tube is formed. The tube is then left free from external traction. We wish to determine the internal state of stress in the tube wall.

A deformation of the form (99) with $\lambda = 2\pi/A$ satisfies the kinematic conditions of the problem. The internal and external radii of the tube are given in terms of the corresponding radii of the segment by $r_a = R_a/\lambda$ and $r_b = R_b/\lambda$, and the ends, initially at $Z_1 = 0$ and $Z_1 = L$, say, are at $z = 0$ and $z = \lambda L$. In the simple tensile state, this deformation is supported by a uniform tensile stress $S_3(0, \lambda)$ on the ends. The stress component σ_{zz} is equal to $S_3(0, \lambda)$ everywhere, and all other stress components are zero.

When the plane faces of the segment are bonded together and the tube is then left free from traction, the deformation is the same as for the simple tensile state, but the stress distribution is different. Since $\sigma_{zz} = S_3 - P$, the condition of no traction on the ends of the tube is satisfied if $P = S_3$ there. Then the equilibrium equation $\partial P/\partial z = 0$ requires that $P = S_3(\lambda, 0)$ everywhere, if the deformation is indeed the same. This appears to present a difficulty, because the normal pressure on the internal and external radii is specified to be zero. However, from the equilibrium conditions (58) and (59) we find that these conditions are satisfied if the fiber tension has the form

$$T = P[-1 - r_a\delta(r - r_a) + r_b\delta(r - r_b)] \qquad (100)$$

Thus, the stress at interior points amounts to an isotropic pressure P in cross-sectional planes. This pressure is contained by a finite tensile force r_bP in the outer boundary and a compressive force r_aP in the inner boundary.

V. Axisymmetric Deformations

We briefly consider axisymmetric deformations without twist, for bodies with an axis of rotational symmetry and fibers initially parallel to the axis.

Such deformations are often kinematically determinate, in the same way that plane problems with sufficient displacement boundary data are. However, analytical solutions are not so easily obtained as in plane problems, because normal lines are not in general straight lines. After setting down some useful geometrical relations in Section V,A, we derive compatibility conditions in Section V,B and from them find an explicit expression for the shapes of normal lines (a result due to T. G. Rogers).

The stress has the same form as for plane deformations superimposed on axial stretching, except that now the direction of stretching is the azimuthal direction (Section V,D). As in plane deformations, the equilibrium equations reduce to two equations in characteristic form, governing the variation of T along fibers and the variation of P along normal lines. It follows that for any kinematically admissible deformation, it is always possible to construct a compatible stress field, just as in plane deformations.

As examples (in Sections V, C, E, and F) we consider only some degenerate cases in which normal lines are straight; expansion and axial shearing of a tube.

A. Geometrical Preliminaries

We consider bodies with an axis of rotational symmetry, with fibers initially straight and parallel to the axis. As the concrete example of greatest interest, we have in mind a tube of internal and external radii R_0 and R_1, respectively, reinforced by fibers in the axial direction. We consider axisymmetric deformations without twist, for which a particle initially at R, ϕ, Z goes to the place

$$\mathbf{x} = \mathbf{i}_r(\phi)r(R, Z) + \mathbf{k}z(R, Z) \tag{101}$$

in the same plane ϕ = constant. Here $\mathbf{i}_r$ is a unit vector in the radial direction, which is expressed in terms of constant vectors $\mathbf{i}$, $\mathbf{j}$, $\mathbf{k}$ by

$$\mathbf{i}_r = \mathbf{i}\cos\phi + \mathbf{j}\sin\phi \tag{102}$$

The unit vector in the azimuthal direction is

$$\mathbf{i}_\phi = -\mathbf{i}\sin\phi + \mathbf{j}\cos\phi \tag{103}$$

The derivatives of these vectors with respect to ϕ are

$$\mathbf{i}_r'(\phi) = \mathbf{i}_\phi(\phi) \qquad \text{and} \qquad \mathbf{i}_\phi'(\phi) = -\mathbf{i}_r(\phi) \tag{104}$$

The directions of fibers and their orthogonal trajectories in meridional planes ϕ = constant are defined initially by

$$\mathbf{a}_0 = \mathbf{k} \qquad \text{and} \qquad \mathbf{n}_0 = \mathbf{i}_r(\phi) \tag{105}$$

After a deformation of the form (101), the field of unit vectors tangent to the fibers is denoted by **a**, and the field of normal vectors by **n**, where

$$\mathbf{a} = \mathbf{k}\cos\theta + \mathbf{i}_r \sin\theta \tag{106}$$

and

$$\mathbf{n} = -\mathbf{k}\sin\theta + \mathbf{i}_r \cos\theta \tag{107}$$

The fiber angle $\theta(R, Z)$ is the angle between the fiber and the symmetry axis. It is independent of ϕ.

Relations (14)–(16) remain valid in the present case. However, (17) and (18) are no longer valid. In the present case **a** and **n** are functions of ϕ, through $\mathbf{i}_r$. Their derivatives with respect to ϕ are

$$\partial\mathbf{a}/\partial\phi = \mathbf{i}_\phi \sin\theta \qquad \text{and} \qquad \partial\mathbf{n}/\partial\phi = \mathbf{i}_\phi \cos\theta \tag{108}$$

Since $\nabla\phi = \mathbf{i}_\phi/r$, then

$$\nabla\cdot\mathbf{a} = \nabla\theta\cdot\partial\mathbf{a}/\partial\theta + \nabla\phi\cdot\partial\mathbf{a}/\partial\phi = (1/r_n) + (\sin\theta)/r \tag{109}$$

and similarly

$$\nabla\cdot\mathbf{n} = (-1/r_a) + (\cos\theta)/r \tag{110}$$

Here r_a and r_n are the fiber and normal radii of curvature, respectively, defined in (15). We note that $\nabla\cdot\mathbf{i}_\phi = 0$.

Since $\nabla r = \mathbf{i}_r$, it is obvious that

$$\mathbf{a}\cdot\nabla r = \sin\theta \qquad \text{and} \qquad \mathbf{n}\cdot\nabla r = \cos\theta \tag{111}$$

These relations connect the function $r(R, Z)$ giving the final radial coordinate of a particle to the fiber angle $\theta(R, Z)$ at that particle.

B. Kinematics

In a deformation of the form (101), an infinitesimal element initially bounded by coordinate surfaces undergoes a locally homogeneous deformation similar to that described in (91) and (92) for a plane deformation superimposed on a stretching orthogonal to the plane of deformation. In the present case, a material curve that is initially a circle of radius R about the symmetry axis becomes a circle of radius $r(R, Z)$ when the body is deformed, and it is thus stretched in the ratio

$$\lambda = r(R, Z)/R \tag{112}$$

The deformation gradients, which describe the deformation locally, have forms analogous to (96):

$$(\mathbf{a}_0 \cdot \nabla_0)\mathbf{x} = \mathbf{a}, \qquad (\mathbf{n}_0 \cdot \nabla_0)\mathbf{x} = (\mathbf{n} + k\mathbf{a})/\lambda, \qquad (\mathbf{i}_\phi \cdot \nabla_0)\mathbf{x} = \lambda \mathbf{i}_\phi \tag{113}$$

The amount of shear k superimposed on the azimuthal stretching is defined by the second of these relations. If we express the directional derivatives in terms of partial derivatives, we obtain

$$\partial\mathbf{x}/\partial Z = \mathbf{a}, \qquad \partial\mathbf{x}/\partial R = (\mathbf{n} + k\mathbf{a})/\lambda, \qquad \partial\mathbf{x}/\partial\phi = R\lambda\mathbf{i}_\phi \tag{114}$$

To obtain compatibility conditions among the functions λ, k, and θ, we eliminate $\mathbf{x}$ from (114) by cross-differentiation. This can be done in three different ways. From the cross-derivatives involving a ϕ derivative, we recover relations equivalent to (111). The remaining relation is

$$\partial\mathbf{a}/\partial R = \partial^2\mathbf{x}/\partial R\,\partial Z = (\partial/\partial Z)[(\mathbf{n} + k\mathbf{a})/\lambda] \tag{115}$$

On carrying out the indicated differentiation, with the help of the results in Section V,A, and then separating the resulting equation into components in the fiber and normal directions, we obtain

$$\theta_R = (k\theta_Z/\lambda) - (\lambda_Z/\lambda^2) \qquad \text{and} \qquad \theta_Z/\lambda = (k/\lambda)_Z \tag{116}$$

where partial derivatives are indicated by subscripts.

To express these relations in terms of derivatives along fibers and normal lines in the deformed state, we first observe from (114) that

$$\partial/\partial Z = \mathbf{a}\cdot\nabla \qquad \text{and} \qquad \partial/\partial R = \lambda^{-1}(\mathbf{n} + k\mathbf{a})\cdot\nabla \tag{117}$$

We also replace λ by r/R in (116). Then, we obtain

$$\mathbf{n}\cdot\nabla\theta = -r^{-1}(\mathbf{a}\cdot\nabla)r \qquad \text{and} \qquad \mathbf{a}\cdot\nabla\theta = r(\mathbf{a}\cdot\nabla)(k/r) \tag{118}$$

By rearranging (118b) and using (111a), we obtain

$$(\mathbf{a}\cdot\nabla)(k - \theta) = (k/r)\sin\theta \tag{119}$$

Thus, it is not possible immediately to state the relation between the amount of shear and the fiber angle, as it was for plane deformations with initially parallel fibers.

The shapes of normal lines can be determined once and for all, but they are not straight, in general. From (118a), with (111), it is easy to verify that

$$r\sin\theta = r_m \tag{120}$$

where r_m is constant on each normal line. The change of position dr, dz, along a normal line satisfies

$$(dr/\cos\theta) + (dz/\sin\theta) = 0 \tag{121}$$

By using (120) to express $\sin\theta$ and $\cos\theta$ in terms of r/r_m in (121), and then integrating, we obtain

$$r = r_m \cosh[(z - z_m)/r_m] \tag{122}$$

The case $r_m = 0$ in (120) yields $\theta = 0$, i.e., the normal line is a straight, radial line. If θ is known along a boundary, the value of r_m for a given normal line can be found by evaluating (120) at the place where that normal line intersects the boundary, and z_m can then be evaluated by using the boundary values of r and z in (122). The entire normal line through that boundary point is then given by (122). When all normal lines have been constructed, the fibers can be constructed as the orthogonal trajectories. This process will ordinarily be done more easily by graphical methods than by analysis. The preceding derivation of the shapes of normal lines is due to T. G. Rogers (unpublished).

When the field of fibers has been constructed, the amount of shear at any point on a fiber can be obtained from (119) when the amount of shear at one point on that fiber is specified. We envisage numerical integration of the equation, except perhaps in special cases.

We record a result that will be useful later. With (15) and (111), we can write the compatibility condition (118a) as

$$(1/r_n) + (\sin\theta)/r = 0 \tag{123}$$

Comparison with (109) shows that this means that the divergence of **a** is zero:

$$\nabla \cdot \mathbf{a} = 0 \tag{124}$$

C. *Example: Expansion and Axial Shearing of a Tube*

As the simplest possible example, we consider a tube whose internal and external radii are initially R_0 and R_1. We suppose that the tube is forced over a rigid cylindrical rod of radius r_0, or is expanded to the radius r_0 by an internal pressure. The fibers along the inner boundary are then straight, with fiber angle $\theta = 0$. On each normal line through the boundary, $r \sin\theta$ is constant, and thus equal to zero, its boundary value. Thus $\theta = 0$ throughout the body. All fibers remain straight and parallel to the axial direction.

Since the material is incompressible, the volume of material between the cylindrical surface r = constant and the inner boundary $r = r_0$ is equal to the corresponding volume occupied by the same material in the undeformed state. Since there can be no stretching in the axial direction, it follows that the final radius $r(R)$ for a particle initially at R is given by

$$\pi(r^2 - r_0^2) = \pi(R^2 - R_0^2) \tag{125}$$

From the condition of inextensibility we see that the axial coordinate z

after deformation of the particle initially at R, Z must be of the form

$$z = Z + u(R) \tag{126}$$

That is, the axial displacement u is constant over each cylinder R = constant. The final place of the particle initially at R, ϕ, Z is

$$\mathbf{x} = \mathbf{i}_r(\phi)r(R) + \mathbf{k}[Z + u(R)] \tag{127}$$

Since

$$\partial\mathbf{x}/\partial R = \mathbf{i}_r r'(R) + \mathbf{k}u'(R) \tag{128}$$

then by noting that $r' = R/r = \lambda^{-1}$ and comparing with (113b), we find that the amount of shear is

$$k = u'(R)/r'(R) = du/dr \tag{129}$$

as expected. If the surface of the tube is left free from traction everywhere except in the contact region between tube and rod, we can anticipate that the amount of shear will be zero everywhere.

D. Stress in Axisymmetric Deformations

The stress must have essentially the same form as for plane deformations superimposed on stretching, which is given in (32) and (34). Now, however, the direction perpendicular to **a** and **n** is the azimuthal direction $\mathbf{i}_\phi$:

$$\boldsymbol{\sigma} = T\mathbf{aa} - P(\mathbf{I} - \mathbf{aa}) + S(\mathbf{an} + \mathbf{na}) + S_3\mathbf{i}_\phi\mathbf{i}_\phi \tag{130}$$

If the response is quasi-elastic, the functions $S(k, \lambda)$ and $S_3(k, \lambda)$ are the same ones that describe material response in plane deformations superimposed on stretching. In axisymmetric deformations these functions vary with position not only because k varies, but also because λ $(=r/R)$ varies with position.

In the equilibrium equation $\nabla\cdot\boldsymbol{\sigma} = \mathbf{0}$, the hoop stress makes a contribution given by

$$\nabla\cdot(S_3\mathbf{i}_\phi\mathbf{i}_\phi) = S_3(\mathbf{i}_\phi\cdot\nabla)\mathbf{i}_\phi = -\mathbf{i}_r S_3/r \tag{131}$$

since S_3 is independent of ϕ and the divergence of $\mathbf{i}_\phi$ is zero. In terms of **a** and **n**, we have

$$\nabla\cdot(S_3\mathbf{i}_\phi\mathbf{i}_\phi) = -(S_3/r)(\mathbf{a}\sin\theta + \mathbf{n}\cos\theta) \tag{132}$$

The contributions of the other terms in (130) to the equilibrium equation are exactly as in (55). In the latter, the divergence of **n** is now given by (110), and the divergence of **a** is zero [see (124)] When the equilibrium equation is separated into components in the fiber, normal, and azimuthal directions, we obtain

$$\mathbf{a}\cdot\nabla T = (2S/r_a) - \mathbf{n}\cdot\nabla S + (S_3 \sin\theta - S\cos\theta)/r \quad (133)$$

$$\mathbf{n}\cdot\nabla P - (P/r_a) = (T/r_a) + \mathbf{a}\cdot\nabla S + (S/r_n) - (S_3\cos\theta)/r \quad (134)$$

and $\partial P/\partial\phi = 0$

If a kinematically admissible deformation has been specified, and the functions S and S_3 are given, then all quantities in these equations except T and P are known. Just as in the case of plane deformations, one equation determines the variation of T along each fiber, and the other determines the variation of P along each normal line, once T has been determined. The value of T at one point on each fiber and the value of P at one point one each normal line must be specified as boundary conditions. One should note that *every kinematically admissible axisymmetric deformation is also statically admissible.* Thus, if the deformation in a given problem can be determined purely kinematically, there is no question that a stress field compatible with the deformation can be constructed.

E. *Example: Inflation of a Tube*

To illustrate the calculation of stress we first consider the pure radial expansion of a tube, for which the deformation has the form (127) with the axial shearing displacement u equal to zero. We suppose that no traction is applied to the outer boundary $r = r_1$ or to the ends of the tube at $z = 0$ and $z = L$, say.

Since there is no shearing, the shearing stress S is zero. Since the fiber angle θ is also zero, (133) shows that the tension is constant along each fiber, and from the boundary conditions at the tube ends it follows that $T = 0$ everywhere. The equation for P, (134), reduces to

$$\partial P/\partial r = -S_3/r \quad (135)$$

With $P = 0$ at the outer boundary $r = r_1$, we obtain

$$P = \int_r^{r_1} S_3/r'\,dr' \quad (136)$$

The hoop stress S_3 depends only on λ since there is no shear, and the dependence of λ on position is found by setting $R = r/\lambda$ in (125):

$$r^2(1 - \lambda^{-2}) = r_0^2(1 - \lambda^{-2}) \quad (137)$$

It is convenient to replace r by λ as the integration variable, by using this expression. We find that at the place where the stretch is λ, the pressure is

$$P = P^*(\lambda) - P^*(\lambda_1) \tag{138}$$

where $\lambda_1 = r_1/R_1$ and the function P^* is defined by

$$P^*(\lambda) = \int_1^\lambda \lambda^{-1}(\lambda^2 - 1)^{-1}S_3(0, \lambda)\, d\lambda \tag{139}$$

F. Example: Axial Shearing

We suppose that a tube of length L is bonded to a rigid shaft at its inner radius and to a rigid cylinder at its outer radius. The outer cylinder is displaced in the axial direction by applying a force F to it while the inner shaft is held fixed.

The deformation has the form (127) with no expansion, i.e., $r = R$, but with a nonzero shearing displacement $u(r)$. The amount of shear is $k = u'(r)$. The shearing stress $S(k, 1)$ is constant over surfaces $r =$ constant, and the resultant shearing force over the entire surface must be equal to the axial force F:

$$F = 2\pi r L S(k, 1) \tag{140}$$

Let $k = K(S)$ be the inverse of the relation $S = S(k, 1)$, i.e., the graph of the latter function with abscissa and ordinate interchanged. Then

$$u'(r) = k = K(F/2\pi r L) \tag{141}$$

and integration yields

$$u(r) = \int_{r_0}^{r} K(F/2\pi r' L)\, dr' \tag{142}$$

By letting the shearing stress be the integration variable, we obtain

$$u(r) = (F/2\pi L) \int_s^{s_0} K(s)/s^2\, ds \tag{143}$$

where

$$s = F/2\pi r L \qquad \text{and} \qquad s_0 = F/2\pi r_0 L$$

For linear shearing response, $K(s) = s/G$ and the displacement is

$$u(r) = (F/2\pi L G) \log(r/r_0) \tag{144}$$

With rS constant, (133) shows that T is constant along each fiber. The condition of no tractions on the ends of the tube implies that $T = kS$ and $P = -S/k$ there. Since kS is also constant along fibers, T is equal to kS everywhere. We omit the calculation of P, which is analogous to the corresponding calculation in the case of simple shearing (Section III,F).

VI. General Three-Dimensional Theory

We have been able to deal with plane problems because of the translational symmetry of such problems in the direction orthogonal to the plane of deformation, and have made some progress toward the solution of axisymmetric problems. We should expect that, as in other branches of mathematical physics, it will not be possible to obtain explicit analytical solutions of problems that have no special symmetry. However, there are symmetrical problems of many kinds that have not been considered here; for example, axisymmetric problems with twist. As a first step toward the solution of other problems, we briefly outline the general theory without specialization to particular geometries.

The kinematic constraint conditions are discussed in Section VI,A, and the equilibrium equations are discussed in Section VI,B. In Section VI,C we record the stress-deformation relation for elastic response, in the form given by Spencer (1972b). The stress response for transversely isotropic, extensible materials is discussed in Section VI,D. The stress-deformation relation for this case, which is due to Ericksen and Rivlin (1954), is likely to be of use in obtaining higher-order approximations taking fiber extensibility to be small but not zero. In Section VI,E we list some problems that have been solved explicitly without assuming fiber inextensibility. For details in these cases, the reader is referred to the book by Green and Adkins (1960).

A. Kinematics

The deformation is described by a mapping $\mathbf{x} = \mathbf{x}(\mathbf{X})$ that specifies the place $\mathbf{x}$ of the particle initially at $\mathbf{X}$. The deformation depends parametrically on time as well, but this need not concern us. The deformation of infinitesimal elements can be regarded as homogeneous, so that elements $d\mathbf{X}$ are mapped linearly to their final configuration $d\mathbf{x} = \mathbf{F}\cdot d\mathbf{X}$. If Cartesian coordinates are used to describe the initial and final states, then $dx_i = x_{i,A}\, dX_A$ ($x_{i,A} = \partial x_i/\partial X_A$), and the *deformation gradient* $\mathbf{F}$ has components $F_{iA} = x_{i,A}$.

The ratio of the volume of a material element in the deformed state to its initial volume is $J = \det \mathbf{F}$. Since we consider the materials to be incompressible, only those deformations for which the determinant J is unity are kinematically admissible.

Let $\mathbf{a}_0(\mathbf{X})$ be the field of fiber directions in the undeformed body. A material element $d\mathbf{X} = \mathbf{a}_0\, dX$ is mapped onto $d\mathbf{x} = \mathbf{F}\cdot\mathbf{a}_0\, dX$ by the deformation. The field $\mathbf{a} = \mathbf{F}\cdot\mathbf{a}_0$ specifies the fiber directions in the deformed body. The length $|\mathbf{F}\cdot\mathbf{a}_0|$ represents the ratio of the length of the deformed element to its initial length dX. Since we consider the fibers to be inextensible, only those deformations for which $|\mathbf{F}\cdot\mathbf{a}_0|$ is unity are kinematically admissible.

The constraint of incompressibility implies that whether fibers are inextensible or not, the divergence $\nabla\cdot\mathbf{a}$ at a given particle has the same value as the divergence $\nabla_0\cdot\mathbf{a}_0$ at the same particle before the deformation (Pipkin and Rogers, 1971a; see also Spencer, 1972b). When the fields $\mathbf{a}$ and $\mathbf{a}_0$ are fields of unit vectors, these divergences are related to the curvatures of curves orthogonal to the fibers. In the special case of plane deformations, the conservation of curvature of normal lines can be derived as a consequence of this more general result concerning conservation of the divergence. The relation defining the shapes of normal lines in axisymmetric deformations can also be viewed as a consequence of this result (Section V,B)

In materials composed of long fibers packed as closely together as possible, the divergence $\nabla_0\cdot\mathbf{a}_0$ is zero initially and thus always zero. This is the generalization to three dimensions of the result for plane deformations that initially parallel fibers remain parallel in any kinematically admissible deformation.

The deformation gradient $\mathbf{F}$ describes the mapping of an infinitesimal element completely. In particular, it includes information about rotations as well as distortions. As a measure of distortion alone, one may use the strain tensor g defined by

$$\mathsf{g} = \mathbf{F}\mathbf{F}^{\mathrm{t}}, \qquad g_{ij} = x_{i,A}x_{j,A} \tag{145}$$

For elastic materials that are transversely isotropic about the fiber direction, the stress can be expressed in terms of g and $\mathbf{a}$. Certain orthogonal invariants of g and $\mathbf{a}$ are of importance:

$$\begin{aligned} J_1 &= \operatorname{tr}\mathsf{g} = g_{ii} \\ J_2 &= \operatorname{tr}\mathsf{g}^{-1} = \tfrac{1}{2}[(\operatorname{tr}\mathsf{g})^2 - \operatorname{tr}(\mathsf{g}^2)] \\ J_3 &= \mathbf{a}\cdot\mathsf{g}\cdot\mathbf{a} = a_i g_{ij} a_j \end{aligned} \tag{146}$$

B. Equilibrium Equations

The stress consists of reactions T and P to the constraint conditions, plus additional terms $\mathbf{S}$ that are specified by constitutive equations:

$$\boldsymbol{\sigma} = T\mathbf{aa} - P(\mathbf{I} - \mathbf{aa}) + \mathbf{S} \tag{147}$$

The equilibrium equation $\nabla \cdot \boldsymbol{\sigma} = \mathbf{0}$ then takes the form

$$\nabla(T + P) \cdot \mathbf{aa} + (T + P)(\nabla \cdot \mathbf{a})\mathbf{a} + (T + P)(\mathbf{a} \cdot \nabla)\mathbf{a} - \nabla P = -\nabla \cdot \mathbf{S} \tag{148}$$

In particular, the component in the fiber direction is

$$\mathbf{a} \cdot \nabla T + (T + P)\nabla \cdot \mathbf{a} = -(\nabla \cdot \mathbf{S}) \cdot \mathbf{a} \tag{149}$$

If all quantities except T are known, for each fiber this is an ordinary differential equation governing the variation of T along the fiber. For the common cases in which $\nabla \cdot \mathbf{a}$ is zero, the equation can be integrated directly and no prior knowledge of P is required.

C. Finite Elastic Stress Response

The stored energy density W for an elastic material is a function of the deformation gradients $x_{i,A}$. From the relation between work and stored energy it can be shown that the extra stress $\mathbf{S}$ can be expressed in terms of derivatives of the potential W by

$$S_{ij} = (\partial W / \partial x_{i,A})\, x_{j,A} \tag{150}$$

We remark that if $\mathbf{S}$ is defined in this way, then T does not represent the total tension in the fiber direction. The component $\mathbf{a} \cdot \mathbf{S} \cdot \mathbf{a}$ can either be omitted from the complete expression for the stress without loss of generality, or retained, as convenience dictates. Similar remarks apply to P.

If the material is transversely isotropic about the fiber direction, then W can be expressed as a function of the three invariants J_1, J_2, and J_3 listed in (146). The expression for $\mathbf{S}$ then becomes

$$\mathbf{S} = 2W_1\mathbf{g} - 2W_2\mathbf{g}^{-1} + 2W_3[\mathbf{a}(\mathbf{g} \cdot \mathbf{a}) + (\mathbf{g} \cdot \mathbf{a})\mathbf{a}] \tag{151}$$

where $W_i = \partial W / \partial J_i$. For moderate deformations it is reasonable to approximate the coefficients W_i by constants. These results are due to Spencer (1972b).

In the theory of constrained isotropic materials due to Adkins and Rivlin (1955) (see also Adkins, 1956a,b, 1958, and Green and Adkins, 1960), the strain energy is taken to be that of an isotropic elastic material rather than a transversely isotropic material. For the case of isotropy, W is independent of J_3, and thus $W_3 = 0$ in the expression for $\mathbf{S}$. As Spencer (1972b) has remarked, this simplification is likely to be appropriate only if fibers are rather sparsely distributed in the material. The same assumption was used by Prager (1969) in connection with plastic response.

For the problems that have been considered in the present paper, the most general deformation of a material element that we have needed to consider is a simple shearing by the amount k superimposed on a perpendicular stretching with extension ratio λ. For such deformations the strain g is

$$\mathsf{g} = \mathbf{aa} + \lambda^{-2}(\mathbf{n} + k\mathbf{a})(\mathbf{n} + k\mathbf{a}) + \lambda^2\mathbf{bb} \tag{152}$$

where $\mathbf{b}$ is a unit vector perpendicular to $\mathbf{a}$ and $\mathbf{n}$, in the direction of stretching. The strain invariants are then given by

$$\begin{aligned} J_1 &= 1 + \lambda^2 + (1/\lambda^2) + (k/\lambda)^2 \\ J_2 &= 1 + \lambda^2 + (1/\lambda^2) + k^2 \\ J_3 &= 1 + (k/\lambda)^2 \end{aligned} \tag{153}$$

The function $W(k, \lambda)$ considered in Section IV,A is obtained from $W(J_1, J_2, J_3)$ by using these expressions for the invariants. We find that

$$S = W_k = 2(W_1\lambda^{-2} + W_2 + W_3\lambda^{-2})k \tag{154}$$

and

$$S_3 = \lambda W_\lambda + kW_k = 2(W_1 + W_2)(\lambda^2 - \lambda^{-2}) + 2W_2k^2 \tag{155}$$

A particularly simple deformation that we have not considered previously is simple shearing across the fiber direction rather than parallel to it. If $\mathbf{a}$, $\mathbf{b}$, $\mathbf{c}$ are unit vectors in the fiber direction, the gradient direction, and the direction of shear, respectively, and the amount of shear is k, so that the deformation is

$$\mathbf{x} = \mathbf{X} + k\mathbf{c}(\mathbf{b}\cdot\mathbf{X}), \tag{156}$$

then the strain g is easily found to be

$$\mathsf{g} = \mathbf{aa} + (\mathbf{b} + k\mathbf{c})(\mathbf{b} + k\mathbf{c}) + \mathbf{cc} \tag{157}$$

The shearing stress is then

$$\mathbf{c}\cdot\mathbf{S}\cdot\mathbf{b} = 2(W_1 + W_2)k \tag{158}$$

The corresponding expression for shearing along the direction of the fibers (with $\lambda = 1$) is $2(W_1 + W_2 + W_3)k$. The extra term involving W_3 represents a difference between the shear moduli for shearing along and across the fibers. In the constrained isotropic theory previously mentioned, it is assumed that there is no difference between these two moduli.

D. Stress Response for Extensible Materials

In some cases the idealized theory, based on treating fibers as inextensible, does not give sufficiently detailed information about stress distributions because it requires specification of stress boundary conditions that may be unknown in the application at hand, or, in the case of closed fibers that do not cross an external boundary, because no boundary condition is available at all. In such cases it may be necessary to seek approximate solutions, with the deformation given by the idealized theory as a lowest-order approximation, for materials with slightly extensible fibers. Insofar as this involves ordinary perturbation methods, the manner in which the next order of approximation can be obtained is indicated by the paper of Green *et al.* (1952) on small deformations superimposed on large deformations. Their formulation is for isotropic materials rather than transversely isotropic materials, but the same methods could be used.

If fibers are extensible we retain the notation $\mathbf{a}_0$ for the field of unit vectors tangential to the fibers in the undeformed state of the body, and still define the field $\mathbf{a}$ in terms of the deformation gradient $\mathbf{F}$ and $\mathbf{a}_0$ by $\mathbf{a} = \mathbf{F}\cdot\mathbf{a}_0$. Now, however, the vectors $\mathbf{a}$ are not constrained to be unit vectors. Instead, the length (squared)

$$J_4 = \mathbf{a}\cdot\mathbf{a} \tag{159}$$

is a new variable upon which the strain energy depends. For an incompressible material, the stress then has the form

$$\boldsymbol{\sigma} = -P\mathbf{I} + \mathbf{S} \tag{160}$$

and in place of a reaction stress proportional to $\mathbf{aa}$, the extra stress $\mathbf{S}$ contains, in addition to the terms in (151), a new term, $2W_4\mathbf{aa}$, where W_4 is the derivative of W with respect to its new argument J_4. The stress-deformation relation for transversely isotropic materials was first derived by Ericksen and Rivlin (1954), and applied to cases in which the direction of transverse isotropy varies with position by Adkins (1955). These results are discussed in the book by Green and Adkins (1960).

E. Solutions for Extensible Materials

There are certain deformations kinematically admissible in incompressible materials that are also statically admissible in every homogeneous, isotropic, incompressible, elastic material. These deformations have been called *controllable* for the class of materials mentioned. In this use of the word, all plane and axisymmetric deformations of ideal fiber-reinforced materials are controllable, since for any kinematically admissible plane or axisymmetric deformation of such a material it is possible to find a compatible stress field satisfying the equilibrium equation with no body force (or a conservative body force).

The deformations that are controllable in materials that are merely incompressible, and not inextensible as well, form a much more limited class. All of these that are known, and these may well be all that are possible, are listed below. In each case, the deformation is described by giving the final position, with coordinates in lower-case letters, in terms of the initial position, with coordinates in capital letters:

(0) homogeneous, isochoric deformations;

(1) expansion of a spherical shell, or a shell first turned inside out:

$$r = \pm(R^3 - R_0^3 \pm r_0^3)^{1/3}, \qquad \theta = \Theta, \quad \phi = \pm\Phi \tag{161}$$

(2) deformations mapping coaxial cylinders onto coaxial cylinders:

$$r^2 = AR^2 + B, \qquad \theta = C\Theta + DZ, \qquad z = E\Theta + FZ \tag{162}$$

(3) deformations mapping parallel planes onto coaxial cylinders:

$$r^2 = 2AX + B, \qquad \theta = CY + DZ, \qquad z = EY + FZ \tag{163}$$

(4) deformations mapping coaxial cylinders onto parallel planes:

$$x = \tfrac{1}{2}AR^2 + B, \qquad y = C\Theta + DZ, \qquad z = E\Theta + FZ \tag{164}$$

(5) Azimuthal shearing and associated deformations:

$$r = AR, \qquad \theta = B \log R + C\Theta, \qquad Z = Z/A^2C \tag{165}$$

In families (2)–(4) the condition of no volume change requires that

$$A(CF - DE) = 1 \tag{166}$$

These deformations are controllable for quasi-elastic response as well as elastic response (for which the stress can be derived from a potential). Moreover, the condition that the material be isotropic can be relaxed in certain ways, depending on which deformation is considered. For each family of deformations there is a family of material membranes called "principal" membranes, such that the deformation involves no shearing deformation between the principal membranes. If the material is inhomogeneous but with constant properties over each principal membrane, the deformation remains controllable. If the material is laminated, with lamination coinciding with the principal membranes, again the deformation is controllable.

For materials reinforced with a single family of straight, parallel fibers, of course every homogeneous deformation is controllable. Among the nontrivial families (1)–(5) those deformations are controllable for which the fibers are initially parallel or perpendicular to the principal membranes. Family (1), the case of spherical principal membranes, is ruled out. In families (2) and (4), the material may be fiber-reinforced with fibers parallel to the axis of the coaxial cylindrical principal membranes. In family (3) the fibers may be either parallel or perpendicular to the principal membranes, which are parallel planes initially. In family (5) the deformations remain controllable for fiber-reinforced materials if the fibers are initially parallel to the axial direction. The use of these deformations in generating solutions for fiber-reinforced or laminated materials is discussed in a little further detail by Pipkin (1970).

These deformations were originally obtained as solutions of problems involving isotropic materials by Rivlin (1948, 1949), Green and Shield (1950), Adkins *et al.* (1953), Ericksen and Rivlin (1954), Ericksen (1954, 1955), Klingbeil and Shield (1966), and Singh and Pipkin (1965). Extensions to transversely isotropic materials and materials with curvilinear aelotropy (i.e., with the preferred direction variable with position in the material) are mainly due to Adkins, and many explicit results for such cases are given in the book by Green and Adkins (1960). There is also a moderately large body of derivative literature involving special cases such as the expansion of a sphere composed of two shells of different materials, for example. We will not attempt to compile a list of such solutions, since no new ideas are involved in obtaining them.

Acknowledgment

This review was prepared under a grant from the Advanced Research Projects Agency to the Materials Science Laboratory at Brown University. This support is gratefully acknowledged.

References

Adkins, J. E. (1955). *Proc. Roy. Soc. London* **A229,** 119.
Adkins, J. E. (1956a). *JRMA* **5,** 189.
Adkins, J. E. (1956b). *Phil. Trans. Roy. Soc. London* **A249,** 125.
Adkins, J. E. (1958). *QJMAM* **11,** 88.
Adkins, J. E., and Rivlin, R. S. (1955). *Phil. Trans. Roy. Soc. London* **A248,** 201.

Adkins, J. E., Green, A. E., and Shield, R. T. (1953). *Phil. Trans. Roy. Soc. London* **A246,** 181.

England, A. H. (1972). *JIMA* **9,** 310.

Ericksen, J. L. (1954). *Z. Angew. Math. Phys.* **5,** 466.

Ericksen, J. L. (1955). *Z. Angew. Math. Mech.* **35,** 381.

Ericksen, J. L., and Rivlin, R. S. (1954). *JRMA* **3,** 281.

Everstine, G. C. (1971). Stress Concentrations and Boundary Layers in Fiber-Reinforced Materials. Ph.D. dissertation, Brown Univ. Univ. Microfilms, Ann Arbor, Michigan.

Everstine, G. C., and Pipkin, A. C. (1971). *Z. Angew. Math. Phys.* **22,** 825.

Everstine, G. C., and Pipkin, A. C. (1973). Boundary Layers in Fiber-Reinforced Materials. *J. Appl. Mech.* **40,** 518.

Everstine, G. C., and Rogers, T. G. (1971). *J. Compos. Mater.* **5,** 94.

Green, A. E., and Adkins, J. E. (1960). "Large Elastic Deformations." Oxford Univ. Press, London and New York.

Green, A. E., and Shield, R. T. (1950). *Proc. Roy. Soc. London* **A202,** 407.

Green, A. E., Rivlin, R. S., and Shield, R. T. (1952). *Proc. Roy. Soc. London* **A211,** 128.

Kao, B. C., and Pipkin, A. C. (1972). *Acta Mech.* **8,** 265.

Klingbeil, W. W., and Shield, R. T. (1966). *Z. Angew. Math. Phys.* **17,** 489.

Lekhnitskii, S. G. (1963). "Theory of Elasticity of an Anisotropic Body." Holden-Day, San Francisco, California.

Lighthill, M. J. (1958). "Fourier Analysis and Generalized Functions." Cambridge Univ. Press, London and New York.

Mulhern, J. F., Rogers, T. G., and Spencer, A. J. M. (1967). *Proc. Roy. Soc. London* **A301,** 473.

Pipkin, A. C. (1970). *In* "Nonlinear Continuum Theories in Mechanics and Physics and their Applications" (R. S. Rivlin, coordinator). Edizioni Cremonese, Rome.

Pipkin, A. C. (1974). *Quart. Appl. Math.*

Pipkin, A. C. and Sanchez, V. M. (1974). *SIAM J. Appl. Math.* **26,** 213.

Pipkin, A. C., and Rogers, T. G. (1971a). *J. Appl. Mech.* **38,** 634.

Pipkin, A. C., and Rogers, T. G. (1971b). *Quart. Appl. Math.* **29,** 151.

Prager, W. (1969). *J. Appl. Mech.* **36,** 542.

Rivlin, R. S. (1948). *Phil. Trans. Roy. Soc. London* **A241,** 379.

Rivlin, R. S. (1949). *Phil. Trans. Roy. Soc. London* **A242,** 173.

Rivlin, R. S. (1955). *ARMA* **4,** 951.

Rivlin, R. S. (1959). *ARMA* **2,** 447.

Rivlin, R. S. (1960). *In* "Structural Mechanics" (J. N. Goodier and N. J. Hoff, eds.). Pergamon, Oxford.

Rivlin, R. S. (1964). *In* "Nonlinear Problems of Engineering." Academic Press, New York.

Rivlin, R. S. (1970). *In* "Nonlinear Continuum Theories in Mechanics and Physics and Their Applications" (R. S. Rivlin, coordinator). Edizioni Cremonese, Rome.

Rogers, T. G., and Pipkin, A. C. (1971a). *J. Appl. Mech.* **38,** 1047.

Rogers, T. G., and Pipkin, A. C. (1971b). *QJMAM* **24,** 311.

Singh, M., and Pipkin, A. C. (1965). *Z. Angew. Math. Phys.* **16,** 706.

Spencer, A. J. M. (1972a). *QJMAM* **25,** 387.

Spencer, A. J. M. (1972b). "Deformations of Fibre-reinforced Materials." Oxford Univ. Press, London and New York.

Treloar, L. R. G. (1958). "The Physics of Rubber Elasticity." Oxford Univ. Press, London and New York.

8

Waves and Vibrations in Directionally Reinforced Composites

J. D. ACHENBACH

The Technological Institute
Northwestern University
Evanston, Illinois

I. Introduction

Materials that are used in engineering applications generally are inhomogeneous. For conventional materials, such as metal alloys, the characteristic lengths of the inhomogeneities usually are, however, very small as compared to lengths characterizing deformations. This is not necessarily the case for materials that are fabricated by compounding reinforcing elements and a matrix material to form a directionally reinforced composite. For such composite materials the lengths characterizing the inhomogeneous structuring are much larger. Indeed, for directionally reinforced composites it is conceivable that for certain loading conditions, particularly of a dynamic nature, the characteristic length of the deformation is of the same order of magnitude as a length characterizing the structuring.

The relative magnitudes of these characteristic lengths significantly affect the mechanical behavior of a composite material. The characteristic lengths must, therefore, play an important role in considerations leading to analytical models for the description of the mechanical behavior of composite materials.

The mechanical response of a composite can adequately be analyzed on the basis of a theory which accounts for the gross mechanical behavior of the composite material if the ratio of the largest characteristic length of the structuring and the smallest length characterizing the deformation is sufficiently smaller than unity. The simplest aspects of the gross mechanical behavior are described by averages of the displacements, the stresses and the strains over representative elements. In the simplest theory for such averaged quantities the averaged stresses are related to averaged strains by means of effective elastic constants. In this theory, which is termed the "effective modulus theory," the mechanical response of the composite is equivalent to the response of a homogeneous but generally anisotropic medium whose "effective moduli" are determined in terms of the elastic moduli of the constituents and the parameters describing the geometrical layout of the composite.

The advantages of the effective modulus theory and of more advanced theories as well are that the actual discrete character of the composite medium is represented within the framework of a homogeneous continuum. Thus, instead of having to deal with numerous sets of field equations (one set for each inhomogeneity), the approximate theory permits us to work with a single set of equations for the composite medium as a whole. For most loading conditions the effective modulus theory is entirely satisfactory. The effective modulus theory becomes, however, less useful for static loading conditions if the interest is focused on local values of the field variables, as for example in the analysis of interface bond failure and other modes of fracture. For dynamic loading conditions the applicability of the effective modulus theory is somewhat restricted even for gross quantities such as frequencies, as will be discussed in Section III.

In this chapter we review some aspects of wave motions and vibrations in directionally reinforced composites for small deformations and linear behavior of the constituents. Some of the basic concepts of the motions of elastic continua are briefly reviewed in Appendixes A and B. An investigation of the propagation of mechanical disturbances is of importance for bodies that are subjected to high-rate loads such as are generated by impact or by explosive charges. For small times after the application of a high-rate load a pattern of transient waves propagates through the medium. These transient waves may interact rather vigorously with the reinforcing elements. For a directionally reinforced composite the character of the dynamic response depends on the direction of propagation of the disturbances. For wave motions propagating in the direction of the reinforcements, the reinforcing elements act as waveguides. If, on the other hand, the wave motion propagates normal to the direction of the reinforcements the reinforcing elements act essentially as obstacles, which reflect and transmit propagating disturbances. As time increases and the disturbances have been reflected several times from the outer boundaries of the body, a pattern of vibrations sets in.

Mechanical disturbances are subjected to attenuation when propagating in composite materials. The attenuation is effected through geometrical dispersion, and dispersion due to other mechanisms, such as are related to inelastic material behavior, delamination, internal voids and cracks, and crushing of composite constituents. From the point of view of structural integrity, dispersion is considered a desirable phenomenon since it reduces the peak intensity of a stress pulse, and hence its potential to do damage. Of the various dispersion mechanisms structural dispersion and dispersion due to inelastic material behavior are best amenable to an analytical treatment.

Geometrical dispersion is the spreading out of pulses caused by interactions at boundaries. The change of shape of a propagating pulse due to geometrical dispersion can be investigated on the basis of an analysis of time-harmonic (sinusoidal) wavetrains. For time-harmonic waves dispersion manifests itself in the dependence of the phase velocity on the wavelength. In a time-harmonic or sinusoidal wave the phase velocity is defined as the rate of advance of points of constant phase angle. An analysis of the propagation of time-harmonic waves results in the derivation of a frequency equation which relates the frequency and the reciprocal wavelength. For linear theories the propagation of a stress pulse can be studied by using a Fourier decomposition of the pulse into modes of time-harmonic wave propagation (see Appendix A). Since the frequency equation governs the phase velocities of the individual Fourier components it provides the essential information on geometrical dispersion.

Geometrical dispersion is a well-known phenomenon for elastic bodies with elongated boundaries such as rods and layers. As an example the propagation of time-harmonic waves in a layer is considered in Appendix B. The Rayleigh–Lamb frequency equation for a layer shows what is known from elementary theories, namely, that at small values of the wavenumber the phase velocity for longitudinal (symmetric) time-harmonic motions varies very little, while the phase velocity of transverse (antisymmetric) motions increases linearly with the wavenumber. On a small scale a directionally reinforced composite essentially is a system of waveguides, and it is thus to be expected that time-harmonic waves, particularly waves that are transverse with respect to the direction of the reinforcing elements, are dispersive while propagating in the composite.

Since the characteristic length of the deformation is expected to be of paramount importance, a first test of approximate theories for the purpose of checking dynamic behavior can be carried out by examining the propagation of time-harmonic waves in a composite medium of infinite extent. Here the characteristic length is the wavelength, which usually enters through the wavenumber $k = 2\pi/\text{wavelength}$. In the presence of dispersion, harmonic waves of different wavelengths propagate at different velocities. This points immediately to a deficiency of the effective modulus theory, because a classical anisotropic continuum model cannot account for the dispersion of free harmonic waves, which should occur in an extended composite medium if the wavelength is of the same order of magnitude as a characteristic length of the structuring. For a layered medium consisting of alternating layers of two elastic materials this restriction was pointed out by Sun *et al.* (1968a).

This chapter includes six sections, two appendixes and a list of references. Appendix A provides background material on the propagation of mechanical disturbances. Some pertinent results on the dynamics of linearly elastic solids are discussed in Appendix B. In Section II we review the effective modulus theory for laminated media and fiber-reinforced composites. A laminated medium consisting of alternating layers of two homogeneous isotropic materials is considered in some detail, and the effective moduli are stated explicitly in terms of the material constants and the thicknesses of the layers. For longitudinal and transverse sinusoidal waves propagating in the direction of the layering the effective modulus theory is used to obtain the constant phase velocities. Next we proceed to discuss the limitations of the effective modulus theory for a laminated medium. The limitations are displayed by a comparison of frequencies and phase velocities with exact results, which are discussed in Section III. Several approximate theories which improve upon the effective modulus theory are discussed in Section IV. In Section V we survey experimental information on waves and vibrations in directionally reinforced composites. To conclude the chapter some pertinent references on topics such as dynamic effects in disordered composites, dynamic fracture, optimization, and nonlinear effects are mentioned in the last section.

In recent related reviews Gurtman (1969) and Peck (1971, 1972) have also provided surveys of theoretical and experimental work in the area of linear and nonlinear wave propagation in composite materials.

II. The Effective Modulus Theory

In the effective modulus theory the mechanical behavior of a composite material is represented by a homogeneous but anisotropic medium. The effective modulus theory, which has been developed to an advanced level of quantitative analysis, is discussed in some detail elsewhere. We will, therefore, limit ourselves to the observation that this theory relates volume averages of the components of the stress tensor (denoted by $\bar{\tau}_{ij}$) to volume averages of the components of the strain tensor (denoted by $\bar{\epsilon}_{kl}$; see Appendix B) by a general anisotropic linear stress–strain relation of the form

$$\bar{\tau}_{ij} = C^*_{ijkl}\bar{\epsilon}_{kl} \tag{1}$$

In Eq. (1) we have employed the usual indicial notation with respect to a Cartesian coordinate system x_i, $i = 1, 2, 3$ (see also Appendix B). In particular it is understood that the tensor summation convention should be invoked whenever indices $i, j, k, l = 1, 2,$ or 3 are repeated. The elastic constants C^*_{ijkl}, which are called the *effective elastic moduli*, are expressions in terms of the material constants of the constituents and the parameters defining the geometrical layout of the composite. The coefficients satisfy the relations

$$C^*_{ijkl} = C^*_{jikl} = C^*_{ijlk} = C^*_{klij} \tag{2}$$

Thus, of the 81 coefficients C^*_{ijkl}, only 21 are independent. It can be shown that C^*_{ijkl} are the components of a fourth-rank tensor. The tensor C^*_{ijkl} is positive definite in the sense that

$$C^*_{ijkl} f_{ij} f_{kl} \geq 0 \tag{3}$$

for all symmetrical f_{ij}, where the equality is satisfied only when $f_{ij} = 0$.

For a fiber-reinforced composite a precise definition of the concept of elastic moduli, and a detailed discussion of means of computing the effective moduli, is given in the work of Hashin (1972).

The number of independent elastic constants is generally much less than 21, due to the existence of symmetries in the structuring of the material. A plane is called a *plane of elastic symmetry* if reflection in the plane leaves the stress–strain law unchanged. If rotation of a coordinate direction about an axis leaves elastic properties unchanged, then the axis is an axis of symmetry. In a composite material the symmetry may be small scale, i.e., for the elastic properties at a point, or it may be a gross property due to the structuring of the composite. Here we consider the case where the constituents of the composite are isotropic, i.e., every axis is an axis of symmetry as far as the individual constituents are concerned, and anisotropy is in evidence only for the gross properties of the medium.

For a fiber-reinforced composite the significant cases with regard to gross anisotropy are orthotropy, square symmetry, and transverse isotropy. An orthotropic elastic body has three mutually perpendicular planes of elastic symmetry. An example for such a material is a fiber-reinforced composite with a rectangular array of identical circular fibers (Fig. 1a). A square symmetric material is an orthotropic material in which the two axis x_2 and x_3 are elastically equivalent. This means that the stress–strain law is insensitive to a 90° rotation around the x_1 axis. The square array of fibers shown in Fig. 1b gives rise to a square symmetric material. For a transversely isotropic material one axis, say the x_2 axis, is an axis of rotational symmetry. Thus, the constitutive relations are insensitive to *any* rotation around the x_2 axis. The hexagonal array of identical fibers shown in Fig. 1c gives rise to transversely isotropic gross properties. Transverse isotropy may also be assumed for a random distribution of unidirectional fibers across the cross section. Another example of gross transverse isotropy is provided by a laminated medium consisting of alternating layers of two isotropic materials, as shown in Fig. 1d.

A. Propagation of Harmonic Waves in Anisotropic Materials

Substitution of Eq. (1) into the stress equation of motion

$$\bar{\tau}_{ij,j} = \bar{\rho}\, \partial^2 \bar{u}_i / \partial t^2 \tag{4}$$

where $\bar{\rho}$ is an effective mass density, yields

$$C^*_{ijkl}\bar{u}_{k,lj} = \bar{\rho}\, \partial^2\bar{u}_i/\partial t^2 \tag{5}$$

In Appendix B we investigate waves in isotropic solids, for which there are only two independent elastic constants. In the present section we will comment briefly on wave propagation in anisotropic materials. This subject has a lengthy history, with some of the first contributions dating back to the middle of the 19th century. In recent years interest has been revived because of interest in the areas of seismology, ultrasonics, and now composite materials.

For a review of waves in anisotropic media we refer to an article by Musgrave (1961). Of particular interest are the results of Synge (1957) for time-harmonic waves propagating in a general anisotropic medium. Let us consider the components of a plane harmonic displacement wave in the form

$$\bar{u}_{\mathrm{m}} = A d_{\mathrm{m}} \exp[i\omega(x_p q_p - t)] \tag{6}$$

where ω is the (real-valued) angular frequency and the components of the

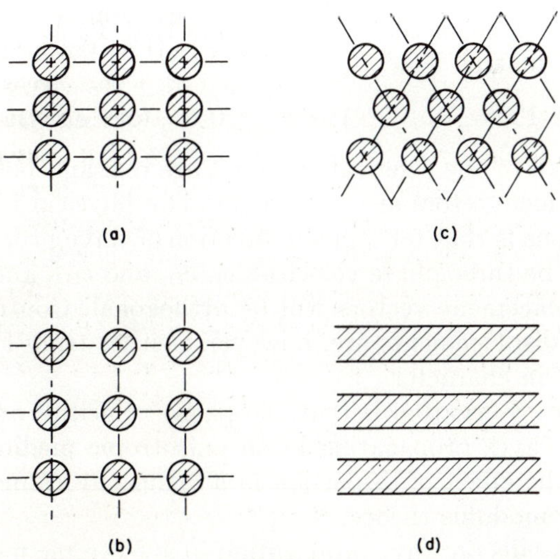

FIG. 1. Directionally reinforced composites: (a) fiber-reinforced composite with rectangular array, (b) square array, (c) hexagonal array, (d) laminated medium.

slowness vector q_j are defined as

$$q_j = p_j/c \tag{7}$$

Here p_j are the components of the vector which defines the direction of propagation and c is the phase velocity. The constant A is the amplitude, and d_m are the components of a vector defining the direction of motion. Upon substitution of (6) into the displacement equation of motion (5) we obtain

$$(C^*_{ijkl}q_jq_l - \bar{\rho}\delta_{ik})d_k = 0, \tag{8}$$

where δ_{ik} is the Kronecker delta. Equation (8) is an equation for the components d_k. For a nontrivial solution the determinant of the coefficients of d_k must vanish, which gives

$$\det | C^*_{ijkl}q_jq_l - \bar{\rho}\delta_{ik} | = 0 \tag{9}$$

If we let

$$\Gamma_{ik} = C^*_{ijkl}p_jp_l \tag{10}$$

Eq. (9) can be rewritten as

$$\det | \Gamma_{ik} - \bar{\rho}c^2\delta_{ik} | = 0 \tag{11}$$

The constants Γ_{ik} are known as the *Christoffel stiffnesses*. If the components p_j are given, Eq. (11) describes three velocity sheets in the space spanned up by p_j.

From the properties of the C^*_{ijkl} it follows that Γ_{ik} is a symmetric and positive definite matrix. That is,

$$\Gamma_{ik} = \Gamma_{ki}, \qquad \Gamma_{ik}d_id_k \geq 0 \quad \text{for} \quad \text{all } d_i$$

It follows that all of the eigenvalues of Γ_{ik} are real and positive and their corresponding eigenvectors are orthogonal. The physical interpretation of these observations is that for a given direction of wave propagation defined by p_i there will be three phase velocities c_{I}, c_{II}, and c_{III}, and the three corresponding displacement vectors will be orthogonal. Contrary to the isotropic case the displacements are, however, neither truly longitudinal nor truly transverse in character.

An important observation is that the phase velocities are constants, so that harmonic waves propagating in an anisotropic medium are not dispersive. Thus, the effect of dispersion in a composite cannot be described by the effective modulus theory.

For further details on wave propagation in anisotropic media we refer to the literature on the subject. An extensive study of the propagation of elastic waves according to the effective modulus theory in materials reinforced by inextensible fibers was recently presented by Weitsman (1972).

B. Transverse Isotropy

Let us consider the case of transverse isotropy in somewhat more detail. A transversely isotropic material is defined by five elastic constants. Choosing the x_2 axis as the axis of symmetry the effective stress–strain relations are

$$\bar{\tau}_1 = C_{11}^*\bar{\epsilon}_1 + C_{12}^*\bar{\epsilon}_2 + C_{13}^*\bar{\epsilon}_3 \tag{12}$$

$$\bar{\tau}_2 = C_{12}^*\bar{\epsilon}_1 + C_{22}^*\bar{\epsilon}_2 + C_{12}^*\bar{\epsilon}_3 \tag{13}$$

$$\bar{\tau}_3 = C_{13}^*\bar{\epsilon}_1 + C_{12}^*\bar{\epsilon}_2 + C_{11}^*\bar{\epsilon}_3 \tag{14}$$

$$\bar{\tau}_4 = C_{44}^*\bar{\epsilon}_4 \tag{15}$$

$$\bar{\tau}_5 = \tfrac{1}{2}(C_{11}^* - C_{13}^*)\bar{\epsilon}_5 \tag{16}$$

$$\bar{\tau}_6 = C_{44}^*\bar{\epsilon}_6 \tag{17}$$

Here we have used the following contracted notation which is often found more convenient, see Hearmon (1961)

$$\bar{\tau}_1 = \bar{\tau}_{11},\quad \bar{\tau}_2 = \bar{\tau}_{22},\quad \bar{\tau}_3 = \bar{\tau}_{33},\quad \bar{\tau}_4 = \bar{\tau}_{23},\quad \bar{\tau}_5 = \bar{\tau}_{13},\quad \bar{\tau}_6 = \bar{\tau}_{12}$$

$$\bar{\epsilon}_1 = \bar{\epsilon}_{11},\quad \bar{\epsilon}_2 = \bar{\epsilon}_{22},\quad \bar{\epsilon}_3 = \bar{\epsilon}_{33},\quad \bar{\epsilon}_4 = 2\bar{\epsilon}_{23},\quad \bar{\epsilon}_5 = 2\bar{\epsilon}_{13},\quad \bar{\epsilon}_6 = 2\bar{\epsilon}_{12}$$

The generalized Hooke's law then becomes

$$\bar{\tau}_q = C_{qr}\bar{\epsilon}_r \tag{18}$$

By comparison of Eqs. (1) and (18) we find

$$C_{11}^* = C_{1111}^*,\quad C_{12}^* = C_{1122}^*,\quad \ldots,\quad C_{16}^* = C_{1112}^*,\quad \text{etc.}$$

C. Laminated Medium

As a specific example of a transversely isotropic medium we will consider a laminated medium consisting of alternating plane, parallel layers of two homogeneous, isotropic, elastic materials. The elastic constants and the thicknesses of the high-modulus reinforcing layers and the low-modulus matrix layers are denoted by λ_f, μ_f, d_f and λ_m, μ_m, d_m, respectively (see Fig. 2). According to the effective modulus theory, the gross elastic behavior of the laminated medium is transversely isotropic with the x_2 axis as the axis of symmetry. The stress–strain relations can then be described by equations of the general form (12)–(17). The effective elastic constants C_{11}^*, etc. have been computed by Riznichenko (1949), Postma (1955), White and Angona (1955), Rytov (1956) and Behrens (1967), both on the basis of static and dynamic considerations. The results are

$$C_{11}^* = (1/D)[(d_f + d_m)^2(\lambda_f + 2\mu_f)(\lambda_m + 2\mu_m) + 4d_f d_m(\mu_f - \mu_m)(\lambda_f + \mu_f - \lambda_m - \mu_m)] \tag{19}$$

$$C_{12}^* = (1/D)(d_f + d_m)[\lambda_f d_f(\lambda_m + 2\mu_m) + \lambda_m d_m(\lambda_f + 2\mu_f)] \tag{20}$$

$$C_{13}^* = (1/D)[(d_f + d_m)^2\lambda_f\lambda_m + 2(\lambda_f d_f + \lambda_m d_m)(\mu_m d_f + \mu_f d_m)] \tag{21}$$

$$C_{22}^* = (1/D)[(d_f + d_m)^2(\lambda_f + 2\mu_f)(\lambda_m + 2\mu_m)] \tag{22}$$

$$C_{44}^* = (d_f + d_m)\mu_f\mu_m/(d_f\mu_m + d_m\mu_f) \tag{23}$$

where

$$D = (d_f + d_m)[d_f(\lambda_m + 2\mu_m) + d_m(\lambda_f + 2\mu_f)] \tag{24}$$

A method to determine the effective elastic constants for more general types of laminations was presented by Chou *et al.* (1972).

As shown earlier in this section it is generally rather complicated to study the propagation of plane time-harmonic waves in an anisotropic medium. For a transversely isotropic material exceptions are waves propagating in the direction of the axis of symmetry, or in a direction normal to the axis of symmetry, which are very easy to analyze. For example, if we consider a transverse wave motion defined by

$$u_2 = A \exp[ik(x_1 - ct)] \tag{25}$$

we easily obtain

$$c = (C_{44}^*/\bar{\rho})^{1/2} \tag{26}$$

where $\bar{\rho}$ is defined as

$$\bar{\rho} = \frac{d_f}{d_f + d_m}\rho_f + \frac{d_m}{d_f + d_m}\rho_m \tag{27}$$

Similarly for a longitudinal wave motion defined by

$$u_1 = A \exp[ik(x_1 - ct)] \tag{28}$$

we find

$$c = (C_{11}^*/\bar{\rho})^{1/2} \tag{29}$$

In Eqs. (26) and (29) the effective elastic constants C_{44}^* and C_{11}^* are defined by Eqs. (23) and (19), respectively.

The effective modulus theory has been used extensively to study the propagation of waves in plates of fiber-reinforced composites; see, e.g., Yang *et al.* (1966) and Chow (1971). Wave surfaces according to the effective modulus theory due to impact on plates were investigated by Moon (1972).

III. Wave Motions in a Laminated Medium—Exact Theory

We consider a laminated medium consisting of alternating plane, parallel layers of two homogeneous, isotropic materials. The Lamé elastic constants and the thicknesses of the high-modulus reinforcing layers and the low-modulus matrix layers are denoted by λ_f, μ_f, d_f, and λ_m, μ_m, d_m, respectively, as previously defined.

A. Time-Harmonic Waves in an Unbounded Medium

For two-dimensional deformations associated with plane strain in the x_3 direction, wave motions representing waves propagating in the direction of the layering (x_1 direction) can be analyzed without much difficulty using

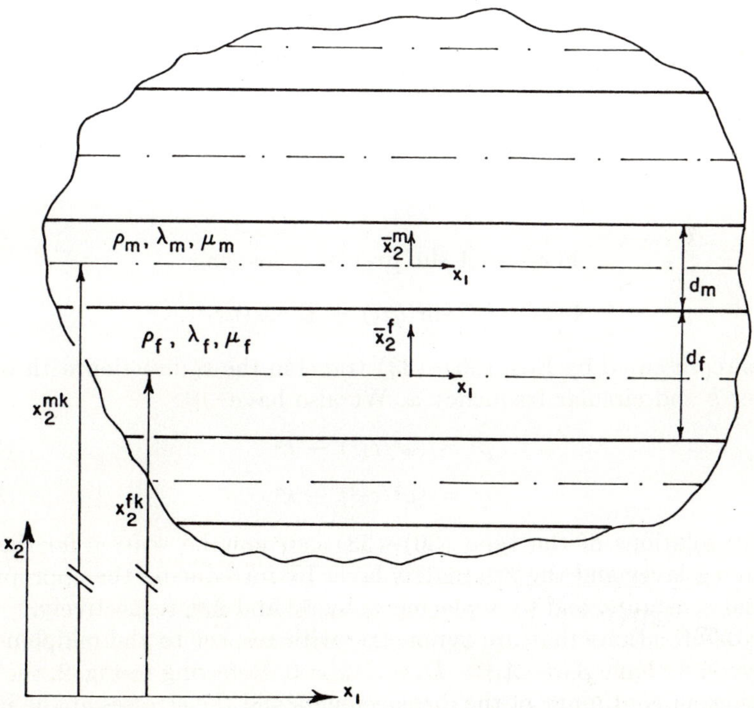

FIG. 2. Laminated composite.

the equations of the theory of elasticity for all layers. For such waves the field quantities are of the form $F(x_2) \exp i(kx_1 - \omega t)$, where the function $F(x_2)$ has the same periodicity as the layering. As a consequence, the deformations are identical in all reinforcing layers and in all matrix layers, respectively. For this relatively simple state of deformation the attention can be restricted to one pair of neighboring layers. We choose the kth pair of reinforcing and matrix layers whose midplane positions are defined by x_2^{fk} and x_2^{mk}, respectively (see Fig. 2). For these two layers we define local coordinate systems $(x_1, \bar{x}_2^{\mathrm{f}}, x_3)$ and $(x_1, \bar{x}_2^{\mathrm{m}}, x_3)$, with axes parallel to x_i, and with origins in the midplanes of the layers. In principle the problem at hand now consists of writing wave solutions of the equations of elasticity for the kth reinforcing layer and the kth matrix layer, and of ascertaining that the stresses and the displacements are continuous at the interfaces, and that the deformation pattern properly reflects the periodic structure of the medium.

Solutions of the equations of elasticity representing plane time-harmonic waves propagating in the x_1 direction, can be found in Appendix B in the form [see Eqs. (63), (68), and (69)]:

$$u_1 = \left[ik\Phi(x_2) + \frac{\partial\Psi(x_2)}{\partial x_2}\right]e^{i(kx_1-\omega t)} \tag{30}$$

$$u_2 = \left[\frac{\partial\Phi(x_2)}{\partial x_2} - ik\Psi(x_2)\right]e^{i(kx_1-\omega t)} \tag{31}$$

where

$$\Phi(x_2) = A\sin(px_2) + B\cos(px_2) \tag{32}$$

$$\Psi(x_2) = C\sin(qx_2) + D\cos(qx_2) \tag{33}$$

Waves represented by Eqs. (30)–(33) travel in the x_1 direction with wavenumber k and circular frequency ω. We also have

$$p^2 = (\omega^2/c_{\mathrm{L}}^2) - k^2 \tag{34a}$$

$$q^2 = (\omega^2/c_{\mathrm{T}}^2) - k^2 \tag{34b}$$

Wave solutions of the type (30)–(33) can now be written for the kth reinforcing layer and the kth matrix layer by introducing the appropriate material constants, and by replacing x_2 by $\bar{x}_2^{\mathrm{f}}$ and $\bar{x}_2^{\mathrm{m}}$, respectively.

For deformations that are symmetric with respect to the midplanes of the layers, we have $A_{\mathrm{f}} = A_{\mathrm{m}} = D_{\mathrm{f}} = D_{\mathrm{m}} = 0$. Referring to Fig. 2, the four conditions of continuity of the displacements and the stresses at the interface defined by $\bar{x}_2^{\mathrm{f}} = \frac{1}{2}d_{\mathrm{f}}$ and $\bar{x}_2^{\mathrm{m}} = -\frac{1}{2}d_{\mathrm{m}}$ yield four homogeneous equations for the four constants B_f, B_m, C_f, and C_m. Because it is possible to consider separately wave motions that are symmetric and antisymmetric with respect to the midplanes of the individual layers, the displacements and the stresses at $\bar{x}_2^m = \frac{1}{2}d_m$ are matched with the corresponding displacements and stresses at $\bar{x}_2^f = -\frac{1}{2}d_f$, and the required periodicity of the deformation is achieved automatically.

The four homogeneous equations for the constants B_f, B_m, C_f, and C_m yield nontrivial solutions only if the determinant of the coefficients vanishes. The latter requirement provides us with a relation between the frequency and the wavenumber, which is the frequency equation. It is a transcendental equation of the general form

$$F(\omega, k) = 0 \tag{35}$$

For symmetric (longitudinal) waves propagating in the direction of the layering, the frequency equation is stated explicitly in the paper by Achenbach and Herrmann (1968b).

For any value of the wavenumber k the frequency equation yields an infinite number of solutions for the frequency ω. A specific wave motion, called a *mode of propagation*, corresponds to each of these frequencies. The curves which display the relationships between the frequency and the wavenumber for the various modes are the branches of the frequency spectrum. The frequencies of the three lowest symmetric modes are shown in Fig. 3 for real-valued wavenumbers. It is to be expected that the frequency equation also yields solutions for wavenumbers that are complex. The corresponding modes, which are of interest in the analysis of transient wave-motions and in the discussion of motions near boundaries, have not yet been analyzed. In Fig. 3 we have plotted the dimensionless frequency Ω, and the dimensionless wavenumber ξ. These quantities are defined as

$$\Omega = \omega d_f/(\mu_m/\rho_m)^{1/2} \qquad \text{and} \qquad \xi = kd_f$$

respectively. The values of the material and the geometrical parameters were chosen as

$$\mu_f/\mu_m = 50, \qquad \rho_f/\rho_m = 3, \qquad d_f/(d_f + d_m) = 0.8$$

and

$$\nu_f = 0.3, \qquad \nu_m = 0.35$$

where μ, ρ, and ν denote a shear modulus, a mass density, and a Poisson's ratio, respectively.

It is of interest to consider some limit cases of the frequency equation. For $\mu_f = \mu_m$ and $\rho_f = \rho_m$, i.e., for a homogeneous isotropic medium, the frequency equation reduces to a simple equation which yields as solutions

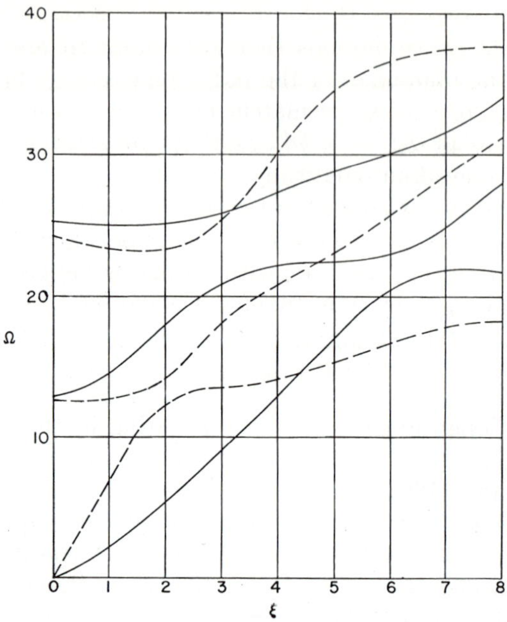

FIG. 3. Branches of the exact frequency spectrum for a laminated medium with $\mu_f/\mu_m = 50$: solid curve for antisymmetric modes and dashed curve for symmetric modes.

the well-known frequencies for longitudinal and transverse waves in an unbounded isotropic medium. In the limit of vanishing ξ, i.e., for infinitely long waves, the frequency equation splits into the product of two equations governing the frequencies of uncoupled symmetric thickness shear and symmetric thickness stretch motions, respectively. In certain ranges of the material constants the cutoff frequency of the lowest symmetric thickness shear mode is less than the cutoff frequency of the lowest symmetric thickness stretch mode; this observation has to be taken into account when constructing approximate theories. For more details on these matters we refer to the work of Achenbach and Herrmann (1968b). At vanishing wavenumber, the frequency of the lowest symmetric mode vanishes, and we can compute the slope of the $\Omega - \xi$ curve at $\xi = 0$ from the frequency equation. This slope is equal to the limiting phase velocity of the lowest mode, and we find

$$c = (C_{11}^*/\bar{\rho})^{1/2} \tag{36}$$

where C_{11}^* and $\bar{\rho}$ are defined by Eqs. (19) and (27). Thus, the phase velocity of the lowest mode at $\xi = 0$ according to the exact theory is just equal to the phase velocity of the single mode of longitudinal motion according to the effective modulus theory.

The analysis runs along very similar lines for wave motions with displacements that are antisymmetric with respect to the midplanes of the layers. The complete frequency equation for this case is presented in a paper by Achenbach (1968). The frequencies for the three lowest antisymmetric modes are also plotted in Fig. 3. In investigating the frequency equation for antisymmetric motion in the limit of vanishing ξ we find equations governing the frequencies of simple antisymmetric thickness stretch and thickness shear motions. The phase velocity at vanishing wavenumber of the lowest antisymmetric mode can be obtained as

$$c = (C_{44}^*/\bar{\rho})^{1/2} \tag{37}$$

where C_{44}^* and $\bar{\rho}$ are defined by Eqs. (23) and (27), respectively. Equation (37) is the phase velocity according to the effective modulus theory for transverse waves propagating in the direction of the layering [see Eq. (26)]. It is illustrative to plot the phase velocity of the lowest antisymmetric mode for various values of μ_f/μ_m versus the dimensionless wavenumber ξ. These phase velocities are shown in Fig. 4. As ξ increases the phase velocities show a considerable deviation from the constant values according to the effective modulus theory, especially for higher values of μ_f/μ_m.

Frequency equations for harmonic waves propagating normal to the direction of the layering can be found in the original paper by Rytov (1956), and also in the book by Brekhovskikh (1960). Plane harmonic waves propagating in an arbitrary direction were examined by Sve (1971). Some of Sve's results are plotted in Fig. 5. The frequency spectrum shown in Fig. 5 clearly reveals the different nature of the propagation of sinusoidal waves at various angles of incidence. For waves propagating normal to the layering there are frequency bands in which no waves with real-valued wavenumbers can propagate. This indicates that the laminated medium acts as a wavefilter for waves propagating normal to the layering. For propagation not normal to the layering, a wave with any frequency can propagate with a real-valued wavenumber.

In summary, the exact analysis of the propagation of time-harmonic waves in a laminated medium of the type shown in Fig. 2 reveals an infinite number of symmetric and antisymmetric modes. The effective modulus theory yields one symmetric (longitudinal) and one antisymmetric (transverse) mode, whose constant phase velocities are equal to the limit-

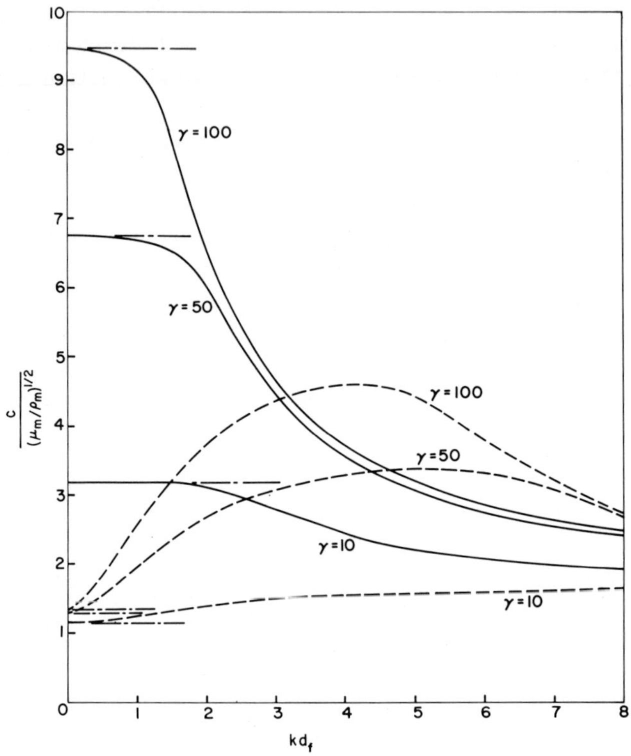

FIG. 4. Phase velocity versus the wavenumber for the lowest symmetric (solid curve) and antisymmetric (dashed curve) modes in a laminated medium with $\gamma = \mu_f/\mu_m$.

ing phase velocities at vanishing wavenumber (infinite wavelength) of the lowest symmetric and antisymmetric modes, respectively, of the exact spectrum.

B. *Time-Harmonic Waves in a Laminated Plate*

If time-harmonic waves propagate in a laminated plate, an additional characteristic length enters, and the phase velocities and the frequencies depend on the wavelength, not only because of the lamination, but also because of the overall thickness of the plate. Sun and Achenbach (1969) have investigated the relative importance of the two types of dispersion. For the lowest flexural and extensional modes the phase velocities and the frequencies were computed as functions of the wavenumber. It was also shown that corresponding results according to the effective modulus theory agree for small values of the wavenumber, for which the dispersive behavior results from the overall thickness of the plate. As the wavenumber increases (wavelength decreases) the dispersion due to the lamination becomes more pronounced, causing the phase velocity to increase as the

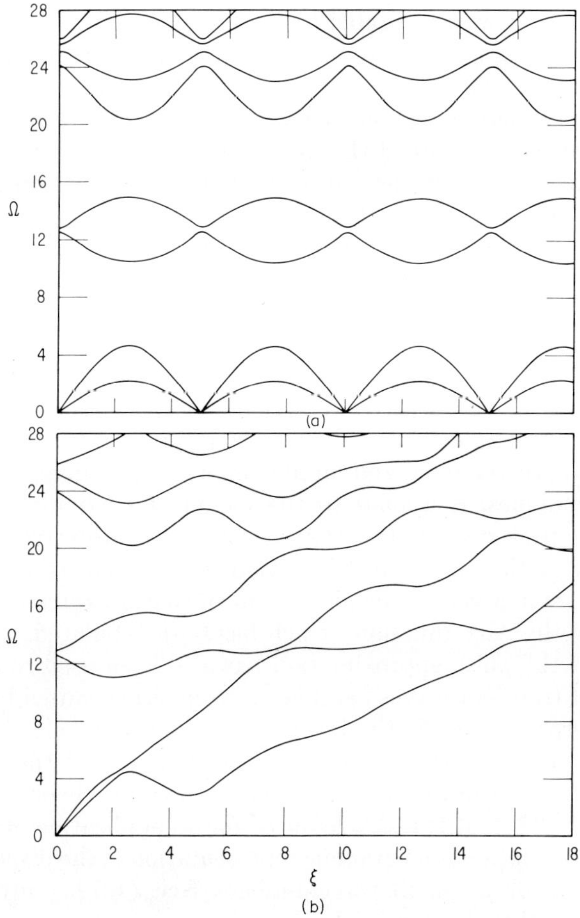

FIG. 5. Branches of the exact frequency spectrum for oblique and normal incidence; $\mu_f/\mu_m = 50$ (from Sve, 1971). (a) $\alpha = 90°$; (b) $\alpha = 75°$.

wavenumber increases. The latter effect cannot be described by the effective modulus theory.

C. Transient Waves

For problems of linear wave propagation the dynamic response to a transient disturbance can be expressed as the superposition of sinusoidal waves, by the application of Fourier integrals. An example of transient analysis of geometric dispersion in laminated media based on Fourier synthesis is the work of Peck and Gurtman (1969), who investigated the propagation of transient waves parallel to the layering in a laminated medium of the type shown in Fig. 2. They considered a semi-infinite medium bounded by the plane $x_1 = 0$. The loading was a uniform normal stress on the $x_1 = 0$ plane with step function dependence on time. The shear stress on the surface was not specified, instead the displacements parallel to the x_2 and x_3 directions were taken to be zero at $x_1 = 0$. These mixed boundary conditions are familiar for mechanical waveguide problems; since they allow the solution to be constructed from sinusoidal infinite wavetrain modes. Exact results for the strain $\partial u_1/\partial x_1$ were given as a superposition of sinusoidal modes in the form of an infinite sum of infinite Fourier integrals. Long-time, far-field, asymptotic approximations to the exact solution were generated using the saddle-point technique. Of particular importance is a contribution from the first mode which is nondecaying in time, so that it dominates the long-time solution. This contribution is known as the *head of the pulse approximation*, and unlike many approximations based on transform techniques, it applies to the region of maximum response. It shows that the peak disturbance propagates at the zero wavenumber limit phase velocity of the first mode. This velocity is given by Eq. (36). The approximation is in a very convenient form since it is expressed in terms of the integral of the Airy function, which has been tabulated.

The head of the pulse approximation has also been used to analyze the propagation of transient waves parallel to the layering caused by impulsive shear loads perpendicular to the layering. Voelker and Achenbach (1969) computed the interface shear stress, and they compared the results with the average stress according to the effective modulus theory. Results are shown in Fig. 6. All that is really required for an application of the head of the pulse approximation is a parabolic representation of the dispersion curve of the lowest mode at small wavenumbers. Sve (1972a) investigated a generalization of this observation by considering complex wavenumbers.

The Fourier synthesis approach is best suited to obtain far-field long-time results. In the near field for short times other methods should be more effective. A problem that has been investigated in some detail deals with transient longitudinal waves propagating normal to the direction of the layering in a laminated medium. The pattern of transient waves is rather complicated due to multiple reflections and refractions at the interfaces of the laminates, as well as due to reflections at the external boundaries. These repeatedly reflected and refracted stress waves may in fact combine to generate rather high tensile stresses in the interior. Some examples showing this phenomenon were given by Chou and Greiff (1968), who employed the method of characteristics. A numerical approach was also described by Chiu (1970). There are in fact several computer codes available to analyze the one-dimensional wave equation for a laminated medium. We mention the TIC and WONDY codes, which are employed by Lundergan and Drumheller (1971a). An analytical study of the reflections and transmissions of stress waves in multiple laminates was carried out by Kinslow (1970).

IV. Advanced Theories and Mathematical Approaches

The limitations of the effective modulus theory have provided a motivation to several investigators to attempt alternative approaches to the

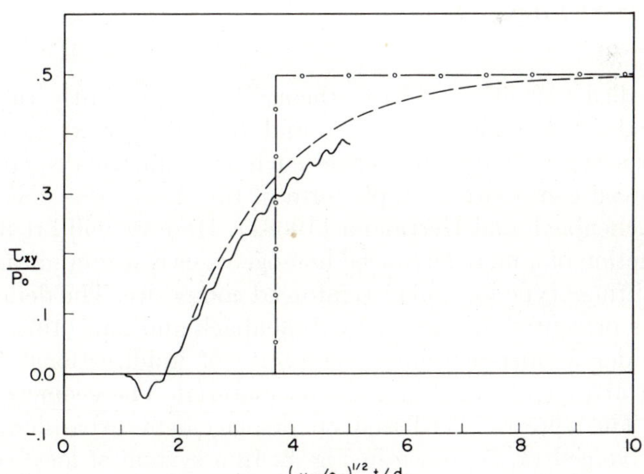

FIG. 6. Interface shear stress in a laminated medium subjected to transverse forces (from Voelker and Achenbach, 1969): (—O—), effective modulus; (—), numerical integration; (- - -) stationary phase.

analysis of the mechanical response of directionally reinforced composites, particularly for dynamic problems. In rough outline these approaches can be separated into two categories, which may be classified as:

(1) the creation of homogeneous continuum models which can account for typical effects due to the discrete structuring of the composite material. Among these we count theories of elasticity with microstructure such as the effective stiffness theory, and also mixture theories and lattice-type theories;

(2) The development of mathematical methods such as asymptotic methods and variational methods to analyze the fields of stress and strain within the actual structuring of the composite.

In this section we will briefly review these approaches.

An apparent advantage of a homogeneous continuum model is that it makes available a set of governing equations together with boundary conditions and initial conditions. As soon as the model has been developed, it can be used for bodies of the composite material of finite dimensions and arbitrary shape. On the other hand, the methods that are based on using elasticity equations for the reinforcing elements and the matrix material together with a direct application of variational methods or asymptotic expansions require a judiciously chosen set of solutions for each particular body.

A. Effective Stiffness Theory

The so-called "effective stiffness theory" was apparently the first continuum model for laminated media and fiber-reinforced composites to account for a typically dynamic effect such as geometric dispersion. For a fiber-reinforced composite a simple form of the theory was developed in a paper by Achenbach and Herrmann (1968a). Here we will briefly describe the construction of a more advanced homogeneous continuum theory of the effective stiffness type for a fiber-reinforced composite. The detailed development was presented in a paper by Achenbach and Sun (1972).

We consider a fiber-reinforced composite of unidirectional fibers in a rectangular array embedded in a matrix material. The geometry is shown in Fig. 7. The approach is based on displacement expansions within a representative cell (k, l) shown in Fig. 8. In a system of local coordinates the following displacement expansions were considered in fiber (k, l) $(r < a)$:

$$u_i^{\mathrm{f}(k,l)} = \bar{u}_i^{(k,l)} + r \cos\theta\, \psi_{2i}^{\mathrm{f}(k,l)} + r \sin\theta\, \psi_{3i}^{\mathrm{f}(k,l)} \tag{38}$$

in the matrix material of cell (k, l) $(r > a)$:

$$u_i^{\mathrm{m}(k,l)} = \bar{u}_i^{(k,l)} + a \cos\theta\, \psi_{2i}^{\mathrm{f}(k,l)} + a \sin\theta\, \psi_{3i}^{\mathrm{f}(k,l)}$$
$$+ (r-a)\cos\theta\, \psi_{2i}^{\mathrm{m}(k,l)} + (r-a)\sin\theta\, \psi_{3i}^{\mathrm{m}(k,l)} \qquad (39)$$

The physical interpretation of the terms clearly is that $\bar{u}_i^{(k,l)}$ is the "gross displacement," while $\psi_{2i}^{\mathrm{f}(k,l)}$, etc. represent "local deformations." Within the actual composite the gross displacements and the local deformations are defined at discrete values of x_2 and x_3, but they are continuous functions of x_1 and t. Since it is not possible to satisfy point by point continuity of the displacements at the interfaces between neighboring cells we impose continuity of the average displacements.

On the basis of these expansions the strain energy and the kinetic energy in the representative cell is subsequently computed. The averages over the cell next yield energy averages which are defined at the centerline of the fiber. Next a transition is achieved from the system of discrete cells to a homogeneous continuum model. The transition is accomplished by defining fields for the kinematical and dynamical variables that are continuous in

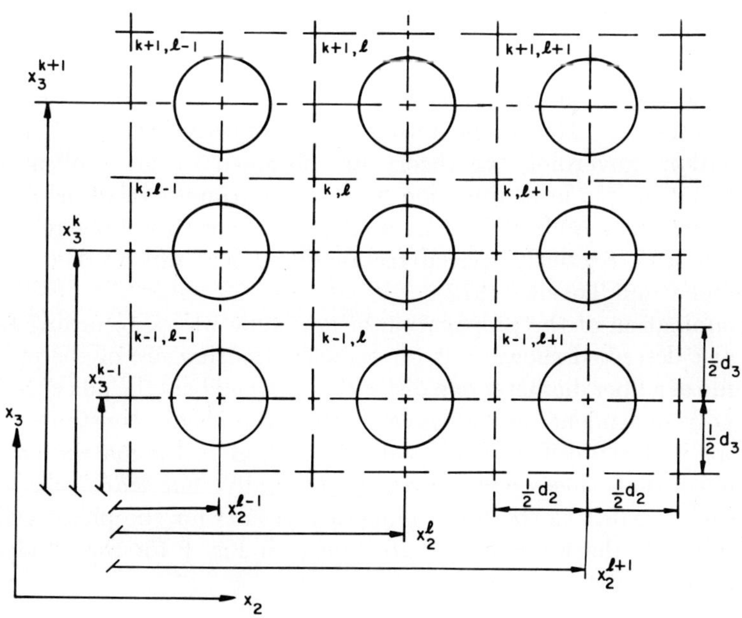

FIG. 7. Fiber-reinforced composite.

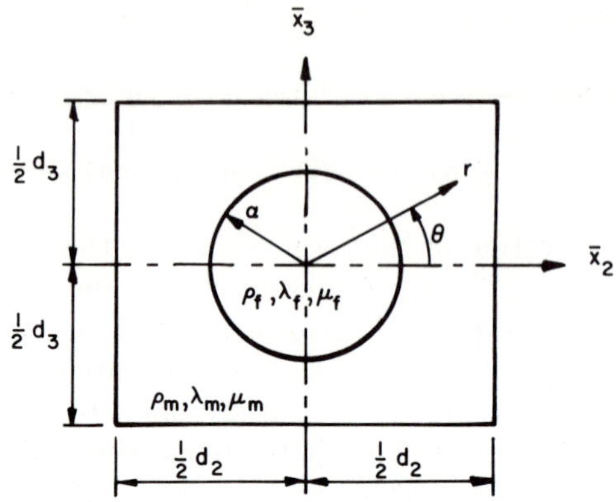

FIG. 8. Cell (k, l) of the fiber-reinforced composite.

all coordinates. Along discrete lines, which are the centerlines of the fibers, the field variables assume the same values as the variables that were defined in the discrete system of cells. The previously computed averaged kinetic and strain energies can then be interpreted as kinetic and strain energy *densities* for the medium. The strain energy density contains not only terms with effective moduli, but also terms with constants which combine the elastic and geometric properties of the constituents into "effective stiffnesses." Hence the terminology "effective stiffness" theory. The equations governing the theory are obtained by an application of Hamilton's principle in conjunction with the continuity relations and the use of Lagrangian multipliers. An analogous theory for fiber-reinforced composites with regularly spaced rectangular fibers was worked out by Bartholomew and Torvik (1972).

An examination of the propagation of harmonic waves according to the theory just described shows that for wavelengths that are of the order of magnitude of a fiber diameter or a distance between fibers the phase velocity is markedly dependent on the wavelength if the elastic constants of the reinforcing material differ substantially from those of the matrix material. This implies that pulses will be dispersed rapidly. For transverse waves propagating parallel to the fibers numerical results for the phase velocity as a function of the wavenumber are shown in Fig. 9 for three values of μ_f/μ_m, namely,

$$\mu_f/\mu_m = 100, \qquad \mu_f/\mu_m = 50, \qquad \mu_f/\mu_m = 10$$

The other material parameters were chosen as

$$\frac{\rho_{\mathrm{f}}}{\rho_{\mathrm{m}}} = 3, \qquad \nu_{\mathrm{f}} = 0.3, \qquad \nu_{\mathrm{m}} = 0.35$$

where ν_{f} and ν_{m} denote Poisson's ratios in the materials of the fibers and the matrix, respectively. The parameters defining the geometrical layout were chosen as

$$d_3/d_2 = 1 \qquad \text{and} \qquad \frac{a}{d_2} = 0.4$$

Note that for $\mu_{\mathrm{f}}/\mu_{\mathrm{m}} = 100$ and at $ka = 1$, which corresponds to a wavelength of π times the diameter of a fiber, the phase velocity is almost three times the phase velocity at $ka = 0$.

For a laminated medium a detailed exposition of the effective stiffness theory can be found in the article by Sun *et al.* (1968b). These authors employed the governing equations to determine curves relating the phase velocity to the wavenumber for waves propagating parallel to and normal to the layering, and they compare these curves with exact results. The lowest antisymmetric mode for waves propagating in the direction of the layering shows the strongest variation of the phase velocity, which is very well described by the approximate theory over a substantial range of wavenumbers. The limiting phase velocities at vanishing wavenumbers agree with those of the effective modulus theory and with the exact limits. The propagation of waves in an arbitrary direction was examined by Sve (1971) who also compared his results with exact curves. The governing equations of the effective stiffness theory were employed by Sun (1970) to study surface waves propagating along the free surface of a layered half-space. It was shown that surface waves are dispersive and that the approximate dispersion curves show good agreement with exact curves.

Further ingredients of a complete theory such as constitutive equations, boundary conditions, uniqueness, etc. were discussed by Achenbach *et al.* (1968b). The theory was refined to include second-order terms in the displacement expansions in a subsequent paper by Achenbach and Herrmann (1968b). Such a refined theory is valid for smaller values of the ratio of the characteristic lengths of the deformation and the structuring. Higher-order refinements were also discussed by Drumheller and Bedford (1973) who developed improved interface conditions and obtained more accurate dispersion curves.

The effective stiffness theory as described above is based on the construction of strain and kinetic energy densities, and the subsequent use of

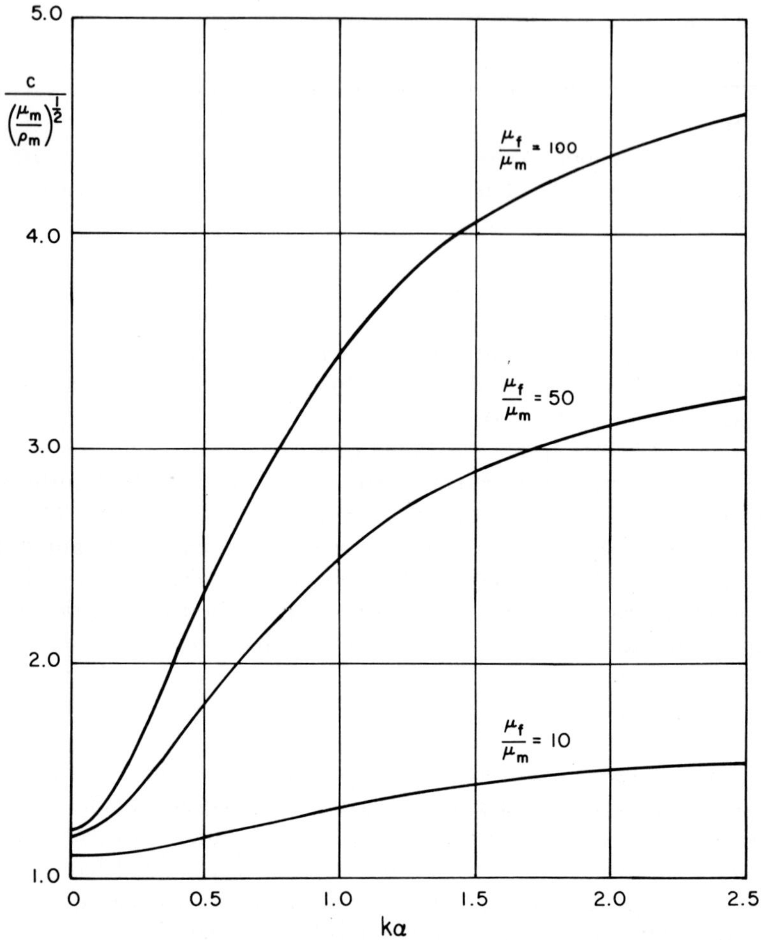

FIG. 9. Dimensionless phase velocity versus dimensionless wavenumber for transverse waves; $d_2/d_3 = 1$, $a/d_2 = 0.4$, $\rho_f/\rho_m = 3$, $\nu_f = 0.3$, $\nu_m = 0.35$.

Hamilton's principle. For constituent materials that are not perfectly elastic this approach encounters difficulties. An alternative way of deriving the equations was presented by Achenbach (1970). This alternative method has the advantage that it can be used for other than elastic behavior of the individual layers. Viscoelastic behavior and temperature effects were included in the work of Grot and Achenbach (1970b). Sve (1972b) has used the effective stiffness theory to examine thermally induced stress waves in a laminated composite. An extension to include large deformations and nonlinear material behavior was also carried out by Grot and Achenbach (1970a). A cylindrical geometry was considered by Chou and Achenbach (1971), who presented field equations governing the mechanical behavior of layered cylinders. In a recent paper Grot (1972) discussed curvilinear laminated composites of a very general type.

Applications of the effective stiffness theory to investigate wave motions in laminated plates and beams were worked out in a series of papers by Sun (1971a, b, 1972).

The various dynamical continuum theories of the effective stiffness type bear a close resemblance to theories of linear elasticity with complex microstructure which were advanced by Mindlin (1964). In the applications to directionally reinforced composites the new material constants of such theories are now directly obtained as functions of the geometrical layout of the composite and of the classical material constants of the constituent materials. This relationship was pointed out for the simplest theory of a laminated medium by Herrmann and Achenbach (1968).

B. Mixture Theories

The basic concepts of continuum theories of mixtures have been studied in considerable detail in the context of modern rational mechanics. In mixture theories two or more media exist at each point in space. While general conservation laws governing the mixture may easily be formulated, the practical application to composite materials encounters difficulties in that it is rather difficult to specify qualitatively the interaction between the constituents on the basis of information concerning the geometrical layout of the composite and the material properties of the reinforcing and matrix materials. Bedford and Stern (1971) have proposed a mixture theory for a laminated medium wherein the interaction parameters were determined on the basis of results of certain simple quasi-static problems. The novel feature of the theory of Bedford and Stern is that each constituent is permitted to have an individual motion, with mechanical coupling between the motions introduced to model constituent interactions in the actual composite. Bedford and Stern (1972) also developed a general thermomechanical theory based on this model, as well as a set of equations applicable to a restricted class of fiber-reinforced materials [see Martin *et al.* (1971)].

C. Lattice-Type Models

In a fiber-reinforced composite the fibers act as waveguides for a wave propagating in the direction of the fibers. For a wave propagating normal to that direction the fibers undergo little deformation and essentially act as obstacles interacting with each other and with the surrounding medium in a manner which is similar to the behavior of the mass particles in a lattice system. These observations have motivated kinematical assumptions regarding the deformations of the reinforcing elements and of the matrix material, which are analogous to those used in waveguides and lattice models, respectively. Thus, the fibers are considered as long and slender structural elements and the matrix is replaced by a system of springs. A three-dimensional theory of this type was worked out by Turhan (1970). A lattice model consisting of a periodic structure of laminated plates which was formed by a redistribution of masses and stiffnesses of fibers was presented by Drumheller and Sutherland (1972).

D. Continuum Theories Based on Asymptotic Expansions

Approximate theories to describe the mechanical behavior of directionally reinforced composites are based on the assumption that the ratio of the characteristic lengths of the structuring and of the deformation is much smaller than unity. In recent years asymptotic methods have been proposed which are explicitly based on the small magnitude of the above mentioned ratio. One approach, utilizing direct asymptotic expansions, has been proposed by Ben-Amoz (1971). The approach appears to be appropriate for the case that the composite material acts as a waveguide. Another technique, utilizing mixed spatial and asymptotic expansions, has been proposed by Hegemier and Nayfeh (1972), who considered the propagation of plane waves normal to the layers of a laminated composite. Truncation of the asymptotic sequences yields a hierarchy of models. To estimate the accuracy the spectrum of phase velocities was investigated. Retention of all terms in the asymptotic sequence yields the exact spectrum (which was discussed in Section III). For the lowest-order dispersive model, accuracy superior to several existing theories was observed. Treatment of a transient pulse showed good correlations with exact data. It was finally shown that the lowest-order dispersive theory can be cast in the standard form of equations for a binary mixture. In a subsequent paper Hegemier (1972) considered extensions for waves propagating parallel to the layers, as well as waves propagating parallel and normal to the fibers of unidirectionally fiber-reinforced composites. Both elastic and viscoelastic material behavior were discussed.

E. Variational Methods

Kohn *et al.* (1972) employed methods based on the Floquet and Bloch theory to study the propagation of harmonic elastic waves through composite media. This approach was also discussed in considerable detail in a paper by Lee (1972). In these methods variational principles are developed in the form of integrals over a single cell of the composite. The variational principles provide means of determining phase velocities and stress distributions in Floquet waves traveling through the composite unchanged in form from cell to cell. Both strain energy and complementary energy principles have been utilized. Because of the sharp change in the elastic constants from the high-modulus reinforcing elements to the low-modulus matrix the analysis must accomodate discontinuities in some stresses and strains across the interfaces. The Rayleigh–Ritz procedure was applied to the solution of the variational equations to calculate dispersion relations. These methods can be used to study time-harmonic waves propagating under an arbitrary angle with the structuring of the composite. In principle the structuring may be rather general and may, for example, consist of parallel fibers in a square or hexagonal array, a three-dimensional array of fibers, and an arbitrary parallelepiped cell. More than one type of inclusion can be considered and the elastic properties and density of the inclusion and matrix can vary with position, as long as they are periodic from cell to cell. At the present time actual results are mostly restricted to laminates, as discussed by Lee (1972). The Floquet wave solutions also provide a means of expressing transient motions for certain geometries. The motion caused by an impulsive pressure variation on the surface of a half-space was obtained in this way by Lee (1972) who also cites other references. Applications of the Floquet and Bloch theory in more than one dimension are briefly discussed by Peck (1972).

Variational methods have also been explored by Wheeler and Mura (1972) and Wu (1971). Instead of using the method of Rayleigh–Ritz, Wu employed, however, the Galerkin method.

Nemat-Nasser (1972a,b) has applied to the Floquet waveproblem more general variational principles than those considered by Kohn *et al.* (1972). He developed general variational principles in which the displacements, the stresses, and the strains in one case, and the displacements and the stresses in another case, are given independent variations, and which include appropriate general boundary and discontinuity conditions. Waves propagating normal to the layering in a laminated medium were analyzed in detail, and dispersion curves were presented. Numerical results showed an extremely rapid convergency to the exact results.

V. Experimental Results

Experimental work on the propagation of waves in directionally reinforced composites has been carried out by means of ultrasonic techniques and by pulse transmission measurements.

A. Ultrasonic Techniques

In principle ultrasonic techniques involve transmission of a pulsed sinusoid of several cycles with a transmitting transducer on one side of the specimen and reception of the modified signal with a receiving transducer on the other side. Results of ultrasonic tests on media consisting of carbon–phenolic laminates reinforced with layers of high-modulus filaments spaced about 0.6 mm apart, which were reported by Asay *et al.* (1968), showed a pronounced variation of the phase velocity with the frequency. Tauchert and Guzelsu (1972) investigated the dispersive behavior of plane harmonic waves in a boron–epoxy composite. They determined the dependence of the group velocity upon the frequency for longitudinal and transverse waves propagating either parallel or perpendicular to the fibers. It was found that transverse waves propagating in the direction of the fibers show a very pronounced dispersive behavior, and that the group velocity increases with the wavenumber.

Sutherland and Lingle (1972) carried out extensive investigations on a fiber-reinforced composite of tungsten wires unidirectionally embedded in an aluminum matrix. This particular composite was selected because (1) it closely simulates several metal–metal composite materials that have engineering applications, (2) both constituents are believed to be linearly elastic in the very low stress range of ultrasonic experiments, and (3) their acoustic impedances differ significantly. The linear elastic behavior of the constituents implies that the dispersion observed in the composite is associated only with the internal geometric array of the fibers, and not with viscoelastic material behavior, as can be suspected in composites with an epoxy matrix. The composite chosen for the experiments was manufactured in two constituent ratios, 2.2 and 22.1% by volume of tungsten, respectively. The dispersive characteristics of these composites were determined for harmonic waves propagating normal to the axis of the fibers by using water-bath techniques with wide-band transducers. The dispersion data obtained demonstrate that for propagation normal to the axis of the fibers, fiber-reinforced composites behave as wavefilters which selectively transmit or reflect periodic waves.

B. Shock-Tube Tests

Whittier and Peck (1969) tested a laminate of carbon–phenolic layers reinforced with layers of high-modulus filaments in a shock tube. A step pressure rise was applied to one side of the specimen by reflection of a gas dynamic shock wave. The spatial average of the free surface velocity over the cross section of the composite was measured with a capacitance gage. The experimental results showed good agreement with corresponding analytical results obtained by Peck and Gurtman (1969). The shock-tube test was found to be an excellent method to characterize the dispersivity of a composite. The stress levels are so low, e.g., 70 psi, that the complications of nonlinearity are certainly avoided.

C. Flyer-Plate Impact Tests

Lundergan and Drumheller (1971a,b,c, 1972) have investigated the dispersion of dilatational stress waves in a laminated composite, both experimentally and analytically. The composite consisted of a number of bilaminar plates, and the propagation of the wave was normal to the plates. A flat flyer plate accelerated by a compressed gas gun was used to induce a rectangular stress pulse. The particle velocity was measured at the opposite end of the composite by an optical interferometer. The experimental results were compared with results obtained by means of computer programs in which the layers are explicitly represented, and which include nonlinear effects. In the first set of experiments (1971a), layers of epoxy and steel were impacted at stress levels on the order of a few kilobars. It was found that the peak amplitude of the transmitted stress wave decreased directly with the width of the input stress pulse. The bulk of the reduction of the stress was attributed to the reflections of the stress wave at the extreme left and right boundaries of the composite. Debonding of the bilaminar plates and dissipation of energy in the matrix material was also thought to have contributed to the stress reduction. Comparisons of the experimental results with numerical results provided an indication of the necessity of considering the nonlinear behavior of the materials.

In subsequent experiments (1971b) laminates of PMMA and stainless steel were used. By varying the number of laminates it was concluded that there was no change in rise time for propagation distances beyond a few unit cells near the side of load application. Another set of experiments (1971c) was carried out on the epoxy–steel laminates with the intention of investigating the response at slightly higher stresses, where the effect of fracture would be more apparent. Subsequently Lundergan and Drumheller (1972) investigated the response of obliquely laminated composites, both experimentally and analytically. Several analytical models and a two-dimensional wave propagation program were used to predict the transmitted waveform. Comparisons were made between the various models and the experimental results. General agreement exists between the predictions of the various models. The experimentally determined first signal velocities and final wave amplitudes agree with the calculations of the models; however, the remaining portion of the experimentally determined stress waves exhibited slower rise times than did those of the calculated waves. It was speculated that the physical conditions existing at the input boundary of the composite are not being adequately incorporated into the models.

VI. Concluding Remarks

This survey of waves and vibrations in directionally reinforced composites was necessarily brief, and the list of references is undoubtedly far from complete. Several important and interesting topics were not at all discussed. Among these we mention dynamic effects in disordered composites, failure mechanisms under dynamic loading conditions, such as fracture of fibers and delamination, optimization of the structuring, and of course the nonlinear dynamic response of directionally reinforced composites. Analytical and experimental work on these topics is available, but this work is mostly of an exploratory nature. Brief discussions of some of these contributions are included in the review articles by Gurtman (1969) and Peck (1971, 1972). It will, however, be of interest to conclude this chapter with a few brief remarks and the listing of some pertinent references on randomness, fracture, optimization and nonlinearity.

The dynamic response of disordered composites was discussed by McCoy (1972), who obtained a formalism that is to be satisfied by the mean (i.e., statistically averaged) field quantities in a statistical sample of heterogeneous, linearly elastic solids. Inertia effects were included in the analysis. Krumhansl (1972) also recently examined randomness and wave propagation in inhomogeneous media. The propagation of waves in randomly layered media has been analyzed by Ziegler (1969).

Fracture under dynamic loading conditions is, of course, a matter of the greatest interest and importance for the experimental work on the propagation of pulses. Failure of the interface bonds is often observed, e.g., Lundergan and Drumheller (1971a,b, 1972). Tensile failure at the interfaces in a composite body was also investigated analytically and experimentally by Achenbach *et al.* (1968a). Spallation of laminated quartz–phenolic composites was investigated by Cohen and Berkowitz (1972), who carried out thin-flyer tests using 5-mil and 15-mil Mylar flyers impacting 0.25-in.-thick composites. They found that spall occurred by delamination with secondary cracks perpendicular to the impact faces.

The study of the stability of flaws at the interfaces of the reinforcing elements and the matrix is also of considerable interest. Generally it is not feasible to manufacture composite bodies so that interface flaws never occur in factory-new composites, or after a period of service. A realistic philosophy is, therefore, to make sure that adequate strength remains if interfaces contain flaws that are smaller than flaws that can readily be found by inspection.

An interface flaw in a stressed composite body gives rise to a significant increase of the local stresses, particularly in the vicinity of the edge of the flaw. If the local stresses reach a sufficiently high level, the flaw becomes unstable and may grow into a sizable debonding surface. It has been found in the analysis of elastodynamic problems that the dynamic stress concentrations are higher than the stress concentrations computed for the corresponding problem of static equilibrium. In view of the dynamic overshoot of the stress concentrations it is possible that extension of an interface flaw may or may not occur, depending on whether the loads are suddenly or gradually applied. A recent study by Brock and Achenbach (1972) has been concerned with the extension of a flaw along the bonded interface of two elastic solids of different elastic constants and mass densities. The extension is caused by stress concentrations that are generated when a system of transient horizontally polarized shear waves strikes the tip of the flaw. It is assumed that rupture of the adhesive bond is preceded by yielding of the adhesive. For various values of the parameters defining the materials and the system of waves, the speed of the leading edge of the zone of yielding was computed. After a critical amount of yielding has taken place the bond ruptures at the trailing edge of the zone of yielding.

Some of the undesirable dynamic effects such as stress amplifications by reflections and refractions can be reduced by proper design of the composite. Investigations of this type fall within the area of optimization. It should of course be realized that the dynamic behavior is only one consideration, and possibly a minor one, in determining the composition of a composite. Some results are available for laminated media. The level of the dynamic stresses in a laminated medium is very sensitive to parameters such as the number of layers, the layer thicknesses, and the material properties. If some of these parameters are prescribed, it is possible to select the remaining parameters to render the tensile stresses as small as possible at certain specified locations and under certain specified loads. Lai and

Achenbach (1972) employed a direct search algorithm for that purpose. Consideration to the optimum design of layered media was also given by Anfinsen (1967) and Hutchinson (1969).

The nonlinear response of directionally reinforced composites requires much further investigation. Deviations from linear behavior can arise because of several mechanisms, among others: large deformations, nonlinear elastic material behavior, plasticity, crushability, and rate-dependent material behavior. Some of the representative theoretical work has centered on the propagation of shocks and the development of Hugoniot relations, e.g., Tsou and Chou (1969) and Torvik (1970). These analyses provide useful insight on the range of response that may be expected under compressive shock loads. Further references to analytical and numerical work can be found in the review papers by Peck (1971, 1972).

For small deformations and for linear behavior of the constitutive materials, an attempt was made to present in this chapter a unified account of theoretical and experimental work on waves and vibrations in directionally reinforced composites. In the recent past good progress has been achieved in the understanding of the peculiarities of the linear dynamic response of composite materials. Indeed, the linear theory is approaching a mature stage with the availability of exact results for laminated media, and several satisfactory approximate theories for both laminated media and fiber-reinforced composites. This body of theoretical knowledge is supplemented by experimental verifications on the propagation of both sinusoidal waves and pulses. It should, however, be noted that there is still a substantial need for further experimental work. Additional efforts are also required for problems of transient dynamic response, particularly with regard to the computation of local field variables such as interface stresses and dynamic stress concentrations.

Appendix A: On the Propagation of Mechanical Disturbances

Wave propagation is a classical subject which has been discussed in great generality in several treatises. A detailed discussion of the propagation of mechanical disturbances, in particularly in elastic solids, can be found in a recent book by Achenbach (1973). For the purpose of reference some of the basic concepts of the propagation of waves are briefly summarized in this appendix.

Mechanical waves originate in the forced motion of a portion of a deformable medium. As elements of the medium are deformed the disturbance is transmitted from one point to the next and the disturbance, or wave, progresses through the medium. In this process the resistance offered to deformation by the consistency of the medium, as well as the resistance to motion offered by inertia, must be overcome. As the disturbance propagates it carries along amounts of energy in the forms of kinetic and potential energies. The transmission of energy is effected because motion is passed on from one particle to the next and not by any sustained bulk motion of the entire medium. Mechanical waves are characterized by the transport of energy through motions of particles about an equilibrium position.

Deformability and inertia are essential properties of a medium for the transmission of mechanical wave motions. If the medium were not deformable, any part of the medium would immediately experience a disturbance in the form of an internal force or an acceleration upon application of a localized excitation. Similarly, if a hypothetical medium were without inertia there would be no delay in the displacement of particles and the transmission of the disturbance from particle to particle would be effected instantaneously to the most distant particle. Indeed, it can be shown analytically that the velocity of propagation of a mechanical disturbance always assumes the form of the square root of the ratio of a parameter defining the resistance to deformation and a parameter defining the inertia of the medium. All real materials are of course deformable and possess mass and thus all real materials transmit mechanical waves.

In mathematical terms the simplest traveling wave in one dimension is defined by an expression of the type $f = f(x - ct)$, where f as a function of the spatial coordinate x and the time t represents a disturbance in the values of some physical quantity. For mechanical waves f generally denotes a displacement, a particle velocity of a stress component. The function $f(x - ct)$ is called a *simple wavefunction* and the argument $x - ct$ is the *phase of the wavefunction*. If t is increased by any value, say Δt, and simultaneously x is increased by $c\,\Delta t$, the value of $f(x - ct)$ is clearly not altered. The function $f(x - ct)$ thus represents a disturbance advancing in the positive x direction with a velocity c. The velocity c is termed the *phase velocity*. The propagating disturbance represented by $f(x - ct)$ is a special wave in that the shape of the disturbance is unaltered as it propagates through the medium.

A. Progressive Harmonic Waves

A progressive or traveling *harmonic* wave is represented by an expression of the form

$$u(x, t) = A \cos[k(x - ct)] \tag{40}$$

where the amplitude A is independent of x and t. Equation (40) is of the general form $f(x - ct)$ and thus clearly represents a traveling wave. The argument $k(x - ct)$ is called the *phase of the wave*; points of constant phase are propagated with the phase velocity c. At any instant t, $u(x, t)$ is a periodic function of x with wavelength Λ, where $\Lambda = 2\pi/k$. The quantity $k = 2\pi/\Lambda$, which counts the number of wavelengths over 2π, is termed the *wavenumber*. At any position the displacement $u(x, t)$ is time-harmonic with time period T, where $T = 2\pi/\omega$. The circular frequency ω follows from Eq. (40) as

$$\omega = kc \tag{41}$$

It follows that an alternative representation of $u(x, t)$ is

$$u(x, t) = A \cos\left[\omega\left(\frac{x}{c} - t\right)\right] \tag{42}$$

Equations (40) and (42) represent trains of sinusoidal waves, which disturb at any instant of time the complete (unbounded) extent of the medium. Harmonic waves are steady-state waves, as opposed to transient waves (pulses).

If the phase velocity is independent of k, very short waves propagate with the same phase velocity as long waves. If the phase velocity does not depend on the wavelength, we say that the system is *nondispersive*. If the material is not purely elastic but displays dissipative behavior, it is found that the phase velocity of harmonic waves depends on the wavelength, and the system is said to be *dispersive*. Dispersion is an important phenomenon because it governs the change of shape of a pulse as it propagates through a dispersive medium. Dispersion occurs not only in inelastic bodies but also in elastic waveguides, as will be discussed in Appendix B.

For mathematical convenience we generally use, instead of Eq. (40), the expression

$$u(x, t) = A \exp[ik(x - ct)] \tag{43}$$

where $i = \sqrt{(-1)}$. Without stating it explicitly, henceforth it is understood that the real part of (43) is to be taken for the physical interpretation of the solution.

B. Standing Waves

Let us consider two displacement waves of the same frequency and wavelength but traveling in opposite directions. Since the wave equation is linear the resultant displacement is

$$u(x, t) = A_+ e^{i(kx - \omega t + \gamma_+)} + A_- e^{i(kx + \omega t + \gamma_-)} \tag{44}$$

where A_+ and A_- are real-valued amplitudes and γ_+ and γ_- are phase angles. If the amplitudes of the two simple harmonic waves are equal, $A_+ = A_- = A$, we can write

$$u(x, t) = 2A \exp[i(kx + \tfrac{1}{2}\gamma_+ + \tfrac{1}{2}\gamma_-)] \cos(\omega t - \tfrac{1}{2}\gamma_+ + \tfrac{1}{2}\gamma_-)$$

The real part of this expression is

$$u(x, t) = 2A \cos(kx + \tfrac{1}{2}\gamma_+ + \tfrac{1}{2}\gamma_-) \cos(\omega t - \tfrac{1}{2}\gamma_+ + \tfrac{1}{2}\gamma_-) \quad (45)$$

Equation (45) represents a *standing wave*, since the shape of the wave does not move. At points where $\cos(kx + \frac{1}{2}\gamma_+ + \frac{1}{2}\gamma_-) = 0$, the two traveling waves always cancel each other and the medium is at rest. These points are called the *nodal points*. Halfway between each pair of nodal points are the antinodes, where the motion has the largest amplitude.

C. Modes of Free Vibration

Standing waves of certain wavelengths form the modes of free vibrations of an elastic body with external boundaries. If we consider, for example, a semi-infinite medium and if we require that the displacement vanishes at $x = 0$, the possible harmonic motions are subject to restrictions. Equation (45) can be used with the angles γ_+ and γ_- chosen so that a nodal point coincides with the boundary $x = 0$, i.e.,

$$\gamma_+ + \gamma_- = \pi$$

and thus

$$u(x, t) = 2A \sin(kx) \sin(\omega t - \gamma_+) \quad (46)$$

When, as a second boundary condition we add $u = 0$ at $x = l$, the harmonic motion is still further limited, for now of all the harmonic motions represented by (46) only those which have a nodal point at $x = l$ can be used. Thus we require

$$\sin(kl) = 0$$

which implies

$$kl = 2\pi l/\Lambda = n\pi, \qquad n = 1, 2, 3, 4, \ldots$$

The distance between nodal points is half the wavelength and thus this distance must be l, $l/2$, $l/3$, etc. The corresponding circular frequencies are

$$\omega = kc = n\pi c/l$$

The circular frequency of the lowest or *fundamental* mode, which is called the *fundamental frequency*, is $\omega_0 = \pi c/l$ rad/sec; in cycles per second the fundamental frequency is $f_0 = c/2l$. The frequencies of the higher modes are in cycles per second, $f_2 = 2c/2l$, $f_3 = 3c/2l$, etc. The higher frequencies are called *overtones*. For the example of the one-dimensional medium with rigidly supported ends the overtones are integral multiples of the fundamental frequency. Overtones with this simple relation to the fundamental are called *harmonics*. Only for the simplest vibrating systems governed by the wave equation are the modes of vibration as elementary as discussed in this section.

D. Propagation of Pulses

As a consequence of the linearity of the wave propagation problems that are under discussion, it is allowable to express the total response to a number of separate excitations as the superposition of the individual responses. Linear superposition, in conjunction with integral representations of forcing functions, provide us with the means of determining solutions to problems of transient elastic wave propagation.

If Eq. (43) satisfies the equation(s) governing wave propagation in the medium for the general case that the phase velocity c depends on the wavenumber (and thus on the frequency by virture of $\omega = kc$), it is evident that

$$u(x, t) = \int_{-\infty}^{\infty} A^*(\omega) e^{i\omega[x/c(\omega)-t]}\, d\omega \tag{47}$$

also satisfies the governing equations. Equation (47) can, however, be interpreted as the solution to a boundary value problem for a semi-infinite medium $x \geq 0$ with boundary condition at $x = 0$

$$u(0, t) = A(t) = \int_{-\infty}^{\infty} A^*(\omega) e^{-i\omega t}\, d\omega \tag{48}$$

If $A(t)$ is known, its "exponential Fourier transform" $A^*(\omega)$ can be obtained from the well-known Fourier integral theorem

$$A(t) = (1/2\pi) \int_{-\infty}^{\infty} e^{-i\omega t}\, d\omega \int_{-\infty}^{\infty} A(s) e^{i\omega s}\, ds \tag{49}$$

i.e.,

$$A^*(\omega) = (1/2\pi) \int_{-\infty}^{\infty} A(s) e^{i\omega s}\, ds \tag{50}$$

Appendix B: Some Results on the Dynamics of Linearly Elastic Solids

For a detailed account of the linearized theory of elasticity we refer to the book by Sokolnikoff (1956). The basic equations are, however, briefly summarized in this appendix for the purpose of reference. Some pertinent results on the vibrations and wave motions in homogeneous, isotropic, linearly elastic solids are discussed in more detail.

Let x_i be a rectangular Cartesian coordinate system with $i = 1, 2, 3$, and let u_i be the displacement vector field. In indicial notation the elastic strain tensor is then defined by

$$\epsilon_{kl} = \tfrac{1}{2}(u_{k,l} + u_{l,k}) \tag{51}$$

where

$$u_{k,l} = \partial u_k / \partial x_l \tag{52}$$

For a homogeneous, isotropic, linearly elastic medium the stress tensor τ_{ij}, where $\tau_{ij} = \tau_{ji}$, is related to the strain tensor by Hooke's law

$$\tau_{ij} = \lambda \epsilon_{kk} \delta_{ij} + 2\mu \epsilon_{ij} \tag{53}$$

where δ_{ij} is the Kronecker delta. Equation (53) contains two elastic constants λ and μ, which are known as *Lamé's elastic constants*. The summation convention should be invoked whenever indices are repeated in the same term. Equations (51)–(53) are supplemented by the equations of motion:

$$\tau_{ij,j} = \rho \ddot{u}_i \tag{54}$$

where a dot denotes a derivative with respect to time. If the strain–displacement relations are substituted into Hooke's law, and the expressions for the stresses are subsequently substituted in the stress equations of motion, we obtain the displacement equations of motion

$$\mu u_{i,jj} + (\lambda + \mu) u_{j,ji} = \rho \ddot{u}_i \tag{55}$$

A. Plane Time-Harmonic Waves

As a simple extension of Eq. (43), a plane harmonic wave propagating in an arbitrary direction in an unbounded medium is represented by

$$u_i = A d_i \exp[ik(p_i x_i - ct)] \tag{56}$$

In Eq. (56), d_i and p_i are the components of unit vectors in the directions of motion and of propagation, respectively. The components of the position vector are denoted by x_i, and $p_i x_i$ = constant represents a plane normal to the unit vector with components p_i. Equation (56) thus represents a plane wave propagating in the direction of the propagation vector. By substituting Eq. (56) into the displacement equations of motion (55) we obtain after some manipulation

$$(\mu - \rho c^2) d_i + (\lambda + \mu)(p_j d_j) p_i = 0 \tag{57}$$

Equation (57) is an equation for the phase velocity c. It is noted that the wavenumber k does not appear in (57), i.e., plane harmonic waves in an unbounded homogeneous isotropic linearly elastic medium are not dispersive.

Since p_i and d_i represent two different unit vectors, Eq. (57) can be satisfied in two ways only:

(a) $p_i = \pm d_i$. Consequently, $p_j d_j = \pm 1$, and Eq. (57) yields

$$c_L^2 = (\lambda + 2\mu)/\rho \tag{58}$$

In this case the motion is parallel to the direction of propagation, and the wave is therefore called a *longitudinal wave*.

(b) If $p_i \neq d_i$, both terms in (57) have to vanish independently, yielding

$$c_T^2 = \mu/\rho \qquad \text{and} \qquad p_j d_j = 0 \tag{59}$$

Now the motion is normal to the direction of propagation, and the wave is called a *transverse wave*.

B. Displacement Potentials

Let us consider a representation of the displacement components in the form

$$u_i = \varphi_{,i} + e_{ijk} \psi_{k,j} \tag{60}$$

where e_{ijk} is the alternating tensor. Substitution of (60) into the displacement equation of motion yields that (55) is satisfied by (60) provided that

$$\varphi_{,ii} = (1/c_L^2)\ddot{\varphi} \tag{61a}$$

and

$$\psi_{k,jj} = (1/c_T^2)\ddot{\psi}_k \tag{61b}$$

where c_L and c_T are given by Eqs. (58) and (59), respectively. Equations (61a) and (61b) are uncoupled wave equations.

Although the scalar potential φ and the vector potential with components ψ_i are generally coupled through the boundary conditions, which still causes substantial mathematical complications, the use of the displacement decomposition generally simplifies the analysis. To determine the solution of a boundary initial value problem one may simply select appropriate particular solutions of Eqs. (61a) and (61b) in terms of arbitrary functions or integrals over arbitrary functions. If these functions can subsequently be chosen so that the boundary conditions and the initial conditions are satisfied, then the solution to the problem has been found. The solution is unique by virtue of the uniqueness theorem of dynamic elasticity.

C. *Waves in Plane Strain in an Elastic Layer*

For motion in plane strain in the x_1x_2 plane we have

$$u_3 \equiv 0, \qquad \frac{\partial}{\partial x_3}(\quad) \equiv 0 \tag{62}$$

Equation (60) then reduces to

$$u_1 = \frac{\partial \varphi}{\partial x_1} + \frac{\partial \psi}{\partial x_2} \tag{63a}$$

$$u_2 = \frac{\partial \varphi}{\partial x_2} - \frac{\partial \psi}{\partial x_1} \tag{63b}$$

For simplicity of notation the subscript 3 has been omited from ψ in (63a,b). The relevant components of the stress tensor follow from Hooke's law as

$$\tau_{21} = \mu\left(2\frac{\partial^2 \varphi}{\partial x_1\,\partial x_2} - \frac{\partial^2 \psi}{\partial x_1^2} + \frac{\partial^2 \psi}{\partial x_2^2}\right) \tag{64}$$

$$\tau_{22} = \lambda\left(\frac{\partial^2 \varphi}{\partial x_1^2} + \frac{\partial^2 \varphi}{\partial x_2^2}\right) + 2\mu\left(\frac{\partial^2 \varphi}{\partial x_2^2} - \frac{\partial^2 \psi}{\partial x_1\,\partial x_2}\right) \tag{65}$$

while the potentials φ and ψ satisfy wave equations, which for plane strain are two-dimensional,

$$\frac{\partial^2 \varphi}{\partial x_1^2} + \frac{\partial^2 \varphi}{\partial x_2^2} = \frac{1}{c_L^2}\frac{\partial^2 \varphi}{\partial t^2} \tag{66}$$

$$\frac{\partial^2 \psi}{\partial x_1^2} + \frac{\partial^2 \psi}{\partial x_2^2} = \frac{1}{c_T^2}\frac{\partial^2 \psi}{\partial t^2} \tag{67}$$

To investigate wave motion in the elastic layer we consider solutions of

(66) and (67) of the forms

$$\varphi = \Phi(x_2) \exp[i(kx_1 - \omega t)] \tag{68}$$

$$\psi = \Psi(x_2) \exp[i(kx_1 - \omega t)] \tag{69}$$

These expressions represent standing waves across the cross section and progressive waves in the direction of the layer. Substituting (68) and (69) into (66) and (67), respectively, the solutions of the resulting equations are obtained as

$$\Phi(x_2) = A \sin(px_2) + B \cos(px_2) \tag{70}$$

$$\Psi(x_2) = C \sin(qx_2) + D \cos(qx_2) \tag{71}$$

where

$$p^2 = (\omega^2/c_L{}^2) - k^2 \tag{72a}$$

$$q^2 = (\omega^2/c_T{}^2) - k^2 \tag{72b}$$

By virtue of Eqs. (63a,b) and Eqs. (70) and (71) the displacement components can be written in terms of elementary functions. For the displacement in the x_1 direction the motion is symmetric (antisymmetric) with regard to $x_2 = 0$, if u_1 contains cosines (sines). The displacement in the x_2 direction is symmetric (antisymmetric) if u_2 contains sines (cosines). The modes of wave propagation in the elastic layer may thus be split up into two systems of symmetric and antisymmetric modes, respectively.

The expressions for the displacements and the stresses must satisfy boundary conditions on the surfaces $x_2 = \pm h$. If the boundaries are free, we have at $x_2 = \pm h$:

$$\tau_{21} = \tau_{22} = 0 \tag{73}$$

These conditions yield the frequency equation, i.e., the equation relating the frequency ω to the wavenumber k. For the symmetric modes we obtain

$$\tan(qh)/\tan(ph) = -4k^2pq/(q^2 - k^2)^2 \tag{74}$$

For the antisymmetric modes the boundary conditions yield

$$\tan(qh)/\tan(ph) = -(q^2 - k^2)^2/4k^2pq \tag{75}$$

Equations (74) and (75) are the well-known Rayleigh–Lamb frequency equations. These transcendental equations look deceptively simple. Although the frequency equations were derived at the end of the 19th century it was not until quite recently that the frequency spectrum was unraveled in complete detail by Mindlin (1960). For a detailed discussion we refer to the book by Achenbach (1973). For an arbitrarily specified value of the wavenumber k, Eqs. (74) and (75) yield an infinite number of solutions for the frequency ω. A specific wave motion of the layer, called a *mode of propagation*, corresponds to each solution satisfying (74) or (75). Thus, the frequency equation yields an infinite number of continuous curves, called *branches*, which display the relationship between the frequency ω and the wavenumber k for a particular mode of wave propagation. The collection of branches constitutes the frequency spectrum.

Acknowledgment

This work was supported by the Office of Naval Research under Contract ONR 064-483 with Northwestern University.

References

Achenbach, J. D. (1968). *J. Acoust. Soc. Amer.* **43,** 1451.

Achenbach, J. D. (1970). The Layered Medium as a Homogeneous Continuum with Microstructure. Tech. Rep. AFML-TR-70-275, Air Force Mater. Lab., Wright Patterson AFB, Ohio.

Achenbach, J. D. (1973). "Wave Propagation in Elastic Solids." Amer. Elsevier, New York.

Achenbach, J. D., and Herrmann, G. (1968a). *AIAA J.* **6,** 1832.

Achenbach, J. D., and Herrmann, G. (1968b). *In* "Dynamics of Structured Solids" (G. Herrmann, ed.), pp. 23–46. Amer. Soc. Mech. Eng., New York.

Achenbach, J. D., and Sun, C. T. (1972). *In* "Dynamics of Composite Materials" (E. H. Lee, ed.), pp. 48–69. Amer. Soc. Mech. Eng., New York.

Achenbach, J. D., Hemann, J. H., and Ziegler, F. (1968a). *AIAA J.* **6,** 2040.

Achenbach, J. D., Sun, C. T., and Herrmann, G. (1968b). *J. Appl. Mech.* **35,** 689.

Anfinsen, L. E. (1967). *J. Appl. Mech.* **34,** 751.

Asay, J. R., Urzendowski, S. R., and Guenther, A. H. (1968). Ultrasonic and Thermal Studies of Selected Plastics, Laminated Materials and Metals. TR No. AFWL-TR-67-91, Air Force Weapons Lab., Kirtland AFB, New Mexico.

Bartholomew, R. A., and Torvik, P. J. (1972). *Int. J. Solids Struct.* **8,** 1389.

Bedford, A., and Stern, M. (1971). *J. Appl. Mech.* **38,** 8.

Bedford, A., and Stern, M. (1972). *Acta Mech.* **15,** 21.

Behrens, E. (1967). *J. Acoust. Soc. Amer.* **42,** 378.

Ben-Amoz, M. (1971). *J. Appl. Phys.* **24,** 5422.

Brekhovskikh, L. M. (1960). "Waves in Layered Media." Academic Press, New York.

Brock, L. M., and Achenbach, J. D. (1973). *Int. J. Solids Struct.* **9,** 53.

Chiu, S. S. (1970). *J. Comput. Phys.* **6,** 17.

Chou, F. H., and Achenbach, J. D. (1970). *AIAA J.* **8,** 1444.

Chou, P. C., Carleone, J., and Hsu, C. M. (1972). *J. Compos. Mater.* **6,** 80.

Chou, S. C., and Greif, R. (1968). *AIAA J.* **6,** 1067.

Chow, T. S. (1971). *J. Compos. Mater.* **5,** 306.

Cohen, L. J., and Berkowitz, H. M. (1972). *J. Franklin Inst.* **294.**

Drumheller, D. S., and Bedford, A. (1973). *J. Appl. Mech.* **40,** 527.
Drumheller, D. S., and Sutherland, H. J. (1973). *J. Appl. Mech.* **40,** 149.
Grot, R. A., *Int. J. Solids Struct.* **8,** 439.
Grot, R. A., and Achenbach, J. D. (1970a). *Int. J. Solids Struct.* **6,** 641.
Grot, R. A., and Achenbach, J. D. (1970b). *Acta Mech.* **9,** 245.
Gurtman, G. A. (1969). Review of Techniques Used to Analyze Stress Wave Propagation in Composites. Systems, Science and Software, La Jolla., California.
Hashin, Z. (1972). Theory of Fiber-Reinforced Materials. NASA CR-1974.
Hearmon, R. F. S. (1961). "Applied Anisotropic Elasticity." Oxford Univ. Press, London and New York.
Hegemier, G. A. (1972). *In* "Dynamics of Composite Materials" (E. H. Lee, ed.), pp. 70-121. Amer. Soc. Mech. Eng., New York.
Hegemier, G. A., and Nayfeh, A. H. (1973). *J. Appl. Mech.* **40,** 503.
Herrmann, G., and Achenbach, J. D. (1968). *In* "Mechanics of Generalized Continua" (E. Kröner, ed.), pp. 69–80. Springer Verlag, Berlin and New York.
Hutchinson, J. R. (1969). *AIAA J.* **7,** 786.
Kinslow, R. (1970). Stress Waves in Multiple Laminates, TTU-ES-70-1. Tennessee Technol. Univ., Cookeville, Tennessee.
Kohn, W., Krumhansl, J. A., and Lee, E. H. (1972). *J. Appl. Mech.* (forthcoming).
Krumhansl, J. A. (1972). *In* "Dynamics of Composite Materials" (E. H. Lee, ed.), pp. 139–150. Amer. Soc. Mech. Eng., New York.
Lai, Y. S., and Achenbach, J. D. (1972). *Comp. Struct.* **3,** 559.
Lee, E. H. (1972). *In* "Dynamics of Composite Materials" (E. H. Lee, ed.), pp. 122–138. Amer. Soc. Mech. Eng., New York.
Lundergan, C. D., and Drumheller, D. S. (1971a). *J. Appl. Phys.* **42,** 669.
Lundergan, C. D., and Drumheller, D. S. (1971b). *In* "Shock Waves and the Mechanical Properties of Solids" (J. J. Burke and V. Weiss, eds.), pp. 141–154, Syracuse Univ. Press, Syracuse, New York.
Lundergan, C. D., and Drumheller, D. S. (1971c). *J. Appl. Phys.* **42,** 4148.
Lundergan, C. D., and Drumheller, D. S. (1972). *In* "Dynamics of Composite Materials" (E. H. Lee, ed.), pp. 35–47. Amer. Soc. Mech. Eng., New York.
Martin, S. E., Bedford, A., and Stern, M. (1971). *In* "Developments in Mechanics" (L. N. H. Lee and A. A. Szewczyk, eds.), Vol. 6, pp. 515–528. Univ. Notre Dame Press, South Bend, Indiana.
McCoy, J. (1973). *J. Appl. Mech.* **40,** 511.
Mindlin, R. D. (1960). *In* "Structural Mechanics" (J. N. Goodier and N. J. Hoff, eds.), pp. 199–232, Pergamon, Oxford.
Mindlin, R. D. (1964). *Arch. Ration. Mech. Anal.* **16,** 51.
Moon, F. C. (1972). *J. Compos. Mater.* **6,** 62.
Musgrave, M. J. P. (1961). *Progr. Solid Mech.* **2,** 61.
Nemat-Nasser, S. N. (1972a). *J. Elast.* **2,** 73.
Nemat-Nasser, S. N. (1972b). *J. Appl. Mech.* **39,** 850.
Peck, J. C. (1971). *In* "Shock Waves and the Mechanical Properties of Solids" (J. J. Burke and V. Weiss, eds.), pp. 155–184. Syracuse Univ. Press, Syracuse, New York.
Peck, J. C. (1972). *In* "Dynamics of Composite Materials" (E. H. Lee, ed.), pp. 8–34. Amer. Soc. Mech. Eng., New York.
Peck, J. C., and Gurtman, G. A. (1969). *J. Appl. Mech.* **36,** 479.
Postma, G. W. (1955). *Geophysics* **20,** 780.
Riznichenko, Yu. V. (1949). *Izv. Akad. Nauk. SSSR Ser. Geofiz. Geogr.* **13,** 518.

Rytov, S. M. (1956). *Sov. Phys. Acoust.* **2,** 68.
Sokolnikoff, I. S. (1956). "Mathematical Theory of Elasticity." McGraw-Hill, New York.
Sun, C. T. (1970). *Bull. Seism. Soc. Amer.* **60,** 345.
Sun, C. T. (1971a). *J. Appl. Mech.* **38,** 231.
Sun, C. T. (1971b). *J. Appl. Mech.* **38,** 947.
Sun, C. T. (1972). *Int. J. Solids Struct.* **8,** 385.
Sun, C. T., and Achenbach, J. D. (1969). Time-Harmonic Transverse and Longitudinal Motions of a Laminated Plate. Northwestern Univ. Struct. Mech. Lab. T.R. No. 69-1, Evanston, Illinois.
Sun, C. T., Achenbach, J. D., and Herrmann, G. (1968a). *J. Appl. Mech.* **35,** 408.
Sun, C. T., Achenbach, J. D., and Herrmann, G. (1968b). *J. Appl. Mech.* **35,** 467.
Sve, C. (1971). *J. Appl. Mech.* **38,** 477.
Sve, C. (1972a). Stress Wave Attenuation in Composite Materials, Aerosp. Rep. No. TR-0172 (2240-30)-4. Aerosp. Corp., Los Angeles, California.
Sve, C. (1972b). *J. Appl. Mech.* **39,** 103.
Sutherland, H. J., and Lingle, R. (1972). Geometric Dispersion of Acoustic Waves by a Fibrous Composite, SC-DR-0488. Sandia Lab., Albuquerque, New Mexico.
Synge, J. L. (1957). *J. Math. Phys.* **35,** 323.
Tauchert, T. R., and Guzelsu, A. N. (1972). *J. Appl. Mech.* **39,** 98.
Torvik, P. J. (1970). *J. Compos. Mater.* **4,** 596.
Tsou, F. K., and Chou, P. C. (1969). *J. Compos. Mater.* **3,** 500.
Turhan, D. (1970). A Dynamic Theory for Directionally Reinforced Composites. Ph.D. dissertation, Northwestern Univ., Evanston, Illinois.
Voelker, L. E., and Achenbach, J. D. (1969). *J. Acoust. Soc. Amer.* **46,** 1213.
Weitsman, Y. (1972). *Int. J. Solids Struct.* **8,** 627.
Wheeler, P., and Mura, T. (1972). *J. Appl. Mech.* **40,** 498.
White, J. E., and Angona, F. A. (1955). *J. Acoust. Soc. Amer.* **27,** 311.
Whittier, J. S., and Peck, J. C. (1969). *J. Appl. Mech.* **36,** 485.
Wu, T. M. (1971). Ph.D. dissertation, Princeton Univ., Princeton, New Jersey.
Yang, P. C., Norris, C. H., and Stavsky, Y. (1966). *Int. J. Solids Struct.* **2,** 665.
Ziegler, F. (1969). *Int. J. Solids Struct.* **5,** 893.

9

Phenomenological Anisotropic Failure Criterion

EDWARD M. WU

Department of Mechanical and Aerospace Engineering
Washington University
St. Louis, Missouri

I. Introduction

In a composite, matrix and reinforcement materials are combined to form a macroscopically multiphase material with the goal of enhancing certain physical properties not realizable by the constituent materials individually. A frequently occurring consequence of such composition (of multiphase materials) is that the physical properties of the resulting materials become highly anisotropic. Two different disciplines may be adopted to determine the physical properties of composites; they are, the mechanistic approach and the phenomenological approach.

The mechanistic approach relates the physical properties of the composite to the properties of the constituent materials through established methods of analysis (e.g., continuum mechanics). This approach is particularly adaptable to the computation of composite properties which characterize the averaged global material responses, such as deformation and conductivity. A particularly useful result from the mechanistic approach is the prediction of composite compliances (Hashin, 1970) from micromechanics analyses. However, for composite properties which are governed by local material responses such as strength and interfacial phenomena, the mechanistic approach is less fruitful. Difficulties arise because of higher precision of analysis and more in-depth knowledge of the physics involved when dealing with local rather than global material responses.

The phenomenological approach treats the heterogeneous composite as a continuum, and a mathematical model is used to correlate the occurrence of the material responses without necessarily explaining the mechanisms which lead to these material responses. If attention is paid to the mathematical requirements of the model, the phenomenological approach is suitable for engineering characterization of the material properties governed by either the averaged global or the local material responses. Treatment of unidirectional composites as homogeneous anisotropic plates (Hearmon, 1952; Lekhnitskii, 1950) exemplifies characterization in a global sense.

While the mechanistic and phenomenological approaches satisfy different needs and possess different appeals, they are by no means self-exclusive. Judicious combination of the two can lead to fruitful results as demonstrated in laminated plate theory (Reissner and Stavsky, 1961) where the treatment of the lamina may be regarded as phenomenological and that of the laminate as mechanistic. The discipline followed in the characterization of failure of composites, whether mechanistic or phenomenological is defined by the level of observation. Figure 1 relates the different characterization disciplines and the plausible linking analytical methods.

It is generally recognized that the failure of a composite originates from local physical processes. Thus, in order to establish a failure criterion from the mechanistic approach, the local irregularity of fiber–matrix geometry must be characterized, detailed analysis methodology beyond the realm of classical continuum mechanics must be established, and the physical mechanism of failure of the individual isotropic phases must be known.

Since these basic ingredients, particularly the last, have yet to be firmly established, the mechanistic approach to composite failure appears to be intractable at present.

Failure characterization as discussed herein is defined at the level where the composite is treated as anisotropic but homogeneous.

A phenomenological failure criterion may be considered as a mathematical model of the material transfer function relating the external excitation to the material responses. "Failure" is interpreted here as the occurrence of any definable discontinuity in the material response. The external excitations can be mechanical, thermal, chemical, magnetic, irradiation, and others. This discussion focusses on the characterization of mechanical failure (e.g., onset of stress–strain nonlinearity or total rupture) due to mechanical excitation (stress, strain, or work). Thus, if the material parameters in the mathematical model are measured in a manner consistent with the discontinuity to be characterized, the failure criterion can be used,

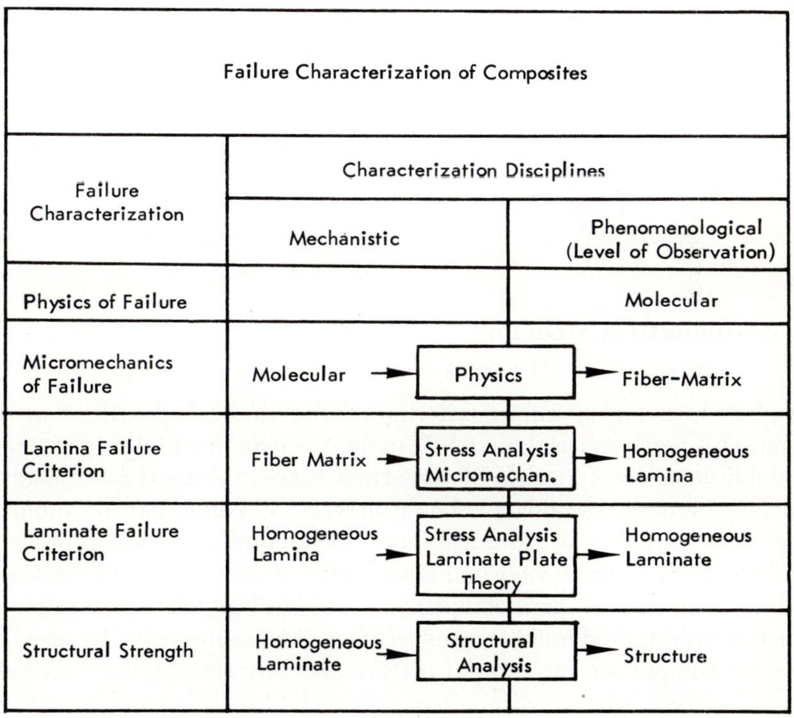

FIG. 1. Classification of composite failure characterization disciplines.

in *isolated* context, to characterize yielding or rupture. The phenomenological failure criterion discussed is intended to be used in the development of composites and in engineering applications. In composite material development, permutations exist for numerous combinations of matrix and reinforcement materials leading to enhancement of different composite properties. For those strength-related properties, a phenomenological failure criterion can supply the feedback in material permutations, provide guidance in fabrication processes, and supply quantitative strength allowables for rational engineering design.

A popular approach for formulation of failure criterion for anisotropic materials has been the generalization of isotropic failure criteria (Jenkins, 1920; Hill, 1948; Norris, 1950; Marin, 1957; Hu, 1958; Azzi and Tsai, 1965; Waddoups, 1967). In many of these adaptations, the original motivating philosophy for the isotropic theory and the underlying assumptions have been overlooked, leading to unnecessarily complex formulations and operational anomalies. The evolution of past formulations will be briefly reviewed† and their limitations clarified. These formulations will be interpreted and their similarities emphasized with the intention of bringing different viewpoints to bear on a failure criterion which is amenable to relatively simple mathematical treatment.

The characterization of strength of anisotropic composites will be treated herein in accordance with the fundamentals of phenomenological characterization, to wit, (1) mathematical modeling, (2) experiments—identification and execution of physical measurements, and (3) data analysis—statistical correlation of measurement results.

II. Mathematical Modeling

A phenomenological failure criterion can be established with or without the aid of a mathematical model. If extensive experimental measurements are made in all components of the stress space, the need for theoretical modeling becomes secondary; the occurrence of failure can be tabulated either in numerical or in graphical form. Such a pure empirical approach is, of course, both costly in time and effort, and cumbersome in application. A mathematical model can complement the failure characterization by reducing the required number of experimental measurements, by providing bases for comparison of material performance, by simplifying data reduc-

† Sendeckyj (1972) has compiled an extensive bibliography of the relevant literatures including many Russian contributions.

tion and documentation, and by improving adaptability in structural analysis applications.

The above advantages of mathematical models are clearly identifiable in the now well-known characterization of linear compliance of anisotropic composites. For a two-dimensional orthotropic composite, the mathematical model (generalized Hooke's law) suggests that the compliance is a fourth-rank tensor thus leading to the fact that only four independent compliance measurements (S_{11}, S_{12}, S_{22}, S_{66}) in the material principal directions will completely determine the six compliances (S'_{11}, S'_{12}, S'_{16}, S'_{22}, S'_{26}, S'_{66}) at any material orientation. Extensive measurements of the six compliances at small incremental changes of material orientation are not required to establish the compliance transformation properties. It also follows that, as is now the practice, the compliance of different composites may be compared through the principal values of the compliance without resorting to the comparison of individual graphs or tabulations of the S_{ij} components versus material axis. In addition, the mathematical structure of the model permits operations in laminate plate analysis (Reissner and Stavsky, 1961) and structural optimization (Waddoups, 1967; Brandmaier, 1970), provides guidance for rational statistical analysis and compaction of experimental data, and facilitates documentation and retrieval in applications (Wu *et al.*, 1973). Such simplification through mathematical modeling will be extended to failure characterization of anisotropic composites with the understanding that:

> The phenomenological mathematical model is intended to aid experimental design; to facilitate interpolation, correlation, and retrieval of experimental observations; it is not intended for extrapolation and interpretation of mechanisms of failure.

A. Theoretical Requirements

In phenomenological strength characterization the mathematical model or failure criterion has to fulfill certain fundamental requirements.

(1) It should be invariant to the definition of the coordinate axis.

(2) It should have sufficient flexibility for modeling any failure envelope without constraint by preconceived assumptions on the mechanism of failure.

(3) It should have only one root for any given radial loading path.

(4) It should be readily simplifiable to provide characterization of failure to the degree of accuracy desired.

(5) It should be mathematically operational, possess unique conversions between the stress space and strain space, and be adaptable for structural analysis (e.g., stress analysis).

(6) It should permit unambiguous definition of the principal directions of strength.

Because the failure criterion discussed herein is phenomenological, it is assumed that any mathematical model which is capable of fulfilling the above requirements is an admissible failure criterion. Obviously, no uniqueness of the formulation of the failure criterion can be established nor is it implied in the ensuing discussion.

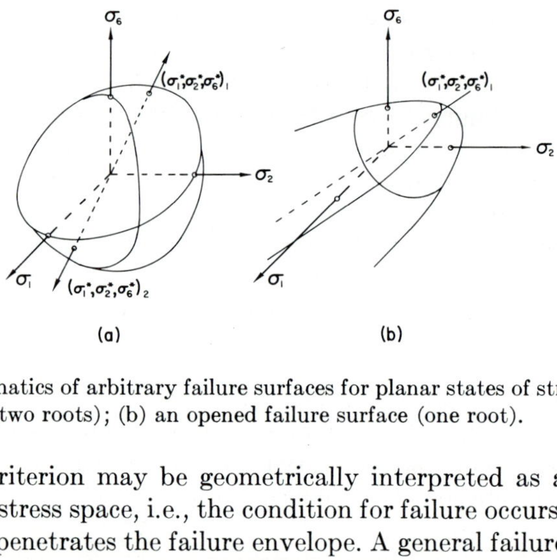

FIG. 2. Schematics of arbitrary failure surfaces for planar states of stress: (a) a closed failure surface (two roots); (b) an opened failure surface (one root).

A failure criterion may be geometrically interpreted as a limiting envelope in the stress space, i.e., the condition for failure occurs when a given stress vector penetrates the failure envelope. A general failure surface for a two-dimensional state of stress is depicted in Fig. 2a,b. This interpretation can be extended to a three-dimensional state of stress by a failure envelope in the six stress component space. The general material parameters for three-dimensional stress will be given but, for the sake of retaining geometric insight, only the two-dimensional state of stress will be discussed.

From the geometric interpretation of the failure envelope, the fundamental requirements for the mathematical model of failure criterion are immediately apparent. Since the state of stress is dependent on the choice of coordinate axis, whereas the condition for failure is intrinsic to the material, the failure criterion must be invariant to coordinate transformation. This requirement of invariancy to coordinate transformation is sometimes considered unnecessary when the material possesses obvious principal directions such as in the case of a unidirectionally reinforced composite lamina. Under these circumstances, it can be argued that the coordinate axis should be oriented along the principal directions of elastic symmetry and all applied stresses may be transformed to such a conventional orientation and examined for failure conditions. Nevertheless, there are numerous practical crystals and composites (e.g., angle-ply laminates) for which the "principal" direction is ambiguous and the invariancy requirement becomes indispensable.

For different anisotropic composites, the respective failure envelope is the manifestation of many different mechanisms of failure and can be of different shapes and sizes, thus the mathematical model must possess sufficient flexibility to model such a failure envelope. Although the failure envelope can be geometrically complex, from inference of classical plasticity theory (Paul, 1968), it is expected to be convex. Even in the absence of convexity (Ashkenazi, 1964), for any given loading condition the failure criterion should have only one root. For example, two colinear radial loading paths may penetrate the failure surface at no more than two points such as shown in Fig. 2a. The case of single root (Fig. 2b), which implies infinite strength under certain stress conditions, is a physically admissible condition although it can be regarded as an engineering rarity.

In any phenomenological failure criterion, the shape of the failure envelope is never completely known until experiments are performed for *all* possible combined states of stress. A failure envelope may appear to be smooth when experimental conditions are spaced sparingly, whereas close experimental mapping may reveal subtle variation in the failure envelope. Past experience involving the discovery of corners in isotropic yield surface is a well-known example. Nevertheless, the degree of accuracy of a failure envelope has to be a compromise between engineering requirements and availability of effort and time. Hence a mathematical model should provide guidance to check the irregularities in the failure surface and should be readily simplifiable for the particular material and application in question.

In engineering analyses, the coordinate axes are usually chosen with regard to the structure so as to simplify the boundary conditions and the solution details may be either in terms of stress or strain. Thus for a failure criterion to be adaptable in engineering analysis, it is desirable that the criterion obey transformation rules and be clearly expressable in stress and/or strain space.

Finally, in the application of a failure criterion such as in structural analysis, in strength optimization, and materials comparison, the principal strength direction should be defined unambiguously (Fischer, 1967; Brandmaier, 1970; Cooper and Wu, 1973).

B. Definition of Failure

The selection of a candidate for a mathematical model of failure criterion can begin with the identification of the excitation parameters and the responses to be characterized. In this mathematical model, mechanical failure response is to be related to mechanical excitation. Mechanical failure is interpreted as the occurrence of any definable discontinuity in mechanical responses. Some of the discontinuities of technical interest are: the onset of nonlinearity in stress–strain relations, the occurrence of irreversible deformations, and the termination of the stress–strain function due to rupture.

The conditions for the occurrence of these phenomena will be broadly referred to in the following as "failure criterion." Ergo, the condition of brittle rupture may be considered as a special case where the failure criterion for yielding coincides with the failure criterion for rupture. Mechanical excitations can be stress, strain, or work; and, in the absence of other external excitations (chemical, thermal, and others), a failure criterion can be expressed as:

$$f(\sigma_{ij}, \mathbf{F}) = 0 \tag{1}$$

$$g(\epsilon_{ij}, \mathbf{G}) = 0 \tag{2}$$

$$h(\sigma_{ij}\epsilon_{ij}, \mathbf{H}) = 0 \tag{3}$$

where **F**, **G**, **H** are material parameters. Fundamentally, Eqs. (1)–(3) each have divergent implications of the physics of failure, i.e., failure is stress actuated, strain actuated, and work actuated. However, in phenomenological characterization, we will only be concerned with their consistency within the framework of continuum mechanics.

In order to maintain uniqueness of conversion between these expressions we assume that the material possesses a strain–energy function up to the onset of the mechanical discontinuity to be characterized; that is, the stress–strain relations may be nonlinear, but they must be single-valued up to the occurrence of "failure." It should be noted this is not a realistic assumption if the failure criterion is to specify the condition for rupture when it is preceded by irreversible deformation. Under these circumstances the application of failure criterion should be limited to radial loading paths only. For materials and applications which approximate the strain–energy function assumption, any one of the three expressions [Eqs. (1)–(3)] may be used for the formulations of a failure criterion provided they comply with the fundamental mathematical requirements.

C. Comparison of Failure Criteria

The stress formulation Eq. (1) will be used for the discussion on the fundamental requirements with the understanding that identical formulation is applicable for the strain formulation and, through the constitutive stress–strain relations, for the work formulation. Clearly, there exists an enormous number of functions in the form of Eq. (1) which are capable of modeling a failure surface. The requirement of invariancy to coordinate definition narrows the choice. Admissible functions must be expressed in terms of stress invariants of the principal stresses or scalar functions of the stresses. In the formulation of failure criterion in terms of principal stresses for isotropic materials, simplifications are possible from the facts that (1) functions must be symmetrical in principal stresses, and (2) the phase relation between the principal stress direction and principal material direction need not be specified. Obviously, neither simplification is possible for anisotropic composites. A formulation of failure criterion in terms of principal stresses must incorporate multiple material parameters to account for the lack of symmetry in strength and to account for the noncoincidence of principal stress directions and principal materials directions. The consequences of overlooking these observations and formulating anisotropic failure criterion in terms of principal stresses are loss of generality and increased complexity. The commonly used maximum strain and maximum stress criteria have the algebraic appearance of formulation in terms of the principal strain and stress tensors. In terms of the stress invariants, an anisotropic failure criterion may be expressed as

$$f(I_1, I_2, I_3, \mathbf{F}) = 0 \tag{4}$$

The distinguishing features of Eq. (4) from the corresponding formulation of isotropic material are: (1) for isotropy, the material parameter **F** consists of only one scalar constant; whereas to account for strength anisotropy, **F** must be multiple parameters of the type invariant to coordinate transformations, (2) simplification through replacement of the stress invariants by deviatoric invariant is not admissible because for anisotropic materials the independency of yielding under hydrostatic stress has not been physically established. Because of these two added generalities needed for anisotropic material, many of the original motives for formulating isotropic failure criterion are no longer applicable. In fact, it will be shown, in Section II,C,5, that formulation as in Eq. (4) results in added algebraic complexity without improvement in flexibility.

Thus, we are left with the formulation in terms of scalar functions of stress as a viable approach. A scalar function with tensor component variables can be formed directly through a tensor polynomial. The formulation of anisotropic failure criterion in the form of a polynomial (in strain) was implicitly anticipated by Voigt (1910) circa 1890 for the characterization of the strength anisotropy of crystals. Contemporary efforts in using similar formulations for anisotropic composites are presented in the works of Ashkenazi (1965), Goldenblat and Kopnov (1965) and others (Malmeyster, 1966; Bogue, 1967; Tsai and Wu, 1971a). In the work of Tsai and Wu (1971a) the tensor polynomial formulations are unified and simplified to the form

$$f(\sigma_{ij}) = F_{ij}\sigma_{ij} + F_{ijkl}\sigma_{ij}\sigma_{kl} + F_{ijklmn}\sigma_{ij}\sigma_{kl}\sigma_{mn} + \cdots = 1 \tag{5a}$$

In this representation, invariancy to coordinate transformation is guaranteed by specifying that each term of the polynomial be a scalar through tensor contractions. In order for this contraction to take place, the material strength parameters F_{ij}, $F_{ijkl}, \ldots$ must be components of tensors of appropriate ranks, i.e., they must obey the transformation rules:

$$F'_{ij} = \frac{\partial x_i'}{\partial x_m}\frac{\partial x_j'}{\partial x_n} F_{mn}$$

$$F'_{ijkl} = \frac{\partial x_i'}{\partial x_m}\frac{\partial x_j'}{\partial x_n}\frac{\partial x_k'}{\partial x_o}\frac{\partial x_l'}{\partial x_p} F_{mnop} \tag{6}$$

$$\vdots$$

The essential features of failure are elucidated (Tsai and Wu, 1971a) through detailed exploration of: determination of the engineering strength parameters, stability conditions, transformational properties, application to three dimensionally reinforced composites, and degenerated cases due to material symmetry. Additional understanding of the tensor polynomial failure formulation may be gained through examination of the fundamental geometric requirements of the failure surfaces. In keeping with the philosophy of phenomenological characterization, a general mathematical model will be established, thus ensuring flexibility with provisions for simplifications through material symmetry considerations and experimental facts. We will begin by examining the flexibility of formulation Eq. (5a) through comparison with other failure criteria based on the first two formulations and show that both those of historical interest and those now employed in engineering practices are special cases of Eq. (5a).

1. Stress Tensor Polynomial

We will refer to the general formulation as tensor polynomials; for conciseness, we rewrite Eq. (5a) in contracted notation†

$$f(\sigma_i) = F_i\sigma_i + F_{ij}\sigma_i\sigma_j + F_{ijk}\sigma_i\sigma_j\sigma_k + \cdots = 1, \qquad i, j, k = 1, 2, \ldots, 6 \tag{5b}$$

For ease in visualization, Eq. (5b) is written in matrix notation for the linear and quadratic terms:

$$\operatorname{tr}\begin{bmatrix} F_1 \\ F_2 \\ F_3 \\ F_4 \\ F_5 \\ F_6 \end{bmatrix}\begin{bmatrix} \sigma_1 \\ \sigma_2 \\ \sigma_3 \\ \sigma_4 \\ \sigma_5 \\ \sigma_6 \end{bmatrix}^{\mathrm{T}} + \begin{bmatrix} F_{11} & F_{12} & F_{13} & F_{14} & F_{15} & F_{16} \\ & F_{22} & F_{23} & F_{24} & F_{25} & F_{26} \\ & & F_{33} & F_{34} & F_{35} & F_{36} \\ & & & F_{44} & F_{45} & F_{46} \\ & & & & F_{55} & F_{56} \\ & & & & & F_{66} \end{bmatrix}$$

$$\times \begin{bmatrix} \sigma_1^2 & \sigma_1\sigma_2 & \sigma_1\sigma_3 & \sigma_1\sigma_4 & \sigma_1\sigma_5 & \sigma_1\sigma_6 \\ & \sigma_2^2 & \sigma_2\sigma_3 & \sigma_2\sigma_4 & \sigma_2\sigma_5 & \sigma_2\sigma_6 \\ & & \sigma_3^2 & \sigma_3\sigma_4 & \sigma_3\sigma_5 & \sigma_3\sigma_6 \\ & & & \sigma_4^2 & \sigma_4\sigma_5 & \sigma_4\sigma_6 \\ & & & & \sigma_5^2 & \sigma_5\sigma_6 \\ & & & & & \sigma_6^2 \end{bmatrix} + \cdots = 1 \tag{5c}$$

F_{ij} is symmetric if we hypothesize the existence of a failure potential, i.e., failure is path-independent, then $f(\sigma_i)$ in Eq. (5b) is of class C2, i.e.,

$$F_{ij} = \partial^2 f/\partial\sigma_i\,\partial\sigma_j = \partial^2 f/\partial\sigma_j\,\partial\sigma_i = F_{ji} \tag{7}$$

Also F_i is symmetric (when used in noncontracted notation) because the stress tensor is symmetric.

† A detailed discussion of contracted notation is given by Tsai and Wu (1971b).

The stress failure tensors F_i and F_{ij} can be expressed in terms of engineering strength measurements by substituting uniaxial experimental strengths into Eq. (5a). For example, for a two-dimensional composite they are (Tsai and Wu, 1971a,b):

$$F_i = \begin{bmatrix} \dfrac{1}{X_1} - \dfrac{1}{X_1'} \\ \\ \dfrac{1}{X_2} - \dfrac{1}{X_2'} \\ \\ 0 \\ 0 \\ 0 \\ \\ \dfrac{1}{X_6} - \dfrac{1}{X_6'} \end{bmatrix}, \qquad F_{ij} = \begin{bmatrix} \dfrac{1}{X_1X_1'} & F_{12} & 0 & 0 & 0 & F_{16} \\ & \dfrac{1}{X_2X_2'} & 0 & 0 & 0 & F_{26} \\ & & 0 & 0 & 0 & 0 \\ & & & 0 & 0 & 0 \\ & & & & 0 & 0 \\ & & & & & \dfrac{1}{X_6X_6'} \end{bmatrix} \tag{8}$$

where X_1, X_1', X_2, X_2' are tension and compression strengths in the 1 and 2 direction, respectively and X_6 and X_6' are the positive and negative shear strengths with respect to the z plane (1, 2 plane). In cases of orthotropy, $X_6 = X_6'$. The components F_{12}, F_{16}, and F_{26} are determined from biaxial tests as outlined by Wu (1972).

With these failure tensors F_i, F_{ij} determined for one material orientation θ, the failure criterion under any complex state of stress can be obtained through substitution of Eq. (8) into (5b) and solution of the resulting algebraic equation. This failure criterion can also be a prediction of failure condition in another (arbitrary) material orientation θ' (Fig. 3). Through tensor transformation rules, Eq. (6), the coefficients F_i, F_{ij} in Eq. (8) can be transformed to F_i', F'_{ij} for orientation θ'. Again, substitution of F_i', F'_{ij}

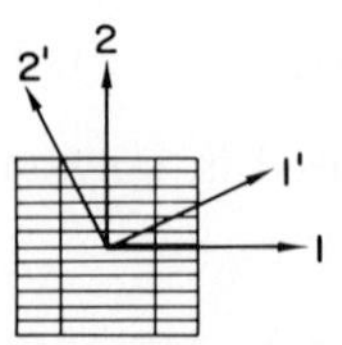

FIG. 3. Transformation of reference coordinates.

into Eq. (5b) provides the failure criterion applicable for orientation θ'. For example, the off-axis strength at θ' when subjected to normal stress σ_1' can be determined as described and the resulting equation is

$$F'_{11}\sigma_1'^2 + F_1'\sigma_1' = 1 \tag{9}$$

The positive root of this equation is the predicted off-axis tension strength; the negative root, the off-axis compressive strength.

Comparison of different formulations will now be made. Contracted notations will be employed unless otherwise specified.

2. Strain Tensor Polynomials

For applications where strain formulation is more convenient, as in laminate plate analysis, a failure criterion of the form Eq. (2) can be expressed as a tensor polynomial analogous to Eq. (5b):

$$G_i\epsilon_i + G_{ij}\epsilon_i\epsilon_j + G_{ijk}\epsilon_i\epsilon_j\epsilon_k + \cdots = 1, \qquad i, j, k = 1, 2, \ldots, 6 \tag{10}$$

where G_i, G_{ij}, etc. are material parameters expressible in terms of the limiting engineering strains. We can express, for example, the strain failure tensors of the linear and quadratic terms of Eq. (10) for a composite plate as

$$G_i = \begin{bmatrix} \dfrac{1}{E_1} - \dfrac{1}{E_1'} \\ \dfrac{1}{E_2} - \dfrac{1}{E_2'} \\ \dfrac{1}{E_3} - \dfrac{1}{E_3'} \\ 0 \\ 0 \\ \dfrac{1}{E_6} - \dfrac{1}{E_6'} \end{bmatrix}, \qquad G_{ij} = \begin{bmatrix} \dfrac{1}{E_1E_1'} & G_{12} & G_{13} & 0 & 0 & G_{16} \\ & \dfrac{1}{E_2E_2'} & G_{23} & 0 & 0 & G_{26} \\ & & \dfrac{1}{E_3E_3'} & 0 & 0 & G_{36} \\ & & & 0 & 0 & 0 \\ & & & & 0 & 0 \\ & & & & & \dfrac{1}{E_6E_6'} \end{bmatrix} \tag{11}$$

where E_i and E_i' are the ultimate extension and compression normal strains in the i direction for $i = 1, 2, 3$. E_j and E_j' are positive and negative shear strain in the j plane for $j = 4, 5, 6$. In the general case, the parameters G_i and G_{ij} are fully populated as in Eq. (5c). Like the stress failure tensors $F_{ij}, \ldots$, the strain failure tensors $G_{ij}, \ldots$ are symmetric if a strength potential is assumed to exist. The transformation rules for the strain failure tensor (in *non*contracted form) are identical to those for the stress-failure tensor indicated by Eq. (6).

The failure surfaces based on stress and strain, respectively, are not independent. The two formulations [Eqs. (5) and (10)] are continuous, thus unique conversion can be made through the stress–strain constitutive relations. If the criteria are used to characterize the onset of nonlinearity of a composite which initially obeys Hooke's Law, then the conversions between these material parameters are

$$\begin{aligned} G_i &= F_m C_{mi} \\ G_{ij} &= F_{mn} C_{mi} C_{nj} \\ G_{ijk} &= F_{mnp} C_{mi} C_{nj} C_{pk} \\ &\vdots \end{aligned} \tag{12}$$

where C_{ij} are components of the elastic stiffness matrix in the generalized Hooke's law for strain: $\epsilon_i = C_{ij}\sigma_j$. Similarly, the conversions from the strain failure tensors to the stress failure tensors are

$$\begin{aligned} F_i &= G_m S_{mi} \\ F_{ij} &= G_{mn} S_{mi} S_{nj} \\ F_{ijk} &= G_{mnp} S_{mi} S_{nj} S_{pk} \\ &\vdots \end{aligned} \tag{13}$$

where S_{mi} are components of the elastic compliance matrix in Hooke's law for stress: $\sigma_i = S_{ij}\epsilon_j$. It is important to note that the above conversion relations, Eqs. (12) and (13), are applicable only for linear stress–strain relations; hence, if the criteria are to be used to characterize rupture which is preceded by nonlinear deformation, the conversion has to be through an appropriate nonlinear constitutive relation.

3. Maximum Strain Criterion

St. Venant's maximum strain theory has been adopted to characterize failure of anisotropic composites. It is pertinent to note that St. Venant's maximum strain theory, even in its original intended form, characterizes isotropic yielding poorly and is generally not used in metal design practices.

Only in highly brittle materials does this criterion provide a measure of description. Perhaps the observation that some polymeric composites are highly "brittle" led to the modification of St. Venant's criterion to anisotropic composites (Waddoups, 1968). St. Venant's maximum strain criterion for isotropic materials may be interpreted as a formulation based on principal strains (i.e., only the maximum principal strain induces failure). In the usual modification of maximum strain criterion for anisotropy, the material is assumed to be orthotropic and the coordinate axes are defined along the material principal directions as illustrated in Fig. 3. Fixing the coordinate system in this manner precludes a transformation to eliminate the shear strain. A criterion applicable to anisotropic material must include a limiting shear strain condition. The modified anisotropic maximum strain criterion can be written in the form

$$-\mathrm{E}_1' \leq \epsilon_1 \leq \mathrm{E}_1, \qquad -\mathrm{E}_2' \leq \epsilon_2 \leq \mathrm{E}_2, \qquad -\mathrm{E}_6' \leq \epsilon_6 \leq \mathrm{E}_6 \tag{14a}$$

where E_i and E_i' follow the same notation defined by Eq. (11). In the case of orthotropy, $\mathrm{E}_6 = \mathrm{E}_6'$ and the failure envelope is symmetrical with respect to the ϵ_6 axis. Geometrically, this formulation produces a rectangular failure envelope in the $\epsilon_1\epsilon_2$ strain space as shown in Fig. 4a. In the usual application, the limiting tensile strain employed is assumed to be E_1. Physical considerations disclose that exclusive use of E_1 as a limiting strain is not strictly correct. A uniaxial tensile stress σ_1 gives rise to a triaxial state of strain from Poisson effect. To avoid possible difficulties in application it was a priori hypothesized [Eq. (14a) and Fig. 4], that the presence of one strain does not affect the limiting values of the other strains. Thus the maximum strain criterion for anisotropic materials as stated in the form of Eq. (14a) is, in fact, a combination of the isotropic criteria of St. Venant's maximum normal strain and Coulomb's maximum shear criterion.

The maximum strain criterion, Eq. (14a), can be rewritten in a single

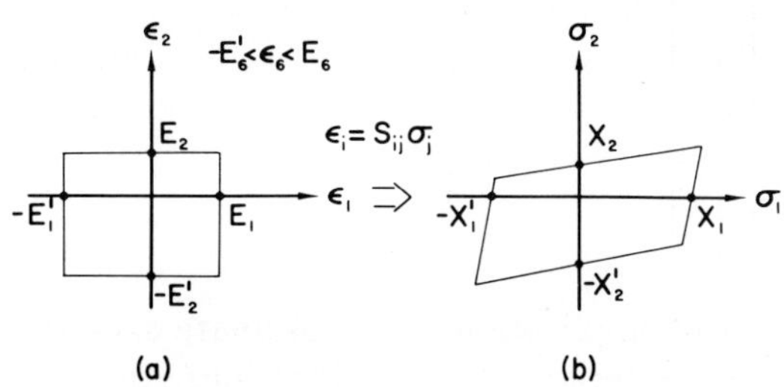

FIG. 4. Maximum strain failure criterion (schematic) in: (a) strain space; (b) stress space.

equation, for the critical state, in the following form:

$$(\epsilon_1 - E_1)(\epsilon_1 + E_1')(\epsilon_2 - E_2)(\epsilon_2 + E_2')(\epsilon_6 - E_6)(\epsilon_6 + E_6') = 0 \qquad (14b)$$

Analytical geometry or direct graphing will verify that Eq. (14b) models a failure surface which is identical to that modeled by Eq. (14a). Expansion of Eq. (14b) produces a polynomial which permits comparison with the strain tensor polynomial formulation, Eq. (10). Simplicity in algebraic details for this comparison will result if we focus our attention on the first strain quadrant. Thus for

$$\epsilon_1 > 0, \quad \epsilon_2 > 0 \qquad \text{and} \qquad -E_6' < \epsilon_6 < E_6$$

Equation (14b) simplifies to be

$$(\epsilon_1 - E_1)(\epsilon_2 - E_2) = 0 \qquad (14c)$$

Expansion of Eq. (14c) yields

$$-\frac{\epsilon_1\epsilon_2}{E_1E_2} + \frac{\epsilon_1}{E_1} + \frac{\epsilon_2}{E_2} = 1 \qquad (14d)$$

Comparing the coefficients of Eq. (14d) to those in the strain tensor polynomial and in the notation of Eq. (11), the maximum strain criterion has the following material coefficients:

$$G_i = \begin{bmatrix} \dfrac{1}{E_1} \\ \dfrac{1}{E_2} \\ 0 \\ 0 \\ 0 \\ 0 \end{bmatrix}, \qquad G_{ij} = \begin{bmatrix} 0 & -\dfrac{1}{2E_1E_2} & 0 & 0 & 0 & 0 \\ & 0 & 0 & 0 & 0 & 0 \\ & & 0 & 0 & 0 & 0 \\ & & & 0 & 0 & 0 \\ & & & & 0 & 0 \\ & & & & & 0 \end{bmatrix} \qquad (15)$$

In current engineering application, the coordinate axes are always defined with respect to the material principal directions, thus the invariancy of Eq. (14a) or (d) to coordinate transformations is not explicitly stated. To check the transformational properties of the maximum strain criterion, we examine the form of Eq. (14d) by first employing the normal practice of transforming the strain for an arbitrary coordinate axis x_i' to the material axis x_i in Fig. 3 by the strain transformation:

$$
\begin{aligned}
\epsilon_1 &= m^2\epsilon_1' + n^2\epsilon_2' - mn\epsilon_6' \\
\epsilon_2 &= n^2\epsilon_1' + m^2\epsilon_2' + mn\epsilon_6' \\
\epsilon_6 &= 2mn\epsilon_1' - 2mn\epsilon_2' + (m^2 - n^2)\epsilon_6'
\end{aligned}
\tag{16}
$$

Upon substitution of Eq. (16) into Eq. (14d) we obtain:

$$
\begin{aligned}
&-\left(\frac{m^2n^2}{E_1E_2}\right)\epsilon_1'^2 - \left(\frac{m^4+n^4}{E_1E_2}\right)\epsilon_1'\epsilon_2' - \left(\frac{m^3n - mn^3}{E_1E_2}\right)\epsilon_1'\epsilon_6' + \left(\frac{m^2}{E_1} + \frac{n^2}{E_2}\right)\epsilon_1' \\
&\qquad -\left(\frac{m^2n^2}{E_1E_2}\right)\epsilon_2^{2\prime} \qquad - \left(\frac{mn^3 - m^3n}{E_1E_2}\right)\epsilon_2'\epsilon_6' + \left(\frac{n^2}{E_1} + \frac{m^2}{E_2}\right)\epsilon_2' \\
&\qquad\qquad + \left(\frac{m^2n^2}{E_1E_2}\right)\epsilon_6'^2 - mn\left(\frac{1}{E_1} - \frac{1}{E_2}\right)\epsilon_6' = 1
\end{aligned}
\tag{17}
$$

Thus Eq. (17) is the maximum strain criterion for a transformed direction x_i'. On the other hand, if we hypothesize that the maximum strain criterion is a special case of the strain tensor polynomial, we can express Eq. (10) in the transformed coordinates by simply writing

$$
\begin{aligned}
&G_{11}'\epsilon_1'^2 + 2G_{12}'\epsilon_1'\epsilon_2' + 2G_{16}'\epsilon_1'\epsilon_6' + G_1'\epsilon_1' \\
&\qquad + G_{22}'\epsilon_2'^2 \quad + 2G_{26}'\epsilon_2\epsilon_6' + G_2'\epsilon_2' \\
&\qquad\qquad + G_{66}'\epsilon_6'^2 \quad + G_6'\epsilon_6' = 1
\end{aligned}
\tag{18}
$$

Comparing the coefficients of Eqs. (17) and (18) and representing them in terms of the strain failure tensor components in Eq. (15), we arrive at

$$
\begin{aligned}
G_{11}' &= 2m^2n^2(-1/2E_1E_2) &&= 2m^2n^2G_{12} \\
G_{12}' &= (m^4+n^4)(-1/2E_1E_2) &&= (m^4+n^4)G_{12} \\
G_{16}' &= (m^3n - mn^3)(-1/2E_1E_2) &&= (m^3n - mn^3)G_{12} \\
G_{22}' &= 2m^2n^2(-1/2E_1E_2) &&= 2m^2n^2G_{12} \\
G_{26}' &= (mn^3 - m^3n)(-1/2E_1E_2) &&= (mn^3 - m^3n)G_{12} \\
G_{66}' &= -2m^2n^2(-1/2E_1E_2) &&= -2m^2n^2G_{12}
\end{aligned}
\tag{19a}
$$

$$G_1' = \frac{m^2}{E_1} + \frac{n^2}{E_2} \qquad = m^2G_1 + n^2G_2$$

$$G_2' = \frac{n^2}{E_1} + \frac{m^2}{E_2} \qquad = n^2G_1 + m^2G_2 \tag{19b}$$

$$G_6' = -mn\left(\frac{1}{E_1} - \frac{1}{E_2}\right) = -mnG_1 + mnG_2$$

It is immediately apparent that Eqs. (19a) and (19b) are, respectively, tensor transformation relations for the fourth and second rank tensors in Eq. (15); thus establishing that:

> The maximum strain criterion expressed in the form of Eq. (14b) is a degenerate case of the strain tensor polynomial formulation Eq. (10), and its coefficients obey tensor transformation algebra. (20)

It is also pertinent to examine the maximum strain criterion with respect to the stress formulation, since in current engineering application the maximum strain criterion is usually presented in the stress domain. We will follow the established practice (Waddoups, 1968) of mapping the failure envelope from the strain domain to the stress domain through the generalized Hooke's law:

$$\epsilon_1 = S_{11}\sigma_1 + S_{12}\sigma_2 + S_{16}\sigma_6 \qquad \epsilon_2 = S_{12}\sigma_1 + S_{22}\sigma_2 + S_{26}\sigma_6$$
$$\epsilon_6 = S_{16}\sigma_1 + S_{26}\sigma_2 + S_{66}\sigma_6 \tag{21}$$

Again, for brevity, we limit our discussion to the first strain quadrant for which the maximum strain failure criterion Eq. (14a) can be restated as: For $\epsilon_1 > 0$, $\epsilon_2 > 0$, $-E_6' < \epsilon_6 < E_6$;

$$E_1 = S_{11}X_1 \geq \epsilon_1, \qquad E_2 = S_{22}X_2 \geq \epsilon_2 \tag{22}$$

Substituting Eq. (22) into Eq. (21) and assuming orthotropic material, i.e., $S_{16} = S_{26} = 0$, we obtain

$$\sigma_1 + \frac{S_{12}}{S_{11}}\sigma_2 \leq X_1, \qquad \sigma_2 + \frac{S_{12}}{S_{22}}\sigma_1 \leq X_2 \tag{23a}$$

Equality in Eqs. (23a) describes the failure surface in stress space. Analogous to Eqs. (14a–c) we can express Eqs. (23a) in the form

$$\left(\sigma_1 + \frac{S_{12}}{S_{11}}\sigma_2 - X_1\right)\left(\sigma_2 + \frac{S_{12}}{S_{22}}\sigma_1 - X_2\right) = 0 \tag{23b}$$

Equation (23b) is the first quadrant of the maximum strain criterion represented in stress space. It may now be regarded as a stress formulation.

Expansion of Eq. (23b) yields

$$-\left(\frac{S_{12}}{S_{22}}\frac{1}{X_1X_2}\right)\sigma_1^{\,2} - \left(1 + \frac{S_{12}^2}{S_{11}S_{22}}\right)\frac{1}{X_1X_2}\sigma_1\sigma_2 - \left(\frac{S_{12}}{S_{11}}\frac{1}{X_1X_2}\right)\sigma_2^{\,2}$$

$$+\left(\frac{1}{X_1} + \frac{S_{12}}{S_{22}}\frac{1}{X_2}\right)\sigma_1 + \left(\frac{1}{X_2} + \frac{S_{12}}{S_{11}}\frac{1}{X_1}\right)\sigma_2 = 1 \qquad (23c)$$

Comparing the coefficients of Eq. (23c) to the stress tensor polynomial Eq. (5b or c) the material coefficients are

$$F_i = \begin{bmatrix} \dfrac{1}{X_1} + \dfrac{S_{12}}{S_{22}}\dfrac{1}{X_2} \\[2ex] \dfrac{1}{X_2} + \dfrac{S_{12}}{S_{11}}\dfrac{1}{X_1} \\[2ex] 0 \\ 0 \\ 0 \\ 0 \end{bmatrix},$$

$$F_{ij} = \begin{bmatrix} -\dfrac{S_{12}}{S_{22}}\dfrac{1}{X_1X_2} & -\dfrac{1}{2}\left(1 + \dfrac{S_{12}^2}{S_{11}S_{22}}\right)\dfrac{1}{X_1X_2} & 0 & 0 & 0 \\[2ex] & -\dfrac{S_{12}}{S_{11}}\dfrac{1}{X_1X_2} & 0 & 0 & 0 \\[2ex] & 0 & 0 & 0 & 0 \\ & & 0 & 0 & 0 \\ & & & 0 & 0 \\ & & & & 0 \end{bmatrix} \qquad (24)$$

Using either Eq. (23c) or substituting the material coefficients in Eq. (24)

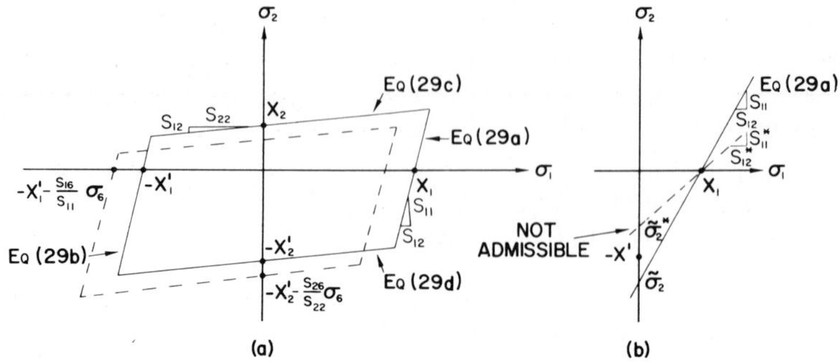

FIG. 5. (a) Maximum strain failure criterion (schematic) in stress space; (b) root for compliance ratio S^*_{11}/S^*_{12} not admissible.

into the stress tensor polynomial, Eq. (5b), it can easily be demonstrated that they represent a failure envelope which is in the familiar form of a parallelogram (Fig. 4b). Since the coefficients in Eqs. (15) and (24) effectively represent the failure envelope constructed from the same (maximum strain) hypothesis, they should be related by Eqs. (12). To verify this, we obtain the stress failure tensors F_i and F_{ij} directly from strain failure tensors G_i and G_{ij} in Eqs. (15) using Eqs. (12). For example:

$$F_1 = G_1 S_{11} + G_2 S_{22} = \frac{1}{\mathrm{E}_1} S_{11} + \frac{1}{\mathrm{E}_2} S_{22} = \frac{1}{X_1} + \frac{S_{12}}{S_{22}} \frac{1}{X_2} \tag{25a}$$

$$\vdots$$

$$F_{11} = G_{11} S_{11}^2 + 2G_{12} S_{11} S_{12} + G_{22} S_{12}^2 = 2\left(\frac{-1}{2\mathrm{E}_1\mathrm{E}_2}\right) S_{11} S_{12} = -\frac{S_{12}}{S_{22}} \frac{1}{X_1 X_2}$$

$$\vdots \tag{25b}$$

Both Eqs. (25a) and (25b) are identical to the respective coefficients listed in Eqs. (24) which were obtained in the conventional manner. Hence we have established that:

> The maximum strain criterion can be expressed as a degenerate form of a stress tensor polynomial through direct conversion from strain failure tensors to stress failure tensors. (26)

This condition is, of course, subject to the stress–strain linearity assumption as stated. Although simplicity was achieved in the above illustration through limiting the criterion to the first strain quadrant, conclusions (20) and (26) hold for the unabridged criterion Eq. (14a or b). Extrapolation can be established either through induction or through implementation of the more complex algebraic details. For example, we can write the maximum strain criterion for the more general case which covers the entire $\epsilon_1\epsilon_2$ space (for $-X_6' < \epsilon_6 < X_6$) by expansion of Eq. (14b) and by comparing to the coefficients of Eq. (10) we obtain

$$G_i = \begin{bmatrix} \frac{1}{E_1} - \frac{1}{E_1'} \\ \frac{1}{E_2} - \frac{1}{E_2'} \\ 0 \\ 0 \\ 0 \\ 0 \end{bmatrix}, \qquad G_{ij} = \begin{bmatrix} \frac{1}{E_1E_1'} & -\frac{G_1G_2}{2} & 0 & 0 & 0 & 0 \\ & \frac{1}{E_2E_2'} & 0 & 0 & 0 & 0 \\ & & 0 & 0 & 0 & 0 \\ & & & 0 & 0 & 0 \\ & & & & 0 & 0 \\ & & & & & 0 \end{bmatrix} \tag{27a}$$

$$G_{ijk}: \begin{cases} G_{112} = \dfrac{-1}{3E_1E_1'}\left(\dfrac{1}{E_2} - \dfrac{1}{E_2'}\right) = \dfrac{-1}{3} G_{11}G_2 \\[2ex] G_{122} = \dfrac{-1}{3E_2E_2'}\left(\dfrac{1}{E_1} - \dfrac{1}{E_1'}\right) = \dfrac{-1}{3} G_{22}G_1 \end{cases} \tag{27b}$$

$$G_{ijkl}: \qquad G_{1122} = \frac{-1}{4}\frac{1}{E_1E_1'}\frac{1}{E_2E_2'} = \frac{-1}{4} G_{11}G_{22} \tag{27c}$$

The transformational properties of Eqs. (27a–c) can be established as in the previous proof and will not be repeated. We note that this intuitively simple criterion actually involves the algebraic complexity of sixth and eighth rank tensors, respectively, Eqs. (27b) and (27c). This penalty of complexity is imposed without the benefit of providing additional material constants since Eqs. (27b) and (27c) depend on constants already incorporated in Eqs. (27a). We also note that compared to Eqs. (27a) the coefficients of the maximum strain criterion in a simplified case [Eqs. (15)] have different algebraic structure involving the engineering strain limits.

This suggests that simplification of the maximum strain criterion entailed more than simple elimination of higher–rank tensors.

The fact that maximum strain criterion results in a piecewise linear failure envelope, Figs. 4a and b, suggests that additional constraints in application may be required to assure that the failure envelope will be physically consistent in the stress space. The maximum strain criterion for a two-dimensional linear anisotropic plate which obeys Hooke's law [Eq. (20)] can be stated in terms of maximum stresses as

$$\mathrm{E}_1 = S_{11}X_1 \geq \epsilon_1 \tag{28a}$$

$$\mathrm{E}_1' = S_{11}X_1' \geq -\epsilon_1 \tag{28b}$$

$$\mathrm{E}_2 = S_{22}X_2 \geq \epsilon_2 \tag{28c}$$

$$\mathrm{E}_2' = S_{22}X_2' \geq -\epsilon_2 \tag{28d}$$

$$\mathrm{E}_6 = S_{66}X_6 \geq \epsilon_6 \tag{28e}$$

$$\mathrm{E}_6' = S_{66}X_6' \geq -\epsilon_6 \tag{28f}$$

Substitution of Eqs. (28a–f) into Eqs. (21) yields the failure envelope in the stress space

$$\sigma_1 = -\frac{S_{12}}{S_{11}}\sigma_2 - \frac{S_{16}}{S_{11}}\sigma_6 + X_1 \tag{29a}$$

$$\sigma_1 = -\frac{S_{12}}{S_{11}}\sigma_2 - \frac{S_{16}}{S_{11}}\sigma_6 - X_1' \tag{29b}$$

$$\sigma_2 = -\frac{S_{12}}{S_{22}}\sigma_1 - \frac{S_{26}}{S_{22}}\sigma_6 + X_2 \tag{29c}$$

$$\sigma_2 = -\frac{S_{12}}{S_{22}}\sigma_1 - \frac{S_{26}}{S_{22}}\sigma_6 - X_2' \tag{29d}$$

$$\sigma_6 = -\frac{S_{16}}{S_{66}}\sigma_1 - \frac{S_{26}}{S_{66}}\sigma_2 + X_6 \tag{29e}$$

$$\sigma_6 = -\frac{S_{16}}{S_{66}}\sigma_1 - \frac{S_{26}}{S_{66}}\sigma_2 - X_6' \tag{29f}$$

The failure envelope in the $\sigma_1\sigma_2$ stress space as described by Eqs. (29a–d) is schematically shown in Fig. 5a for $S_{12} < 0$. The solid envelope represents $\sigma_6 = 0$ and the broken envelope represents $X_6' < \sigma_6 < X_6$. For orthotropic materials for which the maximum strain criterion is frequently used,

$S_{16} = S_{26} = 0$ and the two envelopes coincide, i.e., the presence of σ_6 does not influence the failure envelope in $\sigma_1\sigma_2$ stress space. The failure envelopes in the $\sigma_1\sigma_6$ and $\sigma_2\sigma_6$ stress spaces are represented by the appropriate components of Eqs. (28) and they are similar to that shown in Fig. 5a.

Because of the piecewise linear nature of the failure envelope, additional constraints are required for consistency of conversion between the strain and stress spaces. The constraints on Eqs. (29) can be seen from the Fig. 5b where Eq. (29a) is sketched for two different ratios of S_{11}/S_{12}. It can be seen that the line using S_{11}^*/S_{12}^* ratio is not admissible. The σ_2 axis intercept $\tilde{\sigma}_2{}^*$ predicted by this equation is greater than $-X_2'$ implying that under tension σ_1, the induced Poisson strain in the 2 direction will cause compressive failure at a stress less than the simple compressive strength in the 2 direction. In other words, the simple compressive strength X_2' can never be realized, contradicting Eq. (28d) which states that X_2' is an experimentally measured quantity. In order to avoid this contradiction, the intercept $\tilde{\sigma}_2$ with the σ_2 axis must be less than or equal to $-X_2'$, i.e.,

$$\tilde{\sigma}_2 = -\frac{S_{16}}{S_{12}}\sigma_6 + \frac{S_{11}}{S_{12}}X_1 \leq -X_2' \tag{30}$$

but $-X_6' \leq \sigma_6 \leq X_6$ reduces Eq. (30) to

$$\frac{S_{16}}{S_{12}}\frac{X_6}{X_1} - \frac{S_{11}}{S_{12}} \geq \frac{X_2'}{X_1} \tag{30a}$$

Similarly, consideration for the remainder of Eqs. (28) yields the following constraints:

$$\frac{S_{16}}{S_{12}}\frac{X_6'}{X_1'} - \frac{S_{11}}{S_{12}} \geq \frac{X_2}{X_1'} \tag{30b}$$

$$\frac{S_{26}}{S_{12}}\frac{X_6}{X_2} - \frac{S_{22}}{S_{12}} \geq \frac{X_1'}{X_2} \tag{30c}$$

$$\frac{S_{26}}{S_{12}}\frac{X_6'}{X_2} - \frac{S_{22}}{S_{12}} \geq \frac{X_1}{X_2'} \tag{30d}$$

$$\frac{S_{66}}{S_{16}}\frac{X_6}{X_2} - \frac{S_{26}}{S_{16}} \geq \frac{X_1}{X_2} \tag{30e}$$

$$\frac{S_{66}}{S_{16}}\frac{X_6}{X_2} - \frac{S_{26}}{S_{16}} \geq \frac{X_1'}{X_2'} \tag{30f}$$

The above inequality constraints (30a–f) are written assuming $S_{12} < 0$,

TABLE I

MATERIALS WHICH DO NOT SATISFY ALL CONDITIONS

Material	$-S_{22}/S_{12}$	X_1'/X_2	X_1/X_2'	Condition not satisfied
Glass–epoxy (Scotch-Ply 1002)	13.3	16	5.5	Eq. (30c)
Boron–Al (48% V_f)[a]	6.9	8.1	9.3	Eqs. (30c) and (30d)

[a] Data from Collins and Thomas (1973).

$S_{16} > 0$, and $S_{26} > 0$ for the respective portion of the failure envelope defined by Eqs. (29a–f). In the case of special orthotropy, the constraints by Eqs. (30a–d) can be obtained by setting $S_{16} = S_{26} = 0$; whereas the constraints due to Eqs. (30e,f) vanish.

If the conditions (30a–f) are not met, the conversion between the strain and stress domain is not unique, rendering the physical interpretation of the failure surface ambiguous and the application of the criterion intractable. These constraints are not conditions of only academic interest;

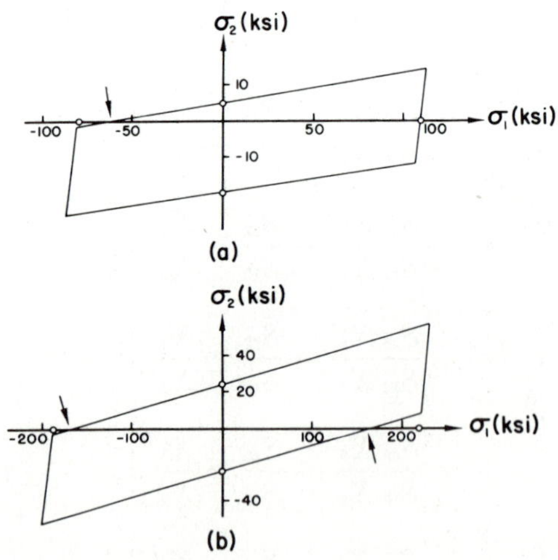

FIG. 6. Maximum strain failure surfaces (actual) for: (a) glass–epoxy composite (Scotch-Ply 1002) and (b) boron–aluminum. Arrows indicate contradictions of experimental facts in the course of conversion from strain space to stress space.

some practical composites possess material constants which do not meet some of the above conditions as shown in Table I and Figs. 6a and b.

We can now place the maximum strain criterion in the perspective of fundamental requirements of a failure criterion.

(1) With proper interpretation [as in Eqs. (14b) and (23b)], it is a degenerate case of either the strain tensor polynomial or the stress tensor polynomial and it is invariant to coordinate transformation.

(2) It lacks the flexibility for modeling failure envelopes of a general form because of the preconceived assumptions that there is no strength coupling between normal strains and between the normal and shear strains. Limited by its mathematical structure, its characterization for composites in general is expected to be compromising.

(3) Although its formulation is intuitively simple it actually involves algebraic complexity of sixth or eighth rank tensors, Eqs. (27b,c). Simplifications lead to complete alteration of the algebraic relation between the material parameters and the engineering strain limits, and entail more than simple elimination of higher-order tensors.

(4) Because it represents a piecewise linear failure envelope, unambiguous conversions between the stress–strain space are subjected to constraints which depend on the strength ratios and the elastic coefficients, Eqs. (30a–f). If the constraint conditions are not met, this criterion can only be used with great caution.

Despite these complexities, the maximum strain criterion is eminently suitable for characterizing the failure of some special composites. For example, an orthogonal bidirectional composite with very low shear modulus (such as cloth with a rubber matrix) is expected to exhibit little coupling between normal strains and normal shear strains. In the absence of such compelling physical considerations, its application to composites in general warrants prudent consideration.

4. Maximum Stress Criterion

Rankin's maximum stress theory for isotropic material was modified by Jenkins (1920) to characterize strength of orthotropic materials. In this formulation the coordinate axis must be oriented along the material principal directions (schematically illustrated in Fig. 3), externally applied stress must be transformed to the coordinates, and failure is hypothesized if the following criterion is not fulfilled:

$$-X_1' \leq \sigma_1 \leq X_1, \qquad -X_2' \leq \sigma_2 \leq X_2 \tag{31a}$$

Whereas the isotropic maximum stress theory can be expressed in terms of maximum principal stress, this is no longer possible for anisotropic materials because the reference coordinates are arbitrarily fixed.

The mathematical structure of this failure criterion is identical to the maximum strain criterion, Eqs. (14a). Thus, using the previous procedures with appropriate modifications, we can readily establish the maximum stress criterion transformational properties and conversion between the stress and strain domains. The important results are presented. As in the previous treatment, we can rewrite the maximum stress criterion in the following form:

$$(\sigma_1 - X_1)(\sigma_1 + X_1')(\sigma_2 - X_2)(\sigma_2 + X_2') = 0 \tag{31b}$$

Both Eqs. (31a) and (31b) will model a rectangular failure envelope in the stress space as shown in Fig. 7. Expansion of Eq. (31b) yields

$$\begin{aligned} &-\frac{\sigma_1^2\sigma_2^2}{X_1X_1'X_2X_2'} - \frac{1}{X_1X_1'}\left(\frac{1}{X_2} - \frac{1}{X_2'}\right)\sigma_1^2\sigma_2 - \frac{1}{X_2X_2'}\left(\frac{1}{X_1} - \frac{1}{X_1'}\right)\sigma_1\sigma_2^2 \\ &+ \frac{\sigma_1^2}{X_1X_1'} + \frac{\sigma_2^2}{X_2X_2'} - \left(\frac{1}{X_1} - \frac{1}{X_1'}\right)\left(\frac{1}{X_2} - \frac{1}{X_2'}\right)\sigma_1\sigma_2 \\ &+ \left(\frac{1}{X_1} - \frac{1}{X_1'}\right)\sigma_1 + \left(\frac{1}{X_2} - \frac{1}{X_2'}\right)\sigma_2 = 1 \end{aligned} \tag{31c}$$

Comparison of Eq. (31c) to the stress tensor polynomial using the notation for engineering strengths in Eqs. (8), the maximum stress criterion has the following material coefficients:

$$F_i = \begin{bmatrix} \frac{1}{X_1} - \frac{1}{X_1'} \\ \frac{1}{X_2} - \frac{1}{X_2'} \\ 0 \\ 0 \\ 0 \\ 0 \end{bmatrix}, \qquad F_{ij} = \begin{bmatrix} \frac{1}{X_1X_1'} & -\frac{1}{2}F_1F_2 & 0 & 0 & 0 & 0 \\ & \frac{1}{X_2X_2'} & 0 & 0 & 0 & 0 \\ & & 0 & 0 & 0 & 0 \\ & & & 0 & 0 & 0 \\ & & & & 0 & 0 \\ & & & & & 0 \end{bmatrix} \tag{32a}$$

$$F_{ijk}: \begin{cases} F_{112} = -\dfrac{1}{3}\dfrac{1}{X_1X_1'}\left(\dfrac{1}{X_2} - \dfrac{1}{X_2'}\right) = -\dfrac{1}{3}F_{11}F_2 \\ \\ F_{122} = -\dfrac{1}{3}\dfrac{1}{X_2X_2'}\left(\dfrac{1}{X_1} - \dfrac{1}{X_1'}\right) = -\dfrac{1}{3}F_{22}F_1 \end{cases} \tag{32b}$$

$$F_{ijkl}: \quad F_{1122} = -\frac{1}{4}\frac{1}{X_1X_1'}\frac{1}{X_2X_2'} = -\frac{1}{4}F_{11}F_{22} \tag{32c}$$

In the form proposed and in the manner it is normally used, Eq. (31a), the maximum stress criterion, is not invariant to coordinate transformation. However, we establish its transformational properties by using the form presented in Eqs. (31c) and (32). The maximum stress criterion [Eq. (31b)] in an arbitrary coordinate x_i can be obtained either by (1) transforming the stresses from σ_i to σ_i' or (2) by transforming the stress failure tensors $F_i, F_{ij}, \ldots$ to $F_i', F'_{ij}, \ldots$ using Eqs. (6). Both of these transformations yield identical results to those shown by Eq. (15)–(19), permitting us to establish that:

> The maximum stress criterion is a degenerate case of the stress tensor polynomial formulation Eq. (5b) and its coefficients obey tensor transformational algebra. (33)

Thus, with the proper interpretation, the maximum stress criterion is invariant to coordinate transformation and can be adapted to mathematical operations in structural analysis.

It is worth noting the similarity between the strength parameters in Eqs. (32) and the strength parameters for the quadratic tensor polynomial in Eqs. (8). The maximum stress criterion is in fact more complex than the

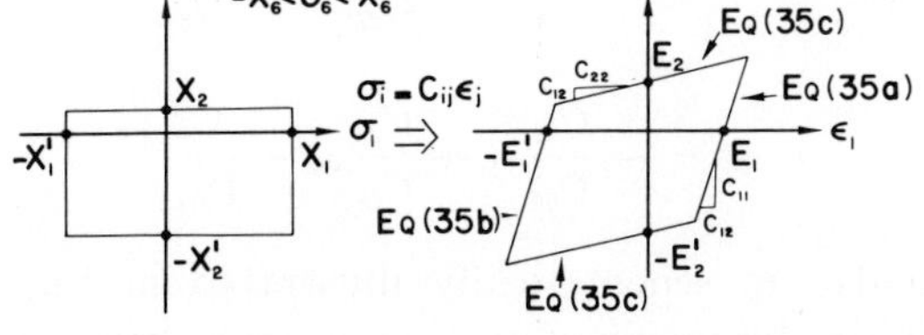

FIG. 7. Maximum stress criterion (schematic): (a) in stress space and (b) in strain space.

quadratic stress tensor polynomial since it contains the algebraic complexity of sixth and eighth rank tensors F_{ijk} and F_{ijkl} without actually introducing new material constants as seen in Eqs. (32a–c). Despite this added complexity, it still has two severe limitations. It does not define the conditions of shear stress induced failure and it assumes that there is no strength coupling between normal stresses and between normal and shear stresses. The first limitation can be overcome by incorporating Coulomb's (1773) type of maximum shear stress failure criterion into Eqs. (31), making a failure criterion of the form:

$$-X_1' \leq \sigma_1 \leq X_1, \qquad -X_2' \leq \sigma_2 \leq X_2, \qquad -X_6' \leq \sigma_6 \leq X_6 \tag{34}$$

This modification is equivalent to including F_6 and F_{66} in terms in Eq. (32a) which serve to make the maximum stress criterion closer to the general form of the stress tensor polynomial.

The failure envelope of the maximum stress criterion which includes the effect of shear stress [Eq. (34)] is represented in the stress space in Fig. 7a. By assuming linearity and through generalized Hooke's law, we can express the maximum stress criterion in the strain space as:

$$\epsilon_1 = -\frac{C_{12}}{C_{11}}\epsilon_2 - \frac{C_{16}}{C_{11}}\epsilon_6 + \frac{X_1}{C_{11}} \tag{35a}$$

$$\epsilon_1 = -\frac{C_{12}}{C_{11}}\epsilon_2 - \frac{C_{16}}{C_{11}}\epsilon_6 - \frac{X_1'}{C_{11}} \tag{35b}$$

$$\epsilon_2 = -\frac{C_{12}}{C_{22}}\epsilon_1 - \frac{C_{26}}{C_{22}}\epsilon_6 + \frac{X_2}{C_{22}} \tag{35c}$$

$$\epsilon_2 = -\frac{C_{12}}{C_{22}}\epsilon_1 - \frac{C_{26}}{C_{22}}\epsilon_6 - \frac{X_2'}{C_{22}} \tag{35d}$$

$$\epsilon_6 = -\frac{C_{16}}{C_{66}}\epsilon_1 - \frac{C_{26}}{C_{22}}\epsilon_6 + \frac{X_6}{C_{66}} \tag{35e}$$

$$\epsilon_6 = -\frac{C_{16}}{C_{66}}\epsilon_1 - \frac{C_{26}}{C_{66}}\epsilon_6 - \frac{X_6'}{C_{66}} \tag{35f}$$

Equations (35a–d) are schematically illustrated in Fig. 7b. It can be readily seen as in the discussion of maximum strain criterion that due to the piecewise linear nature of the respective failure envelopes in the stress any strain spaces, the maximum stress criterion is physically meaningful only when the following conditions are met:

$$\frac{C_{16}C_{22}}{C_{12}C_{66}}\frac{X_6'}{X_1} - \frac{C_{22}}{C_{12}} \geq \frac{X_2'}{X_1} \tag{36a}$$

$$\frac{C_{16}C_{22}}{C_{12}C_{66}}\frac{X_6}{X_1'} - \frac{C_{22}}{C_{12}} \geq \frac{X_1'}{X_2} \tag{36b}$$

$$\frac{C_{26}C_{11}}{C_{12}C_{66}}\frac{X_6'}{X_2} - \frac{C_{11}}{C_{12}} \geq \frac{X_1}{X_2'} \tag{36c}$$

$$\frac{C_{26}C_{11}}{C_{12}C_{66}}\frac{X_6}{X_2'} - \frac{C_{22}}{C_{12}} \geq \frac{X_1}{X_2} \tag{36d}$$

$$\frac{C_{26}C_{22}}{C_{16}C_{66}}\frac{X_6}{X_2} - \frac{C_{26}}{C_{16}} \geq \frac{X_1}{X_2} \tag{36e}$$

$$\frac{C_{26}C_{11}}{C_{16}C_{66}}\frac{X_6}{X_1} - \frac{C_{26}}{C_{16}} \geq \frac{X_1'}{X_2'} \tag{36f}$$

If the conditions of Eqs. (36a–f) are not met, the interpretation of the failure surface becomes ambiguous. For example, if Eq. (34b) is not satisfied, the physical implication is that tensile failure in the 2 direction is not possible; it will be preceded by Poisson effect induced compressive failure in the 1 direction. Such phenomena are hardly observed in real solids which suggests that due to its piecewise linear characterization, the maximum stress criterion must be used with great caution. Although the conditions of Eqs. (36a–f) are derived from the linear elastic case, similar conditions can be derived for the nonlinear materials if a nonlinear constitutive relation is available.

Because the mathematical structure of the maximum stress failure criterion is identical to the maximum strain failure criterion, it possesses the same shortcoming with respect to the fundamental requirements of a mathematical model. We will not repeat the discussions already presented for maximum strain criterion except to restate that the maximum stress criterion is a degenerate case of the tensor polynomial; although it is invariant to coordinate transformation, it entails algebraic complexity with limited flexibility. It appears to be suitable for application to orthogonally bidirectional composites with extremely low moduli. Even in such applications, the strength ratios have to satisfy the constraints stated in Eqs. (36a–f).

5. Criteria in Terms of Invariants

The formulation of anisotropic failure criteria in terms of stress invariants was apparently motivated by the historical development of yield criteria for isotropic materials. The limiting condition of isotropy (directional independency) connotes that all material axes are equally suitable as a reference frame for failure. A convenient vehicle for satisfying imposed invariancy to coordinate transformation is to express failure criteria in terms of scalar functions of the stress invariants. Bridgman (1922) reported that yielding of isotropic material is independent of hydrostatic pressure. Incorporation of this observation allows further simplification of isotropic yield criterion to include only deviatoric invariants.

The adaptation of a similar rationale to the formulation of anisotropic failure criterion can be traced to von Mises (1928) with subsequent revitalization by Hill (1948). Both formulations retain the hypothesis that anisotropic yielding is independent of hydrostatic stress as in the isotropic case. Retention of this hypothesis is in direct contradiction to Bridgman's (1941) latter experiments on anisotropic crystals which demonstrated that yielding occurred under hydrostatic stress. Although it was not explicitly stated, the von Mises–Hill criterion was probably intended for metals which have been cold worked and whose anisotropy ratio is typically small as compared to modern composites. Hence, for its intended application, the erroneous hydrostatic pressure may not be significant. The significance of neglecting the effect of hydrostatic stress for highly anisotropic composites may be examined by reducing the stress tensor polynomial to the von Mises–Hill criterion. Using only the quadratic terms, the stress tensor polynomial, Eq. (5b), failure criterion reduces to

$$F_{ij}\sigma_i\sigma_j = 1 \tag{37}$$

If the failure condition is independent of hydrostatic stress σ_p, then the derivative of Eq. (37) with respect to σ_p must be zero, i.e.,

$$\frac{\partial}{\partial \sigma_p}(F_{ij}\sigma_i\sigma_j) = 0 \tag{38}$$

where

$$\sigma_p = \sigma_1 + \sigma_2 + \sigma_3$$

Solving Eq. (38) we obtain

$$\begin{aligned} F_{11} + F_{12} + F_{13} &= 0, & F_{12} + F_{22} + F_{23} &= 0 \\ F_{13} + F_{23} + F_{33} &= 0, & F_{14} + F_{24} + F_{34} &= 0 \\ F_{15} + F_{25} + F_{35} &= 0, & F_{16} + F_{26} + F_{36} &= 0 \end{aligned} \tag{39}$$

Thus, if failure of a composite is independent of hydrostatic stress, the number of material constants F_{ij} reduces from 21 to 15 by the six subsidiary equations [Eqs. (39)]. For a strength orthotropic composite the number of independent constants for F_{ij} reduces from 9 to 6 by Eqs. (39) and they can be expressed as

$$F_{ij} = \begin{bmatrix} -(F_{12}+F_{13}) & F_{12} & F_{13} & 0 & 0 & 0 \\ & -(F_{12}+F_{23}) & F_{23} & 0 & 0 & 0 \\ & & -(F_{13}+F_{33}) & 0 & 0 & 0 \\ & & & F_{44} & 0 & 0 \\ & & & & F_{55} & 0 \\ & & & & & F_{66} \end{bmatrix} \tag{40}$$

On the other hand, Hill's criterion has the appearance of the second deviatoric stress invariant with modifying materials coefficients

$$F(\sigma_2-\sigma_3)^2 + G(\sigma_3-\sigma_1)^2 + H(\sigma_1-\sigma_2)^2 + 2L\sigma_4^2 + 2M\sigma_5^2 + 2N\sigma_6^2 = 1 \tag{41}$$

Expanding Eq. (41) and comparing coefficients with the stress tensor polynomial, Eq. (5b), we obtain

$$F_i = \begin{bmatrix} 0 \\ 0 \\ 0 \\ 0 \\ 0 \\ 0 \end{bmatrix}, \qquad F_{ij} = \begin{bmatrix} (G+H) & -H & -G & 0 & 0 & 0 \\ & (F+H) & -F & 0 & 0 & 0 \\ & & (F+G) & 0 & 0 & 0 \\ & & & L & 0 & 0 \\ & & & & M & 0 \\ & & & & & N \end{bmatrix} \tag{42}$$

Equations (42) are identical in form to Eq. (40). Thus we conclude that:

The von Mises–Hill criterion is a degenerate case of the stress tensor polynomial with the effect of hydrostatic stress not accounted for. (43a)

It is conceivable that there may exist composites whose failure is independent of hydrostatic stress; for those composites Hill's criterion can be useful. Nevertheless, it is clear that reduction of the independent quadratic material constants from 21 to 15 results in a reduction of flexibility. Experimental data to be presented in the subsequent section indicates that, for graphite–epoxy composites, the conditions of Eqs. (39) are not satisfied. The von Mises–Hill criterion [Eq. (41)] deceptively appears to be a modification of the second deviatoric invariant. In actual fact, the modifying coefficients F, G, H, . . . , etc. are functions of the orientation of the reference axis. Therefore, Eq. (41) is neither an invariant nor can it be generally interpreted as the energy of distortion. Equation (41) was originally proposed for reference axes which coincide with the principal directions of an orthotropic material. Mathematically operational convenience can be gained by applying it in the form of stress tensor polynomials, Eqs. (42), which then allow its transformation to any arbitrary reference axes. The transformational properties of Eqs. (42) can be easily verified by following the proof established previously. We can state that:

The von Mises–Hill criterion, when used in tensor polynomial form, can be expressed in any arbitrary reference coordinate through tensor transformational relationships. (43b)

This statement may appear to be contradictory to examples which show that Hill's failure criterion does not obey tensor transformation. We can readily demonstrate that these so-called "anomalies" resulted from misuse of the criterion. For example, for a composite which is transversely isotropic in strength, i.e.,

$$X_2 = X_3 \tag{44}$$

Eq. (41) can be specialized to the popularly used form

$$\frac{\sigma_1^2}{X_1^2} - \frac{\sigma_1\sigma_2}{X_1^2} + \frac{\sigma_2^2}{X_2^2} + \frac{\sigma_6^2}{X_6^2} = 1 \tag{45}$$

One "anomaly" arises when Eq. (45) is used to characterize the failure of angle-ply laminates where the principal direction is ambiguous. If the reference axes are oriented as shown in Fig. 8a, (p. 388), Eq. (45) is the failure criterion for the x_i reference coordinate. Alternately, if the reference axes are oriented as shown in Fig. 8b, substitution of the same engineering strengths X_1, X_2, X_6 into Eq. (45) yields the failure criterion in the x_i' reference coordinate:

$$\frac{\sigma_1'^2}{X_2^2} - \frac{\sigma_1'\sigma_2'}{X_2^2} + \frac{\sigma_2'^2}{X_1^2} + \frac{\sigma_6'^2}{X_6^2} = 1 \tag{46a}$$

However, if we transform the stress from the x_i' coordinate by replacing $\sigma_1' = \sigma_2$, $\sigma_2' = \sigma_1$, $\sigma_6' = \sigma_6$ in Eq. (46a) we obtain the failure criterion referenced to the x_i coordinate

$$\frac{\sigma_1^2}{X_1^2} - \frac{\sigma_1\sigma_2}{X_2^2} + \frac{\sigma_2^2}{X_2^2} + \frac{\sigma_6^2}{X_6^2} = 1 \tag{46b}$$

Since Eq. (46b) and Eq. (45) are not the same, it appears that Eq. (45) does not obey coordinate transformation rules. The anomaly is not unexpected because Eq. (45) is a specialized form of Hill's criterion where transverse strength isotropy was assumed. Obviously, for the example of angle-ply laminate, $X_2 \neq X_3$ and Eq. (45) is not applicable. If we use Hill's original form, for Fig. 8a the failure criterion is

$$\frac{\sigma_1^2}{X_1^2} - \left(\frac{1}{X_1^2} + \frac{1}{X_2^2} - \frac{1}{X_3^2}\right)\sigma_1\sigma_2 + \frac{\sigma_2^2}{X_2^2} + \frac{\sigma_6^2}{X_6^2} = 1 \tag{47a}$$

For Fig. 8b the failure criterion is

$$\frac{\sigma_1'^2}{X_2^2} - \left(\frac{1}{X_2^2} + \frac{1}{X_1^2} - \frac{1}{X_3^2}\right)\sigma_1'\sigma_2' + \frac{\sigma_2'^2}{X_1^2} + \frac{\sigma_6'^2}{X_6^2} = 1 \tag{47b}$$

It is immediately apparent that transformation of x_i to x_i' through replacement of $\sigma_1' = \sigma_2$, $\sigma_2' = \sigma_1$, $\sigma_6' = \sigma_6$ in Eq. (47b) yields a failure criterion identical to that represented by Eq. (47a). Thus proper application of Hill's criterion does not lead to operational anomalies. Nevertheless, the flexibility of this criterion is inherently limited by the underlying assumptions that the material is orthotropic, that it exhibits no Bauschinger effect, and that it is not affected by hydrostatic stress.

Some of the above limitations can be removed if the failure criterion is formulated in terms of stress invariants instead of deviatoric invariants. The merits of expressing a failure criterion in terms of stress invariant formulations are difficult to assess, because a large number of functional forms exist. Instead, we chose to interpret the stress tensor polynomial Eq. (5b) in the form of stress invariants. The different terms in the stress tensor polynomial can be interpreted separately; denoting the linear terms in stress by f_1, we write

$$f_1 = \mathrm{tr}\begin{bmatrix} F_1 \\ F_2 \\ F_3 \\ F_4 \\ F_5 \\ F_6 \end{bmatrix}\begin{bmatrix} \sigma_1 \\ \sigma_2 \\ \sigma_3 \\ \sigma_4 \\ \sigma_5 \\ \sigma_6 \end{bmatrix}^{\mathrm{T}} = F_1\sigma_1 + F_2\sigma_2 + F_3\sigma_3 + F_4\sigma_4 + F_5\sigma_5 + F_6\sigma_6 \tag{48a}$$

Under the conditions of isotropy, $F_1 = F_2 = \cdots = F$, Eq. (48a) reduces to

$$f_1 = FI_1 + F(\sigma_4 + \sigma_5 + \sigma_6) \tag{48b}$$

In general, we may find an eigenvector and through coordinate transformation express Eq. (48a) in terms of principal stresses $\bar{\sigma}_i$:

$$f_1 = \mathrm{tr}\begin{bmatrix} \bar{F}_1' \\ \bar{F}_2' \\ \bar{F}_3' \\ \bar{F}_4' \\ \bar{F}_5' \\ \bar{F}_6' \end{bmatrix}\begin{bmatrix} \bar{\sigma}_1' \\ \bar{\sigma}_2' \\ \bar{\sigma}_3' \\ 0 \\ 0 \\ 0 \end{bmatrix}^{\mathrm{T}} = \bar{F}_1'\bar{\sigma}_1' + \bar{F}_2'\bar{\sigma}_2' + \bar{F}_3'\bar{\sigma}_3' \tag{49a}$$

where $\bar{x}_i'$ is the principal direction. In the isotropic case since the material coefficient is not a function of direction, i.e., $F_1 = F_2 = \cdots = \bar{F}_1' = \bar{F}_2' = \cdots = F$, Eq. (49a) becomes

$$f_1 = F(\bar{\sigma}_1' + \bar{\sigma}_2' + \bar{\sigma}_3') = FI_1 \tag{49b}$$

However, since the failure condition cannot be a function of reference coordinates, Eq. (48b) and Eq. (49b) must be equal. This requires the second term in Eq. (48b) to vanish implying that for an isotropic material in arbitrary coordinates, $F_4 = F_5 = F_6 = 0$ and the failure criterion involving linear stress terms can be expressed as:

$$f_1 = \operatorname{tr}\begin{bmatrix} F \\ F \\ F \\ 0 \\ 0 \\ 0 \end{bmatrix}\begin{bmatrix} \sigma_1 \\ \sigma_2 \\ \sigma_3 \\ \sigma_4 \\ \sigma_5 \\ \sigma_6 \end{bmatrix}^{\mathrm{T}} = \operatorname{tr}\begin{bmatrix} F \\ F \\ F \\ 0 \\ 0 \\ 0 \end{bmatrix}\begin{bmatrix} \bar{\sigma}_1' \\ \bar{\sigma}_2' \\ \bar{\sigma}_3' \\ 0 \\ 0 \\ 0 \end{bmatrix}^{\mathrm{T}} = FI_1 \tag{50}$$

Hence, the basis for formulating the failure condition in terms of the first stress invariant by the second expression of Eq. (50) can be interpreted as: For isotropy, the eigenvector of the failure tensor F_i always coincides with the eigenvector of the stress tensor.

For anisotropic material, the failure condition is also independent of reference coordinates requiring Eqs. (48a) and (49b) to be equal. However, because of anisotropy, in general,

$$F_1\sigma_1 + F_2\sigma_2 + F_3\sigma_3 \neq \bar{F}_1\bar{\sigma}_1' + \bar{F}_2\bar{\sigma}_2' + \bar{F}_3\bar{\sigma}_3'$$

Thus

$$F_4 \neq 0, \qquad F_5 \neq 0, \qquad F_6 \neq 0 \tag{51}$$

Hence, the anisotropic failure condition will have the algebraic appearance of the first stress invariant only when one of the following conditions are satisfied:

$$f_1 = \operatorname{tr}\begin{bmatrix} \bar{F}_1 \\ \bar{F}_2 \\ \bar{F}_3 \\ 0 \\ 0 \\ 0 \end{bmatrix}\begin{bmatrix} \sigma_1 \\ \sigma_2 \\ \sigma_3 \\ \sigma_4 \\ \sigma_5 \\ \sigma_6 \end{bmatrix}^{\mathrm{T}} = \operatorname{tr}\begin{bmatrix} F_1' \\ F_2' \\ F_3' \\ F_4' \\ F_5' \\ F_6' \end{bmatrix}\begin{bmatrix} \bar{\sigma}_1' \\ \bar{\sigma}_2' \\ \bar{\sigma}_3' \\ 0 \\ 0 \\ 0 \end{bmatrix}^{\mathrm{T}} = \operatorname{tr}\begin{bmatrix} \bar{F}_1 \\ \bar{F}_2 \\ \bar{F}_3 \\ 0 \\ 0 \\ 0 \end{bmatrix}\begin{bmatrix} \bar{\sigma}_1 \\ \bar{\sigma}_2 \\ \bar{\sigma}_3 \\ 0 \\ 0 \\ 0 \end{bmatrix}^{\mathrm{T}} = F_1\sigma_1 + F_2\sigma_2 + F_3\sigma_3 \tag{52}$$

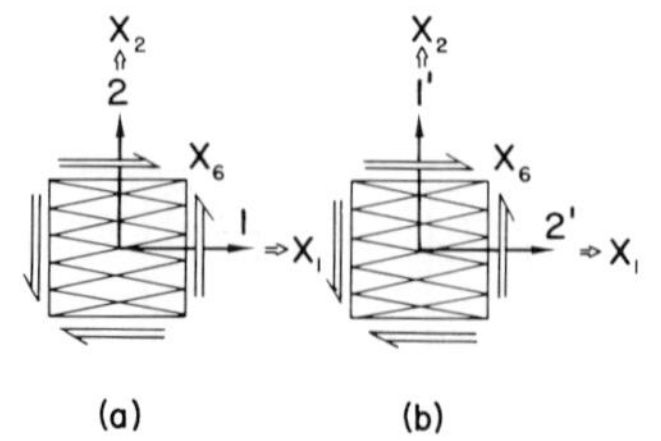

FIG. 8. Two natural definitions of reference axes for angle-ply laminate.

The first condition is when the reference axes coincide with the eigenvector of F_i (the orthotropic material principal direction); the second condition is when the reference axes coincide with the eigenvector of σ_i (the principal stress direction); the third condition is when the eigenvector of the strength tensor and the stress tensor coincide. The second condition is generally not fulfilled and the third condition is satisfied only for isotropy. Thus we conclude that

> With no loss of generality, anisotropic failure criterion formulated by modifying the first stress invariant can be used only for orthotropic materials and when reference coordinates coincide with the material principal directions.

For a general anisotropic composite, the invariant formulation results in a loss of flexibility when the number of independent material constants (for linear effect of stress) is reduced from 6 to 3. Similarly the quadratic terms in the stress tensor polynomial can be written as f_2:

$$f_2 = F_{ij}\sigma_i\sigma_j$$

$$= \mathrm{tr}\begin{bmatrix} F_{11} & F_{12} & F_{13} & F_{14} & F_{15} & F_{16} \\ & F_{22} & F_{23} & F_{24} & F_{25} & F_{26} \\ & & F_{33} & F_{34} & F_{35} & F_{36} \\ & & & F_{44} & F_{45} & F_{46} \\ & & & & F_{55} & F_{56} \\ & & & & & F_{66} \end{bmatrix} \begin{bmatrix} \sigma_1^2 & \sigma_1\sigma_2 & \sigma_1\sigma_3 & \sigma_1\sigma_4 & \sigma_1\sigma_5 & \sigma_1\sigma_6 \\ & \sigma_2^2 & \sigma_2\sigma_3 & \sigma_2\sigma_4 & \sigma_2\sigma_5 & \sigma_2\sigma_6 \\ & & \sigma_3^2 & \sigma_3\sigma_4 & \sigma_3\sigma_5 & \sigma_3\sigma_6 \\ & & & \sigma_4^2 & \sigma_4\sigma_5 & \sigma_4\sigma_6 \\ & & & & \sigma_5^2 & \sigma_5\sigma_6 \\ & & & & & \sigma_6^2 \end{bmatrix} \tag{53}$$

If Eq. (53) is to be used for the isotropic case $F_{11} = F_{22} = \cdots = \hat{F}$, it can be expressed in terms of the second stress invariants as

$$\begin{aligned} f_2 = F_{ij}\sigma_i\sigma_j = \hat{F}I_2 + \hat{F}(&\sigma_1^2 + 2\sigma_1\sigma_4 + 2\sigma_1\sigma_5 + 2\sigma_1\sigma_6 \\ &+ \sigma_2^2 + 2\sigma_2\sigma_4 + 2\sigma_2\sigma_5 + 2\sigma_2\sigma_6 \\ &+ \sigma_3^2 + 2\sigma_3\sigma_4 + 2\sigma_3\sigma_5 + 2\sigma_3\sigma_6 \\ &+ 2\sigma_4\sigma_5 + 2\sigma_4\sigma_6 \\ &+ 2\sigma_5\sigma_6) \end{aligned} \tag{54}$$

For the same isotropic material in another arbitrary coordinate x_i' the resulting failure condition is

$$\begin{aligned} F_{ij}\sigma_i\sigma_j = F_{ij}\sigma_i'\sigma_j' = \hat{F}I_2 + \hat{F}(&\sigma_1'^2 + 2\sigma_1'\sigma_4' + 2\sigma_1'\sigma_5' + 2\sigma_1'\sigma_6' \\ &+ \sigma_2'^2 + 2\sigma_2'\sigma_4' + 2\sigma_2'\sigma_5' + 2\sigma_2'\sigma_6' \\ &+ \sigma_3'^2 + 2\sigma_3'\sigma_4' + 2\sigma_3'\sigma_5' + 2\sigma_3'\sigma_6' \\ &+ 2\sigma_4'\sigma_5' + 2\sigma_4'\sigma_6' \\ &+ 2\sigma_5'\sigma_6') \end{aligned} \tag{55}$$

Since the failure condition for the reference axes x_i and x_i' must be the same, Eq. (55) must be equal to Eq. (54). It follows that the second term of the right-hand side of both equations must be equal or that for isotropy the following 15 materials coefficients must vanish:

$$\begin{aligned} F_{11} = F_{22} = F_{33} &= 0 \\ F_{14} = F_{15} = F_{16} &= 0 \\ F_{24} = F_{25} = F_{26} &= 0 \\ F_{34} = F_{35} = F_{36} &= 0 \\ F_{45} = F_{46} &= 0 \\ F_{56} &= 0 \end{aligned} \tag{56}$$

and Eq. (53) can be expressed as

$$f_2 = \mathrm{tr}\begin{bmatrix} 0 & \frac{\hat{F}}{2} & \frac{\hat{F}}{2} & 0 & 0 & 0 \\ & 0 & \frac{\hat{F}}{2} & 0 & 0 & 0 \\ & & 0 & 0 & 0 & 0 \\ & & & \hat{F} & 0 & 0 \\ & & & & \hat{F} & 0 \\ & & & & & \hat{F} \end{bmatrix}\begin{bmatrix} \sigma_1{}^2 & \sigma_1\sigma_2 & \sigma_1\sigma_3 & \sigma_1\sigma_4 & \sigma_1\sigma_5 & \sigma_1\sigma_6 \\ & \sigma_2{}^2 & \sigma_2\sigma_3 & \sigma_2\sigma_4 & \sigma_2\sigma_5 & \sigma_2\sigma_6 \\ & & \sigma_3{}^2 & \sigma_3\sigma_4 & \sigma_3\sigma_5 & \sigma_3\sigma_6 \\ & & & \sigma_4{}^2 & \sigma_4\sigma_5 & \sigma_4\sigma_6 \\ & & & & \sigma_5{}^2 & \sigma_5\sigma_6 \\ & & & & & \sigma_6{}^2 \end{bmatrix} \tag{57a}$$

Thus Eq. (57a) provides the failure condition in the form of the second stress invariant, i.e.,

$$f_2 = \hat{F} I_2 = 1 \tag{57b}$$

It can be readily seen that the general quadratic terms can be specialized to give it the algebraic appearance of the second stress invariant by writing:

$$f_2 = \mathrm{tr}\begin{bmatrix} 0 & F_{12} & F_{13} & 0 & 0 & 0 \\ & 0 & F_{23} & 0 & 0 & 0 \\ & & 0 & 0 & 0 & 0 \\ & & & F_{44} & 0 & 0 \\ & & & & F_{55} & 0 \\ & & & & & F_{66} \end{bmatrix}\begin{bmatrix} \sigma_1{}^2 & \sigma_1\sigma_2 & \sigma_1\sigma_3 & \sigma_1\sigma_4 & \sigma_1\sigma_5 & \sigma_1\sigma_6 \\ & \sigma_2{}^2 & \sigma_2\sigma_3 & \sigma_2\sigma_4 & \sigma_2\sigma_5 & \sigma_2\sigma_6 \\ & & \sigma_3{}^2 & \sigma_3\sigma_4 & \sigma_3\sigma_5 & \sigma_3\sigma_6 \\ & & & \sigma_4{}^2 & \sigma_4\sigma_5 & \sigma_4\sigma_6 \\ & & & & \sigma_5{}^2 & \sigma_5\sigma_6 \\ & & & & & \sigma_6{}^2 \end{bmatrix} \tag{58}$$

This results in an immediate loss of flexibility because the number of independent coefficients reduces from 21 to 6. The loss of flexibility is least affected in the case of strength orthotropy with reference axes coincident with the material principal directions. Since all shear strengths are symmetrical, Eq. (56) is automatically satisfied, leaving only 9 independent constants If the "invariant" Eq. (57a) is used, the three missing material constants F_{11}, F_{22}, and F_{33} must be added, for example, by adding terms in the form of the square of the first invariant. For example, von Mises' isotropic criterion can be represented as

$$\hat{F} J_2 = \hat{F}\left(\frac{I_1^2}{3} + 3I_2\right) \tag{59}$$

Following the preceding rationale, Eq. (57) can be adopted for the anisotropic case as

$$\operatorname{tr}\left\{\left(\begin{bmatrix} F_1^2 & F_1F_2 & F_1F_3 & 0 & 0 & 0 \\ & F_2^2 & F_2F_3 & 0 & 0 & 0 \\ & & F_3^2 & 0 & 0 & 0 \\ & & & 0 & 0 \\ & & & & 0 & 0 \\ & & & & & 0 \end{bmatrix} + \begin{bmatrix} 0 & F_{12} & F_{13} & 0 & 0 & 0 \\ & 0 & F_{23} & 0 & 0 & 0 \\ & & 0 & 0 & 0 & 0 \\ & & & F_{44} & 0 & 0 \\ & & & & F_{55} & 0 \\ & & & & & F_{66} \end{bmatrix}\right) \times \begin{bmatrix} \sigma_1^2 & \sigma_1\sigma_2 & \sigma_1\sigma_3 & \sigma_1\sigma_4 & \sigma_1\sigma_5 & \sigma_1\sigma_6 \\ & \sigma_2^2 & \sigma_2\sigma_3 & \sigma_2\sigma_4 & \sigma_2\sigma_5 & \sigma_2\sigma_6 \\ & & \sigma_3^2 & \sigma_3\sigma_4 & \sigma_3\sigma_5 & \sigma_3\sigma_6 \\ & & & \sigma_4^2 & \sigma_4\sigma_5 & \sigma_4\sigma_6 \\ & & & & \sigma_5^2 & \sigma_5\sigma_6 \\ & & & & & \sigma_6^2 \end{bmatrix}\right\} = 1 \tag{60}$$

which upon combination is the same as

$$F_{ij}\sigma_i\sigma_j = 1$$

with

$$\begin{aligned} F_{14} = F_{15} = F_{16} &= 0 \\ F_{24} = F_{25} = F_{26} &= 0 \\ F_{34} = F_{35} = F_{36} &= 0 \\ F_{45} = F_{46} &= 0 \\ F_{56} &= 0 \end{aligned} \tag{61}$$

We note that in this modified "invariant" form [Eq. (60)], physical interpretation in terms of distortional energy or octohedral shear stress is no longer applicable. Expressing the criterion in the form of Eq. (60) presents the algebraic appearance of an invariant, but does not provide additional meaningful interpretation of the failure process. In fact, the added algebraic complexity only serves to reduce the flexibility, as the number of independent parameters reduces from 21 to 9.

As can be expected, these undesirable consequences of modeling anisotropic failure criterion after isotropic failure criterion containing only the second invariant are magnified manifoldly when the third invariant is introduced. Historically, several useful isotropic yield criteria which include the second and third deviatoric invariants (J_2 and J_3) have been introduced. Notably, Coulomb's (1773) yield criteria can be expressed as

$$4J_2^3 - 27J_3^2 - 36F^2J_2^2 + 96F^4J_2 - 64F^6 = 0 \tag{62}$$

where F is a material parameter. Another isotropic yield criteria of simpler function in J_2 and J_3 has been proposed by Drucker (1949)

$$J_2^3 + FJ_3^2 = 0 \tag{63}$$

Probably influenced by these yield criteria, Priddy (1971) adhered to the approach of formulating a failure criterion in terms of invariants. While recognizing that the failure of many brittle composites are dependent on hydrostatic pressure, he proposed a failure criterion for orthotropic material of the form:

$$\tilde{\mathrm{II}} = 1 + \tilde{\mathrm{I}} + \mathrm{I}'\mathrm{II} + f\mathrm{III} \tag{64}$$

where $\tilde{\mathrm{I}}$, $\tilde{\mathrm{II}}$ have the essential algebraic form of the first stress invariant modified for anisotropic materials [Eqs. (52) and (58)]. Here f is a scalar constant and II, III are, respectively, the second and third stress invariants. Upon expansion Eq. (64) becomes

$$\begin{aligned} a_1\sigma_6^2 + a_2\sigma_4^2 + a_3\sigma_5^2 + b_1\sigma_1^2 + b_2\sigma_2^2 + b_3\sigma_3^2 + c_1\sigma_1\sigma_2 + c_2\sigma_2\sigma_3 + c_3\sigma_3\sigma_1 \\ = 1 + d_1\sigma_1 + d_2\sigma_2 + d_3\sigma_3 + (e_1\sigma_1 + e_2\sigma_2 + e_3\sigma_3)(\sigma_1\sigma_2 + \sigma_2\sigma_3 + \sigma_3\sigma_1 \\ - \sigma_4^2 - \sigma_5^2 - \sigma_6^2) + f(\sigma_1\sigma_2\sigma_3 + 2\sigma_4\sigma_5\sigma_6 - \sigma_1\sigma_4^2 - \sigma_2\sigma_5^2 - \sigma_3\sigma_6^2) \end{aligned} \tag{65}$$

where a_i, b_i, c_i, d_i, and f are material parameters.

Although further flexibility is introduced through the addition of third order terms in stress, unnecessary algebraic complexity is imposed by adhering to a structure which has the appearance of invariants. For example, redundancy exists where two parameters (e_1 and f) both characterize the interaction effect of $\sigma_1\sigma_2\sigma_3$; indeterminacy occurs where one parameter

(e_1) characterizes two combinations of stress interaction ($\sigma_1^2\sigma_2$ and $\sigma_1^2\sigma_3$) and others. It will be shown in Section II,D,2 that third and higher-order terms can be introduced in a straightforward manner by simply including the desired number of terms in the tensor polynomial, Eqs. (5).

6. *Special Forms of Tensor Polynomials*

We have established that several representative historical anisotropic failure criteria are in fact special and restricted cases of the tensor polynomials, Eqs. (5), regardless of whether the original formulations were motivated by physical intuition or by the form of isotropic invariant yield criteria. On the other hand, there exist numerous formulations of polynomials in terms of tensor scalar products. We will now explore the generality of the tensor polynomial, Eqs. (5), which is but one of the many possible formulations.

Ashkenazi (1964, 1965) and Ashkenazi and Pekker (1970) are active contributors to the characterization of anisotropic failure. The most significant differences in formulation between the tensor polynomial, Eqs. (5), and the criteria proposed by Ashkenazi (1965) lies in the definition of the material parameters which characterize strength. In Eqs. (5) the failure tensors (F_i, F_{ij}, . . . , etc.) form scalar products with the stress tensor σ_{ij} and have the respective dimension of [stress]$^{-1}$, [stress]$^{-2}$, etc. In Ashkenazi's formulation, the material parameters, written in contracted notation, are defined as

$$\alpha_{ij} = 1/\tilde{\sigma}_{ij} \tag{66a}$$

where $\tilde{\sigma}_{ij}$ are ultimate strengths of the material. It is important to note that in noncontracted notation $\tilde{\sigma}_{ij}$ has four free indices $\tilde{\sigma}_{ijkl}$. We emphasize that $\tilde{\sigma}_{ij}$ is not a second rank tensor but is a fourth rank tensor. The physical meaning of $\tilde{\sigma}_{ij}$ is not clearly defined since they have coordinate transformation properties of fourth rank tensors but have dimension of stress (second rank tensor). According to the application details given by Ashkenazi (1965), the material constants α_{ij} are actually defined as

$$\alpha_{ii} = 1/\tilde{\sigma}_i \qquad \text{(no sum)} \tag{66b}$$

and α_{kl}, when $k \neq l$, are determined from tensor transformation relation which can be represented as

$$\alpha'_{ii} = T_{iikl}\alpha_{kl}, \qquad i, j, k, l = 1, 2, \ldots, 6 \quad (i \text{ no sum}) \tag{67}$$

where T_{ijkl} is the standard fourth rank tensor transformation matrix [Eqs. (6)]. In Eq. (67), α'_{ii} and α_{kl} for $k = l$ can be measured physically, respectively, at reference angles θ and θ' in accordance with Eq. (66b) leaving a total of 30 unknowns for $k \neq l$. Since Eq. (67) consists of six independent equations, only 6 out of the 30 unknowns can be determined. Ashkenazi arbitrarily chose to solve for the six constants for $k = l = 1, 2, 3$ and assumed the remaining 24 indeterminant α_{kl} to be zero. Tacitly assuming α_{kl} is symmetrical and drawing upon obvious simplification for transformation at 45°, Ashkenazi (1965) provided formulas for the determination of α_{12}, α_{13}, α_{23} from experimental measurements

$$\alpha_{ij} = \begin{bmatrix} \frac{1}{X_1} & \frac{1}{2}\left(\frac{1}{X_1} + \frac{1}{X_2} - \frac{1}{X_4^{45°}}\right) & \frac{1}{2}\left(\frac{1}{X_1} + \frac{1}{X_3} - \frac{1}{X_5^{45°}}\right) & 0 & 0 & 0 \\ & \frac{1}{X_2} & \frac{1}{2}\left(\frac{1}{X_2} + \frac{1}{X_3} - \frac{1}{X_6^{45°}}\right) & 0 & 0 & 0 \\ & & \frac{1}{X_3} & 0 & 0 & 0 \\ & & & \frac{1}{X_4} & 0 & 0 \\ & & & & \frac{1}{X_5} & 0 \\ & & & & & \frac{1}{X_6} \end{bmatrix} \tag{68}$$

In Eq. (68) the same notation for ultimate strength as in Eq. (8) is used.

Compared to the definition of failure tensors F_{ij}, the formulation of Ashkenazi has four major deficiencies: (1) The symmetry conditions of α_{ij} are arbitrarily assumed and cannot be established as in Eq. (7) for F_{ij}. (2) For $k \neq l$ only three α_{kl} (assuming symmetry) can be determined, the vanishing of the remaining α_{kl} is not from the consideration of material symmetry but by default due to the indeterminacy of the definition. (3) The determination of α_{kl} by Eq. (68) does not account for the sensitivity of the experiments. Normal variations in strength can lead to completely erroneous calculation of α_{kl} (Tsai and Wu, 1972a). (4) Equation (66b) cannot analytically account for differences between tension and compression strengths; a different set of α_{kl} has to be determined by substituting compression strengths into Eq. (68).

The advantage of Ashkenazi's formulation lies in its ability to determine directly strength under one-dimensional states of stress for any arbitrary orientation θ'. For the orientation θ', α'_{ij} can be determined from the transformation equation (67), and inversion of Eq. (66b) immediately provides any uniaxial off-axis strength:

$$\tilde{\sigma}_i' = 1/\alpha'_{ii} \qquad \text{(no sum)} \tag{69}$$

In comparison, uniaxial strength calculation employing the tensor polynomial [e.g., using linear and quadratic terms in Eq. (5b)] requires transformation of F_i and F_{ij} to F_i' and F'_{ij} as in the α_{ij} calculation. In addition, the uniaxial off-axis strength cannot be determined by direct inversion but requires solution of a quadratic algebraic equation as shown in Eq. (9). Ashkenazi, overlooking the stability conditions (Tsai and Wu, 1970a), observed that solution of the quadratic could lead to singularity in strength for certain material orientations. Apparently, this observation motivated the definition of the material parameters in the form of Eq. (66b). This anomaly of singular strength, however, in fact does not exist if F_{12} is determined while taking the sensitivity of the experiments into account (Wu, 1972).

The simplicity of Ashkenazi's formulation in correlating one-dimensional off-axis strength does not carry over to strength under complex states of stress. For the case of general planar stress, Ashkenazi (1965) proposed a failure criterion which can be expressed as

$$\left(\frac{\sigma_1^2}{X_1} + \frac{\sigma_2^2}{X_2} + \gamma\sigma_1\sigma_2 + \frac{\sigma_6^2}{X_6}\right)^2 + \frac{2(\sigma_1\sigma_2 - \sigma_6^2)}{X_6}\left[\frac{\sigma_1^2}{X_1} + \frac{\sigma_2^2}{X_2} + \sigma_1\sigma_2\left(\frac{1}{X_1} + \frac{1}{X_2}\right)\right]$$
$$- (\sigma_1\sigma_2 - \sigma_6^2)[\lambda\sigma_1^2 + \mu\sigma_2^2 + \sigma_1\sigma_2(\lambda + \mu)] + \rho(\sigma_1 + \sigma_2)(\sigma_1\sigma_2 - \sigma_6^2)$$
$$= (\sigma_1^2 + \sigma_2^2 + \sigma_1\sigma_2 + \sigma_6^2) \tag{70a}$$

where

$$\gamma = \left(\frac{4}{X_1^{45°}} - \frac{1}{X_1} - \frac{1}{X_2} - \frac{1}{X_6}\right)$$

and $X_1^{45°}$ is the tensile strength at 45° to the material reference axis and λ, μ, ρ are additional experimentally determined constants. Equation (70a) is valid only for $(\sigma_1 + \sigma_2) \geq 0$; for $(\sigma_1 + \sigma_2) \leq 0$ the tensile strength has to be replaced by compression strength. Thus, to characterize a complete failure surface, two separate criteria are needed requiring a total of 13 constants. The algebraic structure of this criterion can no longer be readily related to the original concept of strength tensors, Eq. (66b). On the contrary, Eq. (70a) has the appearance of equivalent stress type formulations if we rewrite it as

$$\left(\frac{\sigma_1^2}{X_1} + \frac{\sigma_2^2}{X_2} + \gamma\sigma_1\sigma_2 + \frac{\sigma_6^2}{X_6}\right)^2 + I_2\left[\left(\lambda - \frac{2}{X_1X_6}\right)\sigma_1^2 + \left(\mu - \frac{2}{X_2X_6}\right)\sigma_2^2\right.$$

$$\left. + \sigma_1\sigma_2\left(\lambda + \mu - \frac{1}{X_1X_6} - \frac{1}{X_2X_6}\right) - \rho I_1\right] = I_1^2 + I_2 \qquad (70b)$$

More recently, Ashkenazi and Pekker (1970) proposed a simplified version of Eq. (70a) in the form

$$\left(\frac{\sigma_1^2}{X_1} + \frac{\sigma_2^2}{X_2} + \gamma\sigma_1\sigma_2 + \frac{\sigma_6^2}{X_6}\right)^2 = \sigma_1^2 + \sigma_2^2 + \sigma_1\sigma_2 + \sigma_6^2 = I_1^2 + I_2 \qquad (71)$$

The criterion is the direct result of eliminating λ, μ, and ρ as independent constants by assigning their magnitude to be

$$\lambda = 2/X_1X_6, \qquad \mu = 2/X_2X_6, \qquad \rho = 0$$

In this simplified form of Eq. (68), similarity to equivalent stress formulation is even more striking. The complexity of these formulations, particularly in the case of Eq. (70a), prohibits further interpretation of the originator's rationale. Nevertheless, upon expansion both Eqs. (70) and (71) can be written as a polynomial in stress and may be regarded as a special case of Eqs. (5).

In the strength criterion proposed by Goldenblat and Kopnov (1965) the material parameters which characterize anisotropic strength are similar to those in Eqs. (5) where the material parameters are structured to be tensors forming scalar products with the stress tensors. However, the organization of the resulting scalar products in the failure criterion is not the same. The Goldenblat and Kopnov failure criterion can be written as

$$(F_i\sigma_i)^\alpha + (F_{ij}\sigma_i\sigma_j)^\beta + (F_{ijk}\sigma_i\sigma_j\sigma_k)^\gamma + \cdots = 1 \qquad (72)$$

where F_i, F_{ij}, . . . ; α, β, γ, . . . are all experimentally determined parameters. It is apparent that Eq. (72) does not extend the generality of Eq. (5) since the effect of α, β, γ, . . . can all be incorporated in a tensor polynomial of sufficient order. For example, in the special case presented by Goldenblat and Kopnov (1965), the exponents are assigned as

$$\alpha = 1, \qquad \beta = \tfrac{1}{2}, \qquad \gamma = -\infty \qquad (73)$$

For these values, Eq. (72a) can be rewritten as

$$2F_i\sigma_i + F_{ij}\sigma_i\sigma_j - (F_i\sigma_i)^2 = 1 \qquad (74)$$

Equations (74) and (5) both contain 36 independent coefficients (6 for linear terms and 21 for quadratic term). Thus the added algebraic complexity of Eq. (74) results in no more generality. On the contrary, Eq. (74) leads to mixing of linear and quadratic coefficients and subsequent confusion in physically linking failure tensors to engineering strengths. This can be seen by examining the coefficients for Eq. (74) which, for plane stress, are

$$F_i = \begin{bmatrix} \frac{1}{2}\left(\frac{1}{X_1} - \frac{1}{X_1'}\right) \\ \frac{1}{2}\left(\frac{1}{X_2} - \frac{1}{X_2'}\right) \\ 0 \\ 0 \\ 0 \\ \frac{1}{2}\left(\frac{1}{X_6} - \frac{1}{X_6'}\right) \end{bmatrix},$$

$$F_{ij} = \begin{bmatrix} \left(\frac{1}{X_1X_1'} - F_1^2\right) & (F_{12} - F_1^2) & 0 & 0 & 0 & (F_{16} - F_1^2) \\ & \left(\frac{1}{X_2X_2'} - F_2^2\right) & 0 & 0 & 0 & (F_{26} - F_2^2) \\ & & 0 & 0 & 0 & 0 \\ & & & 0 & 0 & 0 \\ & & & & 0 & 0 \\ & & & & & \left(\frac{1}{X_6X_6'} - F_6^2\right) \end{bmatrix} \tag{75}$$

The tensor polynomial has none of these deficiencies [Eq. (8)] and is preferred for operational simplicity. Goldenblat and Kopnov's failure criterion can also be regarded as a special case of the tensor polynomial.

Numerous other failure criteria of various quadratic forms have been proposed by Marin (1957), Hoffman (1967), and others. Marin's criterion can be written as

$$(\sigma_1 - a)^2 + (\sigma_2 - b)^2 + (\sigma_3 - c)^2 + q[(\sigma_1 - a)(\sigma_2 - b) + (\sigma_2 - b)(\sigma_3 - c) + (\sigma_3 - c)(\sigma_1 - a)] = \text{constant} \tag{76}$$

The applicability of this criterion is severely restricted by numerous assumptions which include material orthotropy, equal tensile and compressive strengths, and coincidence of the reference axes, material symmetry axes, and the principal stress directions.

Hoffman's (1967) criterion removes one restrictive assumption by allowing for unequal tensile and compressive strengths. This can be seen from the algebraic structure, an invariant formulation which includes linear terms

$$C_1(\sigma_2 - \sigma_3)^2 + C_2(\sigma_3 - \sigma_1)^2 + C_3(\sigma_1 - \sigma_2)^2 + C_4\sigma_1 + C_5\sigma_2 + C_6\sigma_3 + C_7\sigma_4^2 + C_8\sigma_5^2 + C_9\sigma_6^2 = 1 \tag{77}$$

where C_i are material strength constants. Comparison with the quadratic tensor polynomial [Eq. (5c)], the coefficients in Eq. (77) are

$$F_i = \begin{bmatrix} C_4 \\ C_5 \\ C_6 \\ 0 \\ 0 \\ 0 \end{bmatrix}, \qquad F_{ij} = \begin{bmatrix} (C_2 + C_3) & -C_3 & -C_1 & 0 & 0 & 0 \\ & (C_1 + C_3) & C_2 & 0 & 0 & 0 \\ & & (C_1 + C_2) & 0 & 0 & 0 \\ & & & C_7 & 0 & 0 \\ & & & & C_8 & 0 \\ & & & & & C_9 \end{bmatrix} \tag{78}$$

The flexibility of Hoffman's criterion is greatly improved through inclusion of linear terms, but is still deficient in comparison to the tensor polynomial. The limiting features of Hoffman's criterion are: (1) applicability to only orthotropic materials, i.e., shear coupling coefficients are arbitrarily equated to zero; and (2) the mixing of the coefficients for F_{11}, F_{22}, F_{33} creates the same inconveniences as in Hill's criterion [Eqs. (42)]; coefficients (F_{12}, F_{23}, F_{13}) which characterize normal strength coupling are not independent material constants thus reducing its flexibility.

Other quadratic formulations by Norris (1939), Fischer (1960), Franklin (1968), and Chamis (1969) could be critically examined. However, tedious algebraic manipulations reveal that they are also restricted special cases of the tensor polynomial. We may now expand upon the generality of the tensor polynomial of the form presented in Eqs. (5).

Obviously, Eqs. (5) do not represent the only possible tensor polynomial formulations; there exist many other possibilities. In the tensor polynomial, Eqs. (5), the linear, quadratic, and cubic terms are formed by, respectively, defining failure tensor of second, fourth, and sixth rank to produce scalar products with the stress tensor. Similar scalar products can be formed by restricting failure tensors to second rank only in the form

$$A_i\sigma_i + B_iC_j\sigma_i\sigma_j + \cdots = 1 \tag{79}$$

Since the stress tensor is symmetric, the failure tensors A_i, B_i, and C_i are also symmetric. Compared to the tensor polynomials [Eq. (5c)], the coefficients in Eq. (76) are:

$$F_i = \begin{bmatrix} A_1 \\ A_2 \\ A_3 \\ A_4 \\ A_5 \\ A_6 \end{bmatrix}, \qquad F_{ij} = \begin{bmatrix} B_1C_1 & B_1C_2 & B_1C_3 & B_1C_4 & B_1C_5 & B_1C_6 \\ & B_2C_2 & B_2C_3 & B_2C_4 & B_2C_5 & B_2C_6 \\ & & B_3C_3 & B_3C_4 & B_3C_5 & B_3C_6 \\ & & & B_4C_4 & B_4C_5 & B_4C_6 \\ & & & & B_5C_5 & B_5C_6 \\ & & & & & B_6C_6 \end{bmatrix} \tag{80}$$

In this formulation there are 6 constants for the linear effects of stress which are the same as Eqs. (5); however, there are only 12 independent constants for the quadratic terms compared to 21 in Eqs. (5). In addition to this loss of generality, the material parameters are mixed as can be seen in Eqs. (80). This mixing of the material parameters causes confusion in physical identification, a deficiency shared by the Goldenblat and Kopnov formulation [Eqs. (75)]. Minor variations can be obtained by rewriting Eq. (79) in noncontracted notation and by rearranging the indices, i.e.,

$$A_{ij}\sigma_{ij} + B_{ik}C_{jl}\sigma_{ij}\sigma_{kl} + \cdots = 1 \qquad \text{(noncontracted notation)} \tag{81}$$

In this formulation the number of independent coefficients for the quadratic terms increases from 12 to 18. Nevertheless, compared to Eq. (79) the deficiencies of the lost generality and mixing of coefficient have not been alleviated. Furthermore, symmetry conditions of the material parameters B_{ik} and C_{jl} can no longer be readily identified. Examination of other variations of tensor polynomial formulations, such as letting $C_{jl} = \delta_{jl}$, leads to the conclusions that:

> The tensor polynomial of Eqs. (5) encompasses maximum flexibility without redundancy. It may be applied for one-, two-, or three-dimensional states of stress by simply expanding the indices in accordance with the state of stress in question. (82)

D. Additional Properties of Tensor Polynomials

1. Principal Directions of Strength

Although any surface may be represented by a polynomial in the form of Eqs. (5), not all the surfaces are admissible failure surfaces. Specifically, a failure surface cannot be imaginary and it must consist of a single sheet. The conditions for the coefficients F_i, F_{ij}, . . . , etc. to satisfy these requirements are explored in geometry texts. Geometric interpretation is useful for examining the constraints on F_i, F_{ij}, . . . , etc. and the definition of principal directions. In two-dimensional states of stress the failure surface is three-dimensional since there are three independent stress components σ_1, σ_2, and σ_6. For brevity, the geometric interpretations and the discussion of the roots of Eqs. (5) will be limited to two-dimensional states of stress or three-dimensional failure surfaces. Definition of characteristic directions and orthogonality in n dimensions allows extrapolation of the following results to three-dimensional states of stress or six-dimensional failure surfaces. Expanding Eq. (5b) for two-dimensional states of stress, i.e., $i, j = 1, 2, 6$, we obtain the failure surface in quadratic form:

$$F_{11}\sigma_1^2 + F_{22}\sigma_2^2 + F_{66}\sigma_6^2 + 2F_{12}\sigma_1\sigma_2 + 2F_{16}\sigma_1\sigma_6 + 2F_{26}\sigma_2\sigma_6 + F_1\sigma_1 + F_2\sigma_2 + F_6\sigma_6 = 1 \tag{83}$$

The nature of the surface represented in the above quadratic equation can be examined by reduction into the canonical form. Through coordinate translation and rotation, Eq. (83) may be expressible in terms of one of the 17 canonical forms. Of the 17 different surfaces which can be represented by a general quadratic [Eq. (83)], the admissible ones are those which fulfill the fundamental requirement: a failure surface must possess only one real root for any radial loading path. Thus, imaginary surfaces, surfaces of two sheets, hyperboloid, hyperbolic paraboloid, etc. are not admissible failure surfaces. The only two admissible surfaces are the real ellipsoid and possibly the elliptic paraboloid (for those unusual cases where infinite strength may be exhibited for some state of stress). These two admissible surfaces are shown, respectively, in Fig. 2a and b. The principal directions and the constraints on the coefficients F_i and F_{ij} which would give rise only to these two surfaces can be determined through examination of the two characteristic matrices of Eq. (83):

$$e \equiv \begin{bmatrix} F_{11} & F_{12} & F_{16} \\ F_{12} & F_{22} & F_{26} \\ F_{16} & F_{26} & F_{66} \end{bmatrix}, \qquad \Delta e \equiv | e | \tag{84}$$

$$\mathbf{E} = \begin{bmatrix} F_{11} & F_{12} & F_{16} & \frac{F_1}{2} \\ F_{12} & F_{22} & F_{26} & \frac{F_2}{2} \\ F_{16} & F_{26} & F_{66} & \frac{F_6}{2} \\ \frac{F_1}{2} & \frac{F_2}{2} & \frac{F_6}{2} & -1 \end{bmatrix}, \qquad \Delta\mathbf{E} \equiv | \mathbf{E} | \tag{85}$$

The principal directions l, m, and n in the $\sigma_1\sigma_2\sigma_6$ stress space can be obtained from the eigenvalue λ such that

$$\begin{aligned} (F_{11} - \lambda)l + F_{12}m + F_{16}n &= 0 \\ F_{12}l + (F_{22} - \lambda)m + F_{26}n &= 0 \\ F_{16}l + F_{26}m + (F_{66} - \lambda) &= 0 \end{aligned} \tag{86}$$

Equations (86) can be written in the familiar form:

$$\lambda^3 = \mathrm{I}\lambda^2 + \mathrm{II}\lambda + \mathrm{III} = 0 \tag{87}$$

where

$$\begin{aligned} \mathrm{I} &= F_{11} + F_{22} + F_{66} \\ \mathrm{II} &= F_{11}F_{22} + F_{11}F_{66} + F_{22}F_{66} - F_{16}^2 - F_{26}^2 - F_{12}^2 \\ \mathrm{III} &= \Delta\mathbf{E} \end{aligned} \tag{88}$$

The roots of Eq. (87) are λ_1, λ_2 and λ_6. From established theories of analytical geometry, we determine that for Eq. (83) to be a failure surface it must be in the form of:

real ellipsoid: $[e]$ third rank, $[\mathbf{E}]$ fourth rank, $\Delta\mathbf{E} < 0$ and signs of λ_1, λ_2, λ_6 are the same (89)

elliptical paraboloid: $[e]$ second rank, $[\mathbf{E}]$ fourth rank $\Delta\mathbf{E} < 0$ and signs of λ_1, λ_2, λ_6 are the same (90)

The necessary and sufficient stability condition of a failure criterion for a finite strength composite is represented by Eq. (89), while that for a failure criterion with infinite strength in one stress direction is represented by Eq. (90). A simpler necessary stability condition can be arrived at by considering the trace of the failure surface on alternate planes where $\sigma_i = 0$. For

$$\begin{aligned} \sigma_6 &= 0, \qquad F_{11}F_{22} - F_{12}^2 \geq 0 \\ \sigma_2 &= 0, \qquad F_{11}F_{66} - F_{16}^2 \geq 0 \\ \sigma_1 &= 0, \qquad F_{22}F_{66} - F_{26}^2 \geq 0 \end{aligned} \tag{91}$$

or in general

$$F_{ii}F_{jj} - F_{ij}^2 \geq 0 \tag{92}$$

The principal direction of strength is given by Eqs. (88). Since λ_1, λ_2, and λ_6 are orthogonal to one another, we can conclude that any composite whose failure surface may be characterized by a quadratic in the form of Eq. (83) can be termed orthotropic in strength. The principal direction of strength, however, does not necessarily coincide with the material principal direction as shown schematically in Fig. 9.

The flexibility of the tensor polynomial, Eqs. (5), which lends itself to simplification through appropriate symmetry conditions can be demonstrated for the specially orthotropic composite (Tsai and Wu, 1971b), e.g., a unidirectionally reinforced lamina. If we orient the coordinate axes along the material principal direction as shown in Fig. 9, physical symmetry suggests that a positive shear σ_6^+ [Fig. (10a)] and a negative shear σ_6^- [Fig. (10b)] should be terminated by the same strength. In other words,

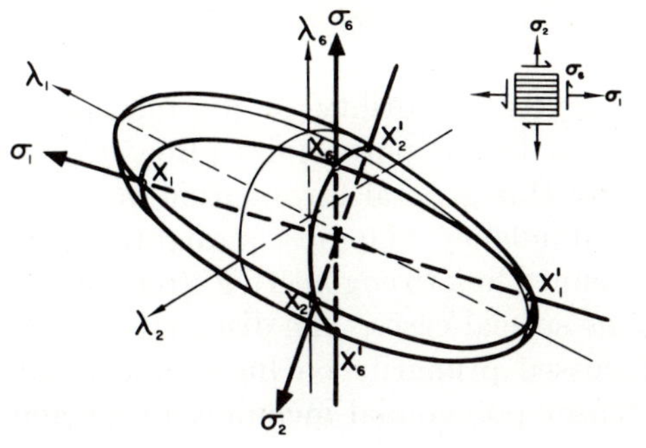

FIG. 9. An ellipsoidal failure surface (quadratic polynomial failure criterion) in stress space depicting the difference between principal directions of elasticity (along σ_1, σ_2, σ_6) and principal directions of strength (λ_1, λ_2, λ_6).

Eqs. (83) or (5) must be symmetrical in σ_6 and may be simplified by assigning zeros for the coefficients for terms which are odd function of σ_6. Thus for the planar case, the strength coefficient in Eq. (5c) reduces to

$$F_i \begin{bmatrix} F_1 \\ F_2 \\ 0 \\ 0 \\ 0 \\ 0 \end{bmatrix}, \qquad F_{ij} = \begin{bmatrix} F_{11} & F_{12} & 0 & 0 & 0 & 0 \\ & F_{22} & 0 & 0 & 0 & 0 \\ & & 0 & 0 & 0 & 0 \\ & & & 0 & 0 & 0 \\ & & & & 0 & 0 \\ & & & & & F_{66} \end{bmatrix} \tag{93}$$

Other cases of symmetry have been discussed by Tsai and Wu (1970a).

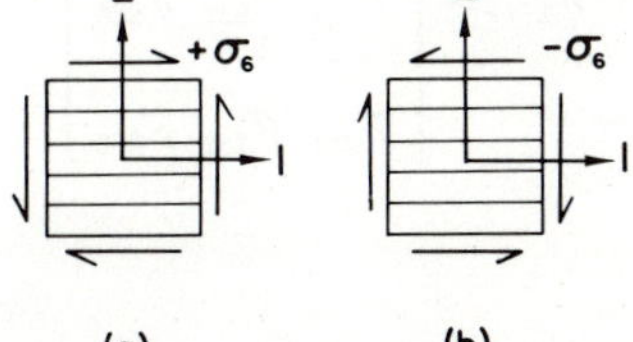

FIG. 10. Strength symmetry in shear for orthotropic materials: (a) positive shear σ_6^+; (b) negative shear σ_6^-.

2. Higher-Order Terms of Tensor Polynomial

Section II,C conveyed the utility of formulating anisotropic failure criterion in terms of a tensor polynomial [Eqs. (5)]. The primary advantages provided by this general formulation are: the greatest possible flexibility without redundancy, the most straightforward operational convenience, a clearly defined link to engineering strengths and direct reduction to existing criteria as special cases. The discussion, for the sake of tractability, has been focussed primarily on linear and quadratic terms. It is obvious that the tensor polynomial including linear and quadratic terms [Eq. (5c) or (83)] is capable of characterizing any anisotropic failure surface expressible as an ellipsoid or paraboloid. It has been observed that the failure surfaces of many crystals and structural materials cannot be closely approximated by a second-order surface [for rock salt, see Sohncke (1869) and Sella and Voigt (1892, 1893); for cast iron, see Priddy (1971); for woven and laminated composites see Ashkenazi (1964) and Wu and Scheublein (1973)]. For these materials, the failure criterion must include higher-order terms of the tensor polynomial. Higher-order terms are subjected to geometric and algebraic constraints to satisfy the fundamental requirements that a failure surface must be of a single sheet and that the failure criterion possess no more than one root for any radial loading path. These constraints can be established through examination of a third-order tensor polynomial, and the results can be extrapolated to fourth- and higher-order terms through induction. A tensor polynomial failure criterion including third-order terms can be written [from Eq. (5b)] as

$$f(\sigma) = F_i\sigma_i + F_{ij}\sigma_i\sigma_j + F_{ijk}\sigma_i\sigma_j\sigma_k = 1 \qquad \text{(contracted notation)} \qquad (94)$$

In addition to the second-order terms expanded in Eq. (5c), the third-order strength parameters for the plane case are

$$F_{1ij} = \begin{bmatrix} F_{111} & F_{112} & F_{116} \\ F_{121} & F_{122} & F_{126} \\ F_{161} & F_{162} & F_{166} \end{bmatrix}, \quad F_{2ij} = \begin{bmatrix} F_{211} & F_{212} & F_{216} \\ F_{221} & F_{222} & F_{226} \\ F_{261} & F_{262} & F_{266} \end{bmatrix}, \quad F_{6ij} = \begin{bmatrix} F_{611} & F_{612} & F_{616} \\ F_{621} & F_{622} & F_{626} \\ F_{661} & F_{662} & F_{666} \end{bmatrix} \qquad (95)$$

In Eqs. (95) the interaction terms F_{ijk} (for $i = j \neq k$, $i = k \neq j$) are symmetrical if we hypothesize that failure is path-independent and $f(\sigma)$

is of class C3, i.e.,

$$
\begin{aligned}
F_{ijk} &= \frac{\partial^3 f}{\partial\sigma_i\,\partial\sigma_j\,\partial\sigma_k} = \frac{\partial^3 f}{\partial\sigma_i\,\partial\sigma_k\,\partial\sigma_j} = F_{ikj} \\
F_{ijk} &= \frac{\partial^3 f}{\partial\sigma_i\,\partial\sigma_j\,\partial\sigma_k} = \frac{\partial^3 f}{\partial\sigma_k\,\partial\sigma_j\,\partial\sigma_i} = F_{kji}
\end{aligned}
\tag{96}
$$

From the symmetry conditions [Eqs. (96)], Eqs. (95) reduce to 11 independent constants. The constraints for these independent constants can be examined, without loss of generality, by examining the special orthotropic case. Inspection of material orthotropic symmetry allows us to deduce as in Section II,D,1 that the constants characterizing odd-order terms in σ_6 must vanish, thus

$$F_{116} = F_{126} = F_{226} = F_{666} = 0 \tag{97}$$

After these simplifications, there remain only 6 admissible independent constants from Eqs. (95) which are

$$F_{111}, \quad F_{222}, \qquad 3F_{112}, \quad 3F_{122}, \qquad 3F_{166}, \quad 3F_{266} \tag{98}$$

The admissibilities of these 6 constants are further constrained by the single-root requirement. The constraint for F_{iii} can be obtained by examining the special case where $\sigma_6 = 0$ and $F_{112} = F_{122} = F_{166} = F_{266} = 0$, upon which Eq. (94) becomes

$$F_{ijk}\sigma_i\sigma_j\sigma_k + F_{ij}\sigma_i\sigma_j + F_i\sigma_i = 1, \qquad i = j = k = 1, 2 \tag{99}$$

In general, the combinations of the roots of Eq. (99) are: (1) one real and two complex roots, (2) three real distinct roots, and (3) three real roots, two of which are equal.

Consideration of the conditions that there exist two distinct roots for two colinear loading paths discloses only combination (3) to be admissible. In order for Eq. (99) to have roots in this combination, the coefficients must satisfy the following constraint:

$$\frac{1}{4}\left[\frac{2}{27}\left(\frac{F_{ii}}{F_{iii}}\right)^3 - \frac{1}{3}\frac{F_{ii}F_i}{F_{iii}} + \frac{1}{F_{iii}}\right]^2 + \frac{1}{27}\left[\frac{F_i}{F_{iii}} - \frac{1}{3}\left(\frac{F_{ii}}{F_{iii}}\right)^2\right]^3 < 0, \qquad i = 1, 2 \quad (\text{no sum}) \tag{100}$$

Since the roots of Eq. (99) are the tensile and compressive strengths in the x_i direction, we can, using Eq. (100), express the coefficients in terms of the engineering strengths:

For third-order polynomial

$$F_i = -\frac{2}{X_i'} + \frac{1}{X_i}, \qquad F_{ii} = \frac{2}{X_i X_i'} - \frac{1}{X_i'^2}, \qquad F_{iii} = \frac{1}{X_i X_i'^2} \tag{101a}$$

For second-order polynomial

$$F_i = \frac{1}{X_i} - \frac{1}{X_i'}, \qquad F_{ii} = \frac{1}{X_i X_i'}, \qquad F_{iii} = 0 \qquad \text{(no sum)} \tag{101b}$$

For comparison, the corresponding relations for a quadratic tensor polynomial are also listed in Eqs. (101). Examination of Eqs. (101) reveals that (1) only two out of the three constants (F_i, F_{ii}, and F_{iii}) are independent, (2) simplification to second-order tensor polynomial form cannot be accomplished by direct elimination of the F_{iii} term without altering the algebraic relation of the F_i and F_{ii} terms. Hence, the third-order terms F_{iii} do not provide additional flexibility, but on the contrary, they inhibit direct simplification procedures. Other than the deficiencies along x_i axis crossings, additional limitations exist. For the $\sigma_1\sigma_2$ plane, the third-order tensor polynomial (for $F_{12} = 0$) can be written as

$$\sigma_2^3 F_{222} + \sigma_2^2 F_{22} + F_2\sigma_2 + W = 1, \qquad W = F_{111}\sigma_1^3 + F_{11}\sigma_1^2 + F_1\sigma_1 \tag{102}$$

Because all the coefficients are already fixed by axis crossing requirements [Eqs. (101)], the term W must be zero, i.e.,

$$F_{111}\sigma_1^3 + F_{11}\sigma_1^2 + F_1\sigma_1 = 0 \tag{103}$$

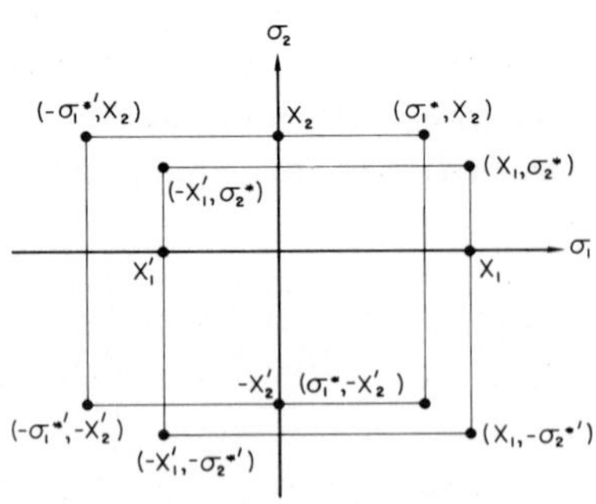

FIG. 11. Sixth rank failure tensor components F_{111} and F_{222} produce 12 distinct roots and cannot admit a continuous single-sheet failure surface.

Equation (103) yields three roots: $\sigma_1 = 0$, $\sigma_1 = \sigma_1{}^*$ and $\sigma_1 = -\sigma_1{}^*$. Alternately, considering Eqs. (102) and (103) and by solving for σ_2, yields three additional roots $\sigma_2 = 0$, $\sigma_2 = +\sigma_2{}^*$, $\sigma_2 = -\sigma_2{}^*$. Thus, as shown in Fig. 11, there exist only 12 distinct roots which satisfy the condition that Eq. (94) has three real roots, two of which are equal. Thus the F_{iii} terms preclude a continuous, single-sheet, failure surface and they are not admissible. Consideration of the planar orthotropic case has eliminated all but four of the third-order coefficients:

$$3F_{112}, \quad 3F_{122}, \qquad 3F_{166}, \quad 3F_{266} \tag{104}$$

These observations can be extrapolated to failure tensors of eighth and higher ranks. The flexibility is characterization introduced by failure tensors of different ranks and can be summarized as (no sum):

F_i: determines center of failure envelope along σ_i axis
F_{ii}: determines the failure envelope intercept of the σ_i axis (or the size of the ellipsoid)
F_{ij}: determines the inclination of the ellipsoid in the $\sigma_i\sigma_j$ plane and characterizes the strength coupling between σ_i and σ_j
F_{iii}: redundant and not admissible
F_{ijk}: determines inclination of failure envelope in the $\sigma_i\sigma_j\sigma_k$ space; vanishes for strength symmetry in σ_k
F_{ijj}: distorts the failure surface from an ellipsoid in the $\sigma_i\sigma_j$ plane (105)
F_{iiii}: redundant and not admissible
F_{ijkl}: determines the inclination in $\sigma_i\sigma_j\sigma_k\sigma_l$ space; vanishes for the planar case
F_{iijk}: determines inclination in $\sigma_i\sigma_j\sigma_k$ space; vanishes for strength symmetry in σ_k
F_{iijj}: distorts failure surface from an ellipsoid in $\sigma_i\sigma_j$ plane
F_{ijjj}: distorts failure surface in the $\sigma_i\sigma_j$ plane
F_{iiiii}: redundant and not admissible
⋮

The individual and combined effects of the failure tensors up to third rank for the $\sigma_1\sigma_2$ plane are illustrated in Fig. 12. The ability of the tensor polynomial to model any arbitrary failure surface is self-evident. For example, Fig. 12d gives the general shape of the failure surface for cast iron (Priddy, 1971) and cross-ply laminates (Wu and Scheublein, 1973). The experimental measurements of these failure tensors are presented in Section III.

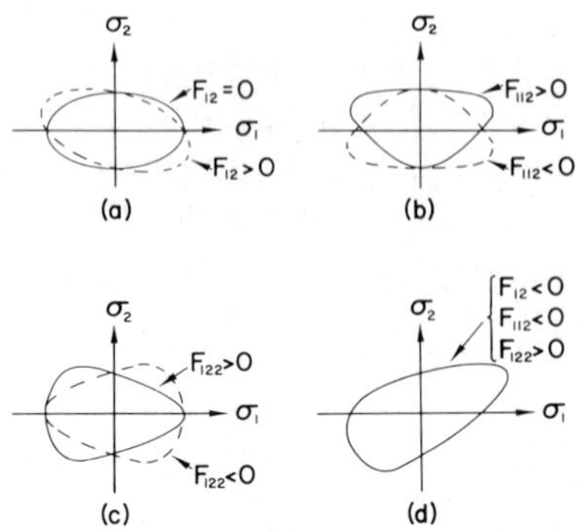

FIG. 12. Individual and combined effects of the failure tensor components which characterize strength interactions.

III. Experiments

Composite strength, i.e., the shape and size of its failure surface, is in general a function of the state of stress, time (creep rupture), stress history (fatigue), service environment, fiber volume fraction, fabrication variables, and others. Herein the primary variable considered is the effect of the state of stress. Even for the simplest case of this limited scope, i.e., the planar case, the experimental mapping of the failure surface in three-dimensional stress space (σ_1, σ_2, σ_6) requires a staggering number of failure experiments. Insight to delineate the shape of the failure surface can narrow the number of tests. Specifically, only strategically meaningful states of stress need be considered. Such insight may come from the mathematical model of a failure criterion.

A salient point in the discussion of mathematical properties of failure criterion is that no uniqueness of any phenomenological failure criterion can be established. Such is a consequence of formulation through modeling rather than through derivation from first principles. Ergo, in the context of phenomenological failure, experiments are not to be interpreted as validification of the theory of failure but rather as confirmation of the rationale of experimentation design and scrutiny of the modeling adequacy of the resulting failure criterion. The objective of experimentation design is to identify the minimum number of independent key experiments required for the determination of a failure surface to the acceptable degree of accuracy. In accordance with the mathematical model, the set of key experiments completely determines the failure surface. Verification of this dictum by additional definitive experiments confirms the characterization capabilities of the failure criterion. Subsequent determination of failure criterion for other composites can follow the established methodology (key experiments only) without the need for extensive experimental confirmation. This rationale is clearly identifiable in the underlying philosophy of phenomenological failure criterion and experimentation design, yet it has been overlooked by many workers in the field. Tenuous data implemented from rudimentary experimentations have been presented in the erroneous perspective of absolute validification of the *theory* of failure. Specifically, the sensitivity of experimentally determined strength parameters to the complex state of stress (the radial loading path) has typically been overlooked. As noted by Tsai and Wu (1971b) a $\pm 3\%$ variation in an off-axis tension test at 45° will completely obscure the determination of the normal stress interaction parameter F_{12}. Unfortunately, off-axis tension tests, because of their simplicity, have been most frequently used for experimental verifications. In view of this lack of sensitivity, it is not surprising that the experimental data in this off-axis tension domain has accomodated almost all proposed criteria. Hence a meaningful confirmation of a failure criterion requires two essential ingredients of experimentation: sound experimental technique and rational guiding key experiments.

A. Experimental Techniques

Sound experimental techniques in the characterization of mechanical properties of materials must satisfy certain fundamental requirements: (1) the stresses in the sample must be computable for the experimental boundary conditions and (2) the stresses in the sample must be uniform. If the mechanical properties to be measured are averaged global material responses, such as deformation, only the first requirement needs to be satisfied. Kavanagh (1972) has demonstrated the use of nonuniform stress samples for the measurement of stiffness constants of anisotropic materials. However, if the mechanical properties to be measured are governed by local material responses, experimentation which satisfies both requirements can lead to significant simplification in the interpretation of results. Failure is a local process and its origin within a sample can seldom be a priori determined. In order not to couple the local material variation with the state of stress, the stresses in samples for failure testing should be both computable and uniform.

For isotropic materials, both of these requirements can be easily met and have been well documented. For anisotropic materials, because of the various deformational couplings, many end-constraints for load introduction are not Saint Venant effects and can cause severe nonuniformity in the gauge section of the sample. Cautions and experimental procedures have been suggested by Halpin *et al.* (1969).

Among the loading conditions required for experimental measurement of strength surfaces are tension, compression, pure shear, and combined loadings. Lenoe (1968) has presented an in-depth review and experimental comparison of many techniques devised for composite testing including: tension, compression, flexture, short beam and solid rod torsion, covering sample geometry, and end-tap configurations. In addition, a wide variety of testing configurations have been proposed to implement shear and biaxial testing in the form of picture frame shear, rail shear, and crossbeam tests. Many of these configurations are geometrically complex and give rise to questionable stress calculations and severe nonuniformity in stress distribution; they are more suitable for stiffness measurement than for strength measurements. Some of these limitations have been examined by Whitney *et al.* (1971). In the case of testing of laminated composites, additional complexity associated with specimens with free edges has been analyzed by Pagano and Pipes (1971) and Whitney and Browning (1972).

One of the most practical geometries for testing anisotropic composites is the uniform thin-walled cylinder. The computation of the state of stress with such a geometry is straightforward; the stress distribution uniformity is high, the gripping forces are self-equilibrated and can be classified as Saint Venant effects; there are no free edges to induce high stress gradients; and any state of planar stress can be obtained through applied axial, torsional, internal pressure and external pressure loading. The design of the geometry of composite cylindrical samples has been delineated by Pagano and Whitney (1970). The major experimental problem associated with testing cylindrical samples lies in the implementation of end-attachments and the extensive data acquisition requirements. Solutions to the end-attachment problem have been proposed by Wu (1972), Cole and Pipes (1972), and Lenoe (1973). The data acquisition requirements have been resolved by incorporation of laboratory real-time computers into the mechanical testing system (Wu and Jerina, 1972). The establishment of the cylindrical test sample and a flexible testing system allows the key guiding experiments to be implemented with confidence.

B. *Guiding Experiments*

On the grounds of generality without redundancy while encompassing currently proposed failure criteria, the tensor polynomial, Eqs. (5), has been affirmed to be the most viable failure criterion. Thus it suffices to concentrate on the establishment of experimentation design based on this mathematical model. The ensuing discussion is for the most basic case of static failure, i.e., short loading time and radial loading paths. The data presented is for graphite fiber- (Moganite II) reinforced epoxy lamina composites produced by Whittaker Corporation.

One of the most prominent features of the tensor polynomial failure criterion, Eq. (5b), is that the relations of the failure tensors (F_i, F_{ij}, . . .) to the engineering strength are not mixed. Hence, any degree of accuracy in characterization can be obtained by systematic inclusion of higher-order terms without recomputation of the constants already determined. Consideration of linear terms alone is obviously insufficient because the resulting failure surface is a plane in the stress space. Progressing to the first- and second-order terms, Eq. (5b) upon expansion becomes

$$F_{11}\sigma_1^2 + F_{22}\sigma_2^2 + F_{66}\sigma_6^2 + 2F_{12}\sigma_1\sigma_2 + 2F_{16}\sigma_1\sigma_6 + 2F_{26}\sigma_2\sigma_6 + F_1\sigma_1 + F_2\sigma_2 + F_6\sigma_6 = 1 \tag{106}$$

By fixing an arbitrary reference axis on a given composite, without necessarily assuming the knowledge or existence of its principal direction, the failure tensors can be determined in the following sequence:

(1) Tension test in 1 direction ($\sigma_2 = \sigma_6 = 0$) to obtain strength: $\sigma_1 = X_1$

(2) Compression test in 1 direction ($\sigma_2 = \sigma_6 = 0$) to obtain strength: $\sigma_1 = -X_1'$

(3) Tension test in 2 direction ($\sigma_1 = \sigma_6 = 0$) to obtain strength: $\sigma_2 = X_2$ (107)

(4) Compression test in 2 direction ($\sigma_1 = \sigma_6 = 0$) to obtain strength: $\sigma_2 = -X_2'$

(5) Pure shear in 1–2 plane ($\sigma_1 = \sigma_2 = 0$) to obtain strength: $\sigma_6 = X_6$

(6) Pure shear in 1–2 plane ($\sigma_1 = \sigma_2 = 0$) to obtain strength: $\sigma_6 = -X_6'$

(7) Biaxial test in 1–2 plane ($\sigma_6 = 0$) to obtain strength:

$$\sigma_1 = \tilde{\sigma}_1, \sigma_2 = \tilde{\sigma}_2$$

(8) Biaxial test in 1–6 plane ($\sigma_2 = 0$) to obtain strength: (107 cont.)

$$\sigma_1 = \tilde{\sigma}_1, \sigma_6 = \tilde{\sigma}_6$$

(9) Biaxial test in the 2–6 plane ($\sigma_1 = 0$) to obtain strength:

$$\sigma_2 = \tilde{\sigma}_2, \sigma_6 = \tilde{\sigma}_6$$

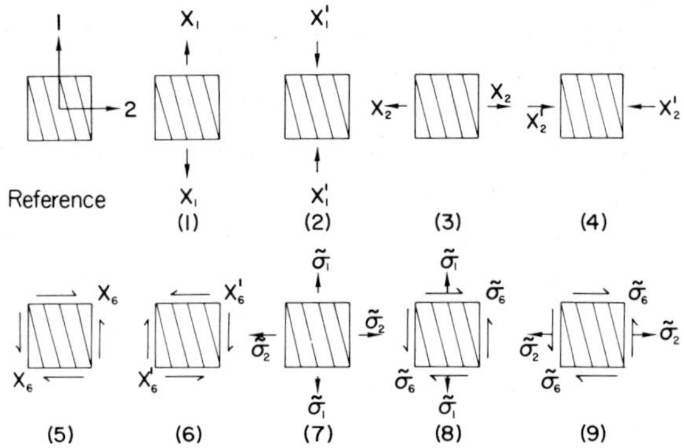

FIG. 13. Reference coordinate definition and 9 guiding experiments for measuring failure tensors in a quadratic polynomial failure criterion.

These basic experiments are schematically illustrated in Fig. 13 and the respective measurements, upon substitution into Eq. (106), yield the following relations for the computation of the failure tensor: For experiments 1–4:

$$F_i = (1/X_i) - (1/X_i') \tag{108}$$

$$F_{ii} = 1/X_iX_i' \qquad \text{(no sum)} \tag{109}$$

For experiments 7–9:

$$(B^2F_{ii} + 2BF_{ij} + F_{jj})\tilde{\sigma}_j^2 + (BF_i + F_j)\tilde{\sigma}_j = 1 \qquad \text{(no sum)} \tag{110}$$

$$B = \tilde{\sigma}_i/\tilde{\sigma}_j, \qquad i, j = 1, 2, 6$$

It has been explicitly pointed out by Wu (1972) that in order to definitively determine F_{12}, F_{16}, and F_{26}, the sensitivity of the experiments must be properly accounted for by choosing the optimal biaxial ratios B which can be computed from the following relations (no sum):

$$\tilde{\sigma}_j = \frac{-(F_iB + F_j) \pm [(F_iB + F_j)^2 + 4(F_{ii}B^2 + 2F_{ij}B + F_{jj})]^{1/2}}{2(F_{ii}B^2 + 2F_{ij}B + F_{jj})} \tag{111}$$

$$\frac{[2\sigma_j(F_{ii}B + F_{ij}) + F_i]}{[2\sigma_j(F_{ii}B^2 - F_{jj}) - F_j]} = -\frac{[2\sigma_j(F_{ii}B^2 + 2F_{ij}B + F_{jj}) + F_iB + F_j]}{B(BF_i + F_j)} \tag{112}$$

Simultaneous solution of Eqs. (111) and (112) yields the respective optimal ratio B for experiments 7–9. The details of these calculations have been presented by Wu (1972). As an example, for determination of F_{12} for graphite–epoxy composite, the optimal biaxial ratio B and the attainable resolution are shown in Figs. 14a and b.

As can be seen from Eqs. (108) and (109), the determination of the strength coefficient F_1, F_2, F_6; F_{11}, F_{22}, F_{66} is direct, i.e., they are not affected by addition or elimination of higher terms. Their resolution can be related directly to the scatter in engineering strengths X_i (Wu, 1972).

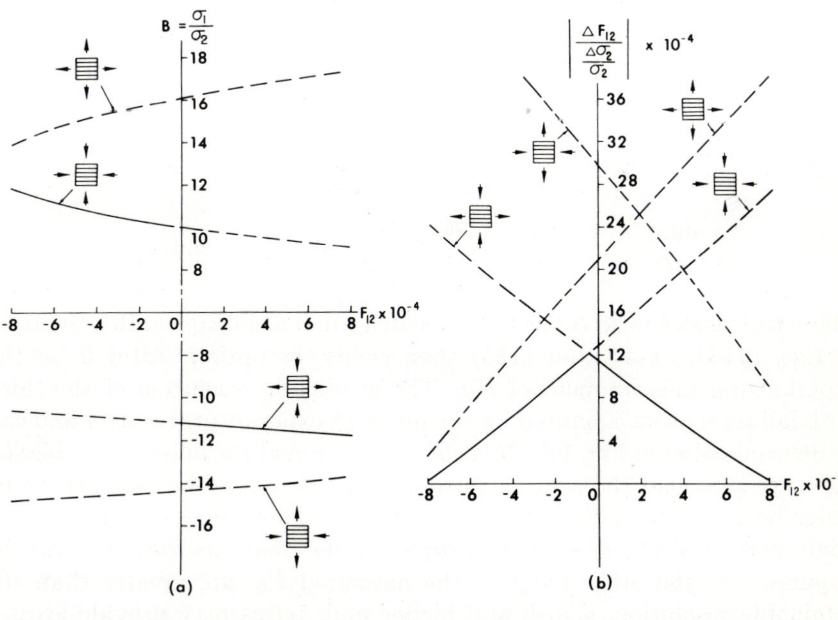

FIG. 14. (a) Optimal biaxial ratio B for measurement of interaction failure tensor F_{12}; (b) attainable resolution of F_{12} for optimal ratio B with experimental scatter $\Delta\sigma_2/\sigma_2$ (Wu, 1972). The values shown are for absolute (solid line) and relative (dashed line) optima.

Thus, if a quadratic failure surface is sufficient to characterize the failure of a composite, only 9 independent experiments [Eqs. (107)] are required. In the case of obvious physical symmetry as exists for unidirectional lamina where symmetry in shear strength is known, experiments 6, 8, and 9 are no longer needed, reducing the minimum number of independent experiments to 6 for orthotropic composites.

If further accuracy in characterization is desired, the third-order terms can be included in the tensor polynomial, i.e.,

$$F_i\sigma_i + F_{ij}\sigma_i\sigma_j + F_{ijk}\sigma_i\sigma_j\sigma_k = 1 \qquad (113)$$

For the simplified case of orthotropic symmetry, we established that the only admissible third-order failure tensor components are F_{112}, F_{122}, and F_{266}. Since these failure tensor components characterize the effect of strength coupling as in F_{12}, F_{16}, and F_{26}, similar caution must be exercised in the determination of the biaxial ratios for experimental measurements. Following the rationale presented by Wu (1972) the optimal biaxial experimental ratios can be determined by defining ψ_{ijj} as the magnifying factor for experimental scatter in the form

$$\Delta F_{ijj}/F_{ijj} = \psi_{ijj}\,\Delta\sigma_j/\tilde{\sigma}_j \qquad \text{(no sum)} \qquad (114)$$

ψ_{ijj} is not a tensor; indicial notation is used for purposes of identification only.

This magnification factor can be minimized with respect to the biaxial ratio through

$$\left.\frac{d\psi_{ijj}}{dB}\right|_{F_{ijj}} = \frac{\partial\psi_{ijj}}{\partial B} + \frac{\partial\psi_{ijj}}{\partial\sigma_j}\frac{\partial\sigma_j}{\partial B} = 0 \qquad \text{(no sum)} \qquad (115)$$

where the biaxial ratio B has the same definition as in Eqs. (110). Solution of Eqs. (113), (114), and (115) then yields the optimal ratio B for the experimental measurement of F_{ijj}. The attainable resolution of the third rank failure tensors is limited by the inherent experimental scatter and can be determined as in Fig. 14b. If the F_{ijj} measured at the appropriate biaxial ratio are less than the attainable resolution, they may be assumed to be insignificant implying that the third-order polynomial does not provide any additional accuracy and terms of higher order than quadratic will not be required. On the other hand, if the measured F_{ijj} are greater than the attainable resolution, eighth and higher rank terms may provide greater accuracy and could, if desired, be measured. Proceeding with this rationale, the lowest-order tensor polynomial failure criterion can be definitely determined.

The procedure for determination of failure surface from minimum key guiding experiments can be definitively confirmed through testing of *highly* anisotropic composites. Failure data reported by Wu (1972) for graphite fiber- (Morganite II) reinforced epoxy composite lamina (fabricated by Whittaker Corporation) can be evaluated from this perspective. These strength data were implemented with care, paying particular attention to boundary conditions of experimentation and stress uniformity with the samples. Compression, pure shear, and biaxial data were obtained from cylindrical specimens. With reference axis 1 oriented along the fiber direction, the averaged results of guiding experiments (1)–(6) are

Test (1):	longitudinal tension,	$X_1 = 149$	ksi	(116)
Test (2):	longitudinal compression,	$X_1' = -103$	ksi	
Test (3):	transverse tension,	$X_2 = 6.3$	ksi	
Test (4):	transverse compression,	$X_2' = -18.2$	ksi	
Test (5):	pure shear,	$X_6 = 10.5$	ksi	
Test (6):	pure shear,	$X_6' = -10.5$	ksi	

Coincidence of shear strength for positive and negative shear in tests (5) and (6) indicates strength orthotropy and precludes the necessity for tests (8) and (9). Tcst (7), the biaxial normal stress experiment, performed at the optimal biaxial ratio for maximum resolution provided the following averaged results:

$$\text{Test (7): biaxial strength} \quad \tilde{\sigma}_1 = 177 \text{ ksi}$$
$$\text{at (biaxial ratio } B = 14) \quad \tilde{\sigma}_2 = -12.8 \text{ ksi} \tag{117}$$

Substitution of results from Tests (1)–(7) into Eqs. (108) to (110), we obtain:

$$\left.\begin{aligned} F_1 &= -0.003 \\ F_2 &= 0.105 \\ F_6 &= 0 \end{aligned}\right\} \{1/\text{ksi}\} \qquad \begin{aligned} \Delta F_1 &= 0.0003 \\ \Delta F_2 &= 0.01 \end{aligned}$$

$$\left.\begin{aligned} F_{11} &= 0.065 \times 10^{-3} \\ F_{22} &= 8.72 \times 10^{-3} \\ F_{66} &= 9.07 \times 10^{-3} \\ F_{12} &= 0.2 \times 10^{-3} \end{aligned}\right\} \{1/\text{ksi}^2\} \qquad \begin{aligned} \Delta F_{11} &= 0.013 \times 10^{-3} \\ \Delta F_{22} &= 1.7 \times 10^{-3} \\ \Delta F_{66} &= 1.8 \times 10^{-3} \\ \Delta F_{12} &= 0.1 \times 10^{-3} \end{aligned} \tag{118}$$

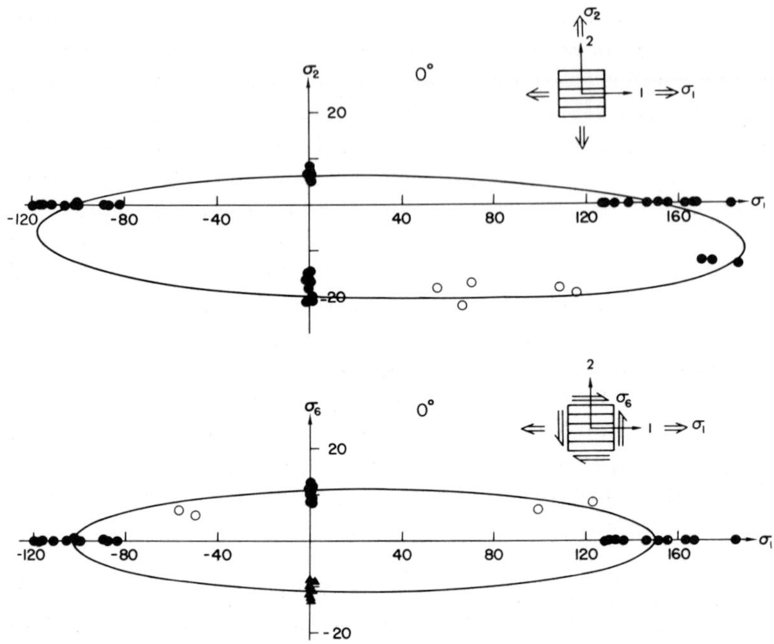

FIG. 15. Traces of failure surfaces determined from guiding experiments (solid points) in material principal direction; stresses in {ksi}.

where ΔF_i, ΔF_{ij} are the theoretical resolutions (Wu, 1972) of the failure tensor based on a 10% scatter in the strength measurements X_i. It can be observed that the measured F_{12} is close to the theoretically attainable resolution suggesting higher terms need not be included.

Substitution of the numerical values from Eqs. (118) into Eq. (106) permits the calculation of an ellipsoidal failure surface as shown in Fig. 15. From Fig. 15 it can be appreciated that geometrically the key guiding experiments provide the necessary number of constants to define the size and shape of the ellipsoidal failure surface. Additional redundant experiments are required to confirm the ability of this failure surface to characterize failure under any arbitrary state of stress. These additional experiments are also reported by Wu (1972). Graphical representation of these data in the three-dimensional $\sigma_1\sigma_2\sigma_6$ space is difficult. Simplicity in presentation will be preserved by considering the trace of the failure surface on the $\sigma_1\sigma_2$ plane and the $\sigma_1\sigma_6$ plane as shown in Fig. 15, where solid circles designate key experiments and open circles designate redundant experiments for confirmation. Other data not on the $\sigma_1\sigma_2$ and $\sigma_1\sigma_6$ planes were obtained from biaxial tests of off-axis cylindrical samples, i.e., helically wrapped lamina tubes. The failure surfaces for these material orientations can be obtained through the transformation of the measured failure tensor [Eq. (18)] to the new reference direction using Eqs. (6). The operational details of such transformation have been illustrated by Tsai and Wu (1971a).

Substitution of these transformed failure tensors F_i' and F'_{ij} into Eqs. (106) or (5b) enables the construction of a failure surface in the θ' orientation. The trace of the failure surface for material orientations where experiments have been conducted are shown in Fig. 16. Open circles again designate redundant experiments for verification. It should be emphasized that all the ellipses in Figs. 15 and 16 are determined from one set of failure tensors [Eq. (118)] and they are geometrically different trace sections of the three-dimensional failure surface in Fig. 9. The ability of the failure surface determined from key experiments to correlate any arbitrary state of stress is evident from Figs. 15 and 16. To achieve further consolidation, the voluminous data in three-dimensional $\sigma_1\sigma_2\sigma_6$ space can be convoluted into the $\sigma_1\sigma_2$ plane for comparison of different failure criteria. In Fig. 17, $f(\sigma_1, \sigma_2, \sigma_6)$ is any given failure criterion to be examined, $(\sigma_1, \sigma_2, \sigma_6)$ is the predicted failure stress for a given radial loading path, and $(\sigma_1{}^*, \sigma_2{}^*, \sigma_6{}^*)$ is an experimental failure data. The deviation of the failure data from the

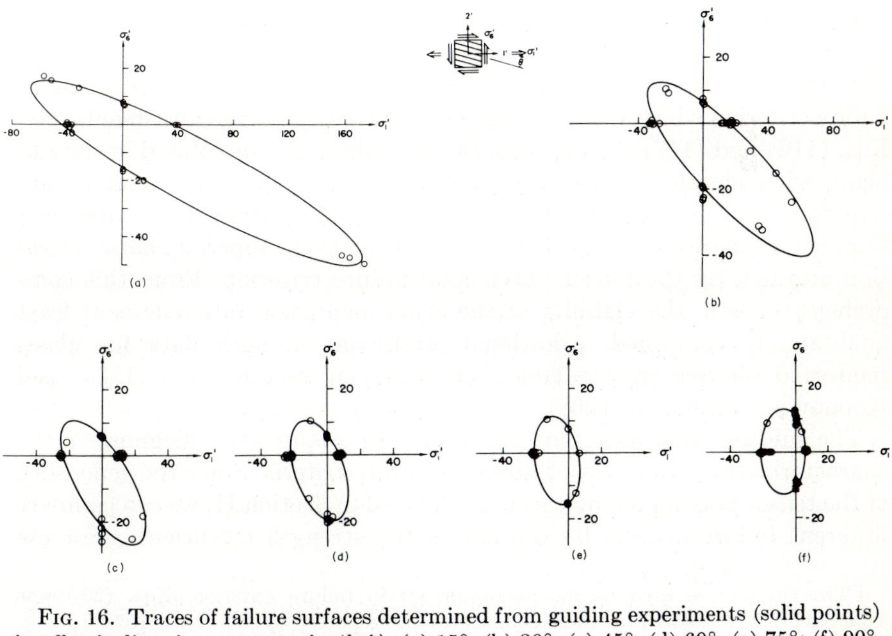

FIG. 16. Traces of failure surfaces determined from guiding experiments (solid points) in off-axis directions; stresses in {ksi}. (a) 15°; (b) 30°; (c) 45°; (d) 60°; (e) 75°; (f) 90°.

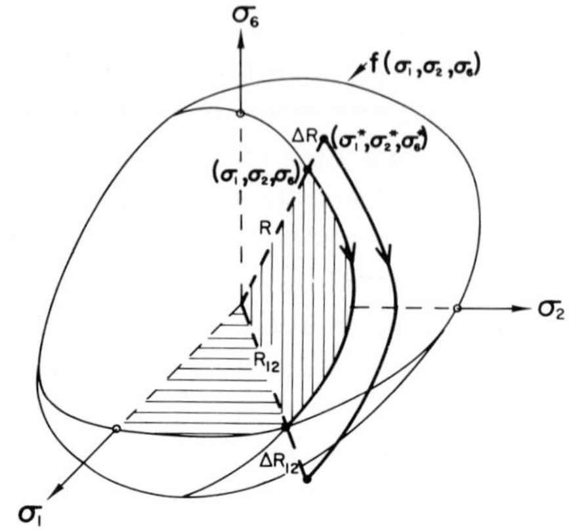

FIG. 17. Convolution of failure data onto a plane by a failure criterion $f(\sigma_1, \sigma_2, \sigma_6)$.

predicted strength is denoted by ΔR in Fig. 17. The relative deviation of the failure data in the $\sigma_1\sigma_2$ plane is defined as ΔR_{12} and can be calculated using the relation

$$\Delta R_{12}/R_{12} = \Delta R/R \tag{119}$$

where R_{12} is the projection of the given loading path in the $\sigma_1\sigma_2$ plane. Using R_{12} and ΔR_{12}, the relative position of the failure data in the three-dimensional $\sigma_1\sigma_2\sigma_6$ space can be represented in the $\sigma_1\sigma_2$ plane. Data shown in Fig. 18a are convoluted onto the $\sigma_1\sigma_2$ plane by the tensor polynomial failure criterion which was determined by the key guiding experiments, i.e., Eqs. (116) and (117). In Fig. 18b, the same data are convoluted in the $\sigma_1\sigma_2$ plane with identical procedures but using the maximum strain failure criterion† and in Fig. 18c by the modified von Mises–Hill criterion.‡ Cursory comparison of Figs. 18(a) to (c) reveals the superior characterization attained by the tensor polynomial failure criterion. From this comprehensive data, the viability of the experimentation rationale is at least qualitatively confirmed. Additional multiaxial strength data for glass-reinforced plastics are available from Protasov and Kopnov (1965) and Kopnov and Protasov (1968).

The guiding experiments are also useful for quantitative diagnosis of the characterization capability of different failure criteria. Since the generality of the tensor polynomial has been established in Section II, we can compare different failure criteria by examining the strength coefficients. For ex-

† The compliance used in the maximum strain failure criterion [Eqs. (29)] was experimentally measured for these composites and reported by Wu *et al.* (1973).

‡ Modified von Mises–Hill criterion was obtained by substituting appropriate tension and compression strengths in Eq. (48b) for the four stress quadrants.

ample, the coefficient F_{12} which characterizes the effect of normal stress coupling is defined by Eqs. (24) for the maximum strain failure criterion; by Eq. (45) for the von Mises–Hill failure criterion; and by Eqs. (78) for the Hoffman failure criterion. The relations for determining F_{12} and their numerical values based on results of key experiments [Eqs. (116)–(118)] are tabulated in Table II. For quantitative comparison of the characterization capabilities, listed in Table II are root mean squares of the difference between all the experimental data (presented in Figs. 15 and 16) and the theoretical prediction by the respective criteria. The root mean square is

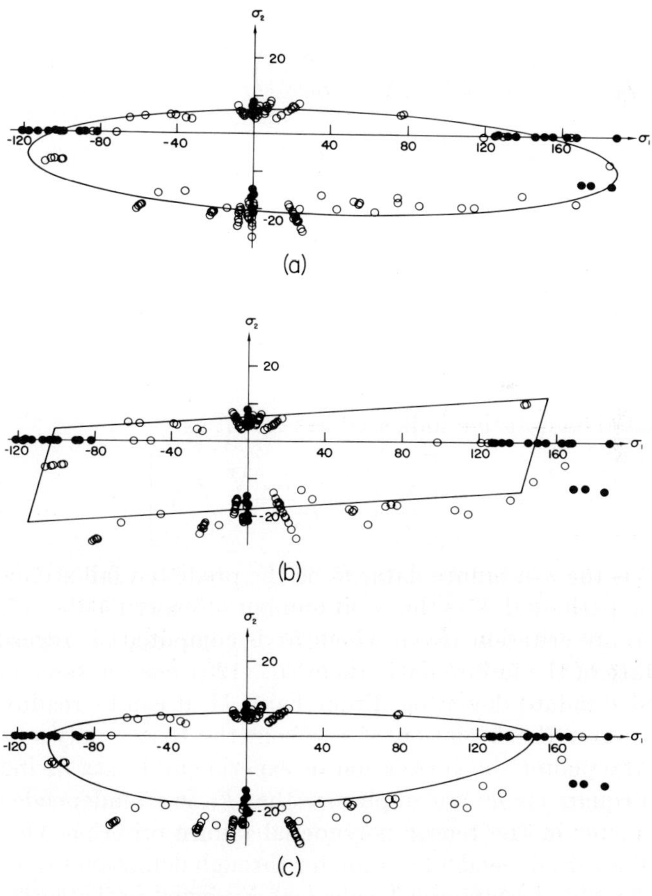

FIG. 18. Failure data convoluted onto $\sigma_1\sigma_2$ plane; stresses in {ksi}: (a) by tensor polynomial failure criterion; (b) by maximum strain failure criterion; (c) by modified von Mises–Hill failure criterion.

TABLE II

DETERMINATION OF F_{12}

Failure criterion	F_{12} (1/ksi)^2 Method of determination	Numerical value ($\times 10^{-4}$)	Correlation of failure data rms
Tensor polynomial Eq. (106)	Measured under optimal biaxial ratio	2.0	0.137
Hoffman Eq. (77)	$F_{12} = -1/2X_1 X_1'$	−0.33	0.160
Maximum strain Eqs. (29)	$F_{12} = \dfrac{-1}{2X_1 X_2}\left(1 + \dfrac{S_{12}^2}{S_{11} S_{22}}\right) \quad \begin{matrix}\sigma_1 > 0,\\ \sigma_2 > 0\end{matrix}$	−5.4	0.200
Modified[a] von Mises–Hill Eqs. (29)	$F_{12} = \dfrac{-1}{2X_1^2}, \quad \sigma_1 > 0$	−0.23	0.226
	$F_{12} = \dfrac{-1}{2X_1'^2}, \quad \sigma_1 < 0$	−0.45	

[a] Appropriate tensorial and compressive strength are used in Eq. (47b) for the four stress quadrants.

defined, based on notation indicated in Fig. 17, as

$$\text{rms} = \left[\sum_n^N \left(\frac{R_n - R_n^*}{R_n} \right)^2 \Big/ N \right]^{\frac{1}{2}} \tag{120}$$

where R_n^* is the nth failure data, R_n is the predicted fail strength for the nth loading path, and N is the total number of experiments (172 here).

If the failure criterion (from which R_n is computed) is regarded as the mean surface of the failure data, then Eq. (120) has the same meaning as normalized standard deviation. From Table II, it can be readily observed that the greater the deviation of F_{12} from the tensor polynomial failure criterion, the poorer the correlation of experimental data as indicated by root mean square error. We emphasize that F_{12} is an independently measured parameter in the tensor polynomial failure criterion. On the other hand, F_{12} is a *fixed* dependent parameter through definition for other failure criteria. Clearly, the numerical values of F_{12} listed in Table II and their deviations from the tensor polynomial value are applicable for the particular graphite–epoxy composite examined. Coincidence of the F_{12} value for a restrictive failure criterion could occur for another composite. However, such coincidence would be purely fortuitous and obviously no general recurrance can be expected.

These quantitative results can be interpreted as an experimental proof of the necessity condition for the tensor polynomial failure criterion and, above all, the rationale of guiding experiments.

IV. Data Analysis

The key for expedient determination of a phenomenological failure criterion for an anisotropic composite is to choose a mathematical model of established generality capable of characterizing a failure surface of any arbitrary shape. Using such a mathematical model as a guide, appropriate numbers of independent excitations can be applied experimentally to measure the corresponding material failure responses. The minimum number of independent key experiments required is equal to the number of the failure tensor components which can be regarded as intrinsic material parameters. The vanishing of higher rank failure tensors from the respective optimal key experiments establishes the lowest order of tensor polynomial which is applicable for the composite examined.

The experimentation rationale delineates the most expedient approach of determining a phenomenological failure criterion with known accuracy through a minimum number of key guiding experiments for one given material orientation. The need for rational data analysis arises: (1) when redundant experimental measurements are performed to improve the confidence of failure surfaces; (2) when the guiding experiments are repeated with respect to different material orientations either for experimental expediency or for verification of the transformational properties of the failure tensors; and (3) when it is desired to reduce the data to compact constants for documentation and retrieval in engineering applications.

A. Analysis of Data from One Material Orientation

If all the experiments, minimum guiding experiments as well as redundant experiments, are performed for only one given material orientation, the failure tensor components $F_i, F_{ij}, \ldots, F_{ijk}, \ldots$ are all scalar quantities. Hence the failure criterion, Eqs. (5), is an algebraic equation with experimentally adjustable coefficients. A second-order tensor polynomial for the planar case has 3 linear coefficients (F_1, F_2, and F_6) and 6 quadratic coefficients (F_{11}, F_{12}, F_{16}, F_{22}, F_{26}, and F_{66}). The magnitude of these 9 coefficients can be optimized for experimental failure data each of which defines an observed failure state ($\sigma_1{}^*$, $\sigma_2{}^*$, $\sigma_6{}^*$).

Optimization can be carried out by minimizing the root mean square error between the mathematical model and the failure data. Two important points must be adhered to in the optimization operation: (1) normalized rms error must be used, and (2) minimization of the rms must be along the respective loading path. Normalization is required because of the extreme differences in strength due to anisotropy which leads to the well-known fact that the absolute scatter is greater for a larger failure strength. A further consequence of strength anisotropy is that the failure surface is elongated along the strong material orientation. Thus, standard numerical techniques which seek optimization along the outward normal to the failure surface are not necessarily in accord with experimental conditions. Data reported herein all followed radial loading paths. The optimization is carried out by minimizing the normalized root mean square error as defined in Eq. (120) through linear regression numerical techniques. The experimental results presented in Fig. 15 and 16 (after transforming all failure stresses to the material principal direction) are thus optimized. Resulting coefficients F_i, F_{ij}, the axis crossing strength, and the rms errors are listed in Table III. From the table it can be seen that (1) the rms error is reduced slightly when all failure data are taken into account, and (2) the basic axis crossing strengths are not drastically changed thus lending further credibility to the guiding experiments.

B. *Analysis of Data from Several Material Orientations*

Ideally, given a flexible mechanical testing system, a general state of plane stress can be applied to a thin-walled cylindrical specimen permitting the experimental mapping of any portion of the failure surface in the three-dimensional $\sigma_1\sigma_2\sigma_6$ space. Although such mechanical testing systems have been described by Wu *et al.* (1972, 1973) and by Cole and Pipes (1972), they are not generally available. Judicious selection of specimen configurations can partially alleviate testing system limitations. For example, an internal pressure system is required in the implementation of biaxial optimal experiments. In lieu of an internal pressure system, a limited combination of biaxial experiment can be achieved through tension–torsion test of an off-axis cylinder where the axis of the sample is not a material principal direction. Under such off-axis testing configurations, the data analysis may be performed by transformation of the failure stresses from the test orientation to one common material reference axis (such as the principal direction), or by analysis of data directly for the testing reference axis. Although the latter approach is not indispensable if the material principal direction is unambiguously defined, it can offer verification for the mathematically operational aspects of the failure criterion. The details of direct analysis of data performed on different material orientations will be discussed.

One of the most important features of the tensor polynomial failure criterion [Eqs. (5)] is that it can be used for any arbitrary reference axis and that it can be transformed from one reference axis to another through tensor transformation rules [Eqs. (6)]. Hence we may, following the guiding experiments, determine the failure tensor F_i, F_{ij}, . . . for a reference orientation θ and repeat similar experiments to determine F_i', F'_{ij}, . . . for another orientation θ'. These two sets of failure tensors are related by the tensor transformation rules, Eqs. (6), which are represented symbolically as

$$F_i' = T_{ij}F_j \tag{121a}$$

$$F'_{ij} = T_{ijkl}F_{kl} \tag{121b}$$

using contracted notation, where T_{ij} and T_{ijkl} are functions of directional cosines for the two reference orientations. In the case where F_i, F_{ij}, . . . ,

TABLE III

Failure tensors	Guiding experiments	All experiments	Axis crossing strengths (ksi)	Guiding experiments	All experiments
Second rank (1/ksi)					
F_1	−0.003	−0.0031	X_1	149	150
F_2	0.105	0.115	X_1'	103	102
F_6	0	0	X_2	6.3	6.1
			X_2'	18.2	20.6
Fourth rank			X_6	10.5	10.9
$(1/\text{ksi})^2 \times 10^{-3}$			X_6'	−10.5	−10.9
F_{11}	0.065	0.065			
F_{12}	0.2	0.1			
F_{16}	0	0			
F_{22}	8.72	7.98			
F_{26}	0	0			
F_{66}	9.07	8.43			
rms	0.137	0.109			

etc. are all determined from a single material orientation, the failure tensors are independent scalar parameters and conventional statistical methods may be used for data averaging of the independent strength measurements X_i. However, in the case under consideration, the failure tensors F_i, $F_{ij}, \ldots$, etc. and F_i', $F'_{ij}, \ldots$, etc. are related by Eqs. (121) and hence are not independent. Their statistical analysis must reflect weighting by the transformation equations [Eqs. (121)].

It has been established (Wu *et al.*, 1973) that data analysis of such tensoral parameters can be performed by direct averaging of the associated scalar invariants. The scalar invariants for the second rank failure tensors are

$$\mathrm{I}^{(2)} = \tfrac{1}{2}(F_1 + F_2) \tag{122a}$$

$$\mathrm{II}^{(2)} = \left[\left(\frac{F_1 - F_2}{2}\right)^2 + F_6\right]^{1/2} \tag{122b}$$

The invariants for orientation θ and θ' can be computed by respectively substituting the measurements F_i and F_i' into Eqs. (121). Since the invariants are scalar quantities independent of reference axis orientations, they may be averaged directly. For example, if F_i are measured for m orientations, the averaged invariants $\bar{\mathrm{I}}^{(2)}$ can be obtained from

$$\bar{\mathrm{I}}^{(2)} = \sum_i^m \alpha_i \mathrm{I}^{(2)}(\theta_i) \Big/ \sum_i^m \alpha_i \tag{123}$$

where α_i's are weighting factors. If the F_i for each orientation are determined from the same number of samples, then the weighting factors become unity. The second-order invariant $\mathrm{II}^{(2)}$ can be averaged using the same relation as in Eq. (123). The two averaged invariants represent the experimental data at different material orientations in consolidated form. Components of the second rank failure tensor at any orientation can be recovered from the transformation equation [Eq. (121a)] represented in terms of the invariants:

$$\begin{bmatrix} F_1' \\ F_2' \\ F_6' \end{bmatrix} = \begin{bmatrix} \mathrm{I}^{(2)} & \mathrm{II}^{(2)} & 0 \\ \mathrm{I}^{(2)} & -\mathrm{II}^{(2)} & 0 \\ 0 & 0 & \mathrm{II}^{(2)} \end{bmatrix} \begin{bmatrix} 1 \\ \cos 2(\theta' - \theta) \\ \sin 2(\theta' - \theta) \end{bmatrix} \tag{124}$$

The scalar invariants for the fourth rank failure tensors are:

$$\mathrm{I}_1^{(4)} = F_{11} + F_{22} + 2F_{12} \tag{125a}$$

$$\mathrm{I}_2^{(4)} = F_{66} - 4F_{12} \tag{125b}$$

$$\mathrm{II}_1^{(4)} = \tfrac{1}{2}[(F_{11} - F_{22})^2 + (F_{16} + F_{26})^2]^{1/2} \tag{125c}$$

$$\mathrm{II}_2^{(4)} = \tfrac{1}{8}[(F_{11} + F_{22} - 2F_{12} - F_{66})^2 + 4(F_{16} - F_{26})^2]^{1/2} \tag{125d}$$

The fourth rank failure invariants can be computed for different orientations and averaged by relations identical to Eq. (123). Retrieval of fourth rank failure tensor components can be obtained from the transformation [Eq. (121b)] expressed in terms of the invariants:

$$\begin{bmatrix} F_{11} \\ F_{12} \\ \dfrac{F_{16}}{2} \\ F_{22} \\ \dfrac{F_{26}}{2} \\ \dfrac{F_{66}}{4} \end{bmatrix} = \begin{bmatrix} U_1 & -\mathrm{II}_1^{(4)} & 0 & -\mathrm{II}_2^{(4)} & 0 \\ -U_4 & 0 & 0 & \mathrm{II}_2^{(4)} & 0 \\ 0 & 0 & \dfrac{\mathrm{II}_1^{(4)}}{2} & 0 & \mathrm{II}_2^{(4)} \\ U_1 & \mathrm{II}_1^{(4)} & 0 & -\mathrm{II}_2^{(4)} & 0 \\ 0 & 0 & \dfrac{\mathrm{II}_1^{(4)}}{2} & 0 & -\mathrm{II}_2^{(4)} \\ U_5 & 0 & 0 & \mathrm{II}_2^{(4)} & 0 \end{bmatrix} \begin{bmatrix} 1 \\ \cos 2(\theta + \delta') \\ \sin 2(\theta + \delta') \\ \cos 4(\theta + \delta) \\ \sin 4(\theta + \delta) \end{bmatrix} \tag{126}$$

where

$$U_1 = \tfrac{1}{8}[3\mathrm{I}_1^{(4)} + \mathrm{I}_2^{(4)}], \qquad U_4 = \tfrac{1}{8}[\mathrm{I}_2^{(4)} - \mathrm{I}_1^{(4)}]$$

$$U_5 = \tfrac{1}{2}[\mathrm{I}_2^{(4)} + \mathrm{I}_1^{(4)}] \tag{127}$$

$$2\delta' = \arctan[(F_{16} + F_{26})/(F_{22} - F_{11})]$$

$$4\delta = \arctan[2(F_{26} - F_{16})/(F_{11} + F_{22} - 2F_{12} - F_{66})]$$

The averaging technique is straightforward for every orientation if the sample sizes are identical for the guiding experiment. If the sample sizes are not the same, the weighting factor in Eq. (123) can be computed by

$$\alpha_i = 1/\mathrm{SEM}\ \mathrm{I}_i^{(2)} \tag{128}$$

where SEM $\mathrm{I}_i^{(2)}$ is the standard error of the mean of $\mathrm{I}_i^{(2)}$ measured at θ_i.

According to mean statistics and Eq. (122a)

$$\mathrm{SEM}^2\, \mathrm{I}^{(2)}(\theta_j) = \left[\frac{\partial \mathrm{I}^{(2)}(\theta_j)}{\partial F_1(\theta_j)}\right]^2 \mathrm{SEM}^2\, F_1(\theta_j) + \left[\frac{\partial \mathrm{I}^{(2)}(\theta_j)}{\partial F_2(\theta_j)}\right]^2 \mathrm{SEM}\, F_2(\theta_j)$$

$$= \tfrac{1}{4}[\mathrm{SEM}^2\, F_1(\theta_j) + \mathrm{SEM}^2\, F_2(\theta_j)] \tag{129}$$

The standard error of the mean of F_1 and F_2 can be derived from their definition, Eq. (108), i.e.,

$$\mathrm{SEM}^2\, F_i = (\partial F_i/\partial \bar{X}_i)^2\, \mathrm{SEM}^2\, \bar{X}_i + (\partial F_i/\partial \bar{X}_i')^2\, \mathrm{SEM}^2\, \bar{X}_i', \qquad i = 1, 2, 6 \tag{130}$$

where $\bar{X}_i$, $\bar{X}_i'$ are the averaged strengths obtained from

$$\mathrm{SEM}\, \bar{X}_i = \mathrm{SD}_{X_i}/(n_{X_i})^{1/2} \tag{131}$$

where SD_{X_i} is the standard deviation of $\bar{X}_i$ for n_{X_i} number of samples:

$$\mathrm{SD}_{X_i} = \left[\sum^{n_{X_i}} (X_i - \bar{X}_i)^2/n_{X_i} - 1\right]^{1/2} \tag{132}$$

Substituting of Eqs. (132) and (131) into (130), we obtain

$$\mathrm{SEM}^2\, F_i = \frac{1}{n_{X_i}}\left(\frac{\mathrm{SD}_{X_i}}{\bar{X}_i}\right)^2 + \frac{1}{n_{X_i}}\left(\frac{\mathrm{SD}_{X_i'}}{\bar{X}_i'}\right)^2 \tag{133}$$

Substitution of Eqs. (133) and (129) into Eq. (128) allows the computation of the necessary weighting factors to be used in the averaging of the first-order invariant for the second rank failure tensor.

In a similar manner the weighting factors can be obtained for the second-order invariant $\mathrm{II}^{(2)}$ of the second rank failure tensor and for the invariants $\mathrm{I}_1^{(4)}$, $\mathrm{I}_2^{(4)}$, $\mathrm{II}_1^{(4)}$, and $\mathrm{II}_2^{(4)}$ of the fourth rank failure tensors. Such calculations, although tedious, are straightforward.

The data shown in Figs. 15 and 16 were obtained at fiber orientations 0, 15, 30, 45, 60, 75, and 90°. Failure tensors computed for these orientations are represented in Figs. 19 and 20, respectively, for the second and fourth rank components. Implementing computational details allow the data to be averaged as described, yielding six invariants (two for F_i' and four for F'_{ij}). The predicted F_i' and F'_{ij} based on these averaged invariants using Eqs. (124) and (125) are shown in Figs. 19 and 20. The exemplary agreement not only verifies the transformational properties of the failure tensors but also suggests that for large sample size, data averaging using the tensor

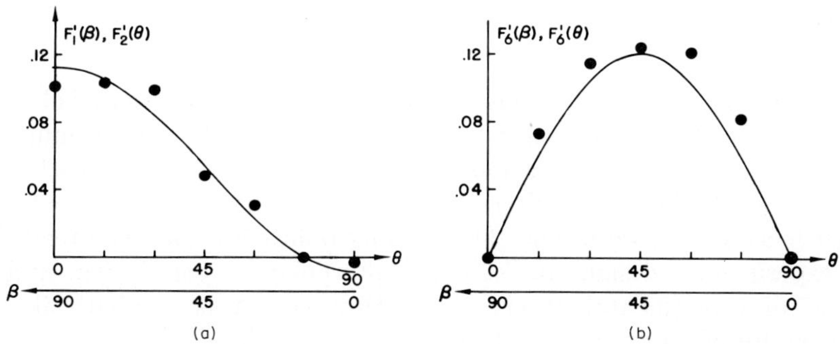

FIG. 19. Failure data in the second rank failure tensor spaces {1/ksi}.

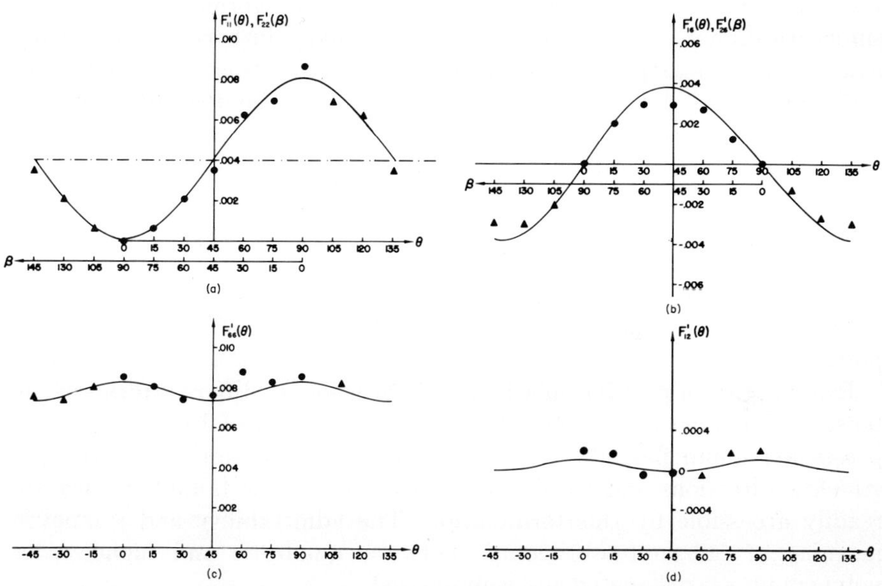

FIG. 20. Failure data in the fourth rank failure tensor spaces {1/(ksi)²}.

invariant is a viable technique. Advantages of the invariant averaging technique are that a large amount of data can be rationally reduced into six compact constants (the invariants) for convenient documentation and, using the transformation equations, they can be retrieved for applications.

Application of invariant averaging to a large amount of data results in compact representation in terms of six constants. Such rational reduction and representation greatly facilitates documentation and retrieval.

V. Summary

The failure of composites has been examined at the lamina level where heterogeneous fiber and matrix are treated as an anisotropic continuum. A phenomenological framework was adopted for the formulation of a mathematical model (the failure criterion) to account for the occurrence of failure under mechanical excitations. The ensuing failure criterion is intended to aid experimental design; to facilitate interpolation, correlation, and retrieval of experimental observations. However, it is emphatically not intended for interpretation of the mechanism of failure.

Arbitrary failure surfaces may be readily expressed in terms of the stress or strain failure tensor polynomials which meet certain fundamental mathematical requirements. Past formulations of anisotropic failure criteria were cast in terms of tensor polynomials. Such reformulation revealed similarities among the various criteria and many hitherto unknown desirable operational properties including the transformation of reference coordinates and direct conversion between stress and strain domains. Contrary to the general belief, several intuitively simple formulations (e.g., maximum strain criterion and maximum stress criterion) are mathematically complex. Their piecewise linear nature necessitates additional constraints to ensure consistency of conversion between stress and strain, and precludes application to some engineering composites. Formulations modeled after isotropic invariant characteristics were found to be limited to special orthotropy and are, therefore, degenerate cases of the failure tensor polynomials.

Examination of past formulations and variations of different forms of the tensor polynomial reveal that the tensor polynomial of Eqs. (5) encompasses maximum flexibility without redundancy. Definition of principal strength directions and the stability condition of the failure surface are readily accessible by this formulation. The admissibility and geometric meaning of failure tensor components for quadratic and higher-order polynomials are delineated and summarized.

Rational experimental design based on the failure tensor polynomial has reduced the monumental experimental task to a limited number of guiding experiments. These experiments provide definitive bases for determining the lowest order of failure criterion which will adequately characterize failure within inherent material scatter.

Experimental data for graphite–epoxy composites confirmed the viability of the experimental design. Tensor failure polynomial criterion, determined by the guiding experiments, provided good correlation of comprehensive complex stress data. In addition, the penalties for employing current restrictive failure criteria were quantatively compared.

Statistical methods for data reduction employing scalar invariants of the failure tensors were presented. Scalar invariants provide useful links for rational analysis of strength data measured at different material orientations, and serve as convenient parameters for compacting experimental data for documentation and ready retrieval in engineering design applications.

Unification of failure criteria for anisotropic composites in tensor polynomial form enables the selection of failure criterion to be based on analysis of characterization accuracy desired, rather than advocation of arbitrary preferences.

Acknowledgments

The author wishes to express his gratitude to Dr. M. J. Michno, Jr., Washington University, for his invaluable comments and suggestions and to acknowledge the financial support by the Nonmetallic Materials Division of the Air Force Materials Laboratory through Contract AF-33615-71-C-1440.

List of Symbols

σ_i,	$i = 1, 2, \ldots, 6$ Stress tensor in contracted notation
ϵ_i	$i = 1, 2, \ldots, 6$ Strain tensor in contracted notation
S_{ij}	$i, j = 1, 2, \ldots, 6$ Linear compliance coefficients
C_{ij}	$i, j = 1, 2, \ldots, 6$ Linear stiffness coefficients
$F_i, F_{ij}, F_{ijk}, \ldots$, etc.	$i, j, k, \ldots = 1, 2, \ldots 6$ Stress failure tensor of second, fourth, sixth rank, etc.
$G_i, G_{ij}, G_{ijk}, \ldots$, etc.	$i, j, k, \ldots = 1, 2, \ldots, 6$ Strain failure tensor of second, fourth, sixth rank, etc.
X_i	$i = 1, 2, \ldots, 6$ Ultimate strength of positive uniaxial stress
X_1	Tensile strength in 1 direction
X_2	Tensile strength in 2 direction
X_6	Positive shear strength
X_i'	$i = 1, 2, \ldots, 6$ Limit strength of negative uniaxial stress
X_1'	Compressive strength in 1 direction
X_2'	Compressive strength in 2 direction
X_6'	Negative shear strength
$\mathbf{E}_i$	$i = 1, 2, \ldots, 6$ Limit strain of positive uniaxial strain
$\mathbf{E}_i'$	$i = 1, 2, \ldots, 6$ Limit strain of negative uniaxial strain
f	Failure criterion function
f_1	Linear terms in failure criterion function
f_2	Quadratic terms in failure criterion function

$\psi_i, \psi_{ij}, \psi_{ijk}, \ldots$, etc.	Magnification factors (not tensors) of experimental scatter in the computation of F_i, F_{ij}, F_{ijk}, etc.
e	Characteristic matrix of F_{ij}
$\mathbf{E}$	Characteristic matrix of F_i, F_{ij}
λ_i	$i = 1, 2, \ldots, 6$ Principal strength directions
B	Biaxial stress ratio defined in Eq. (110)
I_1, I_2, I_3	First, second, and third stress invariants
J_1, J_2, J_3	First, second, and third deviatoric stress invariants
$I_{(i)}^{(j)}$	ith first-order invariant of a jth rank tensor
$II_{(i)}^{(j)}$	ith second-order invariant of a jth rank tensor

References

Ashkenazi, E. K. (1964). *Zavod. Lab.* **30,** No. 2, 285–287.

Ashkenazi, E. K. (1965). *Mekh. Polim.* **1,** No. 2, 60–70.

Ashkenazi, E. K., and Pekker, F. P. (1970). *Mekh. Polim.* **6,** No. 2, 284–294.

Azzi, V. D., and Tsai, S. W. (1965). *Exp. Mech.* **5,** 283–288.

Bogue, D. C. (1967). The Yield Stress and Plastic Strain Theory for Anisotropic Materials. Oak Ridge Nat. Lab. Rep. ORNL-TM-1869.

Brandmaier, H. E. (1970). *J. Compos. Mater.* **3,** 422–425.

Bridgman, P. W. (1922). *Proc. Amer. Acad. Arts Sci.* **58,** 165–242.

Bridgman, P. W. (1941). *J. Appl. Phys.* 461–469.

Chamis, C. C. (1969). *In* "Composite Materials: Testing and Design." ASTM-STP 460, pp. 336–351.

Cole, B. W., and Pipes, R. B. (1972). *In Proc. Conf. Fiberous Compos. Flight Vehicle Design.* Air Force Flight Dynamics Lab. Rep. AFFDL-TR-72-130.

Collins, B. R., and Thomas, R. L. (1973). General Plane Stress Failure for Boron-Aluminum Composite. Air Force Mater. Lab. Rep. No. AFML-TR-73.

Cooper, A. A., and Wu, E. M. (1973). *In* "Structural Composite Reinforced with Short Fibers." Dekker, New York.

Coulomb, C. A. (1773). *Mem. Math. Phys.* **7,** 343.

Drucker, D. C. (1949). *J. Appl. Mech. Trans. ASME* **71,** 349–357.

Fischer, L. (1960). *Mod. Plast.* **37,** No. 10, 120–203.

Fischer, L. (1967). *Trans. ASME* **89,** B, No. 3, 399–402.

Franklin, H. G. (1968). *Fiber Sci. Tech.* **1,** 137–141.

Goldenblat, I. I., and Kopnov, V. A. (1965). *Mekh. Polim.* **1,** No. 2, 54–58.

Halpin, J. C., Pagano, N. J., Whitney, J. M., and Wu, E. M. (1969). *In* "Composite Materials: Testing and Design," ASTM-STP 460, pp. 37–47. Amer. Soc. Testing and Mater.

Hashin, Z. (1970). *In* "Mechanics of Composite Materials" (F. W. Wendt, H. Liebowitz, and N. Perrone, eds.). Pergamon, Oxford.

Hearmon, R. F. S. (1952). *Brit. J. Appl. Phys. Ser. 1* **3,** 120–124.

Hill, R. (1948). *Proc. Roy. Soc. A* **193,** 281–297.

Hoffman, O. (1967). *J. Compos. Mater.* **1,** 200–206.

Hu, L. W. (1958). *J. Franklin Inst.* **265,** 187–204.

Jenkins, C. F. (1920). *In* Report on Materials of Construction Used in Aircraft and Aircraft Engines. Great Britain Aeronaut. Res. Committee.

Kavanagh, K. T. (1972). *Exp. Mech.* **12,** 50–56.
Kopnov, V. A., and Protasov, V. D. (1968). *Mekh. Polim.* **4,** No. 1, 64–66.
Lekhnitskii, S. G. (1950). "Theory of Elasticity of an Anisotropic Elastic Body." Holden-Day, San Francisco.
Lenoe, E. M. (1968). Evaluation of Test Techniques for Advanced Composites. Air Force Mater. Lab. Rep., AFML-TR-68-166, I to III.
Lenoe, E. M. (1973). Evaluation of a Cylindrical Specimen Testing Technique. Army Mater. and Mech. Res. Center Rep. AMMRC-TN-73-1.
Malmeyster, A. (1966). *Mekh. Polim.* **2,** No. 4 324–331.
Marin, J. (1957). *J. Aeronaut. Sci.* **24,** 4.
Norris, C. B. (1939). *Trans. Amer. Soc. Mech. Eng.,* **61,** 259–261.
Norris, C. B. (1950). Forest Products Lab. Rep. No. 1816.
Pagano, N. J., and Pipes, R. B. (1970). *J. Compos. Mater.* **4,** 538–548.
Pagano, N. J., and Pipes, R. B. (1971). *J. Compos. Mater.* **5,** 50–57.
Pagano, N. J., and Whitney, J. M. (1970). *J. Compos. Mater.* **4,** 360–378.
Paul, B. (1968). *In* "Fracture" (H. Liebowitz, ed.), Vol. II, p. 313. Academic Press, New York.
Priddy, T. G. (1971). A Fracture Theory for Brittle Anisotropic Materials. Sandia Lab. Rep. SC-DR-71 0764.
Protasov, V. D., and Kopnov, V. A. (1965). *Mekh. Polim.* **1,** No. 5, 26–28.
Reissner, E., and Stavsky, Y. (1961). *J. Appl. Mech.* **28,** 402–408.
Sella, A., and Voigt, W. (1892). *Gott Nachr.* 494.
Sella, A., and Voigt, W. (1893). *Wied. Ann. Bd.* **48,** 636.
Sendeckyj, G. P. (1972). *In Compos. Maters. Testing Design, 2nd Conf.* ASTM-STP 497, pp. 41–51. Amer. Soc. for Testing and Mater.
Sohncke, L. (1869). *Pogg. Ann. Bd.* **137,** 177–185.
Tsai, S. W., and Wu, E. M. (1971a). *J. Compos. Mater.* **5,** 58–80.
Tsai, S. W., and Wu, E. M. (1971b). A General Theory of Strength for Anisotropic Materials. Air Force Mater. Lab. Rep. AFML-TR-71-12.
Voigt, W. (1910). "Lehrbuch der Kristallphsyik." Teubner, Leipzig.
von Mises, R. (1928). Zeitschrift fur Angewandte Mathematik und Mechanik Vol. 8, pp. 161–185.
Waddoups, M. E. (1968). *In* "Composite Materials Workshop" (S. W. Tsai, N. J. Pagano, J. C. Halpin, eds.). Technomic, Westport, Connecticut.
Whitney, J. M., and Browning, C. E. (1972). *J. Compos. Mater.* **6,** 300–303.
Whitney, J. M., Stansbarger, D. L., and Howell, J. M. (1971). *J. Compos. Mater.* **5,** 24–35.
Wu, E. M. (1972). *J. Compos. Mater.* **6,** 473–489.
Wu, E. M., and Jerina, K. L. (1972). *Mater. Res. Std.* **12,** No. 2, 13–18.
Wu, E. M., and Scheublein, J. (1973). *In* Composite Materials: Testing and Design, ASTM-STP 546. Amer. Soc. for Testing and Mater.
Wu, E. M., Jerina, K. L., and Brissey, F. L. Jr. (1972). Deformation and Strength Characterization of Graphite Reinforced Epoxy Composites. I. Guiding Experiments, Experimental Techniques and Computer Implementation. Air Force Mater. Lab. Rep.
Wu, E. M., Jerina, K. L., and Lavengood, R. E. (1973). *In* Analysis of the Test Methods of High Modulus Fibers and Composites, ASTM-STP 521. Amer. Soc. for Testing and Mater.

10

Photoelastic Investigation of Composites

I. M. DANIEL

IIT Research Institute
Chicago, Illinois

I. Introduction

Composite materials, especially filamentary composites, possess unique properties of high strength, low density, good fatigue properties, and adaptability to the required function of the structure. Mechanics of composites deals with their mechanical response under load, i.e., stress and strain distributions from which the deformation and strength of a structure can be evaluated. Depending on the scale of observation, this study is classified as macromechanics or micromechanics. In the former case, the material is treated as composed of homogeneous, albeit anisotropic, plies and the quantities of interest are the average stress, strain, and strength. A composite laminate under this study can be characterized adequately by two Young's moduli, two Poisson's ratios, a shear modulus, and three strength values.

The micromechanical point of view recognizes the fact that the behavior of a composite material or structure is intimately related to the internal stress level and distribution and to the load transfer between the constituent parts. The field of micromechanics comprises the study of these internal stresses and the mechanics of internal reactions and interactions of the constituent parts due to imposed forces. This knowledge provides a rational basis for evaluating and designing the macroscopic response of the material, understanding failure modes, and establishing failure criteria.

Possible failure modes include tensile fracture of the matrix, bond failure at the fiber matrix interface, and tensile failure of the fibers. These modes are not independent but they can interact and trigger each other. Failure initiation is obviously related to the internal state of stress which is determined by the geometry, loading, and properties of the constituent materials. The state of stress may be very complex and difficult to determine by analysis, thus, experimental means are essential and sometimes indispensable. Experimental methods applicable to the study of mechanics of composites include photoelasticity, strain gages, moiré, and holography. Of these, the photoelastic method offers versatility in dealing with a variety of problems and is especially effective in the study of micromechanics.

A critical situation arises when a unidirectionally reinforced composite is loaded transversely to the fibers. In this case stiffness reaches a minimum and stresses and strains in the matrix become the governing failure criteria. Most of the relevant micromechanical studies have been analytical (Chamis and Sendeckyj, 1968). Some analyses deal with the average (macroscopic) response and give expressions for transverse moduli and thermal expansion coefficients of the composite. Some of these analyses are based on energy considerations using variational principles (Paul, 1960; Hill, 1963; Hashin and Rosen, 1964; Hashin, 1965; Rosen *et al.*, 1964; Schapery, 1968), others give approximate empirical expressions (Greszczuk, 1966). Adams and Doner (1967), Adams *et al.* (1967), and Foye (1966, 1967) gave expressions based on exact elasticity solutions.

Stresses in the matrix are of great importance when loads act in a direction normal to the fibers, since in that case initial failure is governed by the stress and strain concentration in the matrix. Kies (1962) conducted simple analyses and calculated the strain magnifications existing at the fiber–matrix interface. Herrmann and Pister (1963) obtained an exact elasticity solution using numerical methods. Detailed solutions of the same problem have been given by Wilson and Hill (1965) by a conformal mapping approach, Foye (1967), who used a discrete element method, Adams and Doner (1967), who followed finite-difference procedures, and Clausen and Leissa (1967), who applied a point matching technique. Analytical methods have also been developed to account for the effects of anisotropy (Whitney, 1967), viscoelasticity (Hashin, 1966), plasticity (Shaffer, 1968), and random filament packing (Adams and Tsai, 1968).

Related experimental work is relatively limited and it has consisted primarily of two-dimensional studies. Shrinkage stresses around inclusions have been analyzed by Daniel and Durelli (1961, 1962) and Koufopoulos and Theocaris (1969) by means of two-dimensional models. Sampson (1962) and Daniel (1966, 1969) have studied stress distributions in two-dimensional photoelastic specimens containing arrays of inclusions and subjected to external load and matrix shrinkage. Durelli *et al.* (1970) conducted two- and three-dimensional studies of shrinkage and loading stresses in matrices with various types of inserts and Marloff and Daniel (1969) conducted an extensive three-dimensional photoelastic study of a realistic fiber-reinforced composite model.

Another important problem in the micromechanical study of composites is the load transfer between matrix and fiber when the load acts parallel or at an angle to the fiber. A substantial amount of experimental photoelastic work has been reported dealing with matrix stresses in general, interfacial stress distributions and concentrations near fiber ends and breaks, and failure modes and fracture propagation. Most of this work is qualitative in nature. Microphotoelastic experiments using models with actual prototype fibers have been described by Schuster and Scala (1964), who studied quantitatively the stress interactions near the tip of a single sapphire whisker, Alexander *et al.* (1965), Haener (1966), and Caputo and Hilzinger (1969), who reported primarily qualitative studies. Load transfer and especially stress distributions and concentrations near a fiber end have been the subject of many investigations using two-dimensional models such as

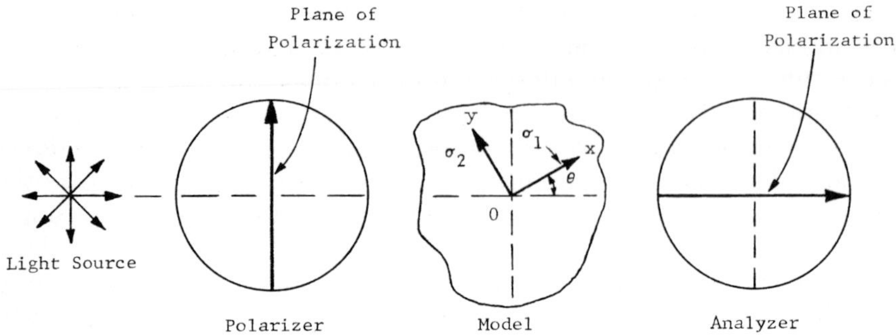

FIG. 1. Schematic arrangement of the elements of a plane polariscope. (σ_1, σ_2 are principal stresses.)

those reported by Lewicki and Pindera (1956), Tyson and Davies (1965), MacLaughlin (1966, 1968), Stone (1967), Allison and Hollaway (1967), Pih and Sutliff (1968), and Chen and Lavengood (1969).

The application of photoelasticity to the study of dynamic phenomena in composites has been very limited to date. It has been applied to the study of crack propagation by the author (Daniel, 1969, 1970) and wave propagation by Hunter (1970).

In addition to the micromechanical applications, photoelastic methods have been used in macromechanical studies of composites. The photoelastic coating method, yielding average surface strains, has been used successfully by Dally and Alfirevich (1969) and the author (Rowlands *et al.*, 1973; Daniel *et al.*, 1973). A new approach introduced recently by Pih and Knight (1969), Sampson (1970), and Dally and Prabhakaran (1971) makes use of transparent birefringent filamentary composite models with anisotropic elastic and optical properties.

Photoelasticity, in general, offers a wide variety of approaches to the study of mechanics of composites. The present chapter deals with a number of these techniques, such as microphotoelasticity, two-dimensional modeling of transverse and longitudinal sections, three-dimensional modeling, and dynamic photoelasticity. Applications of these techniques to specific problems are illustrated and discussed.

II. Fundamentals of Photoelasticity

Photoelasticity is based on the phenomenon of temporary double refraction (optical anisotropy) which certain isotropic transparent materials exhibit when subjected to stress. This optical phenomenon, when viewed in a field of polarized monochromatic light is manifested in the form of interference fringes or alternate dark and light bands. These fringes, referred to as "isochromatic fringes," are ordered according to the number of darkness–brightness cycles that appear at a point as the load is increased from zero to its final value. The fringe order represents relative light retardation in wavelengths. The principal optical axes, as Maxwell found, coincide with the principal stress directions.

Photoelastic models are examined in the polariscope. The simplest one, the plane polariscope, consists of a light source, a polarizer, and an analyzer. The last two elements convert ordinary light to plane polarized. If a stressed model is placed in the field (Fig. 1) with one principal stress direction at an angle θ with the axis of the polarizer, the light intensity at a point O is given by

$$I = 2a^2 \sin^2 2\theta \sin^2 \alpha/2 \tag{1}$$

where a is the amplitude of the plane polarized light emerging from the polarizer and α is the retardation introduced by the stressed model. Extinction occurs either when $\theta = 0, \pi/2$, or when $\alpha = 2n\pi$, i.e., the retardation is an integral number of wavelengths (integral fringe orders). These points of extinction form two families of fringes, *isoclinics* (loci of points of constant inclination of principal stresses) and *isochromatics* (loci of points of constant number of wavelengths of retardation).

The isoclinic family is eliminated in the so-called "circular" polariscope by introducing two properly oriented quarter-wave plates (Fig. 2).

The optical transformations that take place in this polariscope can be described as follows:

(1) The polarizer transforms the ordinary light into plane polarized.

(2) The first quarter-wave plate transforms the plane polarized light into circularly polarized.

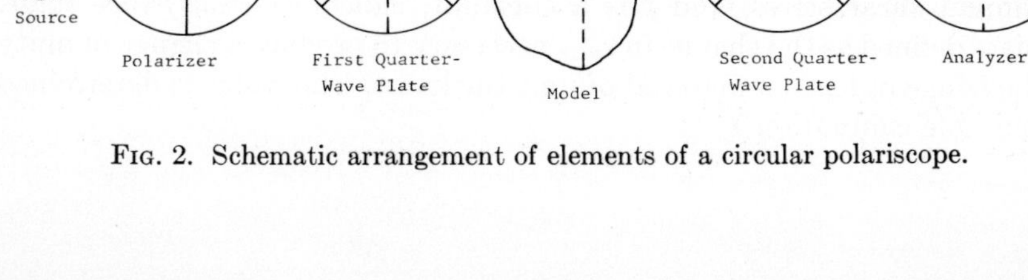

FIG. 2. Schematic arrangement of elements of a circular polariscope.

(3) The second quarter-wave plate restores the original plane polarized light or rotates it through 90° depending on whether it is crossed or parallel to the first plate.

(4) The analyzer blocks or lets through the plane polarized light according to the desired background.

Isochromatic fringes are related to the state of stress by means of the stress-optic law. Motivated from the stress–strain relations

$$\begin{aligned} \epsilon_1 &= (1/E)[\sigma_1 - \nu(\sigma_2 + \sigma_3)] \\ \epsilon_2 &= (1/E)[\sigma_2 - \nu(\sigma_1 + \sigma_3)] \\ \epsilon_3 &= (1/E)[\sigma_3 - \nu(\sigma_1 + \sigma_2)] \end{aligned} \tag{2}$$

we can write

$$\begin{aligned} n_1 - n_0 &= C_1\sigma_1 + C_2(\sigma_2 + \sigma_3) \\ n_2 - n_0 &= C_1\sigma_2 + C_2(\sigma_3 + \sigma_1) \\ n_3 - n_0 &= C_1\sigma_3 + C_2(\sigma_1 + \sigma_2) \end{aligned} \tag{3}$$

where C_1, C_2 are the stress-optic coefficients, n_1, n_2, n_3 are the principal indices of refraction, and n_0 is the initial index of refraction in the isotropic unstressed body.

Due to the difficulties involved in measuring the principal indices and the principal optical directions, photoelasticity is usually confined to measuring relative birefringence. From Eqs. (3),

$$n_1 - n_2 = (C_1 - C_2)(\sigma_1 - \sigma_2)$$

or

$$\begin{aligned} n &= (n_1 - n_2)t/\lambda \\ &= (C_1 - C_2)(t/\lambda)(\sigma_1 - \sigma_2) \\ &= (t/f)(\sigma_1 - \sigma_2)/2 = (t/f)\tau_{\max} \end{aligned} \tag{4}$$

or

$$\sigma_1 - \sigma_2 = 2nf/t \tag{5}$$

where n is the fringe order, t is the length of optical path and, in some cases, thickness of specimen, λ is the wavelength of light, σ_1, σ_2 are the principal stresses acting on planes parallel to axis of light propagation, $\tau_{\max}$ is the maximum shear stress, and f is a constant, called *material fringe value.* (This is defined as the change in $\tau_{\max}$ necessary to produce a change of unity in the fringe order in a material of unit thickness. This value is determined by suitable calibration.)

In the general case where the principal planes in the model are not parallel to the axis of light propagation, the stress-optic law is expressed in terms of the "secondary" principal stresses. These are the maximum and minimum of those normal stresses acting on planes parallel to the direction of propagation. These stresses may vary across the thickness of the model but only their average values influence the optical effect.

When the stress field is uniaxial, as it is on free boundaries, one of the principal stresses is zero and the nonzero component is determined directly from the stress-optic law above. In general, separation of principal stresses requires knowledge of some complementary information or the use of some auxiliary numerical operations.

Photoelastic techniques are applicable to two- and three-dimensional problems. A two-dimensional analysis is justified when the state of stress in the structure can be approximated by a plane stress or a generalized plane stress condition. In such cases, a model is fabricated from a sheet of transparent plastic known to possess the desired photoelastic properties. The model is machined to be geometrically similar to the prototype and is subjected to loads simulating those acting on the prototype. The loaded model is viewed in a field of circularly polarized light and the observed fringe patterns usually give a direct indication of regions of high and low stress. The fringes represent loci of points of constant maximum shear.

For three-dimensional problems, a three-dimensional scaled model of the prototype is machined out of certain epoxies having the desired "stress-freezing" properties. In this so-called "frozen-stress" method, the loaded model is slowly heated to a "critical" temperature, held there for some time, and, finally, slowly cooled to room temperature. This process freezes or "locks-in" the state of deformation corresponding to the elastic state of stress at the elevated temperature. This state of stress is not disturbed by careful slicing of the model. Thin slices are removed from the model wherever the stress distribution is required. The material is calibrated by subjecting a loaded calibration specimen (usually a disk under diametral loading) to the same temperature cycle as the model and observing the frozen fringe pattern.

The fringe patterns from the slices can be interpreted in much the same manner as in the case of a two-dimensional model. Shear stress distributions throughout the slice and principal stresses on free and pressure-loaded boundaries can be determined directly from the patterns. The determination of principal stresses at interior points is considerably more complicated and requires the use of auxiliary methods.

Another three-dimensional method is the "scattered-light" method. The method is nondestructive, does not require stress freezing, and the experiments can be conducted at room temperature where model material properties, such as Poisson's ratio, closer simulate those of prototype materials. When an intense, monochromatic, polarized beam, such as that emitted by a laser, enters a transparent stressed medium it produces a scattered-light fringe pattern visible when viewed in a direction perpendicular to the primary beam.

According to the stress-optic law for scattered light the difference of secondary principal stresses on a plane perpendicular to the direction of light propagation is proportional to the rate of change of birefringence along the light path:

$$\sigma_1' - \sigma_2' = (1/c)\, dn/ds \tag{6}$$

where c is the stress-optic coefficient and dn/ds is the gradient of birefringence along primary light path.

To resolve the stress components from photoelastic data, some auxiliary method, such as the shear difference method, is used. This method is based on the numerical integration of the equations of equilibrium. When integrating along a line parallel to the x axis use is made of the equilibrium equation

$$\frac{\partial \sigma_x}{\partial x} + \frac{\partial \tau_{yx}}{\partial y} + \frac{\partial \tau_{zx}}{\partial z} = 0 \tag{7}$$

which, when integrated, yields the following equation in finite difference form:

$$\sigma_x = \sigma_{x_0} - \sum_{x0}^{x} \Delta\tau_{yx}\, \Delta x/\Delta y - \sum_{x0}^{x} \Delta\tau_{zx}\, \Delta x/\Delta z \tag{8}$$

The shear differences $\Delta\tau_{yx}$ and $\Delta\tau_{zx}$ on the xy and xz planes are determined from the shear stresses along two pairs of parallel auxiliary lines lying on either side of the line of integration. These shear stresses are determined by

$$\tau_{yx} = -n'(f/t') \sin 2\theta_{1-x}' \tag{9}$$

$$\tau_{zx} = -n''(f/t'') \sin 2\theta_{1-x}'' \tag{10}$$

where θ_{1-x} is the angle between the x axis and maximum secondary principal stress measured positive counterclockwise. The symbols $'$ and $''$ refer to the xy plane and xz plane, respectively.

Thus, if σ_x is known at a point x_0, σ_x at any point is given by Eq. (8). Then, σ_y and σ_z are given by

$$\sigma_y = \sigma_x - 2n'(f/t') \cos 2\theta_{1-x}' \tag{11}$$

$$\sigma_z = \sigma_x - 2n''(f/t'') \cos 2\theta_{1-x}'' \tag{12}$$

The above stress components σ_x, σ_y, σ_z, τ_{yx}, and τ_{zx} can be determined by analyzing two slices of a frozen-stress model in the xy and xz planes. The remaining component τ_{yz} can be determined from oblique incidence measurements or from an additional subslice in the yz plane.

III. Two-Dimensional Model Analyses of Stresses on a Transverse Plane

A. Introduction

Matrix stresses in a unidirectional composite on a transverse plane arise from a variety of causes: (1) matrix shrinkage during curing, (2) temperature variations and resulting differential thermal expansions, (3) axial loading and resulting unequal transverse deformations in matrix and reinforcement, and (4) transverse loading. The first three types of stresses are similar in nature since they arise from uniform differential transverse deformation of the matrix and reinforcement. The usual approach in studying these stress distributions is by making a two-dimensional photoelastic model of the transverse cross section of the composite material. This is accomplished by casting the epoxy matrix around an array of stiff inclusions. One problem associated with two-dimensional composite models results from the out-of-plane restraint at the inclusion–matrix interface. Thus, a three-dimensional state of stress exists at the interface and it changes gradually into a two-dimensional state of plane stress away from the interface. This phenomenon, referred to as "pinching" effect, has been discussed by Parks and Durelli (1965) and Durelli and Parks (1967) who proposed a knife-edge configuration at the interface that renders the plane stress photoelastic analysis valid. Rhines (1968) proposed a general approach to the solution of problems with out of plane restraint by extending the concept of generalized plane stress. In this approach, the average transverse normal stress is not ignored but is assumed to decay exponentially with distance from the interface. Another approach to the problem of out of plane restraint is to insert stiff inclusions with lubricated edges into undersized holes in a soft matrix (Daniel and Durelli, 1962).

B. Shrinkage Stresses

In early attempts to determine shrinkage stresses, Paraplex P-43 polyester resin was cast around glass disk inclusions and the isochromatic

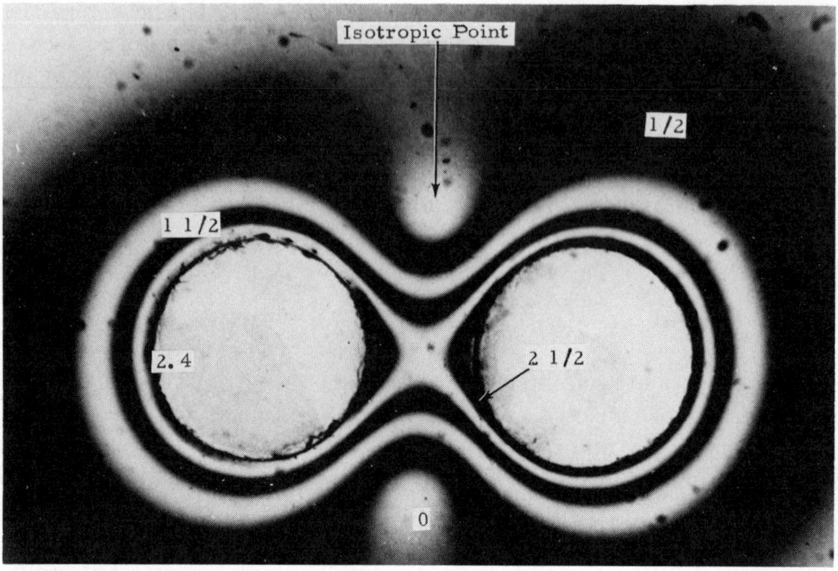

FIG. 3. Isochromatic fringe pattern in Paraplex P-43 around two circular glass inclusions $\frac{1}{4}$ in. in diameter and spaced $\frac{3}{8}$ in. on centers.

patterns analyzed (Daniel and Durelli, 1961). One of the simplest problems solved was that of shrinkage stresses around two inclusions (Fig. 3). In interpreting such fringe patterns it must be assured that the proper fringe value is used. A fringe value can be obtained by testing a calibration specimen of cured unreinforced material, but it is not certain if this value interprets correctly fringe patterns formed during curing when the material properties are changing. This problem can be by-passed if all results are expressed in dimensionless form with the nominal stress equal to the compression around a single isolated inclusion.

The modulus of the reinforcing fibers is usually so much higher than that of the matrix that the reinforcement can be considered infinitely rigid. The problem of matrix shrinkage around rigid inclusions is equivalent to the second boundary value problem of circular holes subjected to constant radial displacements. This suggests the alternative approach of applying mechanically these displacements by inserting rigid inclusions in undersized holes in a soft matrix (Daniel and Durelli, 1962) (see Fig. 4). This approach is valid in cases where all radial interface stresses are compressive.

Fringe patterns in the cast model (Fig. 3) were analyzed using the shear difference method in two dimensions [Eqs. (8), (11), (12)]. Results along the axis of symmetry normal to the axis through the centers of the inclusions are shown in Fig. 5. Deformations in the rubber model (Fig. 4) due to insertion of the inclusions were measured with a comparator microscope and the computed strains were used along with the photoelastic data to separate stresses. Results along the same axis of symmetry expressed in dimensionless form are in good agreement with those obtained from the cast model (Fig. 5).

The state of stress in the neighborhood of a group of inclusions can be described by the stress level and the dimensionless stress distribution. The stress level, defined by the nominal interface pressure around a single inclusion, depends on the effective shrinkage and the elastic constants of the material.

The dimensionless stress distribution is the stress distribution obtained by plotting the actual stresses divided by the nominal versus a dimensionless abscissa. This stress distribution is independent of size of inclusion, matrix material properties, and amount of effective shrinkage. For a given geometry of inclusion arrangement, the dimensionless stress distribution in one matrix material is sufficient to give the state of stress in any other material, provided the nominal stress is known for both materials.

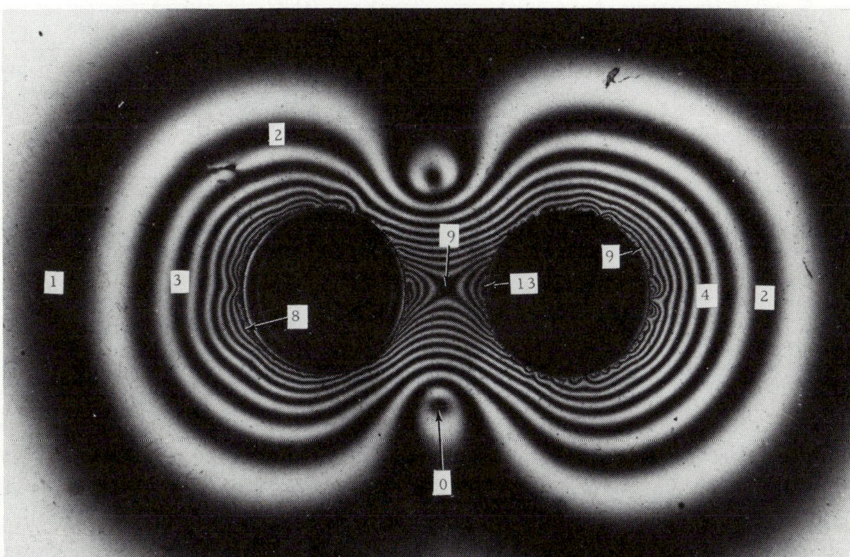

FIG. 4. Isochromatic fringe pattern in Hysol 8705 around two slightly oversized $\frac{1}{2}$ in. diameter glass inclusions spaced $\frac{3}{4}$ in. center to center (mechanical simulation of shrinkage stresses).

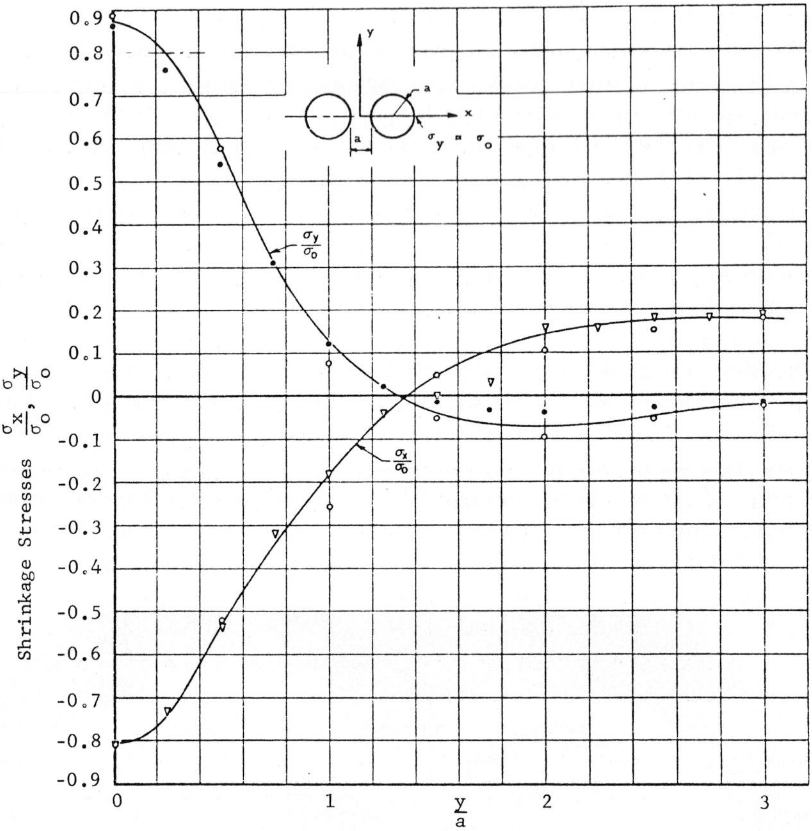

FIG. 5. Shrinkage stresses along y axis between two circular inclusions in Hysol 8705 with $\frac{1}{2}$-in. diameter inclusions (○) and Paraplex P-43 plates with $\frac{1}{4}$-in. diameter inclusions (●, △).

More realistic composite models can be made by casting the matrix around larger arrays of inclusions. Figure 6 shows shrinkage fringe patterns around square arrays of inclusions of various spacings simulating different fiber volume ratios.

Koufopoulos and Theocaris (1969) have described a two-dimensional shrinkage stress analysis around square arrays of inclusions. They simulated two modulus ratios of $E_f/E_m = 2.4$ and $E_f/E_m = 11.4$ and three fiber volume fractions, $v_f = 0.35$, $v_f = 0.55$, and $v_f = 0.71$. They analyzed the models by accepting the out of plane restraint as a realistic condition existing in the prototype fibrous composite and they made no attempt to correct for it. They separated stresses along axes of symmetry using the shear difference method and expressed the results in dimensionless form by dividing the stresses by the radial compression at the interface of a single inclusion. It was found that the stress distribution around the inclusions depended on the elastic moduli of the two phases as well as on the amount of shrinkage. It was shown that for close packing of the inclusions, i.e., high fiber volume fraction, large compressive stresses are developed at the smallest interfiber distance, while tensile stresses at the center of the largest interfiber distance could extend to parts of the interface. These tensile stresses may initiate cracking in the area. The maximum compressive stress ratio at the interface increases with E_f/E_m ratio, from 2.6 for $E_f/E_m = 2.4$ to 3.9 for $E_f/E_m = 11.4$. The matrix area subjected to biaxial tensile stresses increases with decreasing E_f/E_m and increasing v_f.

Knowledge of shrinkage stresses is essential since they are superimposed to the stresses induced by external loading. In some cases, the combination of the two states of stress may be deleterious.

In addition to the shrinkage stresses, thermal stresses arise in composites from the difference in thermal expansion of the constituents when the material is exposed to a temperature change. When the temperature variation is uniform throughout the composite the state of stress produced is similar to that produced by shrinkage. If the temperature varies nonuniformly through the composite material, other experimental or analytical techniques are required for stress analysis. Some two-dimensional photothermoelastic techniques for studying thermal stresses in composite materials were reported by Blejwas *et al.* (1969).

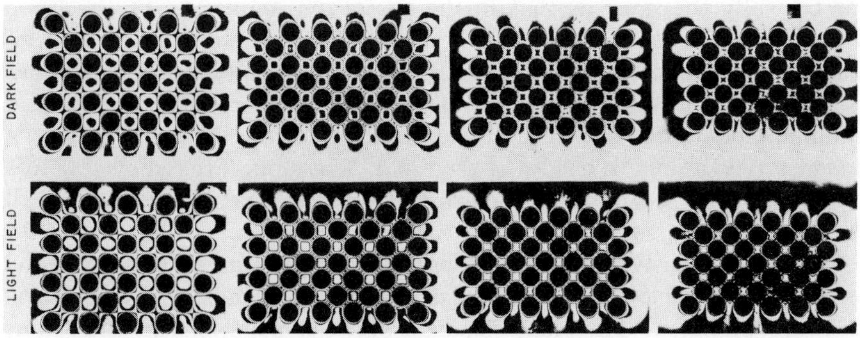

FIG. 6. Isochromatic fringe patterns due to shrinkage stresses in fiber-reinforced composite models.

C. *Mechanical Loading Stresses*

1. *Introduction*

Stresses in the matrix are of great importance when loads act in a direction normal to the fibers, since in this case initial failure is usually governed by the strain concentration in the matrix. Kies (1962) conducted simple analyses and calculated the strain magnifications existing in a matrix under transverse loading of the composite. For a square array, for example, he obtained

$$k_\epsilon = \bar{\epsilon}_{\max}/\epsilon_0 = [2 + (\Delta/R)]/[(\Delta/R) + (2E_{\mathrm{m}}/E_{\mathrm{f}})] \qquad (13)$$

where R is the fiber radius, Δ is the matrix space between fibers, E_{m}, E_{f} are the elastic moduli of matrix and fiber, respectively, $\bar{\epsilon}_{\max}$ is the maximum strain in the matrix in the x direction (direction of applied load and parallel to the side of elementary square), and ϵ_0 is the average strain. This analysis involves some gross approximations of the stress and material distribution. Herrmann and Pister (1963) obtained an exact elasticity solution using numerical methods. Detailed solutions have also been developed by Wilson and Hill (1965), Foye (1966), Adams and Doner (1967), and Clausen and Leissa (1967). Related experimental work using two-dimensional models has been reported by Sampson (1962) and Daniel (1966, 1969). Some of the latter work is described briefly below.

2. *Experimental Procedure*

Transverse sections through the fibers were simulated by casting the matrix resin around a group of rigid inclusions. The matrix material used was basically the same as that in the prototype composite. It was Epon 828 cured with 8 pph Diethylene Triamine at room temperature. The inclusions were steel disks $\frac{1}{2}$ in. in diameter and $\frac{1}{4}$ in. thick. Closed Plexiglas molds lined with mylar were used. The inclusions were held in place and manipulated slightly from the outside to free entrapped air by means of pins inserted through the sides of the mold. These pins are removed before complete curing to allow the inclusions to "float" and follow the overall shrinkage of the resin.

The most extensively studied case was the square array under parallel normal loading. One set of four models containing from 35 to 45 inclusions each was prepared. The clear spacings between inclusions in these four models were 1, $\frac{1}{2}$, $\frac{1}{4}$, and $\frac{1}{8}$ of the inclusion radius. The model dimensions were 5.25 in. × 11.50 in. × 0.25 in. In addition to these models, a calibration specimen of the same overall dimensions but with only one inclusion

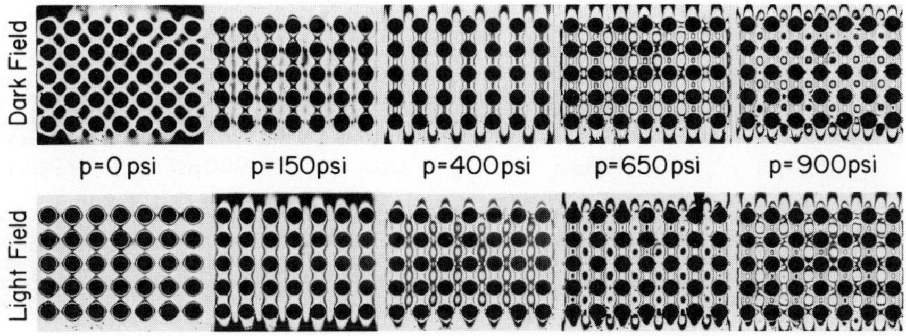

FIG. 7. Isochromatic fringe patterns for square array under parallel normal loading (inclusion spacing $\Delta/R = 1$).

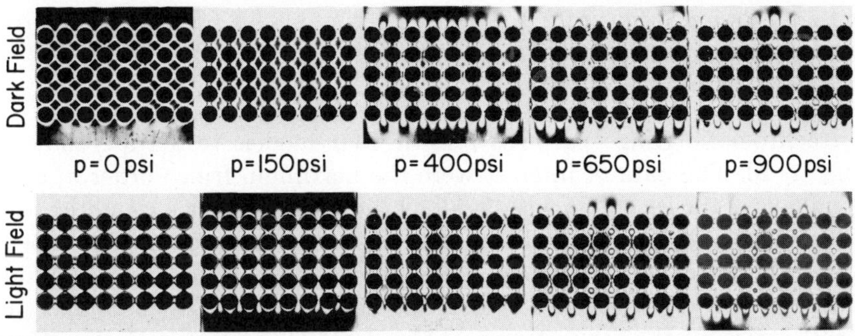

FIG. 8. Isochromatic fringe patterns for square array under parallel normal loading (inclusion spacing $\Delta/R = \frac{1}{2}$).

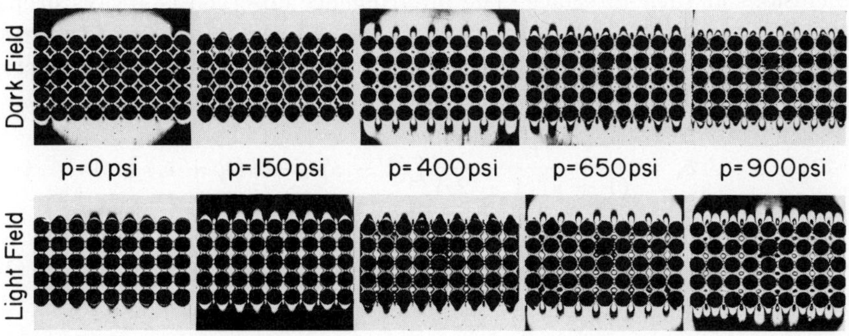

FIG. 9. Isochromatic fringe patterns for square array under parallel normal loading (inclusion spacing $\Delta/R = \frac{1}{4}$).

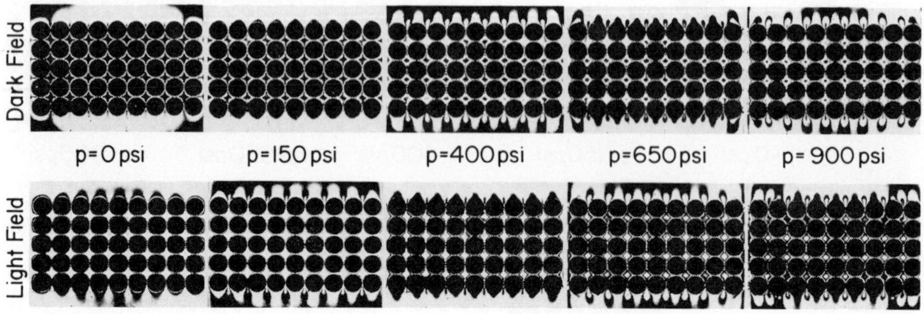

FIG. 10. Isochromatic fringe patterns for square array under parallel normal loading (inclusion spacing $\Delta/R = \frac{1}{8}$).

was prepared. The specimens were loaded in uniaxial compression in a specially designed fixture.

3. *Stress Concentration Factor*

Isochromatic fringe patterns for the four models tested are shown in Figs. 7–10. The data of interest were the maximum fringe order n_1 at the boundary of the central inclusions along the vertical diameter and the fringe order n_0 at a point sufficiently removed from the inclusions and the loaded edge. The latter was used as a check on the stress fringe value, obtained independently by a calibration test, and as a measure of the nominal (uniform) stress for the computation of the stress concentration factor. Figure 11 shows a typical plot of the variation of n_1 and n_0 with load.

As discussed earlier, the true state of stress in these models is a complex three-dimensional one due to the out of plane restraint introduced by the inclusions at their interface. The birefringence observed is always proportional to the in-plane stresses averaged through the thickness of the specimen. Stress–strain relations used must be expressed in terms of average values through the thickness:

$$\bar{\sigma}_x = \frac{E}{(1+\nu)(1-2\nu)}[(1-\nu)\bar{\epsilon}_x + \nu(\bar{\epsilon}_y + \bar{\epsilon}_z)]$$

$$\bar{\sigma}_y = \frac{E}{(1+\nu)(1-2\nu)}[(1-\nu)\bar{\epsilon}_y + \nu(\bar{\epsilon}_x + \bar{\epsilon}_z)] \qquad (14)$$

$$\bar{\sigma}_z = \frac{E}{(1+\nu)(1-2\nu)}[(1-\nu)\bar{\epsilon}_z + \nu(\bar{\epsilon}_x + \bar{\epsilon}_y)]$$

where $\bar{\sigma}_x$, $\bar{\sigma}_y$, $\bar{\epsilon}_x$, $\bar{\epsilon}_y$ are in-plane stresses and strains normal and parallel to the loading direction and averaged through the thickness of the specimen, and $\bar{\sigma}_z$ and $\bar{\epsilon}_z$ are the average out of plane stress and strain.

The inclusions can be considered rigid compared with the matrix due to the high ratio of moduli. Conditions of perfect bond at the interface and inclusion rigidity require that

$$\bar{\epsilon}_x = \bar{\epsilon}_z = 0 \tag{15}$$

From Eqs. (14) it follows that

$$\bar{\sigma}_x = \bar{\sigma}_z = [\nu/(1 - \nu)]\bar{\sigma}_y \tag{16}$$

Using the stress-optic law

$$\bar{\sigma}_x - \bar{\sigma}_y = 2n_1 f/2 \tag{17}$$

we obtain the average stress components along an axis of symmetry

$$\bar{\sigma}_x = \bar{\sigma}_z = \frac{\nu}{1 - 2\nu} \cdot \frac{2n_1 f}{t} \tag{18}$$

$$\bar{\sigma}_y = \left(\frac{1 - \nu}{1 - 2\nu}\right)\frac{2n_1 f}{t} \tag{19}$$

To determine the stress concentration factor, the average stress was calcu-

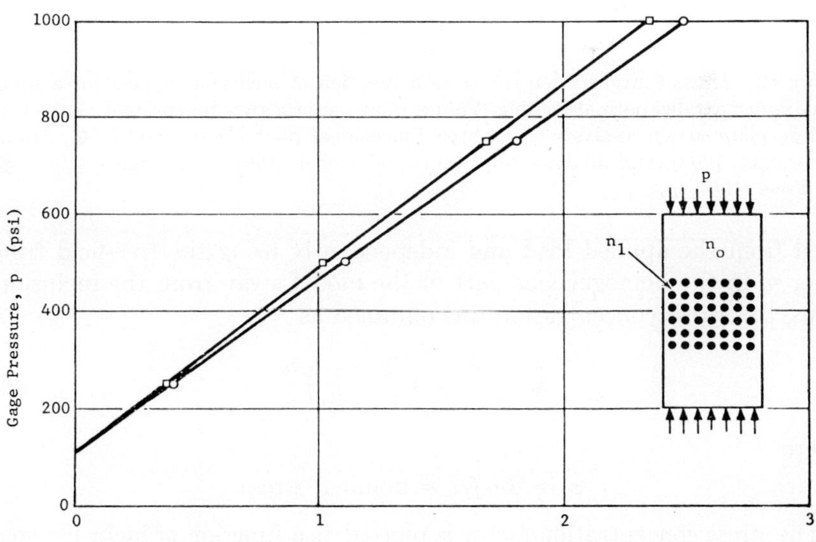

FIG. 11. Birefringence variation as a function of applied pressure (inclusion spacing $\Delta/R = \frac{1}{4}$). Values shown are for n_0 (□) and n_1 (○).

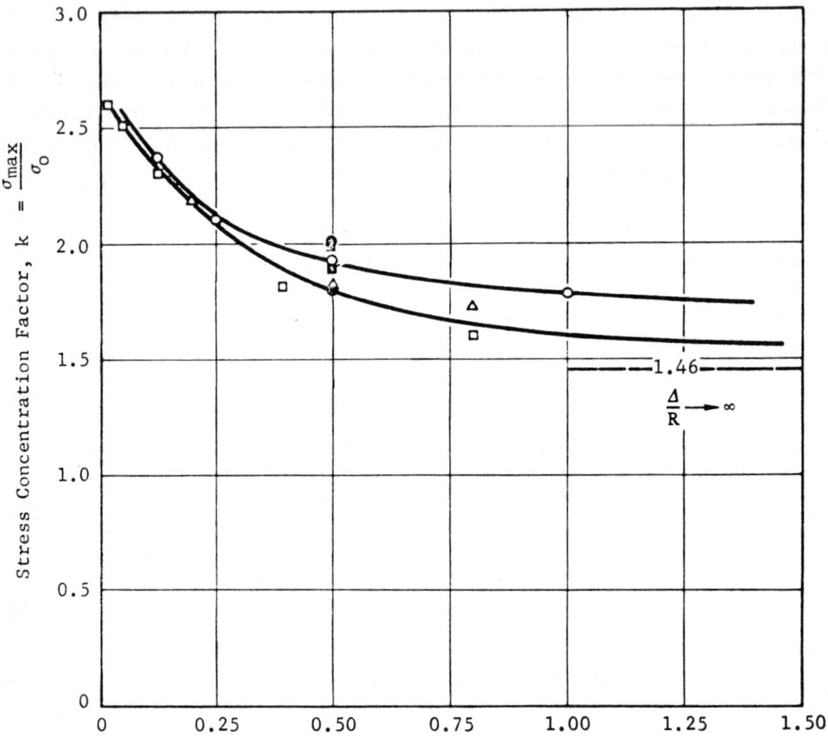

FIG. 12. Stress concentration factor as a function of inclusion spacing for a square array under parallel normal loading. Values shown are for two-dimensional photoelastic models, plane strain analysis (○), three-dimensional photoelastic model (●, Marloff and Daniel, 1969), and analysis by Adams and Doner (1967, □), Adams (1968, ■), and Foye (1967, △).

lated from the applied load and independently using the free-field fringe order n_0 in the homogeneous part of the model away from the inclusions. Thus, the stress concentration was obtained as

$$k_\sigma = \frac{\bar{\sigma}_y}{\sigma_0} = \left(\frac{1-\nu}{1-2\nu}\right)\frac{n_1}{n_0} \tag{20}$$

where

$$\sigma_0 = 2n_0 f/t = \text{nominal stress}$$

The stress concentration factor is plotted as a function of inclusion spacing in Fig. 12. Photoelastic results are compared with those by Foye (1967) and Adams and Doner (1967). Also shown in this figure are results from a subsequent three-dimensional photoelastic analysis (Marloff and Daniel, 1969) and a newer analysis by Adams (1968) based on a complex variables approach. The plane strain analysis of the photoelastic data above, which is perfectly valid at the interface, gives somewhat higher values for the stress concentration factor since conditions of plane strain do not hold away from the interface. However, for fiber volume fractions higher than 0.50 (inclusion spacing ratios $\Delta/R < 0.5$) the discrepancies become very small. Therefore, the closer the inclusion array the more valid is the plane strain analysis of two-dimensional photoelastic data.

4. Strain Concentration Factor

The quantity of importance in transversely loaded composites is the strain concentration factor. This factor is defined as the ratio of the maximum interface radial strain to the average strain in the composite model

$$k_\epsilon = \bar{\epsilon}_y/\epsilon_0 \tag{21}$$

From the relation

$$\bar{\epsilon}_x - \bar{\epsilon}_y = [(1+\nu)/E](\bar{\sigma}_x - \bar{\sigma}_y) \tag{22}$$

the boundary condition

$$\bar{\epsilon}_x = 0 \tag{23}$$

and Eqs. (18) and (19), it follows that the maximum strain is

$$\bar{\epsilon}_y = [(1+\nu)(1-2\nu)/(1-\nu)E]\bar{\sigma}_y \tag{24}$$

The nominal strain is

$$\epsilon_0 = \sigma_0/E_c \tag{25}$$

where σ_0 is the average stress and E_c is the transverse composite modulus. Then, the strain concentration factor is

$$k_\epsilon = \frac{\bar{\epsilon}_y}{\epsilon_0} = k_\sigma\left(\frac{E_c}{E_m}\right)\frac{(1+\nu)(1-2\nu)}{(1-\nu)} \tag{26}$$

where E_m is the matrix modulus and ν is the matrix Poisson's ratio. For $\nu = 0.36$,

$$k_\epsilon = 0.595k_\sigma(E_c/E_m) \tag{27}$$

Of the quantities entering expression (26), k_σ is obtained from photoelastic data only, E_m and ν are obtained from calibration of the matrix material, and E_c is either calculated or measured directly in the model. The composite modulus in the present case was determined directly by measuring the relative displacement of inclusions under load. This displacement

FIG. 13. Experimental setup for measurement of transverse composite modulus.

was measured by means of linear differential transformers. Figure 13 shows the experimental setup and Fig. 14 shows a plot of stress versus relative inclusion displacement. The composite modulus measured in this manner is plotted versus inclusion spacing in Fig. 15 where it is compared with theoretical results by Adams and Doner (1967) and Foye (1967). The agreement seems to be very satisfactory.

On the basis of these results and using Eq. (27), the strain concentration factor was computed for the four models tested. Results are tabulated in Table I and plotted in Fig. 16 as a function of inclusion spacing. It is seen that the strain concentration factor rises sharply with increasing fiber volume fraction. From a value of k_ϵ of approximately 4 at $v_f = 0.50$, it increases to approximately 12 at $v_f = 0.70$. The limiting value of k_ϵ for contiguous inclusions in the case of the present models is $(k_\epsilon)_{max} \simeq 21$.

5. *Prediction of Transverse Strength*

In using the results obtained from model tests for boron-reinforced plastic composites it must be noted that the stress concentration factor obtained from the models applies without any significant modification to the proto-

TABLE I

STRESS AND STRAIN CONCENTRATION FACTORS FOR PHOTOELASTIC MODELS TESTED

Inclusion spacing Δ/R	Ratio of composite to matrix modulus E_c/E_m	Stress concentration factor k_σ	Strain concentration factor k_ϵ
∞[a]	1.00	1.46	0.87
1	2.11	1.78	2.23
0.500	3.31	1.91	3.76
0.250	5.36	2.11	6.73
0.125	8.18	2.37	11.53
0	14.5	2.4	21.0

[a] Single inclusion.

type composite. To obtain the strain concentration in the boron prototype, note must be made of the dependence of the composite modulus on the modulus ratio between filament and matrix. This ratio for the prototype is 100 compared to 55 for the model. To compute the strain concentration factor in the prototype boron composite, one would have to introduce the

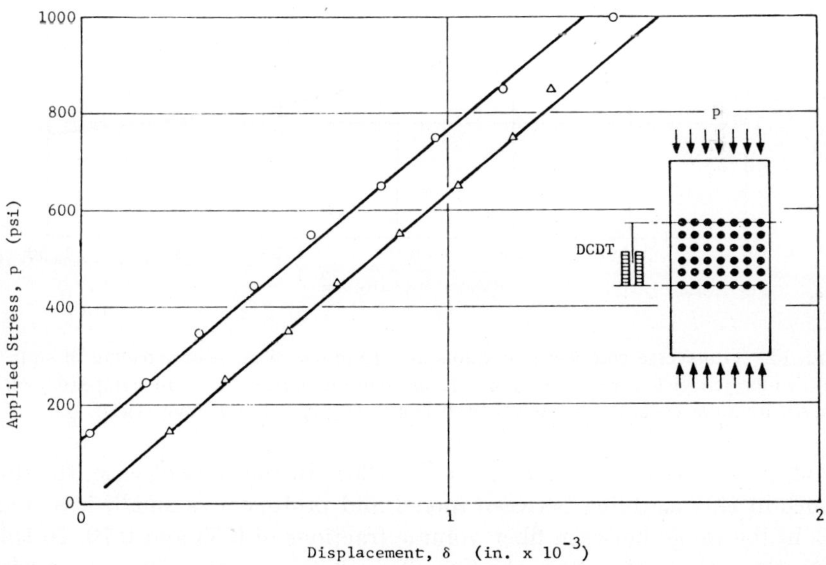

FIG. 14. Pressure versus relative displacement between inclusions in longitudinal direction for loading (○) and unloading (△), ($\Delta/R = 1$).

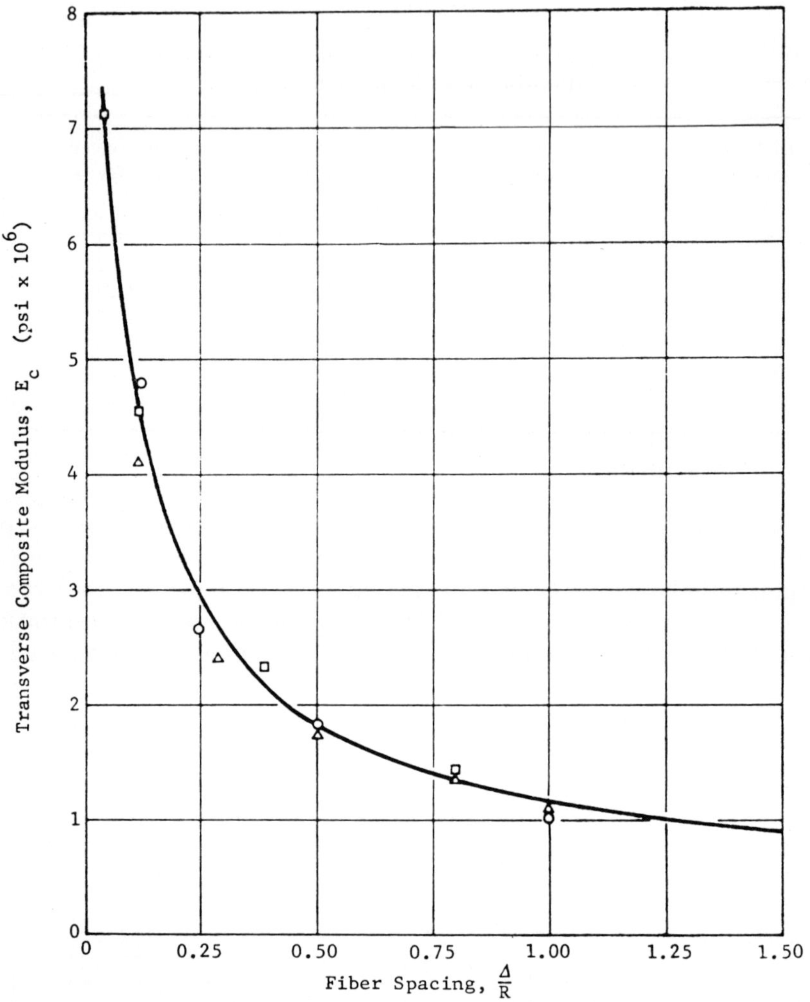

FIG. 15. Transverse composite modulus as a function of inclusion spacing in square array under parallel normal loading. Experimental values (○) are compared with theoretical values obtained by Adams and Doner (1967, □) and Foye (1967, △).

prototype composite modulus E_c in Eq. (26). In the present case, the difference in this modulus between model and prototype is small, less than 10% in the range between fiber volume fractions of 0.50 and 0.70. In this range, then, the strain concentration varies between approximately 4 and 12. This fact points out that the governing criterion in the case of transverse loading is the ultimate strain of the matrix material. Considering the fact that in commonly used boron composites the longitudinal modulus is approximately ten times the transverse modulus, it follows that the same load produces a maximum strain in the matrix 40–120 times higher when the load acts in the transverse direction than when it acts in the longitudinal direction.

In designing a composite it may seem advantageous to use the highest possible fiber volume fraction. This would give a high longitudinal strength but a relatively low transverse tensile strength due to the resulting high

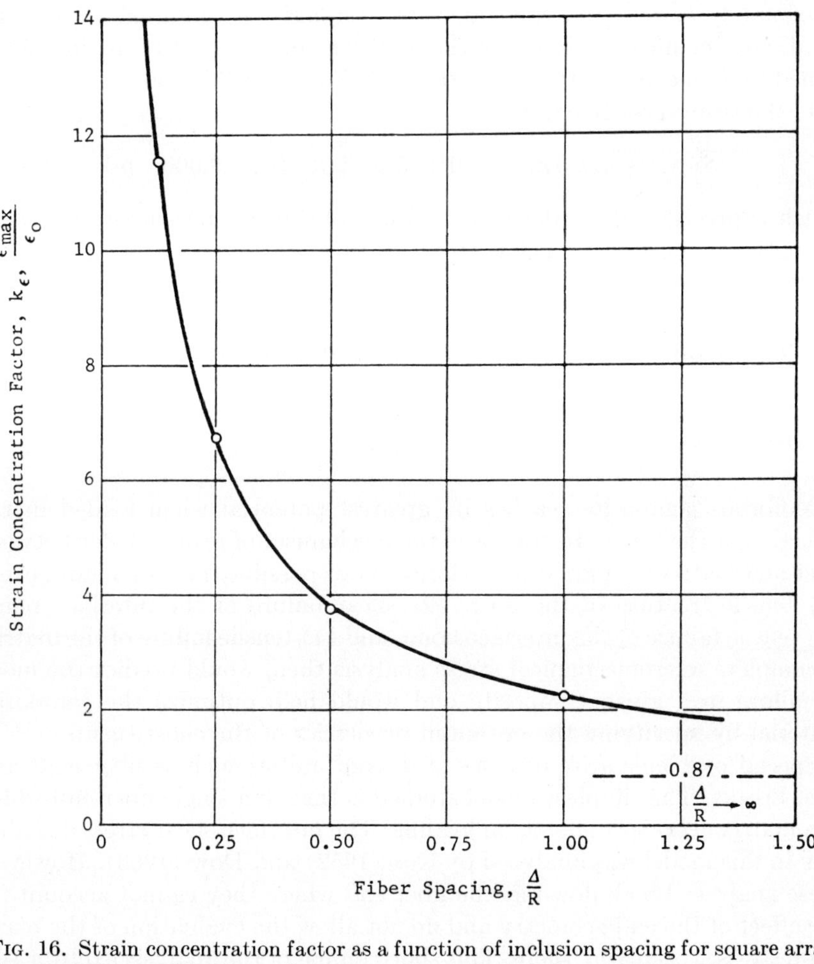

FIG. 16. Strain concentration factor as a function of inclusion spacing for square array under parallel normal loading, where $E_f = 30 \times 10^6$ psi, $E_m = 0.55 \times 10^6$ psi, $\bar{\epsilon}_{max}$ is the maximum strain in matrix, and ϵ_0 is the average strain.

strain concentration. The shape of the curve of Fig. 16 suggests that an optimum fiber volume fraction might be between 0.50 and 0.60 which corresponds roughly to the transition between the shallow and the steep portions of the curve. Fortuitously, the most commonly used fiber volume fraction in boron–epoxy composites is $v_f = 0.55$.

The results of the two-dimensional tests discussed here also help explain and are in agreement with results from boron–epoxy composites. A typical composite of fiber volume fraction 0.55 has an axial strength of 220,000 psi and an ultimate strain (both matrix and fiber) of 0.007 in./in. The corresponding strain concentration factor in the transverse direction is approximately $k_\epsilon = 5$ and the transverse composite modulus $E_c = 3 \times 10^6$ psi. Using the ultimate strain of the matrix resin $\epsilon_{ult} = 0.015$ in./in., determined by independent testing of the bulk Narmco 5505 resin, we can compute the transverse strength as

$$S_T = (\epsilon_{ult}/k_\epsilon) E_c = 0.015/5 \times 3 \times 10^6 = 9{,}000 \quad \text{psi}$$

which is precisely the order of magnitude of directly measured transverse strength in boron–epoxy composites.

IV. Two-Dimensional Model Analyses of Stresses on an Axial Plane

A. Introduction

A fibrous composite reaches its greatest potential when loaded in the direction of the fibers. In this case, the mechanism of load transfer between fiber and matrix is of great importance. Four possible fracture modes exist: (1) tensile fracture of the fiber, (2) shear failure of the interface bond, (3) tensile failure of the interface bond, and (4) tensile failure of the matrix. A complete micromechanical stress analysis then, would predict the mode of failure in a given composite and would help optimize the composite material by specifying the optimum properties of the constituents.

Special problems arise in cases of discontinuities such as fiber ends and fiber breaks. The simplest model studied is that of a single fiber embedded in a matrix cylinder under axial loading. The interface shear stress distribution in this model was analyzed by Cox (1952) and Dow (1963). However, these analyses break down at the fiber end where they cannot account for the effect of the end geometry and do not allow the evaluation of the maximum stresses occurring there. This shortcoming of the analysis led to a series of photoelastic investigations.

Tyson and Davies (1964) tested a model consisting of an aluminum strip of rectangular cross section cemented in a slot of a sheet of photoelastic material of the same thickness. With the model loaded in the direction of the fiber, maximum shear stresses of 2.5 times the nominal matrix tension were observed close to the square end of the fiber. The distributions of the maximum shear stress and interface shear stress were obtained along the length of the fiber. The photoelastic results agreed with the analytical results at distances from the fiber end greater than two fiber diameters.

More comprehensive photoelastic investigations including separation of stresses and complete stress distributions in the fiber and surrounding matrix were undertaken by Allison and Hollaway (1967) and Stone (1967). They employed similar models using Araldite CT200 as the matrix material and a Dural or steel strip as the reinforcing fiber. Allison and Hollaway studied both a square-ended and a round-ended fiber. The metal strip simulating the fiber was inserted and bonded into a slot in a sheet of the matrix material. The matrix and fiber were loaded independently in such a way as to produce equal axial strains in both away from the fiber end.

B. Maximum Shear Stress

The maximum shear stress in the matrix is significant for matrix materials with viscoelastic and plastic behavior under shear. This value, which can be readily obtained from the isochromatic fringes, peaks near the end of the fiber and is greatly dependent on the shape of the fiber end. A number of investigators have reported results on maximum shear concentration factors but it is very difficult to compare them because they used a variety of models and loading conditions and gave different definitions of the concentration factor. Allison and Hollaway (1967) reported values of

$$\tau_{\max} = 4.2\sigma_0 \quad \text{for the round-ended fiber}$$

$$\tau_{\max} = 2.3\sigma_0 \quad \text{for the square-ended fiber}$$

where σ_0 is the nominal matrix tensile stress. Stone (1967) found $\tau_{\max} = 3.4\sigma_0$ for the round-ended specimen but for a slightly different model. These results do not agree with those given by MacLaughlin (1966) and Pih and Sutliff (1968) who predicted higher maximum shear around a square end than around a semicircular end. It must be noted that they used a different model preparation procedure by casting the matrix around the fiber and they introduced the entire load through the fiber.

MacLaughlin (1966, 1968) studied the effects on maximum shear stress in the matrix of fiber end geometry including square, semicircular, and v-shaped ends. He also studied the effects of a gap at the end of a fiber, an open or closed gap between two collinear fibers, lateral spacing between fibers (representing 0.16 and 0.45 fiber volume fractions), eccentricity of fibers, and overlap of fiber ends. Maximum shear stress concentration factors, defined as τ_{max}/τ_{ave}, as high as 13 were measured in geometries where two fiber ends were in close proximity (up to one fiber diameter apart). These concentrations were approximately 50% higher than the maximum concentrations around single fiber ends.

Pih and Sutliff (1968) also studied the effects of end geometry and found maximum shear concentration factors as high as 12 for a 30° wedge-ended fiber. They also found that the maximum shear concentration dropped with increasing angle between fiber and load direction up to 45°.

C. Interface Shear Stress

The interfacial shear stress is significant from the point of view of shear failure of the interface bond and the critical aspect ratio or ineffective length of the fiber. This quantity is computed from the isochromatic and isoclinic data as

$$\tau_{xy} = \tfrac{1}{2}(\sigma_1 - \sigma_2) \sin 2\theta = (nf/t) \sin 2\theta \tag{28}$$

where θ is the isoclinic angle, or inclination of σ_1 with respect to the fiber (x) direction. This quantity reaches a peak near the end of a fiber and drops off rapidly along the fiber length. Allison and Hollaway (1967) reported peak values of interfacial shear

$$(\tau_{xy})_{max} = 1.93\sigma_0 \qquad \text{for a round end}$$

$$(\tau_{xy})_{max} = 2.0\sigma_0 \qquad \text{for a square end}$$

which are in agreement with results by Stone (1967) and Tyson and Davies (1964). Bond failure does not imply catastrophic failure of the composite. The resulting stress redistribution is more acceptable than fracture of the matrix itself. Another important result is the point at which this shear stress becomes zero, which corresponds to the point where the fiber load becomes maximum. This occurred at a distance of 20 diameters from the end (Allison and Hollaway, 1967), indicating a critical fiber aspect ratio of 40. MacLaughlin found this ratio to vary between 40 and 80 for multifiber models, depending on the lateral spacing of the fibers.

D. Maximum Tensile Stress in Matrix

Tensile failure in the matrix is governed by the maximum tensile stress σ_1. Both Allison and Hollaway (1967) and Stone (1967) separated the principal stresses from the photoelastic data. The former found that

$$(\sigma_1)_{\max} = 7\sigma_0 \qquad \text{for} \quad \text{a round end}$$

$$(\sigma_1)_{\max} = 4\sigma_0 \qquad \text{for} \quad \text{a square end}$$

compared to the latter's value of $(\sigma_1)_{\max} \simeq 5.8\sigma_0$ for the round end. For a brittle elastic matrix, it is quite probable that the maximum tensile stress above will control fracture initiation.

E. Maximum Fiber Tensile Stress

From the matrix stresses and the boundary conditions at the interface it is possible to compute stresses in the fiber. The equality of tangential strains at the interface implies tangential stresses in the fiber much higher than the other stress components which are transferred unchanged from matrix to fiber. Thus, the maximum fiber stress would occur near the tip rather than at the location of maximum fiber load. Allison and Hollaway (1967) gave values of $172\sigma_0$ and $85\sigma_0$ for the maximum fiber tension near a round and a square end, respectively. For an elastic fiber the ratio of the maximum fiber to matrix stress would tend to increase as the matrix yields.

V. Microphotoelastic Investigations

Microphotoelastic experiments utilize models with prototype reinforcing fibers, such as glass, boron, sapphire whiskers, etc. The models simulate the prototype composite further in that they retain the three-dimensional stress conditions and they reproduce the inherent bond characteristics between matrix and reinforcement. Microphotoelastic tests have been conducted to study stress distributions and concentrations around fiber ends, fiber breaks, and fiber debonding to determine ineffective fiber length and to investigate modes and propagation of microfracture.

The closer simulation of prototype conditions in microphotoelastic experiments is achieved at the expense of accurate quantitative analysis. Specimens usually consist of a thin layer of epoxy matrix with one or more fibers embedded in it. The layer is usually a few fiber diameters thick. The nature of the photoelastic method is such that only the average through the thickness effects are observed. Except in a few instances, such as the case of a single fiber, the individual stress components cannot be separated and only average effects can be studied.

Several microphotoelastic investigations have been described dealing with different objectives. One of the earliest ones was reported by Schuster and Scala (1964) who studied stresses around high-strength Sapphire (α-Al_2O_3) whiskers. A procedure was outlined whereby the average through the thickness value of the principal stress difference was corrected to give the principal stress difference on the plane through the axis of the whisker. They assumed a linear variation of this stress difference with radial distance from the whisker–matrix interface to a radius where free-field conditions prevail. A maximum shear concentration of 2.5 was obtained for a square-ended whisker, which is in good agreement with results of two-dimensional photoelastic studies (Tyson and Davies, 1964; Allison and Hollaway, 1967). The stress concentration in whiskers with tapered ends was much lower. Matrix reinforcement was observed to within 5 diameters from the whisker tip. The highest stress concentrations were observed at points of whisker fracture occurring after whisker embedment. This causes transverse cracking in the matrix. No quantitative analysis was attempted of the state of stress in the vicinity of a fiber break.

Haener (1966) conducted microphotoelastic tests to obtain qualitative correlations with theoretical stress distributions in unidirectional multifiber composites under matrix shrinkage and external loading.

Qualitative studies of fracture mode and propagation by means of microphotoelastic models were conducted by Rosen (1964), Alexander *et al.* (1965), and Caputo and Hilzinger (1969). The latter used one- and two-ply microphotoelastic glass–epoxy and graphite–epoxy models with widely spaced fibers. They used both still photography and high-speed photography at a rate of 3000 frames/sec. This framing rate, however, is too slow for the study of real dynamic phenomena. They studied the influence on fracture of the fiber type, fiber orientation, type of flaw (cut fiber, unbonded fiber, matrix void, exposed edges), and loading condition (static tension, biaxial compression–tension, biaxial tension–tension, cyclic tension, and impact). The role of the matrix in distributing the load among the fibers was clearly evidenced as was also the importance of the fiber–matrix interfacial bond. Failure modes in simple tension, in biaxial compression–tension and tension–tension, and in cyclic loading indicated that fibers oriented at an angle to the principal tensile stress constitute flaws. The interfacial bond plays a major role in these failures, as illustrated by the characteristic shear mode in $\pm 45°$ configurations. Broken fibers in the $0°$ direction affected strength and failure mode more drastically than did the introduction of isolated matrix flaws or unbonded regions. These results are in agreement with Rosen's (1964) observations. Fracture of $0°/90°$ specimens was initiated at transverse fiber bundles and was accompanied

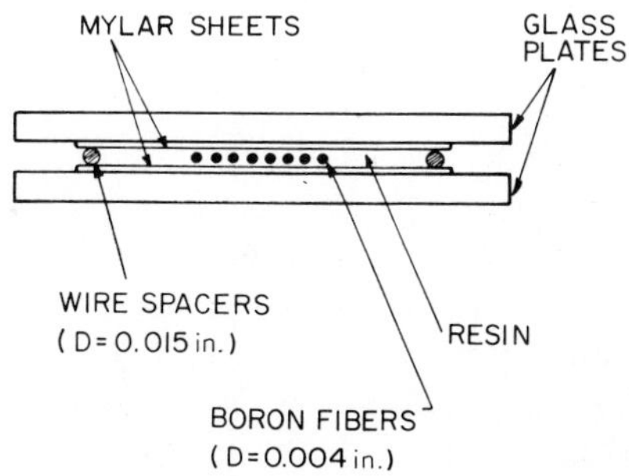

FIG. 17. Arrangement for preparation of microphotoelastic specimens.

by stacked cracks and debonding. Initial flaws did not have any appreciable effect on these specimens. Specimens with ±45° fiber orientation failed in shear along the fibers with failure initiation at crossover points and near the edges of the specimen. These specimens exhibited extreme sensitivity to the presence of voids.

Microphotoelastic experiments were reported by Daniel (1969) who studied the effects on matrix stress concentration of fiber orientation, fiber

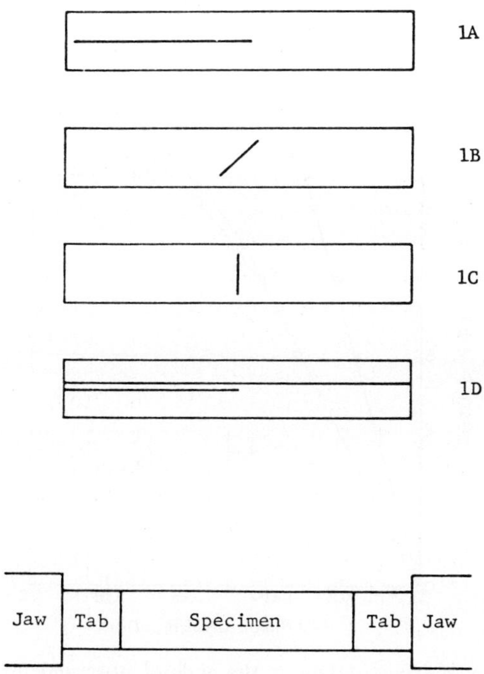

FIG. 18. Microphotoelastic specimens and loading arrangement.

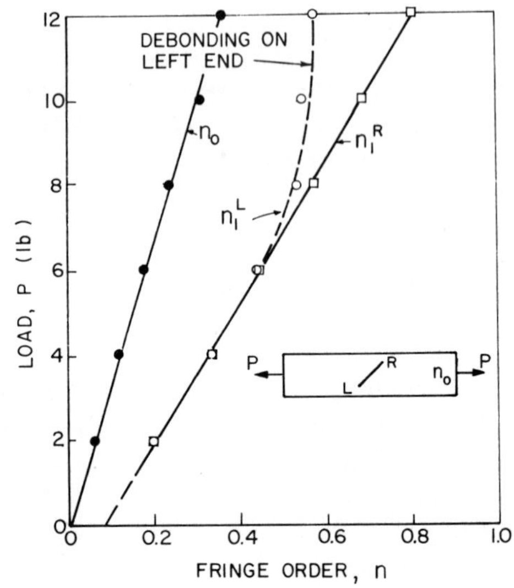

FIG. 19. Variation of fringe orders at the ends of fiber and in the free field with applied load (specimen 1B).

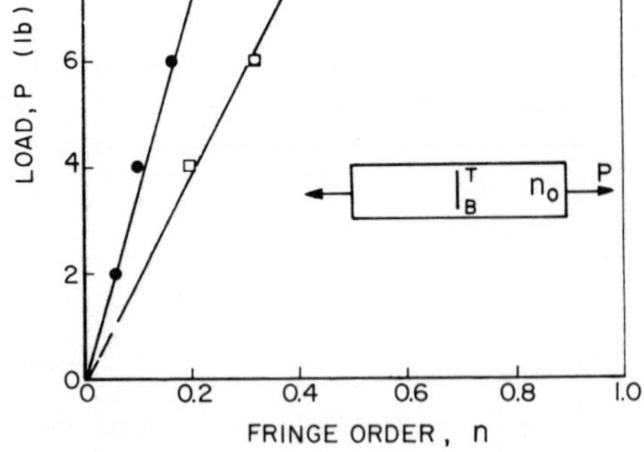

FIG. 20. Variation of fringe orders at the ends of fiber and in the free field with applied load (specimen 1C).

debonding, and matrix cracking at the interface. The arrangement and procedure used for preparation of the specimens is illustrated in Fig. 17. They were cast between glass plates lined with mylar sheets. Steel wires, 0.015 in. in diameter, were used as spacers. Boron fibers were taped at the ends with spacers so that they would be centered. The group of specimens tested and the loading arrangement is shown in Fig. 18. The specimen cross section was 0.50 × 0.015 in. In specimen 1A, debonding occurred at a low level of load. The fiber was fully debonded at a load of 10 lb. The progress of debonding was followed easily by the movement of the maximum fringe location away from the free end. Specimen 1B also failed by debonding starting at one end. Maximum fringe orders at the two ends of the fiber and the average fringe order away from the fiber are plotted as a function of load (Fig. 19). Birefringence variation with load for specimen 1C shows a nonlinearity at a load level of about 8 lb (Fig. 20). Contrary to the nonlinearity associated with debonding a rapid increase in fringe order with load is noted here. This increase may be associated with yielding and incipient cracking in the matrix. One conclusion reached from these three models was that the "average" maximum shear concentration factor at the fiber end decreased from a value of 2.5 for the axial fiber to 2.1 for the 45° fiber and 1.8 for the 90° fiber. This finding is in agreement with results of two-dimensional photoelastic tests. In specimen 1D, shrinkage cracks normal to the fiber direction were developed. Birefringence at the fiber end, in the free field, and at the crack was measured

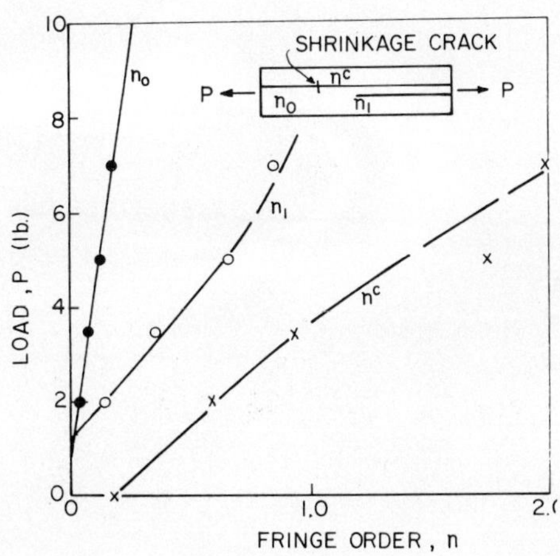

FIG. 21. Variation of fringe orders at fiber end, shrinkage crack and in the free field (specimen 1D).

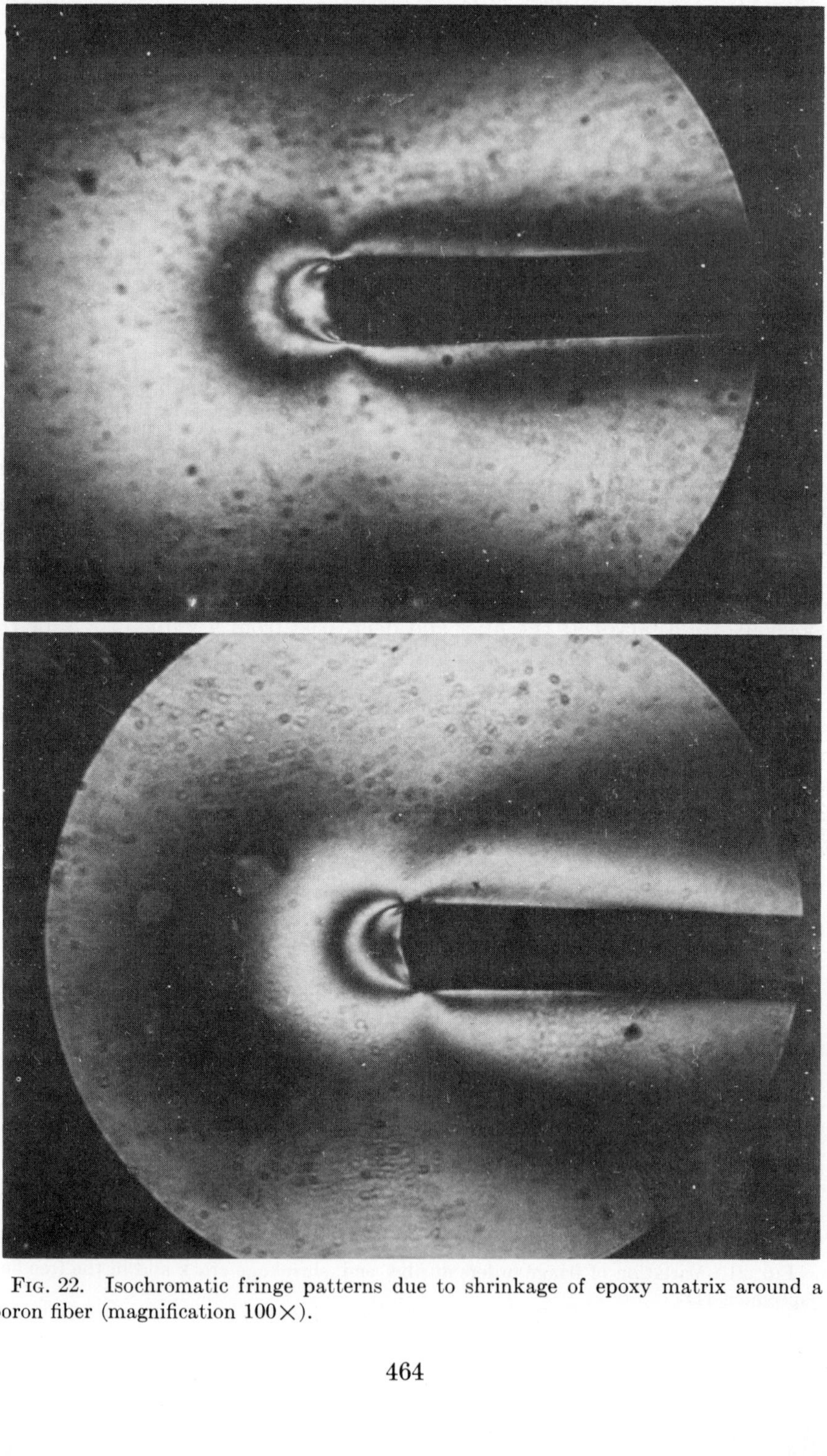

FIG. 22. Isochromatic fringe patterns due to shrinkage of epoxy matrix around a boron fiber (magnification 100×).

and plotted as a function of load (Fig. 21). The curve for fringe order at the end of the fiber is concave upward which indicates debonding. The curve for birefringence at the crack is concave downward as in the case of specimen 1C.

In all cases discussed heretofore matrix shrinkage was minimized by using room temperature curing resins. In the prototype, however, which is cured at an elevated temperature, shrinkage stresses may be appreciable. An illustration of such a shrinkage stress field around a boron fiber end is given in Fig. 22. High stress concentrations at the edges of the fiber end are manifested.

VI. Three-Dimensional Model Studies

A. Introduction

Three-dimensional photoelastic investigations have been very limited. Durelli *et al.* (1970) and Parks *et al.* (1970) studied stress distributions around inserts of various geometries in matrices subjected to shrinkage and mechanical loading. They used a low-temperature curing epoxy cast around Plexiglas or other epoxy inserts. They used standard stress-freezing and slicing techniques and assumed the material to be incompressible since Poisson's ratio is very close to 0.5 at the critical temperature. Their analysis was primarily limited to the interface of single isolated inserts, such as bars. Some experiments with single bar inclusions using the scattered-light technique were reported by Pih and Sutliff (1968) and Jenkins (1968). The first extensive stress analysis of a realistic three-dimensional composite model was reported by Marloff and Daniel (1969). They used the standard stress-freezing technique to determine stress distributions in the matrix of a unidirectionally reinforced composite model subjected to matrix shrinkage and normal transverse loading. The model consisted of reinforcing and matrix components having a modulus ratio simulating that of a boron-reinforced plastic.

B. Experimental Procedure

Two identical models of dimensions shown in Fig. 23 were prepared. The matrix was an epoxy formulation suggested by Sampson (1963) with a relatively low critical temperature (175°F) and rather small polymerization shrinkage. It consisted of the following:

2 parts by weight ERL-2774

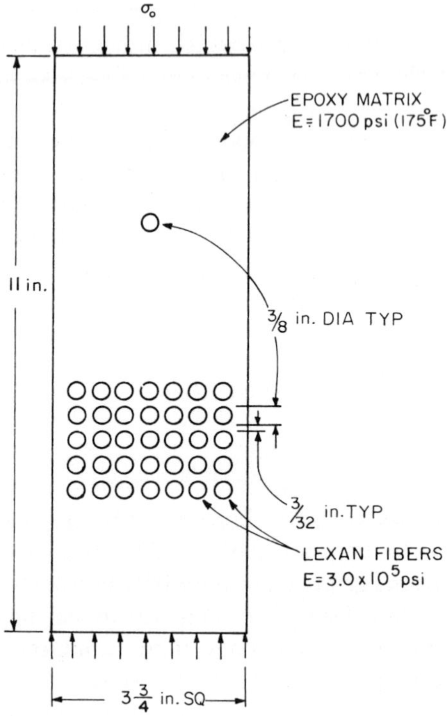

FIG. 23. Photoelastic model of fiber-reinforced composite.

2 parts by weight ERL-2795

1 part by weight ZZL-0803

At the critical temperature, this material exhibits an elastic modulus of 1700 psi. Lexan (polycarbonate) rods, $\frac{3}{8}$ in. in diameter, were used as the reinforcing elements. This material exhibits excellent stability at the critical temperature of the matrix and a relatively constant modulus of 3×10^5 psi up to 280°F. This combination resulted in a modulus ratio of 175. No consideration was given to the effect of Poisson's ratio of the two materials. The clear spacing of $\frac{1}{2}$ the fiber radius provided a volume fraction of 0.5 for a square array. A single isolated fiber was included for calibration purposes. The fiber array was located sufficiently far away from the ends of the specimen to ensure uniformity of loading.

The casting procedure consisted of heating both epoxy resins to 220°F, degassing for 10 min, and then cooling to room temperature. The hardener was added and the resulting system was mixed thoroughly. The epoxy was slowly poured into the mold and allowed to gell under a controlled temperature of 60°F for 24 hr. Final cure was accomplished by heating the mold and specimen to 175°F at a rate of 2°F/hr, soaking at this temperature for 24 hr, and then cooling to room temperature at a rate of 2°F/hr. The specimen was then removed from the mold and machined to its final size.

One of the specimens was subjected to a uniform axial compression of 75 psi by means of air pressure distributed over the ends. The other specimen was left unloaded. Both specimens were subjected to the stress-freezing cycle, i.e., raising the temperature to 175°F at a rate of 2°F/hr, maintaining this temperature for 8 hr, and finally lowering it to room temperature at a rate of 2°F/hr.

C. Analysis

Figures 24 and 25 show isochromatic fringe patterns corresponding to shrinkage and combined shrinkage and external loading in slices normal to the rods removed from the center of each specimen. Birefringence and isoclinic data were obtained in the regions of interest on a comparator polariscope giving a tenfold magnification of the patterns on the screen.

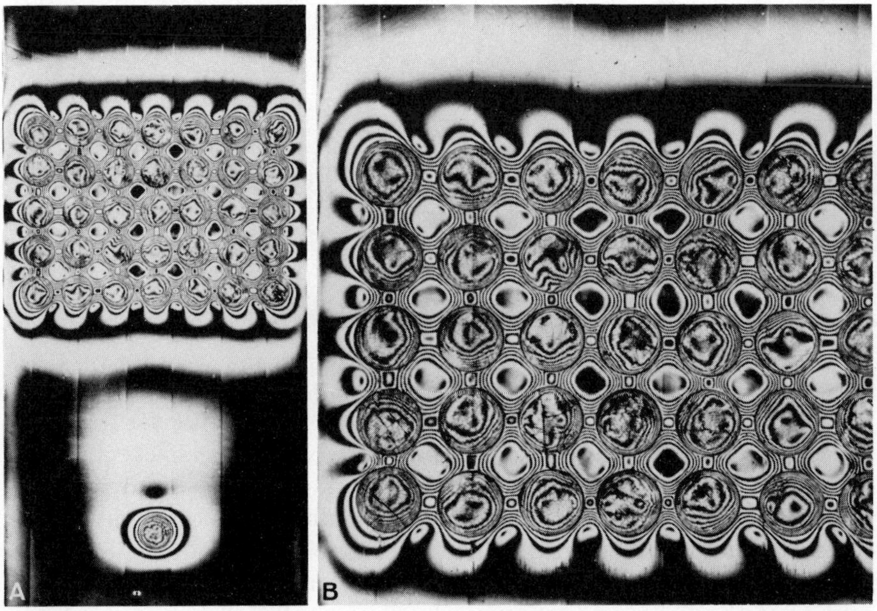

FIG. 24. Isochromatic fringe patterns in model subjected to matrix shrinkage: (A) overall view; (B) enlarged view through region of analysis.

FIG. 25. Isochromatic fringe patterns in model subjected to external loading and matrix shrinkage: (A) overall view; (B) enlarged view through region of analysis.

The principle of superposition was used to separate the effects of shrinkage from those of combined shrinkage and loading. Complete stress separation was obtained in both cases and corresponding stress components were subtracted from each other. Because of the large number of rods simulating an infinite array, there are four such axes of symmetry, two connecting the centers of the rod cross sections and two along the matrix segments between rods.

The method used to separate principal stresses was a numerical integration procedure based on Filon's transformation of the Lamé–Maxwell equilibrium equations for an axis of symmetry (Frocht, 1941). This technique has been applied successfully by Sampson (1963) for a similar two-dimensional specimen. If a principal stress σ_x or σ_y is known at a given point of coordinate x_0 or y_0, the stress at any point along the axis of symmetry is determined from the following equations:

$$\sigma_{x_i} = \sigma_{x_0} - \sum_0^i (\sigma_x - \sigma_y)\ (\Delta\theta/\Delta y)\ \Delta x \tag{29}$$

$$\sigma_{y_i} = \sigma_{y_0} - \sum_0^i (\sigma_x - \sigma_y)\ (\Delta\theta/\Delta x)\ \Delta y \tag{30}$$

where $\sigma_x - \sigma_y$ is the principal stress difference given by the fringe order and $\Delta\theta$ is the change in isoclinic parameter over a distance Δy or Δx from the line of integration.

The starting point for the integration was the intersection of two axes of symmetry shown as point O in Fig. 26. Using Eq. (30), the integration was carried out to $y = a$ giving the distribution of $\sigma_y - \sigma_{y_0}$ along the y

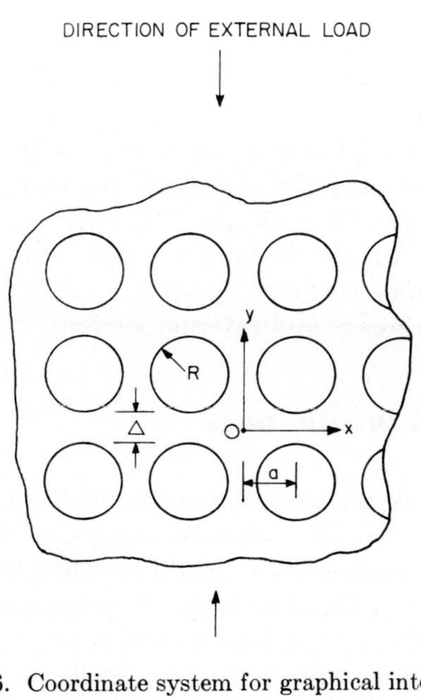

Fig. 26. Coordinate system for graphical integration.

axis. Using the relation

$$\sigma_x - \sigma_{y_0} = (\sigma_x - \sigma_y) + (\sigma_y - \sigma_{y_0}) \tag{31}$$

the distribution of $\sigma_x - \sigma_{y_0}$ can be determined. From static equilibrium

$$\int_0^a \sigma_x \, dy = 0 \tag{32}$$

where the x direction for the loaded specimen is normal to the direction of the applied load. Using this equation, the initial value σ_{y_0} can be determined from

$$-\sigma_{y_0} = \frac{1}{a} \int_0^a (\sigma_x - \sigma_{y_0}) \, dy \tag{33}$$

Equations (30) and (31) were then used to obtain the components σ_y and σ_x at all points along the y axis. Equation (29) then gives the values of σ_x along x axis from which the values of σ_y are computed using the fringe-order readings along the same axis. Similar integration procedures were used to obtain principal stress components along the other axes of symmetry. The procedure above was identical for both the shrinkage and the combined shrinkage and external load conditions.

Algebraic subtraction of corresponding stress components for the two loading conditions gave the effect of external load only. Subslices were removed in the direction of the fiber to allow determination of the stress component parallel to the fiber axes (σ_z). Since the faces of the subslice correspond very closely to principal planes, the isochromatic fringe patterns viewed in the rz plane gave the difference between the axial and radial stresses at any point across the section. Using the predetermined values of the radial stresses, the distributions of the axial stresses were determined. Again, superposition was used to separate the effect of shrinkage from the combined effect of shrinkage and external loading.

D. *Shrinkage Stress Distributions*

The distribution of shrinkage stress components along the ligament centerline is shown in Fig. 27. The values shown represent the average of two orthogonal ligaments which ideally should give identical results. Figure 28 shows the distribution of the shrinkage stress components across the section between fibers. The components σ_r and σ_θ are normal and parallel to the centerline. The stress component σ_z is parallel to the fiber axis. Values at $r/\Delta = 0.5$ represent the interface stresses. Reduction of the stresses to dimensionless form was accomplished by dividing the computed stresses by the shrinkage pressure at the interface of the isolated fiber present in the specimen. As shown by Daniel and Durelli (1961), the shrinkage stress distribution around an isolated circular inclusion is essentially identical to that produced around a circular hole subjected to a boundary pressure p. At the interface, the radial and tangential stresses are

$$\sigma_r = -p, \qquad \sigma_\theta = p \tag{34}$$

In terms of the photoelastic fringe order, the interface pressure is

$$p = nf/t \tag{35}$$

From Fig. 28, it is observed that the circumferential stress σ_θ has a maximum value at the center of the minimum section of the ligament and it decreases toward the interface. The radial stress σ_r is compressive across the same section and nearly constant in magnitude with only a slight increase at the interface. The stress σ_z in the direction of the fiber is tensile,

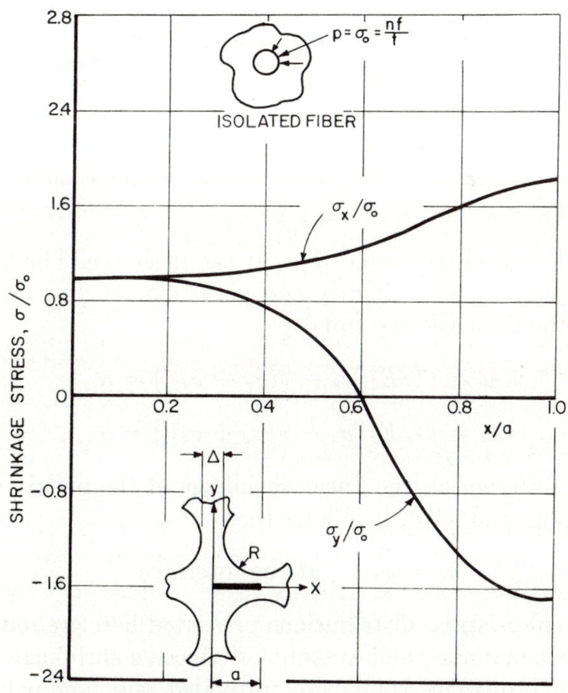

FIG. 27. Distribution of shrinkage stresses along ligament centerline, $\Delta/R = 0.50$.

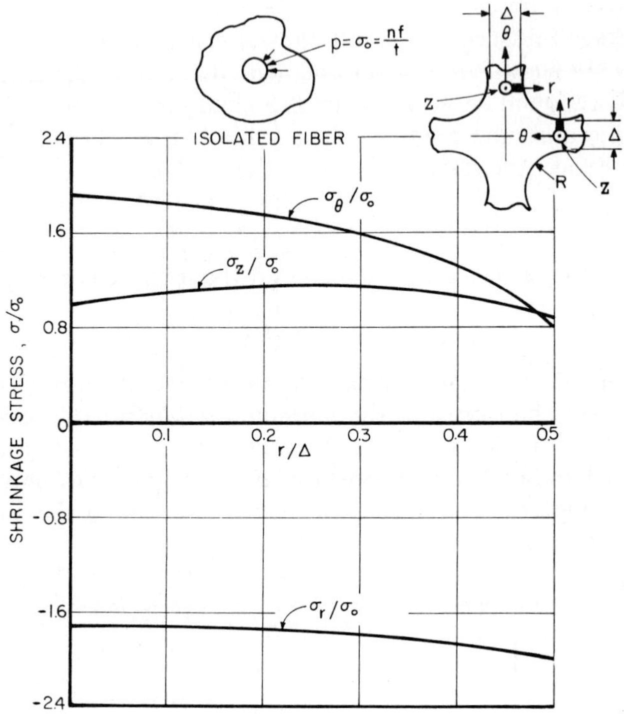

FIG. 28. Distribution of shrinkage stresses across section between fibers, $\Delta/R = 0.50$.

nearly constant, and nearly equal to σ_θ at the interface. The latter results from the conditions of fiber rigidity and perfect bond. For this condition, the strains on the interface are equal:

$$\epsilon_\theta = (1/E)[\sigma_\theta - \nu(\sigma_r + \sigma_z)] = \alpha \tag{36}$$

$$\epsilon_z = (1/E)[\sigma_z - \nu(\sigma_r + \sigma_\theta)] = \alpha \tag{37}$$

where α is the differential free linear shrinkage of the matrix and fiber.

From Eqs. (36) and (37), it follows that

$$\sigma_\theta = \sigma_z \qquad \text{at the interface} \tag{38}$$

The dimensionless stress distributions presented here are independent of fiber size, matrix material, and amount of effective shrinkage and are applicable to the prototype composite, provided the nominal stress p is known. For a single isolated fiber, this nominal stress away from the ends is given by

$$p = E\alpha \tag{39}$$

where E is the modulus of matrix material.

E. External Load Stress Distributions

The stress distributions due to external load along ligament center lines are plotted in Figs. 29 and 30, and across sections between fibers in Figs. 31 and 32. The results are presented in dimensionless form by dividing the stress components by the average stress σ_0 through the composite section. The latter was obtained by integrating the σ_y component along the x axis. This value was within 2% of the average stress computed from the applied load.

Inspection of Figs. 31 and 32 shows that the principal stresses at the interfaces are equal. This is a result of the conditions for a rigid fiber bonded to a matrix and subjected to mechanical loading only (no shrinkage effects). For this condition, all interface tangential strain components are zero, re-

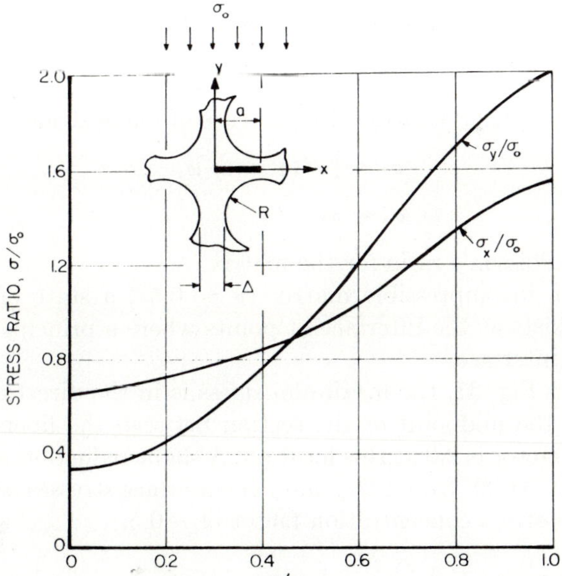

FIG. 29. Distribution of external load stresses along ligament normal to direction of applied load, $\Delta/R = 0.50$.

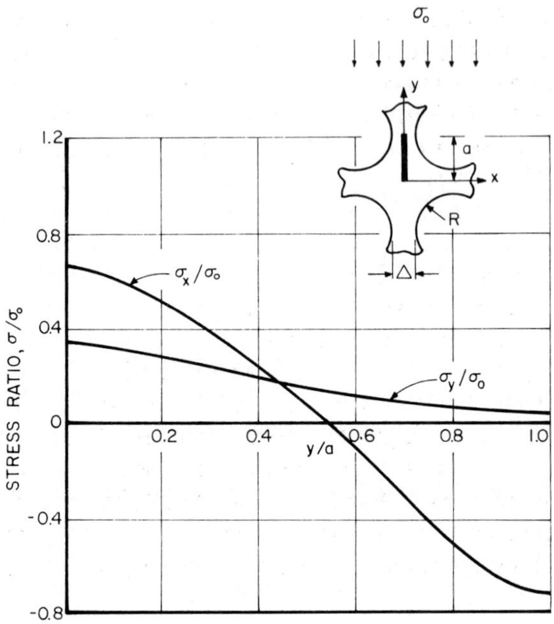

FIG. 30. Distribution of external load stresses along ligament parallel to direction of applied load, $\Delta/R = 0.50$.

sulting in the following relations for the case of plane strain:

$$\sigma_\theta = [\nu/(1 - \nu)]\sigma_r \tag{40}$$

$$\sigma_z = \sigma_\theta \tag{41}$$

where ν is the Poisson's ratio for the matrix.

Thus, for an incompressible matrix ($\nu = 0.5$), a state of hydrostatic compression exists at the interface at points where a principal direction is normal to the interface.

As seen from Fig. 31, the maximum stress is in the direction of loading and occurs at the midpoint of the section between the fibers. The stress concentration factor is 2.0 at this location. A slight reduction to 1.80 occurs at the interface. At 90° from this point, the interface stresses become tensile and result in a stress concentration factor of −0.5.

F. Discussion

Most of the relevant theoretical results available are limited to maximum stress concentration factors for specific values of Poisson's ratio for the components, viz., 0.35 for the matrix and 0.20 for the fiber. A limited comparison of the present results with the analysis of Adams and Doner (1967) for a boron–epoxy composite has shown reasonable agreement. For a filament volume fraction of 55% they show stress concentration factors of 1.86 and −0.5 for the two interfaces compared with 1.80 and −0.5 for the present work. The theory does not predict the increase in stress concentration factor at the ligament midpoint, however. Goree (1967) investigated the case of in-plane loading of two rigid cylindrical fibers in an infinite matrix using a complex variable technique and pointed out a variation in maximum stress location with inclusion spacing. For wide spacings, the maximum principal stress occurs at the interface, but for spacings less than one-fiber radius, the location moves to the midpoint. He also indicates a considerable influence of Poisson's ratio of the matrix with the incompressible matrix producing the highest stress for a given spacing. For example, for a clear spacing of one-half the fiber radius, a Poisson's ratio of 0.5 produces a 40% increase in stress over that for a ratio of 0.25. In general, the influence of Poisson's ratio of the component materials has not been thoroughly investigated. The rather good agreement of the experimental results above

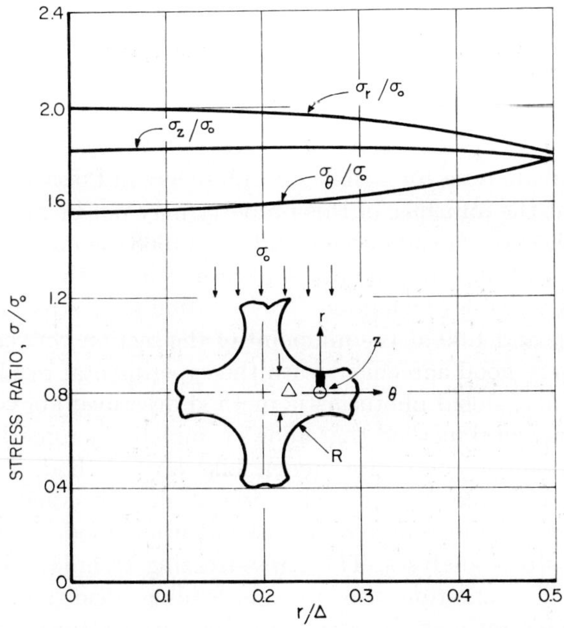

FIG. 31. Distribution of external load stresses across section between fibers parallel to direction of applied load, $\Delta/R = 0.50$.

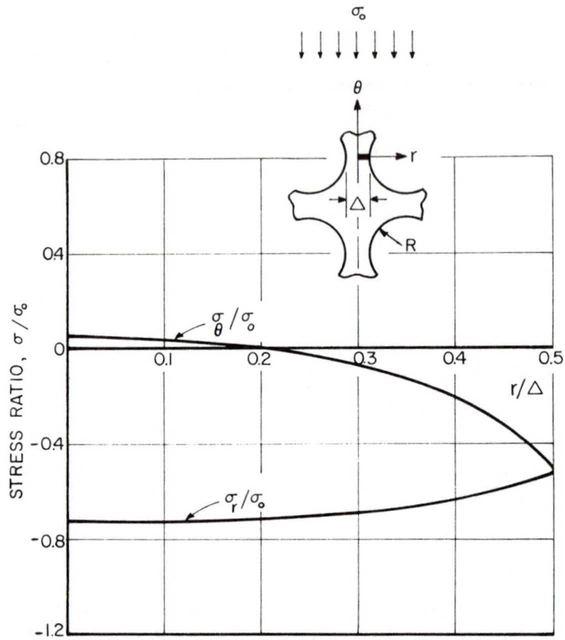

Fig. 32. Distribution of external load stresses across section between fibers normal to direction of applied load, $\Delta/R = 0.50$.

with theoretical analyses for significant differences in Poisson's ratios seems to indicate that the influence of this property may not be appreciable.

Following the experiments above, Adams (1968) communicated to the author a new solution using the material properties of the model used here and a complex variables technique. This solution gave stress ratios of 1.89 at the interface and 1.99 at the midpoint of the section between the fibers, which are in very good agreement with the experimental results above.

The three-dimensional photoelastic method described above is probably the only experimental method that allows complete micromechanical stress analysis of a realistic composite model. Like any method, it has its difficulties and undesirable features. The separation of residual stresses from those due to external loading requires two identical models and essentially two complete stress analyses. The stress-freezing technique is destructive and allows only a one-time use, for one loading condition, of expensive models. Poisson's ratio at the stress-freezing temperature is near 0.5 and it does not simulate that of prototype composites (0.35). This discrepancy may be a source of errors. Several attempts to overcome these difficulties have been made. Edelman (1969) reported on a procedure for producing shrinkage-free photoelastic models. Jenkins (1968), Pih and Sutliff (1968), and the author have attempted to apply scattered-light techniques, which are nondestructive and allow testing at room temperature where the matrix Poisson's ratio is the same as that of the prototype. Figure 33 shows a simple composite model consisting of a glass rod embedded in an epoxy matrix and loaded in compression in a scattered-light polariscope using a laser light source. Figure 34 shows a scattered-light fringe pattern obtained when the laser beam is directed along the interface parallel to the fiber axis.

VII. Dynamic Investigations

A. Introduction

Dynamic photoelastic studies of composites are relatively rare. Hunter (1970) presented a preliminary dynamic photoelastic study of wave propa-

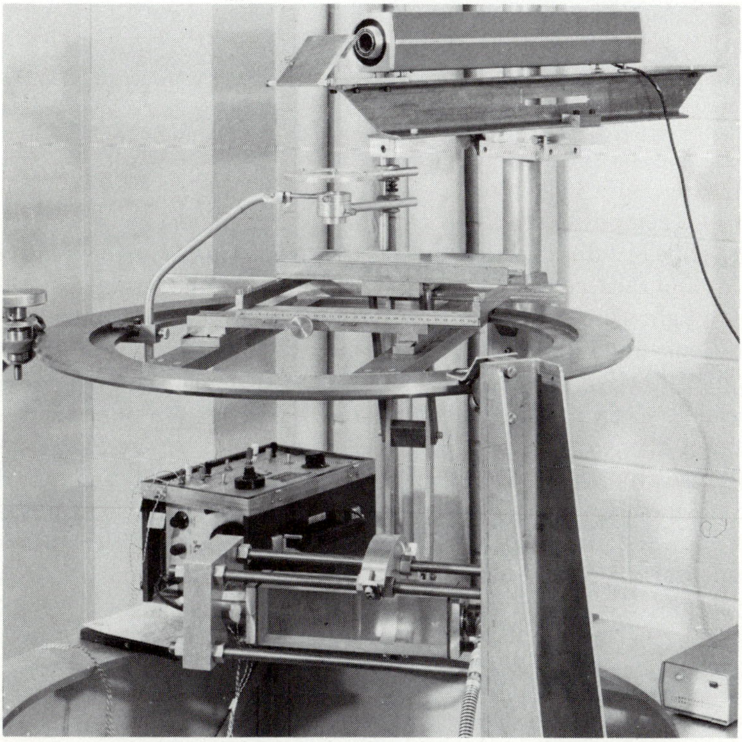

FIG. 33. Composite model in scattered-light polariscope.

FIG. 34. Scattered-light fringe pattern along fiber–matrix interface in the direction of fiber axis.

gation in a composite model. A two-dimensional model consisting of alternating strips of "fiber" and "matrix" materials was loaded explosively on one edge and the dynamic fringe patterns photographed using a Q-switched laser as a light source. The study was qualitative and the modeling not very realistic since the ratio of dynamic moduli of the fiber and matrix materials was only 1.61. The author (Daniel, 1969, 1970) has reported a dynamic photoelastic study of the dynamics of crack propagation in a more realistic model of a filamentary composite. The objective of this task was to study the fracture propagation in the matrix of a unidirectionally reinforced composite when one interior fiber breaks. The energy released suddenly usually causes a crack to propagate in the matrix toward the adjacent fibers. The experimental procedure and results of this investigation are described briefly below.

B. Experimental Procedure

The models were idealized composites consisting of alternating strips of reinforcing and matrix materials. The models were plastic plates $\frac{1}{4}$ in.

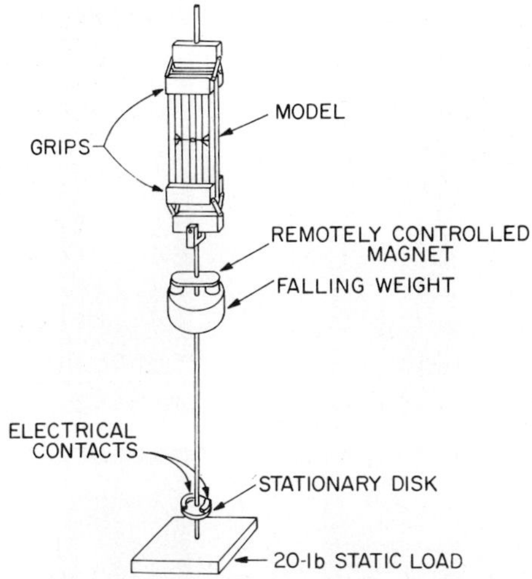

FIG. 35. Fixture for dynamic tensile loading of composite models.

thick containing parallel and equidistant glass bars of $\frac{1}{4} \times \frac{1}{4}$ in. cross section. Polyester resin (Homalite 100) strips were cemented to the glass bars to produce the composite model. Plexiglas pads were cemented to the ends of the specimen. Small holes were drilled and crack configurations were machined ultrasonically in the middle of the central glass bar to ensure and localize crack initiation.

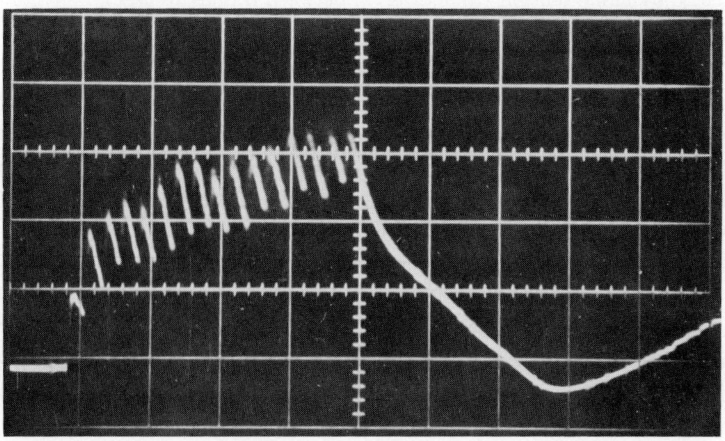

FIG. 36. Oscilloscope record showing time delay between trigger pulse and first spark and time sequence of 16 sparks (20 μsec/division and 2 V/division).

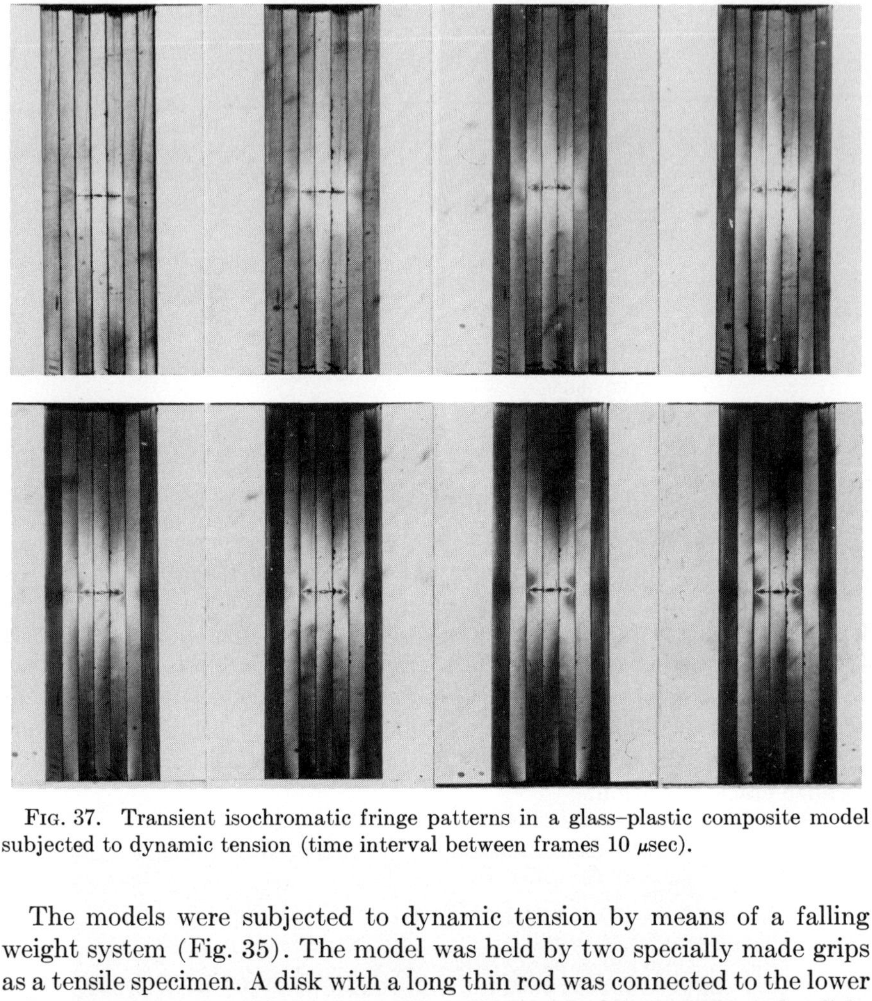

FIG. 37. Transient isochromatic fringe patterns in a glass–plastic composite model subjected to dynamic tension (time interval between frames 10 μsec).

The models were subjected to dynamic tension by means of a falling weight system (Fig. 35). The model was held by two specially made grips as a tensile specimen. A disk with a long thin rod was connected to the lower grip. A perforated weight (6 lb) slid on this rod and impacted on the disk. A 20-lb static load was added at the bottom of the rod to eliminate time delays caused by slack in the system.

The dynamic fringe patterns and the propagating crack were recorded with a multiple spark Cranz–Schardin camera operating at a rate of 200,000 frames/sec. The oscilloscope and the camera were triggered by the closing of a circuit upon contact of the falling weight with the stationary disk. A photomultiplier device was used to sense individually the 16 sparks of the camera. An oscilloscope record giving the exact time and sequence of the sparks is shown in Fig. 36. A time delay between loading and triggering of

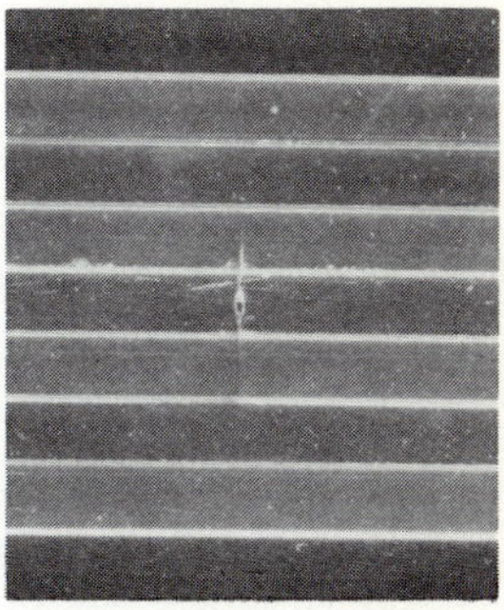

FIG. 38. Composite model after dynamic tensile loading showing crack propagation and arrest in the matrix adjacent to central glass bar.

the camera was introduced to coordinate fracture propagation and the camera recording.

C. *Results and Discussion*

Figure 37 shows a sequence of eight frames from the Schardin camera record of the first test. The crack propagated from the central glass bar into the adjacent matrix layers and stopped near the interface of the next two glass bars on either side. This initial crack propagation and arrest is shown in Figs. 38 and 39. Although the dynamic load was sufficiently high to initiate the crack, the short duration of the load did not provide sufficient energy for further propagation. Other factors tending to arrest the crack are

FIG. 39. Detailed view of arrested crack in composite model after dynamic tensile loading.

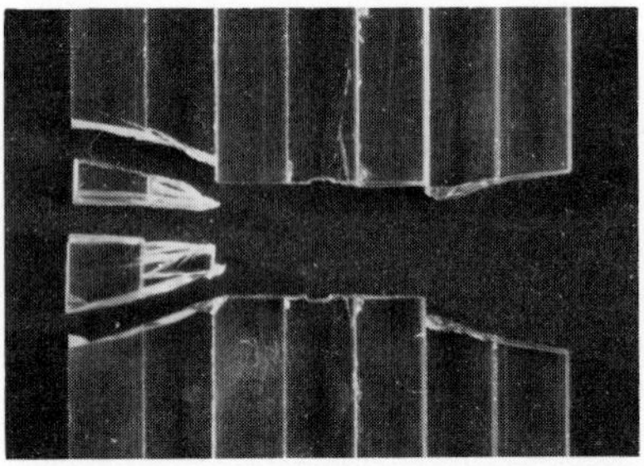

Fig. 40. Composite model after second dynamic tensile loading.

nonlinear and plastic deformation at the tip of the crack causing "crack blunting" as discussed by Irwin (1964) and reflections from the glass–plastic interfaces of transverse stress waves emanating from the crack border as discussed by Schardin (1959). An arrested or near stationary crack in a strain rate sensitive material tends to increase the specimen compliance by developing large nonlinear strains near its border, thus cushioning somewhat the effect of the dynamic load (Irwin, 1964).

One characteristic phenomenon illustrated in Fig. 39 is the shape of the arrested crack. Its pronounced horseshoe shape shows a higher velocity near the free surfaces of the specimen. This is contrary to previously observed phenomena and to expectations (Irwin, 1969). One might conjecture that it is a result of the dynamic type of loading. The high stresses produced tend to break bonds and promote cracking; however, the short duration of the load keeps the energy low and therefore tends to limit the development of crack surfaces. It may be then that this shape represents the maximum crack propagation with a minimum of energy expenditure.

The record of Fig. 37 shows some hairline traces extending beyond the ends of the more visible crack. These traces are more visible in the first frames of the record and disappear subsequently. Upon examination of the specimen after crack arrest no traces of cracks were visible in this region. The model was loaded anew in dynamic tension. This time the crack propagated farther into the other two glass bars where it bifurcated and finally resulted in complete separation of the model (Fig. 40). Comparison of Figs. 37 and 40 shows that the crack followed exactly along the indicated initial hairline traces. One proposed explanation of this phenomenon is that as a result of the initial loading stress crazing or microcracking is produced beyond the visible crack and is mostly concentrated on the surface of the specimen. Their optical effect on transmitted light seems to be appreciable enough to show in the photographic records. Molecular bonds may be broken by the high dynamic stress but the energy input is not sufficiently high to open the cracks and create new surfaces. This phenomenon may be related to and considered an extreme case of the horseshoe shape effect discussed above. The fact that these microcracks appear to close may be due to the short duration of the loading. However, there is some residual damage which accumulates and results in complete failure upon reloading.

In subsequent tests, the spacing of the glass bars was increased by using twice as wide plastic bars. The sequence of photoelastic fringe patterns of Fig. 41 illustrates the large stress concentration at the tip of the propagating crack. One important characteristic observed in these fringe patterns is the inclination of the loops formed by the fringes at the tip of the crack. This angle of inclination here is larger than 90° which is appreciably different from previous results in homogeneous materials. Gerberich (1962) observed angles of 45 and 60° for slowly growing internal and edge cracks, respectively. Wells and Post (1958) indicate values of up to 80° for running cracks. The angle of inclination of an isochromatic loop θ_m, the maximum radial span r_m to this loop, and the fringe order (or equivalently the maximum shear stress τ_m) are related to the stress intensity factor K or the crack extension force $\mathcal{G}$ as shown by Irwin (1958). It has been pointed out that the latter quantity is very sensitive to changes in the angle of inclination. The difference in this angle observed in this test points to a substantial difference in the crack extension force for the composite model.

The rate of crack growth in this test is shown in Fig. 42. The maximum velocity measured is 9600 in./sec, which is much lower than the theoretical limiting value of 28,000 in./sec. The crack propagated at this constant velocity and then it was momentarily arrested as shown by the horizontal portion of the plot. The specimen, however, did fail and separated completely during the test.

The slow velocity and near arrest of the crack can be attributed as before to "crack blunting" due to inelastic effects and to reflected stress waves. After passage of the growth inhibiting stress wave the crack was able to resume its course to complete failure. Such interactions between cracks and stress waves have been observed by Rader (1967). Another inhibiting factor may be the type of stress singularity present. Generally, for a homogeneous material, the stress singularity near the crack is of the form r^{-n} with $n = \frac{1}{2}$. However, for a crack reaching the interface between two dissimilar materials, $n > \frac{1}{2}$ when the crack enters from the hard medium into the soft one and $n < \frac{1}{2}$ for the reverse case (Zak and Williams, 1962). Thus, a more severe type of singularity exists in the former than in the latter case.

Figure 43 shows another sequence of isochromatic fringe patterns depicting a later stage of the same crack propagation phenomenon in an identical model. The failure pattern for this model is shown in Fig. 44. As observed earlier, failure followed along the precursor traces visible in Fig. 43. An interesting phenomenon, observed in the previous test as well, is the bifurcation occurring when the crack reaches the interface with the glass bar

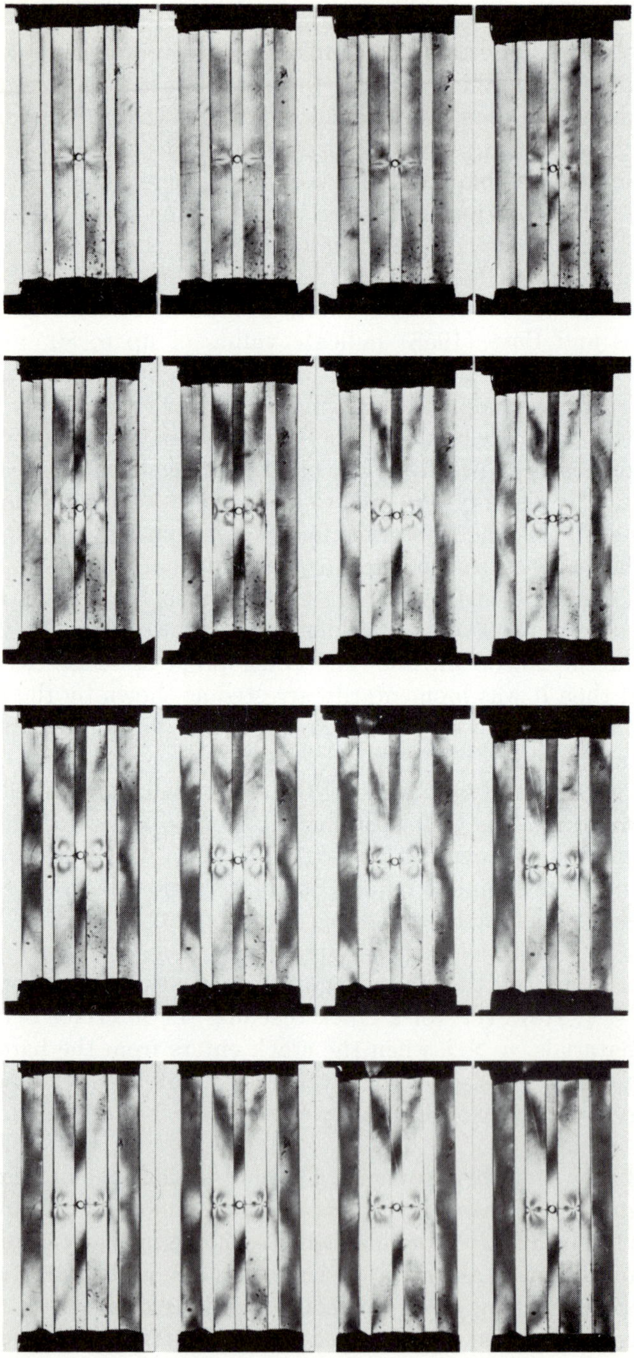

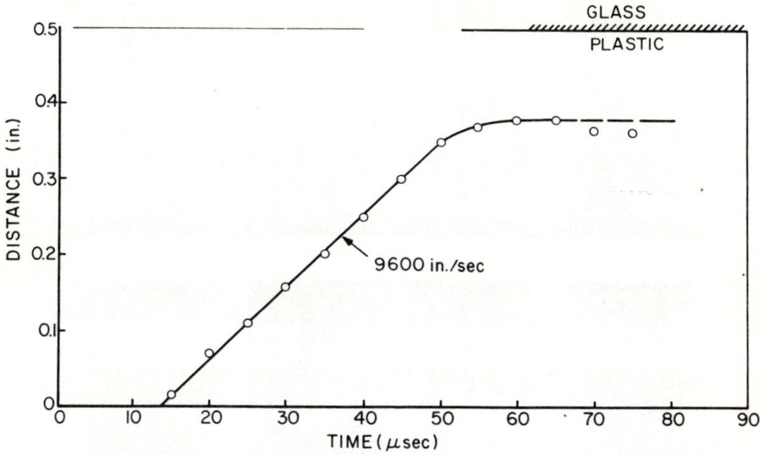

FIG. 42. Rate of crack propagation in glass–plastic composite model subjected to dynamic tension.

despite the fact that the propagation velocity at this point was far below the limiting value. One possible reason is the horseshoe shape of the crack mentioned earlier. This causes crack initiation in the outer glass bars at two separate points near the surface where the crack arrives sooner than at the interior points.

Results of this photoelastic investigation were compared with those of Rosen's (1964), who considered a unidirectional fibrous composite with a statistical distribution of flaws loaded in uniaxial tension in the direction of the fibers. In the photoelastic model discussed here, the crack propagated across the entire specimen whereas Rosen's model failed after accumulation of many randomly distributed fiber breaks. In the latter case, fiber breaks occur at weak points or preexisting flaws in the fibers which tend to be randomly distributed and of more or less comparable severity. These flaws were such that no breaks occurred before half of the ultimate load was exceeded. Once a fiber breaks at a point, the crack extends into the surrounding matrix but tends to be arrested by the adjacent fibers. The flaw distribution and severity in the fibers is such that other fiber breaks occur at other points before the stress concentration at the arrested crack can cause crack extension. In other words, a crack would occur at a weak point of the fiber rather than a strong point that happens to be in contact with cracked matrix material.

FIG. 41. Transient isochromatic fringe patterns surrounding a running crack in a glass–plastic composite model subjected to dynamic tension (camera speed 200,000 frames/sec).

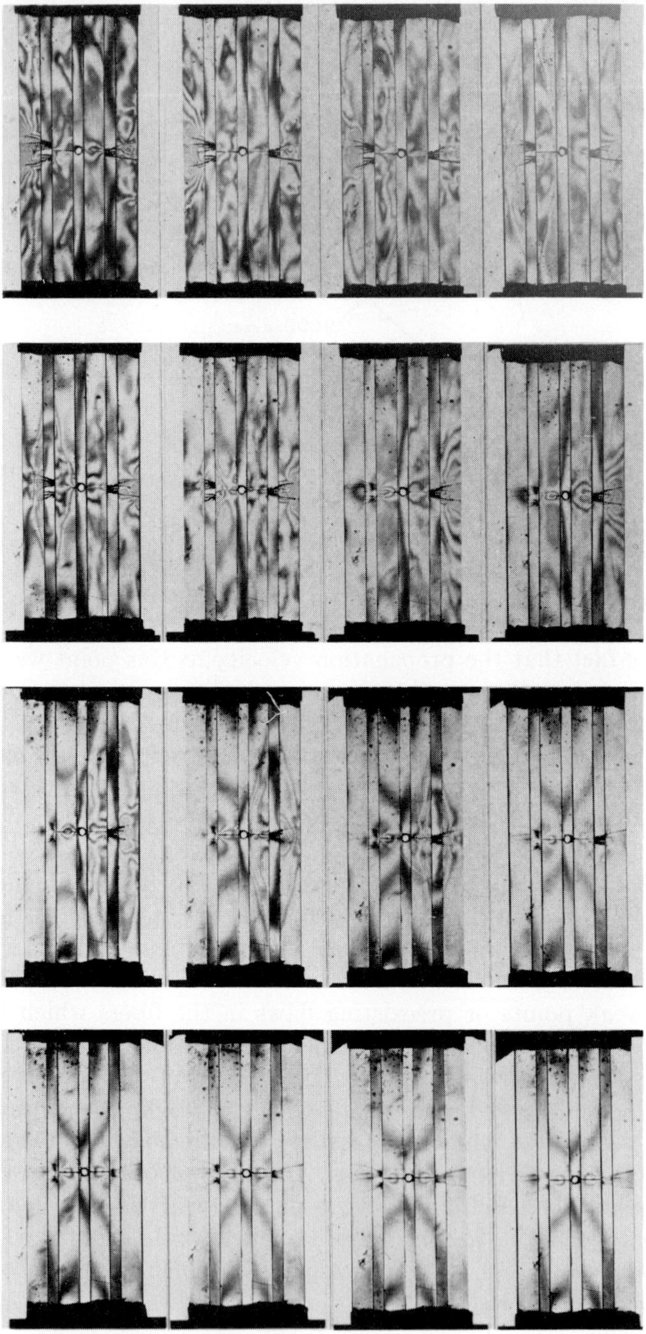

FIG. 43. Transient isochromatic fringe patterns in a glass–plastic composite model subjected to dynamic tension (camera speed 200,000 frames/sec).

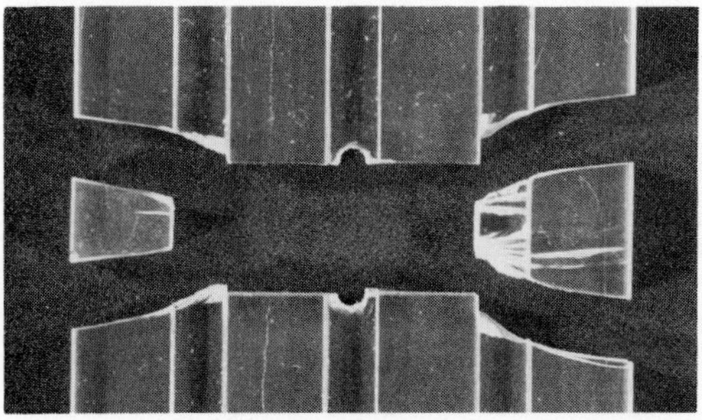

FIG. 44. Failure pattern in model of Fig. 43.

In the present investigation, the mechanics of only one single crack, or fiber break, was studied. The presence of one single severe flaw or stress raiser dominated the entire fracture and started crack propagation at a much lower load than would be required without this stress raiser. However, the observed reluctance of the initial crack to propagate across the glass bars adjacent to the central one corroborates the findings in Rosen's (1964) model. The stress concentration at the tip of the crack reaching the adjacent glass bars is more severe than any existing flaws at other points of the reinforcing bars and thus the crack continues before any new fiber breaks occur. Other factors that may contribute to this phenomenon here are the dynamic nature of the load and the small number of reinforcing bars in the model.

References

Adams, D. F. (1968). Private communication.
Adams, D. F., and Doner, D. R. (1967). *J. Compos. Mater.* **1,** 152.
Adams, D. F., and Tsai, S. W. (1968). Rand Corp. RM-5608-PR.
Adams, D. F., Doner, D. R., and Thomas, R. L. (1967). AFML-TR-67-96.
Alexander, E. L., Caputo, A. A., Prado, M. E., and Hilzinger, J. E. (1965). AFML-TR-65-283.
Allison, I. M., and Hollaway, L. C. (1967). *Brit. J. Appl. Phys.* **18,** 979.
Blejwas, T. E., Lauffer, W. D., and Tauchert, T. R. (1969). *Exp. Mech.* **9,** 478.
Caputo, A. A., and Hilzinger, J. E. (1969). USAAVLABS Tech. Rep. 69-42.
Chamis, C. C., and Sendeckyj, G. P. (1968). *J. Compos. Mater.* **2,** 332.
Chen, P. E., and Lavengood, R. E. (1969). Monsanto/Washington Univ. Ass., Rep. HPC-68-60.
Clausen, W. E., and Leissa, A. W. (1967). AFML-TR-67-151.
Cox, H. L. (1952). *Brit. J. Appl. Phys.* **3,** 72.
Dally, J. W., and Alfirevich, I. (1969). *Exp. Mech.* **9,** 97.

Dally, J. W., and Prabhakaran, R. (1971). *Exp. Mech.* **11,** 346.
Daniel, I. M. (1966). IIT Res. Inst. Rep. M6132.
Daniel, I. M. (1969). AFML TR-69-101, Vol. II.
Daniel, I. M. (1970). *J. Compos. Mater.* **4,** 178.
Daniel, I. M., and Durelli, A. J. (1961). *Proc. of SPI* Sect. 19-A, 1.
Daniel, I. M., and Durelli, A. J. (1962). *Exp. Mech.* **2,** 240.
Daniel, I. M., Rowlands, R. E., and Whiteside, J. B. (1973). *ASTM STP* **521,** 143.
Dow, N. F. (1963). G. E. Rep. R63SD61.
Durelli, A. J., and Parks, V. J. (1967). *Exp. Mech.* **7,** 381.
Durelli, A. J., Parks, V. J., Feng, H. C., and Chiang, F. (1970). *In* "Mechanics of Composite Materials" (J. W. Wendt, H. Liebovitz, and N. Perrone, eds.), pp. 265–335. Pergamon, Oxford.
Edelman, W. E. (1969). *Exp. Mech.* **9,** 380.
Foye, R. L. (1966). *SAMPE Symp., 10th, San Diego, California.*
Foye, R. L. (1966, 1967). North Amer. Aviat., Columbus Div., Quart. Rep. No. 1, 2, 3 on Contract No. AF33(615)-5150.
Frocht, M. M. (1941). "Photoelasticity," Vol. I. Wiley, New York.
Gerberich, W. (1962). *Exp. Mech.* **2,** 359.
Goree, J. G. (1967). *J. Compos. Mater.* **1,** 404.
Greszczuk, L. B. (1966). *Proc. SPI* Sect. 8A.
Haener, J. (1966). USAAVLABS Rep. No. 66-62.
Hashin, Z. (1965). *J. Mech. Phys. Solids* **13,** 119.
Hashin, Z. (1966). *AIAA J.* **4,** 1411.
Hashin, Z., and Rosen, B. W. (1964). *J. Appl. Mech.* **31,** 223.
Herrmann, L. R., and Pister, K. S. (1963). ASME Paper No. 63-WA-239.
Hill, R. (1963). *J. Mech. Phys. Solids* **11,** 357.
Hunter, A. R. (1970). Presented at SESA Fall Meeting, Boston, Massachusetts.
Irwin, G. R. (1958). *Proc. SESA* **16,** 1, 69.
Irwin, G. R. (1964). *J. Eng. Power* **86,** 444.
Irwin, G. R. (1969). Private communication.
Jenkins, D. R. (1968). *Exp. Mech.* **8,** 467.
Kies, J. A. (1962). NRL Rep. 5752.
Koufopoulos, T., and Theocaris, P. S. (1969). *J. Compos. Mater.* **3,** 308.
Lewicki, B., and Pindera, J. T. (1956). *Polish Acad. Sci. II,* No. 4.
MacLaughlin, T. F. (1966). *Exp. Mech.* **6,** 481.
MacLaughlin, T. F. (1968). *J. Compos. Mater.* **2,** 44.
Marloff, R. H., and Daniel, I. M. (1969). *Exp. Mech.* **9,** 156.
Parks, V. J., and Durelli, A. J. (1965). *J. Appl. Mech.* **32,** 504.
Parks, V. J., Durelli, A. J., Chandrashekhara, K., and Chen, T. L. (1970). *J. Appl. Mech.* **37,** 578.
Paul, B. (1960). *Trans. Met. Soc. AIME* **218,** 36.
Pih, H., and Knight, C. E. (1969). *J. Compos. Mater.* **3,** 94.
Pih, H., and Sutliff, D. R. (1968). AFML-TR-68-380.
Rader, D. (1967). *Exp. Mech.* **7,** 160.
Rhines, W. J. (1968). *Exp. Mech.* **8,** 481.
Rosen, B. W. (1964). *AIAA J.* **2,** 1985.
Rosen, B. W., Dow, N. F., and Hashin, Z. (1964). NASA CR-31.
Rowlands, R. E., Daniel, I. M., and Whiteside, J. B. (1973). *Exp. Mech.* **13,** 31.

Sampson, R. C. (1962). Aerojet-General Corp. Rep. No. 0411-10F.
Sampson, R. C. (1963). *Exp. Mech.* **3,** 225.
Sampson, R. C. (1970). *Exp. Mech.* **10,** 210.
Schapery, R. A. (1968). *J. Compos. Mater.* **2,** 380.
Schardin, H. (1959). *Proc. Int. Conf. Mech. Fracture, Swampscott, Massachusetts.*
Schuster, D. M., and Scala, E. (1964). *Trans. Metall. Soc. AIME* **230,** 1635.
Shaffer, B. S. (1968). *AIAA J.* **6,** 2316.
Stone, D. E. W. (1967). Roy. Aircraft Establishment, Tech. Rep. 67015.
Tyson, W. R., and Davies, G. J. (1965). *Brit. J. Appl. Phys.* **16,** 199.
Wells, A. A., and Post, D. (1958). *Proc. SESA* **16,** 69.
Whitney, J. M. (1967). *J. Compos. Mater.* **1,** 188.
Wilson, H. B., and Hill, J. L. (1965). *In* Mathematical Studies of Composite Materials, II. Rohm and Haas Rep. No. S-50, p. 39.
Zak, A. R., and Williams, M. L. (1962). California Inst. Technol., GALCIT SM42-1.

Author Index

Numbers in italics refer to the pages on which the complete references are listed.

G

H

I

J

T

U

V

W

Y

Z

Subject Index

F

G

H

I

K

L

M

N

O

A 4
B 5
C 6
D 7
E 8
F 9
G 0
H 1
I 2
J 3